Lecture Notes in Biomathematics

ctd. on inside back cover

Lecture Notes in Biomathematics

Managing Editor: S. Levin

89

W. Alt   G. Hoffmann (Eds.)

Biological Motion

Proceedings of a Workshop
held in Königswinter, Germany, March 16–19, 1989

 Springer-Verlag Berlin Heidelberg GmbH

Editors

Wolfgang Alt
Abteilung Theoretische Biologie, Universität Bonn
5300 Bonn, FRG

Gerhard Hoffmann
Zoologisches Institut II, Universität Würzburg
8700 Würzburg, FRG

Mathematics Subject Classification (1980): 92-06, 92A06, 92A08, 92A09, 92A18

ISBN 978-3-540-53520-1   ISBN 978-3-642-51664-1 (eBook)
DOI 10.1007/978-3-642-51664-1

© Springer-Verlag Berlin Heidelberg 1990

2146/3140-543210 – Printed on acid-free paper

PREFACE

"...behavior is not, what an organism *does itself, but to what we point. Therefore, whether a type of behavior of an organism is adequate as a certain configuration of movements, will depend on the environment in which we describe it."*
(Humberto Maturana, Francisco Varela: El arbol del conocimiento, 1984)

"A thorough analysis of behavior must result in a scheme, that shows all regularities that are to be found between the sensorical input and the motorical output of an animal. This scheme is an abstract representation of the brain."
(Valentin Braitenberg: Gehirngespinste, 1973)

During the 70ies, when Biomathematics (beyond Biomedical Statistics and Computing) became more popular at universities and research institutes, the problems dealt with came mainly from the general fields of 'Population Biology' and 'Complex Systems Analysis' such as epidemics, ecosystems analysis, morphogenesis, genetics, immunology and neurology (see the first series of Springer Lecture Notes in Biomathematics). Since then, the picture has not considerably changed, and it seems that "a thorough analysis of behavior" of single organisms and, moreover, of their mutual interactions, is far from being understood. On the contrary, mathematical modellers and analysts have been well-advised to restrict their investigations to specific aspects of 'biological behavior', one of which is 'biological motion'. Until now, only a few Conference Proceedings or Lecture Notes have paid attention to this important aspect, some of the earlier examples being Vol.24: 'The measurement of biological shape and shape changes' (1978) or Vol.38: 'Biological Growth and Spread' (1980). Nevertheless, several remarkable monographs on particular topics in theoretical modeling or mathematical analysis of (loco)motion have been published, for example, by Werner Nachtigall: 'Insects in Flight' (1968), by James L. Lighthill: 'Flagellar hydrodynamics' (1976) or by Hermann Schoene: 'Orientierung im Raum' (1980).

The idea for this Workshop, the Proceedings of which we now present, arose between 1987 and 1988, when Gerhard Hoffmann from Würzburg and Hans Scharstein from Köln had reported on their work on Animal Physiology at the Biomathematical Colloquium organized by three colleagues in Bonn. In response, one of these, Wolfgang Alt, visited the Zoological Insitute at Köln (group of Gernot Wendler) and was fascinated by a colloquium talk on ciliary motion presented by Hans Machemer from the Cell Physiology Department in Bochum, in particular by the demonstration of his famous plastic flagella model. Finally, coming into mind as participant of an interesting meeting on chemotaxis of leukocytes, Hans Gruler from the Biophysics Department in Ulm joint us and completed the group of five organizers. We soon developed the concept of a four-day Workshop on "Modeling, Analysis and Simulation of Biological Motion" and requested various scientists to assemble for a discussion of our main topic: "Mechanisms and Control of Locomotion". The selection of motile species taken into consideration was clearly influenced by the organizers' own research fields: predominantly unicellular organisms, blood and tissue cells, but also insects, crustaceans and even fish were placed 'on the table'.

The task of conducting the discussion in the seven working groups and of writing introductions and conclusions for the seven **Sections** of these Proceedings (as well as the important job of acting as referees for the other contributions) was performed by some of the organizers and, furthermore, by Wolfram Zarnack (Zoological Institute, Göttingen), Michael Vicker (Cell Biology, Bremen), Graham Dunn (Cell Biophysics, London) and by Leah Edelstein-Keshet (now at the Mathematics Department in Vancouver). Due to their commitment and enthusiasm the atmosphere during the Workshop was exciting, and it stimulated a continuous work from Thursday evening, March 16, until Sunday morning, March 19, 1989, in the comfortable 'Stegerwaldhaus' of the Jakob-Kaiser-Stiftung placed at the little wine-growing village Königswinter on the Rhine, opposite to Bonn and at the foot of the 'Drachenfels'. Those who never can finish a discussion concluded the Workshop by a walk through the forest around this pleasent hill.

It is obvious, that the success of this meeting has been caused by the 'synergetic' effort of all participants coming from overseas (Japan, Canada, USA), from various countries of Western Europe and from different regions of (Western) Germany, but also from the local region around Köln/Bonn. In particular, the group of Wolfgang Briegleb from the DLR (German Aero-Space Research Center) showed their interest in the behavior of microorganisms under microgravity. In general, the constellation of participants, with their different educational backgrounds, has been very inspiring for an interdisciplinary approach to the discussions and the envisaged research projects. Among the 47 participants, half of them theorists and half experimentalists, there were 5 medical physiologists, 19 biologists, 12 'theoreticians' (originating in physics or chemical engineering) and 11 mathematicians, all willing to learn from each other and to combine differing aspects and methods which are effective in Biomathematics: modeling (Theoretical Biology), data and image analysis (Statistics, Informatics), analysis of dynamical systems or stochastic processes and their simulation (Applied and Numerical Mathematics).

In particular, the complex dynamical systems inducing the 'miracles' of naturally evolved 'biological motors' (on the individual side) and resulting cooperative movements of organisms (on the population side) require adequate mathematical theories and techniques to give insight into the underlying highly non-linear world of spatio-temporal phenomena. Therefore, it has been very fortunate for the Workshop to obtain generous financial, material and personnel support by the Research Program SFB 256 on "Nonlinear Partial Differential Equations" sponsored by the German Research Fond (DFG). Special acknowledgments for doing the tedious organizational work during the Workshop can be given to the secretaries Marianne and Anke Thiedemann, to the 'assistants' Beate Pfistner and Thomas Pohl as well as to the other doctorands of the Division of Theoretical Biology in Bonn.

The originally not intended long period of preparation, collection and revision of the articles for these Proceedings has unfortunately postponed the publication of several readily produced results. Nevertheless, it has enabled a substantial maturation of some other results which had been 'freshly' presented during the Workshop and which lead to fruitful discussions.

Finally, the editors wish to thank all participants and contributors for their cooperation and patience as well as Annedore Buhler at the Zoological Institute in Würzburg for the extensive correspondence and editing work, and the team at the Springer Verlag for their critical support.

Wolfgang Alt (Bonn) and Gerhard Hoffmann (Würzburg)

Hans Machemer explaining his '9 + 2 microtubular doublets' model of a eukaryotic flagella by rotating it in a 'natural' way.

Hugh Crenshaw demonstrating his colorful cardboard model of a microorganism and bringing it onto a helical path by rotation about two different 'artificial' body axes.

CONTENTS

INTRODUCTION

Motion is one of the essential requisites for life, since besides motility all (other) characterizing properties of living organisms as self-reproduction, mutability and metabolism require some kind of movement, occuring on different organizational levels: Displacement or locomotion of whole organisms (cells, plants, animals) and their primary offsprings (spores, seeds, gametes) as well as movement of their organs (hearts, legs, antennæ) and of various organelles (flagella, protoplasma).

At first glance all such examples of 'biological motion' might be classified by the distinction, whether they represent an 'active' or a 'passive' motion. Let us take, for instance, a causative viewpoint: *Passive movement* of biological entities, just as of physical particles, is caused by molecular motion and mainly driven by thermal energy, pressure or other potential gradients (e.g. plasmids transported through gap junctions, spores carried by wind or plankton floating in water). In this view, the physico-chemical 'constraints' surrounding a body give it a (passive) mobility which, in appropriate mathematical models can be quantified by parameters as diffusivity or drift velocity. However, if passivity means lack of endogenous, energy consuming activity, then what defines, for example, the passive motion of magnetic bacteria in a magnetic field (which orients their movement just by carrying suitable particles in their body) to be 'biological'? One might answer 'nothing', but one could also rely on the functional importance of this property beyond its purely physical description: *Biological evolution* probably has selected these types of bacteria and their specific morphology, supporting a better 'search' for favorable living conditions.

Active movement of biological entities (e.g. chemotaxis of *E.coli* bacteria) clearly provides such functional abilities in much more plentiful and effective ways. However, from the viewpoint of modern computer and roboter technology, any 'automobile' vehicle or 'automatic' apparatus is a functioning, albeit artificial, example for active motion as long as driving energy is (self-)supplied and some steering or (self-)regulating mechanisms are implemented. Then, what defines these types of *technical motion* to be 'non-biological'? A clear answer is difficult, since one calls 'bionics' the (more or less successful) technique, which tries to copy or at least simulate *natural examples* of active biological motion (e.g. flying, walking or fish navigation). The problems still remaining or newly arising (in fulfilling some of those 'dreams of mankind') can be substantiated by the growing amount of unresolved questions in understanding various 'biological motors' (flagella, contractile filaments or whole muscles) and the ways how they perform complex movement patterns (ciliate swimming, cell division, amoeboid motion, neural control of locomotion). Moreover, research on biological motion and the development of statistical descriptions or mathematical models has not been restricted to these motory aspects, but has always tried to include sensory aspects in order to explain the functionally important facets of motion behavior (e.g. orientation of beatles, organization of ants or fish swarming).

In any case, what ensures the surprising efficiency and adabtability of 'natural' motile system, making them 'superior' to physico-chemical or technical systems, seem to be the following characteristic properties:

- Self-maintenance and endogenous regulation of <u>motility</u> within a wide range of environmental influences;

- Signal perception, transduction to the motor control system and sensory adaptation enabling <u>motory responses</u> to external stimuli and interactive phenomena of <u>collective motion</u>;

- Learning capacity and evolutionary variability inducing 'optimal' performance of functional <u>motion behavior</u>.

Again, mathematical models should help to quantify analogous concepts as (active) <u>motility</u> or <u>taxis</u> on different temporal or spatial scales. Indeed, often quite simple analogs of physico-chemical theories have been carried over to biological situations. However, for describing and explaining the whole range of phenomena in biological motion indicated above, additional modeling efforts are requested calling for close collaboration between theorists and experimentalists. Adequate theoretical concepts and experimental assays used or to be developed touch such diverse fields as biochemistry, biophysics, physiology or behavioral sciences. The interdisciplinary task then is to formulate and analyze corresponding mathematical models, perform computer simulations and compare the 'numerical results' with the previously evaluated and analyzed data, call them 'biological results'. This comparison aims at estimating the model parameters, it eventually requires variation of the proposed mathematical model and might necessitate a revision of the underlying theoretical concepts, thereby influencing the performance of (new) experiments.

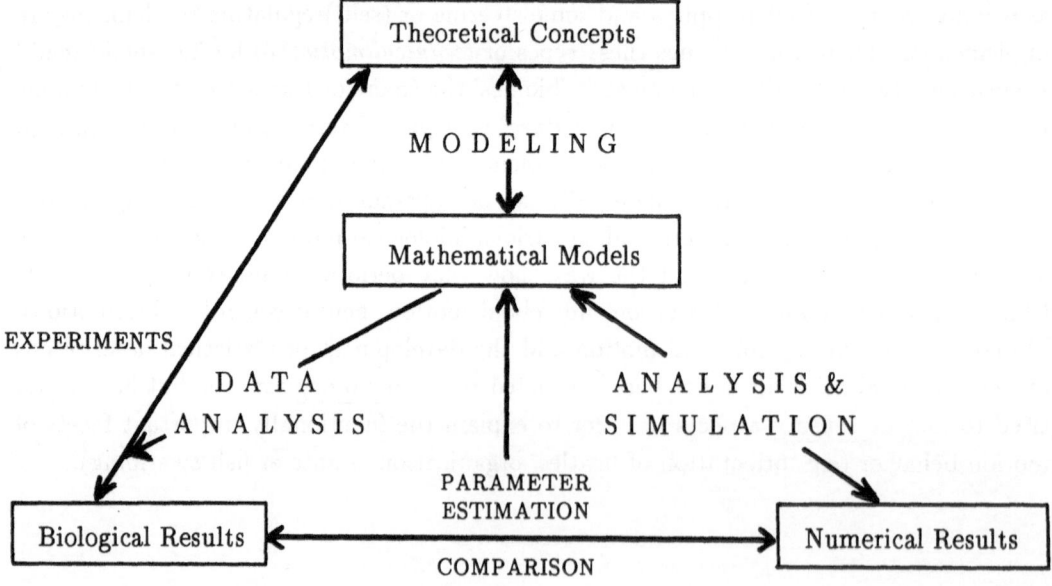

The Workshop on 'Modeling, Analysis and Simulation of Biological Motion' intended to (at least partially) realize this ideal framework of biomathematical reasearch on a specific field, which obviously attracted a bunch of scientists with various methodological and mental scopes: More than forty mathematicians, physicists and biologists congregated in several plenary sessions, discussion and working groups for exchanging their experiences and ideas about seven selected topics: Movement paths and search behavior, neuromotoric aspects of animal movement, collective behavior and swarming, kinesis and/or taxis, microtubuli and cilia motion, cellular motion and shape changing, and also image analysis. The subdivision of these Proceedings, roughly following those topics, combines relevant presentations given at the Workshop together with some spontaneous contributions during the course of disputations. The coordinators of each section give a brief (for section I/2 and part III more extensive) introduction into the topic and an overwiev over the articles. At the end of each section, they present the discussion results, draw conclusions, compare different approaches and outline perspectives for further study and research. Certain relevant mathematical methods and theoretical concepts (including a brief glossary) of more general interest for other topics have been summarized in a series of framed BOXES, placed within or between the articles.

The three sections of **PART I: Motion of cell or body parts** are mainly concerned with the motory aspects, whereas the following three sections of **PART II: Locomotion of single organisms** take up the sensory as well as the functional aspects when investigating the structures of individual paths and their distribution in time and space. Finally, **PART III: Collective motion** incorporates possible synergetic interactions between moving individuals.

For the elementary biological units, single cells, the authors in **PART I** describe, with the aid of image analysis as well as moment and differential equations, the physico-chemical properties of intracellular contractile systems leading to **cytoplasmic streaming** or **deformations of the cell periphery (section I/1)**, and the wonderful dynamics of **cilia and flagella** which, as highly sensitive parts of the plasmalemma, contain actively sliding filametous structures (**section I/2**). For insects, birds or fishes some possibilites of **motoneural control of muscle contraction** in legs, wings or tails are addressed in (**section I/3**). It is amazing and hopefully promising to detect and work up strong analogies in the physiological and biochemical principles of motor control on all these organizational levels: Electric potentials, ions (e.g. Ca^{++}) and other primary/secondary etc. messengers and their interactions with membrane bound receptors and specific 'contractile' filaments (e.g. myosin, dynein, centrin) seem to play the major role in movement coordination of different parts of a cell (e.g. cilia of *Paramecium*) as well as of a body (e.g. legs of crayfish).

The resulting patterns of motor activity and subsequent locomotion have since long been investigated on a more *descriptive* level, with strong support by physical and mathematical modeling, in particular by nonlinear (partial) differential equations. This applies to many of the contributions in these Proceedings, but also to other, well-known classical or recent theories as biorheology, fluid dynamics of swimming and flying, sceletal and heart muscle dynamics, optomotorics and insect orientation, pattern formation in embryogenesis and, not to forget, the related dynamics of plant growth. Although most of these important topics and corresponding scientists could not at all be included into the scope of our small Workshop (we ask for kind understanding), it should be emphasized that during the last years reasearch on biological motion has moved more and more to a *physiological* level. Thus, there is a strong desire for complementary theories and suitable mathematical models which are able to elucidate, as far as possible, the whole (network of) regulatory pathways between sensory inputs and (loco)motory outputs of a moving organism.

The authors in **PART II** take up this challenging ploblem and try, from various viewpoints, to approach the central question, how to characterize differing locomotion patterns and their efficiency in response to external stimuli. The basic difficulties become apparent when reformulating this question: How can we detect or distinguish 'mechanisms,' or at least some principles, by which an organism detects its environment ('where it is') and decides upon its motile behavior ('where to go'); formulating it even more principally (thereby indicating the modeller's limitations): How do we know (from observations of its motion and environment) what an organism 'knows' (its 'internal' informations)?

In order to detect possible 'search plans' some authors try to define and describe observable **characteristics of search paths** by carefully analyzing empirical track data and/or by performing simulations of 'model random walker' obeying certain (more or less abstract, mostly stochastic) decision rules (**section II/1**). Then, in some cases, the search success or degree of orientation can be evaluated and traced back to certain properties of the locomotion path (e.g. turning) and their regulation. For the particular, but typical situation of **orientation responses in gradients** several examples of microorganisms, blood and tissue cells, and one example of social insects (ants) are presented in detail (**section II/2**). Again, as in other parts of these Proceedings, the used mathematical methods range from (2-D and 3-D) image tracking over statistic analyses to (deterministic and stochastic) ordinary or partial differential equations (please, have a look over the broad spectrum of methodological BOXES).

Regarding the diversity of biological situations (still to be) modelled and of mathematical techniques (still to be) applied, it is not surprising that 'language problems' appear and that these are not (so easily) resolvable, as the outstanding problem of **identifying taxis and kinesis** (**section II/3**). Two experimental assays for quantifying (transient) accumulation and/or chemotactic drift of migrating cell polulations are well suited to stimulate a discussion which had started, as usually also on the Workshop, with a 'struggle for words', but which subsequently used the chance to move from controversal general statements to more constructive formulations about details of observed or proposed orientation mechanisms, corresponding to the details of biomathematical analysis offered in the preceding sections.

The further evolution of biomathematics and theoretical biology will probably go hand in hand with acute problems in basic and applied fields of biology, one of which belongs to the facet of synergetics and comprises such important phenomena as the formation and movement of **cohesive patterns of individually active and mutually interacting members** of a population (swarms, herds and other, higher order social organizations). A glance of the large demands and difficulties in creating and treating suitable mathematical models sparkles through the final **PART III**. Besides some more elementary methods, the authors mainly choose Monte-Carlo simulation or nonlinear (partial) differential (integral) equations, where they implement certain 'suggested' interaction rules, in order to reproduce at least some features of the 'natural' communication process. As in the case of myxobacteria swarming, our rare knowledge of biological details seems to demand any thorough analysis of such 'mathematical model swarms'.

PART I

MOTION OF CELL OR BODY PARTS

INTRACELLULAR STRUCTURES AND CELL SHAPE

(Coordinator: Graham Dunn)

Introduction

Even in its humble role of quantitative description, the power of mathematics should not be underestimated. Phrases describing cells moving like 'ships without rudders' or like 'pigs at a food trough' may be graphic and poetically appealing but, like poetry, they create different images in different minds. The simplest numerical description, if well designed, can have three invaluable properties besides that of mere quantitation: it can be objective, repeatable and unambiguous. But numerical description should also be readily interpretable and versatile if it is to compete with verbal description. Ideally, it might also provide a new and meaningful way of looking at the problems of biological motion. By analogy with a good high-level computer language, it should provide an interface between the way a moving cell operates and the way a human mind operates.

The first three contributions in this section deal with the problems of quantitative description of cellular shape and motion. **Dunn and Brown** have tried to overcome one of the main objections to numerical descriptions of changes in cell shape: that they ignore the features of real interest. We describe simple measures of cell shape based on the moment transform that are not only quite easy to interpret in verbal terms but that also seem to have biological significance. The contribution of **Keller, Eisele, Zimmermann & Lackie** demonstrates the importance of understanding cellular shape changes in a practical application of these techniques. **Peter Noble's** contribution takes a quite different but complementary approach and describes the use of the median axis transform as an unambiguous means of visualising and identifying the protrusions of a moving cell.

The spectacular success of the molecular description of cell motility in recent years leaves no need for me to extol its virtues. But, at the risk of overstretching the analogy with computer operation, the molecular description is only one of the chain of languages linking mechanism and understanding: it is the machine language. A knowledge of the assembly language, intermediate between the machine- and high-levels, is also required for any satisfactorily complete explanation of cell motility. It seems that, in the headlong rush to unravel the molecular complexities of the cell's motile machinery, many biologists have forgotten that they are dealing with an essentially mechanical system and that a satisfactory explanation of its functioning will eventually be written as much in biophysical terms as in biomolecular terms.

The remaining four papers in this section deal with the cell's motile machinery at the assembly language level. **Jürgen Bereiter-Hahn** has been a pathfinder in the field of understanding the physicochemical principles of cell motility and his contribution with **Braun and Voth** is again a pioneering exploration of largely unknown territory: the grosser mechanics of the moving cell. **Pohl's** paper is for me a glimpse into the future in which mathematics and physics combine with observation in demonstrating how molecular mechanism might be translated into the complex behaviour of the whole cell. The key to understanding cell motility must lie in finding the relationships between intracellular dynamics and translocation and the last two papers remind us that the actomyosin transducer is not the only motor available to the cell. **Weiss and Langford** demonstrate how state-of-the-art microscopy can combine with sophisticated motion analysis techniques to elucidate the molecular mechanism of microtubule-based intracellular motility. Finally, **Melkonian's** contribution discusses a recently discovered mechanism of contraction based not on sliding filaments but on Ca^{2+}-mediated coiling and supercoiling of the protein centrin. The ciliary and flagellar locomotion of some protozoans is assisted by rapid contractions of this protein.

QUANTIFYING CELLULAR SHAPE USING MOMENT INVARIANTS

Graham A. Dunn and Alastair F. Brown
MRC Cell Biophysics Unit, King's College London,
26-29 Drury Lane, LONDON WC2B 5RL, UK

Abstract

Two-dimensional moments are useful for describing the size, location, orientation and shape of isolated cells in culture. Physical moments can characterise the distribution of matter (dry mass) within the cell whereas geometrical moments can characterise the geometrical properties of the cell outline. Here we describe the calculation of geometrical moments from either digitised images or outlines of isolated cells. We explain why the lower orders of moments are a useful summary of cell shape and how two interpretable shape factors, elongation and dispersion, can be derived from these. These shape factors appear to have a biological significance since certain experimental treatments can affect either one without affecting the other. Finally, we briefly discuss how the higher orders of moments may be used to describe cell shape in more detail and we speculate on the use of third order moments for detecting any asymmetry in cell shape that may be induced by chemotactic gradients.

1. Introduction

The mechanism and control of cellular motility present problems of fundamental importance to understanding such basic processes as embryonic development, wound healing and malignant invasion. In this Workshop, the working group on Cellular Motility recognised the need to develop methods for describing and quantifying cellular motility before the more general methods of analysis and model testing described in this volume can be applied to this specialised area of biological motion.

One problem that is peculiar to the study of crawling motility in tissue cells is their rapidly changing shape and structure.

Special methods are needed even to define the position and orientation of a motile cell with sufficient consistency and objectivity for detailed motion analysis. At the meeting, the problem of describing the shape of moving cells gave rise to much discussion and a radically different but complementary approach to the one given here was presented by Noble (this section).

A general theme that emerged in the working group is that the key to understanding the mechanism of cellular motion must lie in relating intracellular dynamics (including changes in shape and molecular structure) to translocative dynamics. Our approach to this problem has been to develop a method of using interferometric microscopy coupled with digital image processing and analysis to record the dynamic changes in mass distribution within single living cells in culture (Brown and Dunn 1989). Interferometric microscopy is ideally suited to this task since, with little further processing, the image can be calibrated to yield a <u>two-dimensional map of the dry mass density</u> of cellular material (Figure 1).

Figure 1.

Contour map of the distribution of dry matter in a living fibroblast. The contours are calibrated in tenths of a picogram per square micrometre.

Our initial aim is to build up a machine-readable database of the spontaneous variability in fibroblast behaviour. It is envisaged that databases of this type will be used primarily to obtain a numerical description of the major characteristics of cell motility (see also Noble's article). The static pattern of mass distribution within a single cell can be characterised to any degree of accuracy using the mathematical quantities known as the moment series. These also serve to define the centroid of the mass distribution which provides a very consistent estimate of cell position for characterising translocative motility. Time-series analysis can be used to obtain the autocorrelation function for displacement of the centroid and thus to obtain an almost complete description of the statistical properties of cell translocative behaviour (Dunn and Brown 1987, and see articles by Alt and Scharstein in this volume). In addition, the dynamics of the changing mass distribution can be quantified using finite element analysis to yield the minimum velocities of mass transport within the moving cell (Brown and Dunn 1989). The kinetic energy of the total transport of cytoplasmic mass provides the first generally useful measure of intracellular motility. Sensitive techniques of this type will prove essential for characterising the effects of molecular modification or genetic manipulation of the cellular motile machinery and for testing models of the motile mechanism (see Pohl, this section).

In this article we will concentrate on the use of moments as purely geometrical descriptions of cell outlines. Geometrical moments are increasingly being used to describe the location, orientation and shape of cells in tissue culture (Keller et al., this section) and have already proved effective for detecting and quantifying such diverse effects as those of tumour promoting factors (Brown et al., 1989), microtubule disrupting drugs (Middleton et al., 1988, 1989), and micro-alignment in the culture substratum (Dunn and Brown, 1986). We will describe how to calculate geometrical moments and will discuss their future potential and their advantages and disadvantages over other methods of deriving shape factors.

2. What are moments?

Moments are used by physicists and engineers to describe certain basic properties of the distribution of matter (mass) in space. For example, the zeroth, first and second order moments contain complete information about the total mass, the location of the mass centroid or centre of gravity and the rotational inertia of the mass distribution. These moments of the lower orders describe the more fundamental aspects of the mass distribution. However, the moment series can be extended indefinitely to describe the mass distribution to any required degree of accuracy.

Moments need not be three-dimensional: one-dimensional moments are commonly used by statisticians to describe the basic properties of a univariate statistical distribution. In fact, the familiar measures of a sample's location and dispersion, known as mean and variance, are derived from the zeroth, first and second order of one-dimensional moments. Third and fourth order moments are required to extend the description to include measures of a sample's skewness and kurtosis. Two-dimensional moments, besides describing bivariate statistical distributions, are frequently used in image analysis to describe the distribution of image intensity or brightness. The lower order moments are very insensitive to the resolution of the image and are therefore useful for automatic recognition of simple patterns such as printed characters (Hu 1962) where the fine details of the pattern are unimportant. Increasing orders of moments contain information about increasingly fine detail and it is interesting to note that the first twenty or so orders of moments can be used to reconstruct a blurred but usually recognisable image (Boyce and Hossack 1983).

Most forms of microscopy provide information in only two dimensions and so for describing images of cultured cells we need only consider <u>two-dimensional moments</u> in the plane of x,y-coordinates. The remainder of this article will deal exclusively with two-dimensional moments. These are defined as:

$$(1) \qquad m_{jk} = \int_{-\infty}^{\infty} \int_{-\infty}^{\infty} x^j y^k f(x,y) \, dxdy$$

where j and k can be any non-negative integers and the integral
is taken over the whole image plane. The <u>order of each moment</u> is
simply the sum $j+k$. Thus there is one two-dimensional moment of
the zeroth order; there are two of the first order, three of the
second order and so on. In image analysis, the function $f(x,y)$ is
conventionally the image intensity at the point (x,y). In the
case of interferometric microscopy, however, true <u>physical
moments</u> of the distribution of dry mass can be calculated by
assigning a value for the dry mass density of the cell to $f(x,y)$
which can be calculated from the image intensity at (x,y) (Brown
and Dunn, 1989). But here we will concentrate on the use of
moments as a purely geometrical description of a cell's outline.
In this case, $f(x,y)$ takes a value of 1 if (x,y) lies within the
outline and a value of 0 otherwise. These <u>geometrical moments</u> can
be envisaged as physical moments if matter is assumed to be
distributed at unit density within the cell outline - as if the
cell shape were cut from a sheet of material or lamina of uniform
thickness. The <u>zeroth order moment</u> then describes the <u>spread area</u>
of the cell, the two <u>first order moments</u> contain further
information needed to calculate the x and y locations of its
<u>geometrical centroid</u> (i.e. the centre of gravity or balance point
of the horizontal lamina) and the three <u>second order moments</u> are
needed to describe the cell's <u>orientation</u> and its basic <u>shape</u>
(the moments of inertia of the lamina).

3. Moments as shape factors

It is well known that the infinite series of geometrical moments
is a complete description of a binary image, it is simply a
transform of the information contained in the image (Teague,
1980). Roughly speaking, if as many numbers are used in a
description of a cell's shape using a <u>truncated moment series</u> as
in, say, a <u>polygonal approximation</u> to the cell outline, then the
two descriptions are approximately equivalent in the detail that
they contain. Other descriptions of cell shape also have this
property, such as the <u>median axis transform</u> (Blum, 1973 and

Noble, this section) and various <u>Fourier transforms of a plane closed curve</u> (Lord and Wilson, 1984). But the main advantage of using moments is that the lower orders give a particularly simple description of the fundamental properties of shape as well as containing valuable information about size, location, and orientation. This is why they are so commonly used by statisticians to describe the basic properties of a distribution.

The first three orders of moments, six numbers in all, can be manipulated to yield four numbers describing size, *x*-location, *y*-location and orientation and two numbers which are <u>shape factors</u>. These shape factors have the advantage of ignoring fine detail whereas many simple shape factors are sensitive to detail at the expense of ignoring the basic properties of the shape. This is particularly true of the commonly used shape factors based on <u>perimeter to area ratios</u> or those based on <u>'caliper' diameters</u> of cells. Of course, the final choice of shape factors depends on what features the investigator wants to measure but it should be remembered that measures that are sensitive to fine detail are also sensitive to the resolution of the measuring technique. Such quantities often carry as much information about the limitations of the measuring technique as they do about cell shape.

For analysing the basic properties of cell shape using moments, it is often unnecessary to go beyond the second order. However, the four <u>third order moments</u> carry information about <u>asymmetry of shape</u> which may prove useful for examining the asymmetry that almost inevitably accompanies the translocation of a cell. The higher order moments become much more difficult to interpret intuitively but can be used empirically as purely descriptive numbers to classify shapes in <u>automatic cell recognition systems</u> or for assaying the effect of an experimental treatment.

4. Calculating moments from a cell's image

The method of calculating moments depends on how the cell's outline data are stored. If these take the form of a <u>digitised image</u> as in a video frame store, then the arithmetic of

calculating the moments is very simple. If the outline is
described by a <u>polygonal approximation</u>, such as a sequence of
coordinates entered into a computer from a photomicrograph using
a digitising tablet, then the calculation becomes a little more
complicated.

The former approach has the advantage that data collection can be
fully automatic but its disadvantage is that, for phase contrast
or Nomarski microscopy of living cells, a good deal of
preprocessing of the image is often necessary in order for the
computer to 'recognise' the cell outline. This is greatly
simplified if interferometric microscopy is used (Brown and Dunn,
1989) or if the living cells can be made to fluoresce or can be
fixed and heavily stained. Otherwise, humans are usually much
better and quicker at interpreting and tracing cell outlines than
computers.

a) Digitised images:

In calculating the moments of a digitised image, the two-
dimensional integral of formula (1) is replaced by a discrete
summation over all the picture elements or <u>pixels</u> of the image:

$$(2) \qquad m_{jk} \quad = \quad \Sigma x^j y^k f(x,y)$$

For calculating the moments of the outline of an isolated cell,
the function $f(x,y)$ is given a value of zero if the pixel whose
central coordinates are (x,y) (in true cellular dimensions) lies
outside the outline. In practice, therefore, the summation need
only be taken over a portion of the x,y-plane containing the
outline of interest and it should be restricted in this way if
the image contains more than one cell outline. If the pixel (x,y)
lies within the outline then $f(x,y)$ takes a value corresponding
to the area of a pixel in true cellular dimensions.

Second and higher order moments calculated by discrete summation
in this way are not strictly accurate (even after allowing for
the fact that the digitisation is itself an approximation). This

is a well known source of error in calculating the variance of statistical data that are grouped into discrete intervals and can be corrected by a method due to Sheppard (see Kendall and Stuart, 1977, Vol. I, p 86). Fortunately, the error is small enough to be ignored if the cell area is many times larger than the area of a pixel.

The moments are more easily calculated in pixel address coordinates and converted to true cellular coordinates later. If the pixels are square and the moments are only to be used to calculate orientation and shape factors, then they need not be converted. Taking summations over the n pixels within a cell outline, the first three orders of moments are then simply:

0th: $m_{00} = n$

1st: $m_{10} = \Sigma x, \qquad m_{01} = \Sigma y$

2nd: $m_{20} = \Sigma x^2, \qquad m_{11} = \Sigma xy, \qquad m_{02} = \Sigma y^2$

b) Polygonal outlines:

According to Routh (1905), any geometrical moment of the form m_{j0} for a triangle with vertices $(0,0)$, (x_{i-1}, y_{i-1}), (x_i, y_i), taken in clockwise order, is given by:

$$(3) \qquad m_{j0} = \frac{2a_i}{(j+1)(j+2)} \cdot \frac{x_{i-1}^{j+1} - x_i^{j+1}}{x_{i-1} - x_i}$$

where $a_i = \frac{1}{2}(x_i y_{i-1} - x_{i-1} y_i)$ is the area of the triangle.

Other moments can be found by considering a rotation of the coordinate axes through 45° (Routh 1905). Since a_i does not change on rotation, it is not difficult to show that the sum of all moments of order p, with binomial coefficients, is given by:

$$(4) \qquad \sum_{k=0}^{p} \begin{bmatrix} p \\ k \end{bmatrix} m_{(p-k)k} = \frac{2a_i}{(p+1)(p+2)} \cdot \frac{(x_{i-1}+y_{i-1})^{p+1} - (x_i+y_i)^{p+1}}{(x_{i-1}+y_{i-1}) - (x_i+y_i)}$$

To find an expression for the moment m_{jk} it is necessary to substitute for p (= $j+k$) in the expression on the right and then, after dividing and expanding, to delete all terms that are not multiples of j x's and k y's, and finally to divide by the binomial coefficient p over k. This can be written as:

$$(5) \qquad \begin{bmatrix} p \\ k \end{bmatrix} m_{jk} \;=\; \frac{2a_i b_{ijk}}{(p+1)(p+2)}$$

where $b_{ijk} \;=\; x_{i-1} b_{i(j-1)k} + y_{i-1} b_{ij(k-1)} + \begin{bmatrix} p \\ k \end{bmatrix} x_i^j y_i^k$

and $x_i^0 = y_i^0 = 1$ (for all x_i, y_i) and $b_{i(-1)k} = b_{ij(-1)} = 0$

If a polygonal approximation to a cell outline is represented by a sequence of coordinate pairs (in true cellular dimensions) taken in clockwise order: (x_0, y_0), $(x_1, y_1) \ldots (x_n, y_n)$, where $(x_0, y_0) = (x_n, y_n)$, then it follows from the additive property of moments that any moment m_{jk} of the polygon can be found by summing the moment for n of these triangles over i = 1 to n (moments for the anticlockwise triangles automatically cancel those for the clockwise triangles outside the polygon):

$$(6) \qquad \begin{bmatrix} p \\ k \end{bmatrix} m_{jk} \;=\; \frac{2}{(p+1)(p+2)} \sum_{i=1}^{n} a_i b_{ijk}$$

This gives expressions for the first three orders of moments as:

$$(7) \qquad m_{00} = \sum_{i=1}^{n} a_i \qquad\qquad (8) \qquad m_{10} = \frac{1}{3} \sum_{i=1}^{n} a_i (x_{i-1} + x_i)$$

$$(9) \qquad m_{20} = \frac{1}{6} \sum_{i=1}^{n} a_i (x_{i-1}^2 + x_{i-1} x_i + x_i^2)$$

$$(10) \qquad m_{11} = \frac{1}{12} \sum_{i=1}^{n} a_i (2x_{i-1} y_{i-1} + x_{i-1} y_i + x_i y_{i-1} + 2x_i y_i)$$

The moments m_{01} and m_{02} are found by substituting y for x in (8) and (9) (keeping the same subscripts).

5. Moment invariants

The moments m_{jk}, which may be called <u>raw moments</u>, suffer from the problem that the different aspects of information about cell size, location, orientation and shape are distributed among them in an inconvenient jumble. For example, the values of all raw moments are dependent on the area of the cell. This problem is easily overcome, however, by designing <u>moment invariants</u>. Described crudely, the technique is to extract the useful information contained in each moment and then to transform all the remaining moments to eliminate this information.

This procedure can be envisaged as a sequence of <u>geometrical transformations</u> of the cell outline. First the outline is moved parallel to the x-axis until the transformed m_{10}, denoted by $'m_{10}$, is equal to zero. The amount of this shift is an inverse measure of the x-location of the outline's geometrical centroid. This is repeated for the y-axis until $'m_{01} = 0$. These two transformations generally change the values of all other moments apart from the zeroth order moment. The new moments $'m_{jk}$ are known as <u>central moments</u> and have the property that they are <u>invariant to translation</u>. Next the outline is magnified or reduced in size until the new moment $''m_{00} = 1$. The amount of this transformation is obviously an inverse measure of the outline's area. The moments $''m_{jk}$ are now called <u>normalised central moments</u> and are <u>invariant to dilation</u> (i.e. to change of scale). Next the outline is rotated until $'''m_{11} = 0$ and $'''m_{20} > '''m_{02}$. The amount of this rotation, which is always in the range $\pm\pi/2$, is an inverse measure of the orientation of the original outline. The two new moments $'''m_{20}$ and $'''m_{02}$ are <u>rotational invariants</u> and their values are entirely independent of the choice of coordinate axes. They are pure measures of shape and are often simply called moment invariants.

6. Calculating moment invariants

First the relevant information is extracted from the first two orders of raw moments:

<u>spread area</u> = a = m_{00}

<u>x-location of geometrical centroid</u> = $\bar{x}$ = m_{10}/m_{00}

<u>y-location of geometrical centroid</u> = $\bar{y}$ = m_{01}/m_{00}

Central moments are defined as:

(11) $'m_{jk} = \displaystyle\int_{-\infty}^{\infty} \int_{-\infty}^{\infty} (x-\bar{x})^j (y-\bar{y})^k f(x,y)\,dxdy$

For the second order, the raw moments may be transformed into central moments as follows:

$'m_{20} = m_{20} - a\bar{x}^2$ $'m_{11} = m_{11} - a\bar{x}\bar{y}$ $'m_{02} = m_{02} - a\bar{y}^2$

Higher order formulae for this are derived easily by substituting for j and k in (11) and expanding. Hu (1962) gives the third order formulae.

Normalised central moments are defined by Teague (1980) as:

(12) $''m_{jk} = 'm_{jk}/a^{(j+k+2)/2}$

Greater accuracy can usually be obtained in calculating the normalised central moments by transforming the outline data instead of the raw moments. The outline can be normalised and centralised after calculating the zeroth and first order raw moments.

The <u>orientation</u>, ϕ, is given by:

(13) $\phi = \dfrac{1}{2} \tan^{-1}\left[\dfrac{2''m_{11}}{''m_{20}-''m_{02}}\right]$

The value of the arctangent function should be chosen so that ϕ is the tilt angle of the long axis of the outline with respect to the x-axis and $-\pi/2 \leq \phi \leq \pi/2$. Teague (1980) gives a method of doing this which is equivalent to using the dividend and divisor as the first and second arguments for the ATAN2 function in Fortran 77. The orientation is undefined if both arguments are zero and ϕ should be set to zero for the following in this case.

The second order rotational invariants are the two <u>principal</u> <u>moments</u> obtained by rotating the outline through the angle $-\phi$. They are calculated as:

(14) $'''m_{20} = ''m_{20}c^2 + 2''m_{11}cs + ''m_{02}s^2$

(15) $'''m_{02} = ''m_{20}s^2 - 2''m_{11}cs + ''m_{02}c^2$

where: $c = \cos(\phi)$ $s = \sin(\phi)$

7. Interpreting the second order moments

a) Area, location and orientation:

Assuming that the cell outline is represented by a thin lamina of uniform unit density, it is always possible to construct a unique elliptical lamina with density ≤ 1 that exactly matches all the true physical moments of the cell lamina through the second order (Teague 1980). This is known as the <u>Legendre or equimomental</u> <u>ellipse</u>. It is a model of the size, position, orientation and shape of the cell as far as can be described by the geometrical moments of the first three orders. The elliptical lamina's mass equals the area of the cell lamina since the masses of the two laminas must be equal (in order to match zeroth order physical moments) and the cell lamina has unit density. The centroid of the ellipse is obviously at its centre and so this coincides with the centroid of the cell lamina. This geometrical centroid is an unambiguous description of the location of the cell and a sequence of such locations is very useful for describing the translocation of a living cell (although the true physical centroid of dry mass described earlier is probably preferable for precise work). The long axis of the ellipse is the axis about which its <u>moment of inertia</u> is a minimum and this serves to define the <u>long axis</u> of the cell lamina; the angle that this makes with the x-axis is therefore a useful description of the orientation of the cell. Orientation is obviously not definable if the ellipse happens to be a circle and we can then say that the cell outline has no unique long axis. Two numbers complete the description of the ellipse, the lengths of its two principal radii, and these contain information about the shape of the cell.

b) Elongation:

Consider transforming the equimomental ellipse in the same way as we transformed the cell outline. It now has unit mass and lies with its centre on the origin and its long axis coinciding with the x-axis. All its moments of the first three orders still match the moments of the transformed cell lamina. Its long and short radii, a and b, are simply related to the two moment invariants: $a = 2\sqrt{'''m_{20}}$, $b = 2\sqrt{'''m_{02}}$ (Routh, 1905). The problem is how to make two convenient measures of shape from these two numbers. Dunn and Brown (1986) proposed a simple measure of the _elongation_ of the cell which is equivalent to $\log_2(a/b)$. _Elongation_ takes a minimum value of zero if the cell is not elongated, i.e. if the equimomental ellipse is a circle, and it takes a value of 1 if the ellipse is twice as long as it is wide. This is quite easy to interpret since an approximately elliptical cell with an _elongation_ of 2, for example, is approximately four times as long as it is wide.

Elongation can be seen as a measure of yet another transformation in the sequence of transformations. This consists of compressing the cell outline along the x-axis (and stretching it along the y-axis to maintain unit area) until the new moments $''''m_{20}$ and $''''m_{02}$ are equal. Thus five transformations yield five descriptions of the outline's geometrical properties and leave one item of information from the first three orders of moments which is invariant to all five transformations.

c) Dispersion:

Consider the moment of inertia of the transformed cell lamina about the z-axis perpendicular to the x,y-plane. The last transformation has reduced this moment as far as possible by making the cell shape as compact and circular as possible. But, in general, the cell lamina is still not circular (unless the original cell shape was elliptical) and any protrusions will have the effect of dispersing its mass away from the centroid and increasing this moment of inertia - much as a spinning ice skater extends limbs to achieve the same effect. After the last transformation, the equimomental ellipse is now a circle whose area is still πab. Since it still matches all moments of the cell lamina through the second order it has unit mass and the only way that it can disperse this mass to match the moment of inertia of the cell lamina is by having an area greater than unity (recall that the equimomental ellipse need not have unit density). The area of the normalised equimomental ellipse is therefore a measure of this dispersion of mass away from the centroid. Dunn and Brown (1986) proposed _dispersion_ as a measure of how much a cell shape differs from an ellipse and it can be shown that this is equivalent to $\log_2(\pi ab)$. A cell of elliptical shape has the minimum _dispersion_ of zero. A cell shape with a _dispersion_ of 1 has a non-normalised equimomental ellipse of twice the cell's area and a density of 1/2. Figure 2 shows how it is possible to use this theory of equimomental ellipses to construct a simple shape with any given _elongation_ and _dispersion_.

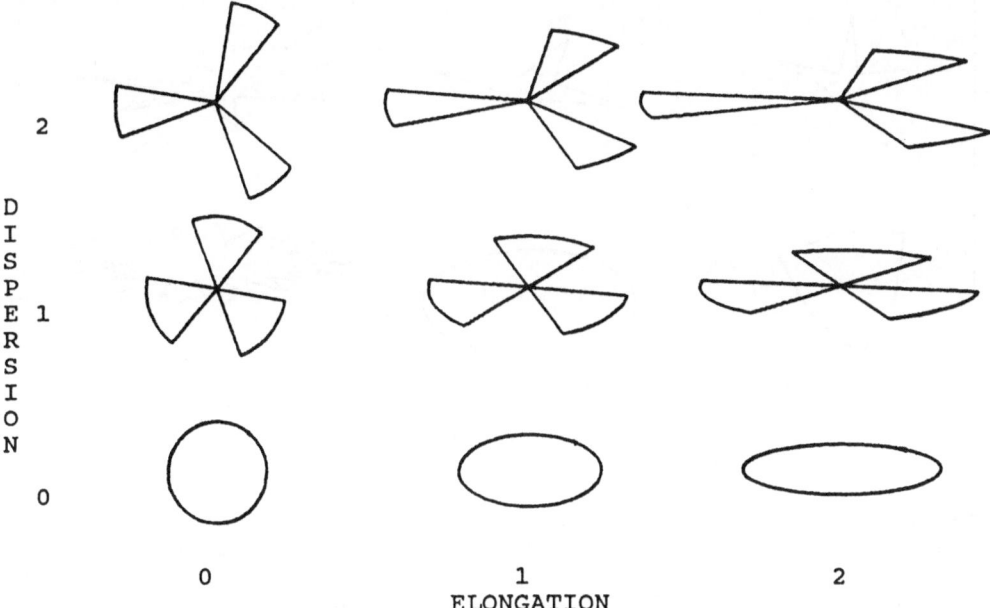

D
I
S
P
E
R
S
I
O
N

2

1

0

0   1   2
ELONGATION

Figure 2. Each outline has unit area. Each of the two upper outlines in the first column is created by removing three (or more) uniformly spaced segments from a circle of area equal to $2^{dispersion}$ so that the remainder has unit area. Each remaining column is created by applying a parallel projection transformation to the first column so that the ellipse on the bottom has a ratio of long to short radius equal to $2^{elongation}$.

d) Extension:

In some circumstances it is convenient to have a single measure
to describe the lack of compactness of a cell shape. Dunn and
Brown (1986) suggested a measure called *extension* which is the
sum of *elongation* and *dispersion*. This takes the minimum value of
zero only if the cell shape is circular. Figure 3 summarises the
three shape measures applied to various shapes. With practice, it
is not difficult to make a good guess at the extension,
elongation and dispersion of a cell outline.

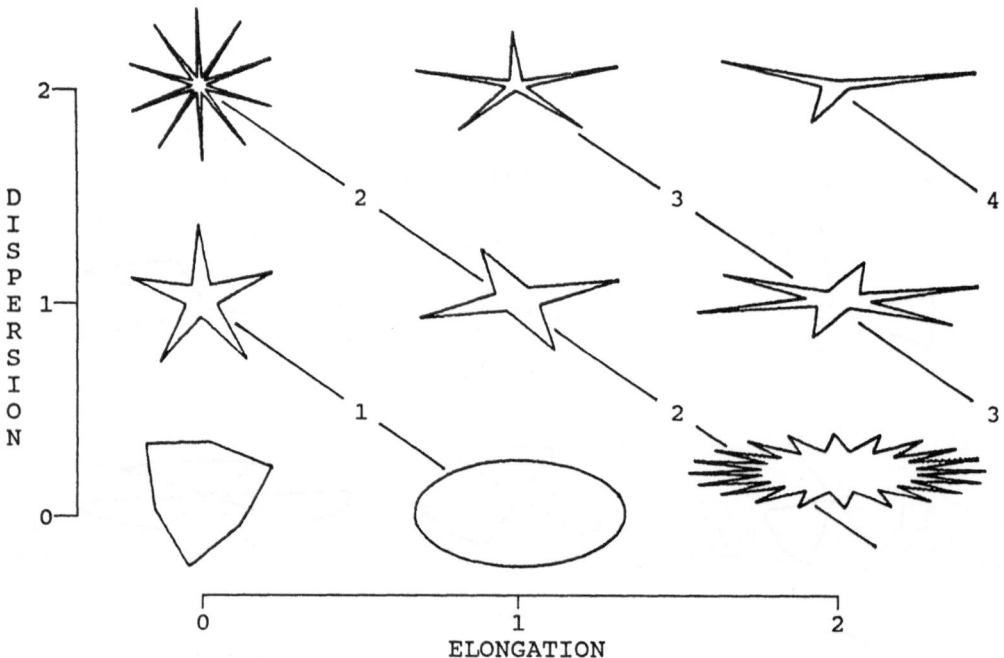

Figure 3. *Extension, dispersion* and *elongation* are shown for
various geometrical shapes. Each outline is positioned so that
its centroid coincides with the point determined by its
elongation and *dispersion*. *Extension* increases diagonally away
from the origin and values of *extension* are shown on the diagonal
lines. Note that the outline in the bottom right hand corner has
a low *dispersion* (0.2) even though its perimeter to area ratio is
very high. This illustrates that *dispersion* is relatively
insensitive to fine detail and is affected mainly by the grosser
features of shape.

8. Biological significance of elongation and dispersion

Although we designed *elongation* and *dispersion* to be readily interpretable as measures of cell shape, it is interesting that they also seem to have biological significance. In one case, at least, this significance appears to arise from the simple transformation properties of the shape factors. *Dispersion* is invariant to stretching (parallel projection) of the cell shape as well as to the other transformations and so it is invariant to all the so-called <u>affine transformations</u>. Dunn and Brown (1986) observed that culturing fibroblasts on a finely grooved substratum affects their mean *elongation* but does not detectably affect their mean *dispersion* (Figures 4a, 4b). This absence of effect on *dispersion* is strong evidence that the biological effect of the substratum is to impose a parallel projection or stretch transformation on the cell shape (and possibly on the cell translocation although this remains to be studied). Applying the inverse of the calculated transformation to the experimental outlines results in a sample of outlines that is statistically indistinguishable from the control sample in terms of *dispersion* and residual *elongation* (Figure 4c). Further testing of whether the biological effect is truly a transformation could be done by comparing the shapes of the transformed outlines of Figure 4c with those of the control outlines of Figure 4a using higher order moment invariants (see next section).

Another strong indication that these shape factors have a biological relevance is that the tumour promoter TPA affects the mean *dispersion* of fibroblasts without significantly affecting their mean *elongation* (Brown et al. 1989) as shown in Figure 5. In this case there is very little overlap between the control and experimental samples. Thus each shape factor can be unaffected by an experimental treatment that strongly affects the other which suggests that the two shape factors are measures of biologically distinct processes. Not all treatments, however, affect only one of the shape factors and Middleton et al. (1988) showed that treatment with either colcemid or nocodazole reduces both *elongation* and *dispersion* in secondary fibroblasts.

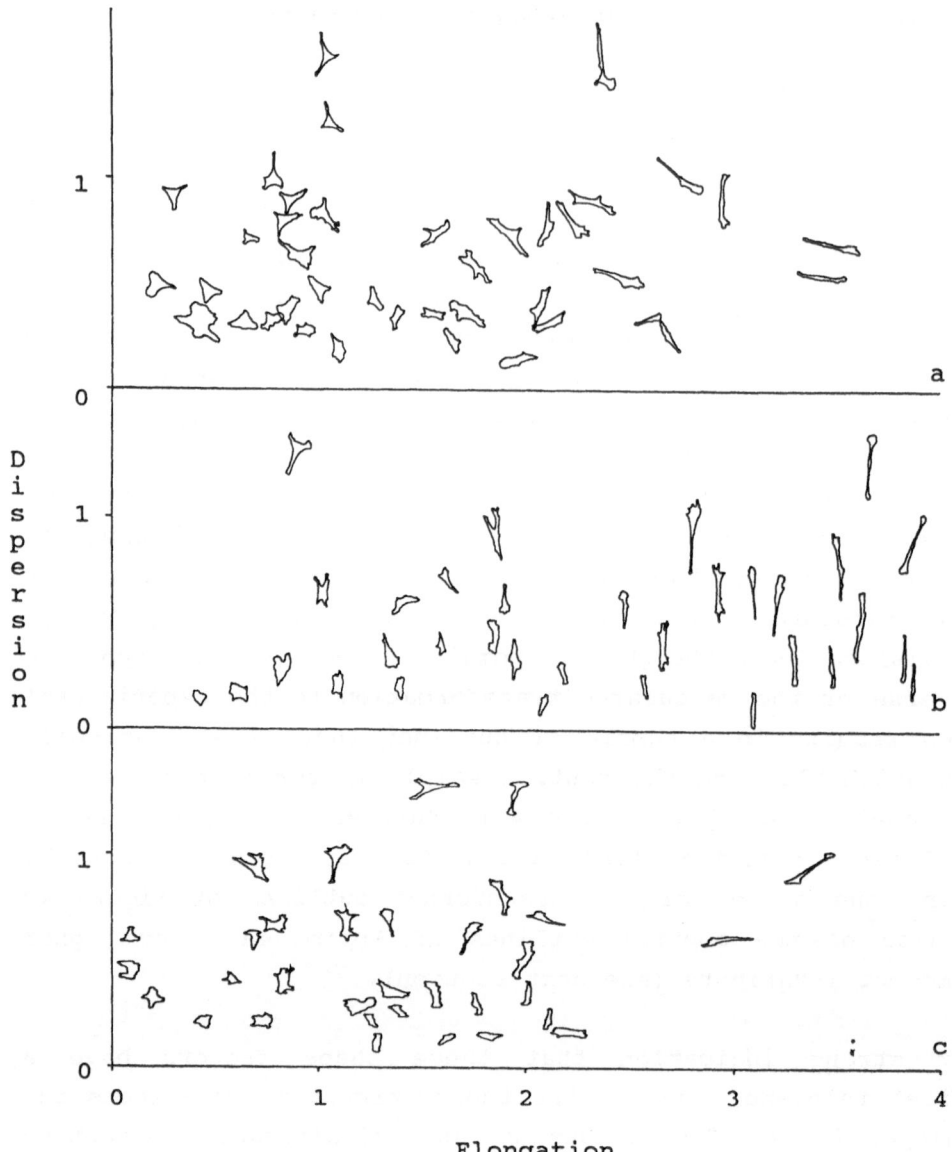

Elongation

Fig.4 Outlines selected from samples of fibroblasts cultured on
planar (a) and grooved substrata (b). The inverse of the mean
transformation imposed by the grooves was applied to the outlines
in (b) resulting in the outlines in (c). The centroid of each
outline coincides with the point determined by its *elongation* and
dispersion. (see text for details.)

	Sample mean *elongation* ± sem	Sample mean *dispersion* ± sem	
(a)	1.5874 ± 0.0701	0.5453 ± 0.0270	n = 146
(b)	2.3858 ± 0.1001	0.4552 ± 0.0267	n = 116
(c)	1.5254 ± 0.0724	0.4552 ± 0.0267	n = 116

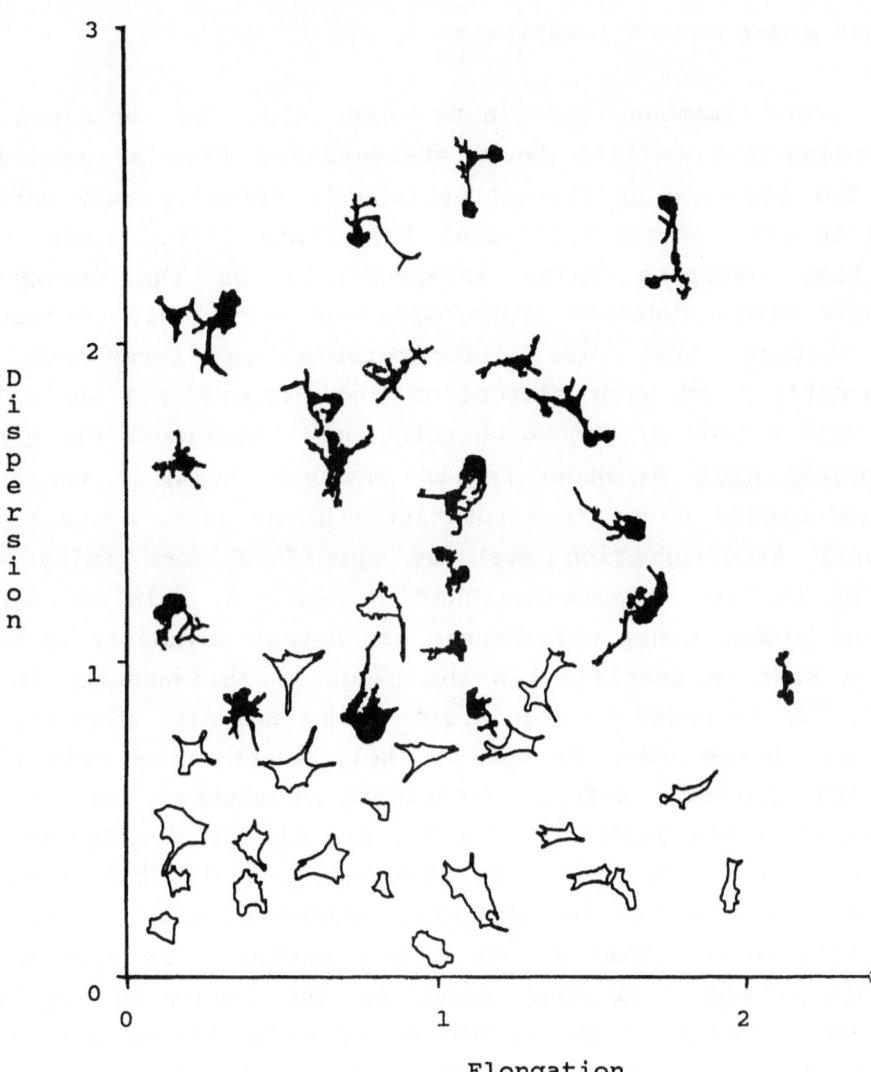

Fig.5 Outlines selected from samples of mouse fibroblasts of the CAK-7 line cultured in control medium (open outlines) and medium containing 100ng/ml 12-o-tetradecanoyl phorbol-13-acetate (solid outlines). The centroid of each outline coincides with the point determined by the *elongation* and *dispersion* of the cell's shape. (see Brown et al 1989 for full details.)

	Sample mean *elongation* ± sem	Sample mean *dispersion* ± sem	
Control	0.8304 ± 0.0676	0.4223 ± 0.0335	n = 57
TPA-treated	0.8226 ± 0.0673	1.4455 ± 0.0784	n = 50

9. Higher order moment invariants

Higher order moment invariants can also be obtained by transforming the outline to a standardised form as described above. The sequence of transformations is normally only carried through to produce the rotational invariants $'''m_{jk}$ since it is these that describe shape independently of the choice of coordinate axes. (Further transformations serve only to produce shape factors that are interpretable in terms of the transformations as with *elongation* and *dispersion*.) Hu (1962) called this method of producing rotational invariants the Method of Principal Axes. In order for the moments $'''m_{jk}$ of the third order and higher to be true rotational invariants, however, the rotational transformation must be specified more fully than before by further stipulating that $'''m_{30} > 0$. This is because third and higher order odd moments can detect asymmetry in shape and so ϕ must be specified in the range $\pm\pi$ (orientation in the full circle) in order to standardise these moments. Even so, the method can break down in the unlikely event of a cell shape possessing perfect n-fold rotational symmetry because the principal axes are undefined if $n > 2$ and all third order moments are zero if $n = 2$ or $n > 3$. In some cases, it is still possible to patch up the method and Hu (1962) suggested how to deal with 3-fold rotational symmetry. But a more satisfactory approach is to design moment invariants based on the theory of algebraic forms. Hu called this the method of Absolute Moment Invariants and gave the formulae for seven such invariants (the seventh detects reflections of the shape) as functions of the first four orders of normalised central moments. Teague (1980) modified this approach and extended it to include higher orders using Zernike moments. The number of absolute moment invariants that can be constructed is now practically unlimited.

10. Interpreting higher order moments

a) Shape vectors and phase space:

Third and higher order moment invariants become increasingly difficult to interpret as the order increases. One approach to this problem is not to attempt to interpret them but to treat them as a purely empirical description of cellular shape. Higher order moment invariants can be used to construct a shape vector for a cell outline. This is simply a list of values of moment invariants which serves as a description of the shape of the outline up to a chosen level of detail. If there are N shape factors in the list, then the shape vector can be thought of as specifying the position of the cell in an imaginary N-dimensional phase space. The difference between two shapes can then conveniently be described by a single number which is simply the distance between the two representative points in phase space. If an experimental treatment affects cell shape, then we can think of the control and experimental samples as two clouds of points in phase space (a multidimensional version of Figure 5) and the question of how effective the chosen shape vector is for quantifying the response reduces to the question of to what extent the clouds overlap in phase space.

An extension of this approach is to consider how the shape vector for a cell changes during the treatment. The shape vector for each cell now describes a curvilinear trajectory in phase space and the clustering and parallelism of these trajectories is an indication of the uniformity in the effect of the treatment. The spontaneous shape changes of 'randomly' moving cells may also be described by such trajectories and the identification and characterisation of well worn tracks in phase space may eventually lead to a deeper understanding of the nature of common shape changes in moving cells. Regions of track divergence, for example, may correspond to states of the cells in which they are particularly susceptible to external influences which may control their motility.

As mentioned earlier, the empirical description of shape is quite adequate for the recognition and classification of cell shape by computer and for assaying the effect of an experimental treatment. In these cases it may be possible to use Principal Component Analysis to reduce or collapse the dimensions of phase

space, while retaining most of the discriminating power of the
shape vector for a particular application, by constructing new
moment invariants from linear combinations of the old ones. If
three or less of these new invariants still have adequate
discriminating power, the phase space becomes much easier to
visualise and to treat statistically.

b) The inverse moment problem:

An obvious approach to interpreting the higher order moment
invariants is to construct a simple outline with the same shape
vector as the cell outline. If this could be done it would be
much easier to understand intuitively which features of the cell
outline are described by the shape vector. This is known as the
inverse moment problem. Unfortunately, there is no unique and
theoretically satisfying simple outline that can match any given
outline even for the first three orders of moments. While it is
true that the equimomental ellipse matches a cell lamina for all
physical moments of the first three orders, this is not a valid
outline since it has less than unit density. Certain simple
outlines, such as those in Figures 2 and 3, can be contrived to
match the first three orders of moments of any outline but their
arbitrariness makes them generally unsatisfactory as summaries of
the features of the cell outline.

The inverse moment problem has a much more satisfactory solution
for reconstructing an image of graded intensity from its
truncated moment series (Teague 1980, Boyce and Hossack 1983).
This method can be applied to a cell outline if the outline is
treated as a silhouette or binary image with two intensities: say
black background and white cell. The image reconstructed from the
truncated moment series is not binary but graded in intensity.
Teague suggests that a binary image may be obtained from the
reconstructed image by thresholding or taking the contour of a
particular intensity. The reconstructed binary image does not,
however, exactly match the moments of the original within the
chosen orders and it suffers from the further defect that the
reconstructed outline may consist of more than one closed curve.

Nevertheless, if a suitable threshold is chosen, the method gives quite a good impression of the information contained in the moments and Teague shows examples of the letters E and F reconstructed through various orders of moments.

c) Higher order shape factors:

Yet a third approach is to construct more readily interpretable shape factors from the moment invariants in the same way that elongation and dispersion are constructed from the second order invariants. But, before appealing to the gymnastic abilities of the mathematicians, we should have some ideas about what features of cell shape we wish the shape factors to describe and discriminate. This problem is really a vicious circle because we do not yet understand the biological significance of various shape changes and we need some sophisticated methods for describing and quantifying shape before we can discover any general patterns. There are, however, some general biological principles that might guide this search for a universal language of cell shape.

One fairly well accepted principle of cell motility is that the number of active processes (protrusions or pseudopodia) and/or passive processes (retraction fibres and 'tails') is indicative of the commitment of a cell to move in a particular direction. Many subjective classifications of the shapes of motile cells are therefore based on counting processes. These subjective classifications vary enormously with different observers and it may, therefore, be useful to design moment invariants as process counters. It is quite likely, however, that Fourier transforms of a closed curve will be more adaptable than moments for this since the magnitude of the nth harmonic is an obvious candidate for measuring n-polarity. The median axis transform is also useful for visualising the mulitpolarity of a cell (Noble, this section) but deriving shape factors for process counting using this approach presents some difficulties. A problem with both approaches is that of distinguishing between active and passive processes and it is here that moments may prove to be more effective. For example, a combination of second and third order moment invariants would probably be effective for distinguishing a bipolar cell (two active processes) from a monopolar cell (one active and one passive process) because the third order invariants would be able to detect any asymmetry due to a difference in shape between the active and the passive process.

Another principle of biological motion is that the motile organism or cell tends to proceed in a direction along the axis of greatest bilateral symmetry. We have just seen that the third order moments are well suited for detecting and measuring any asymmetry in a cell's shape. For a centralised and normalised cell outline, consider rotating the coordinate axes through a full circle. The moment m_{30} as a function of the angle θ through which the axes are rotated is given by:

(16) $m_{30}(\theta) = ''m_{30}c^3 + 3''m_{21}c^2s + 3''m_{12}cs^2 + ''m_{03}s^3$
       where: $c = \cos(\theta)$ $s = \sin(\theta)$

Plotted as a polar curve, this moment generally (but not invariably) shows three positive lobes which exactly coincide with three negative lobes (since $m_{30}(\theta) = -m_{30}(\theta+\pi)$). Plotting the curve superimposed on a cell outline reveals the pattern of asymmetry in the outline (Figure 6). If the cell has a single passive process or tail, this very commonly coincides with the largest lobe (see bottom two rows in Figure 6). In analysing the outlines of moving cells, we find that the largest of the three lobes is generally oriented in the opposite direction to the direction of translocation of the cell. This approach is already proving useful for detecting the direction and magnitude of any asymmetry in the response of cells to a putative chemotactic field.

As a method of visualising and summarising the third order moments, the polar curve of $m_{30}(\theta)$ is loosely analogous to the equimomental ellipse of the second order moments (a polar curve of $2\sqrt{m_{20}(\theta)}$ is an oval shape not very different from the equimomental ellipse). This suggests that it may be possible to

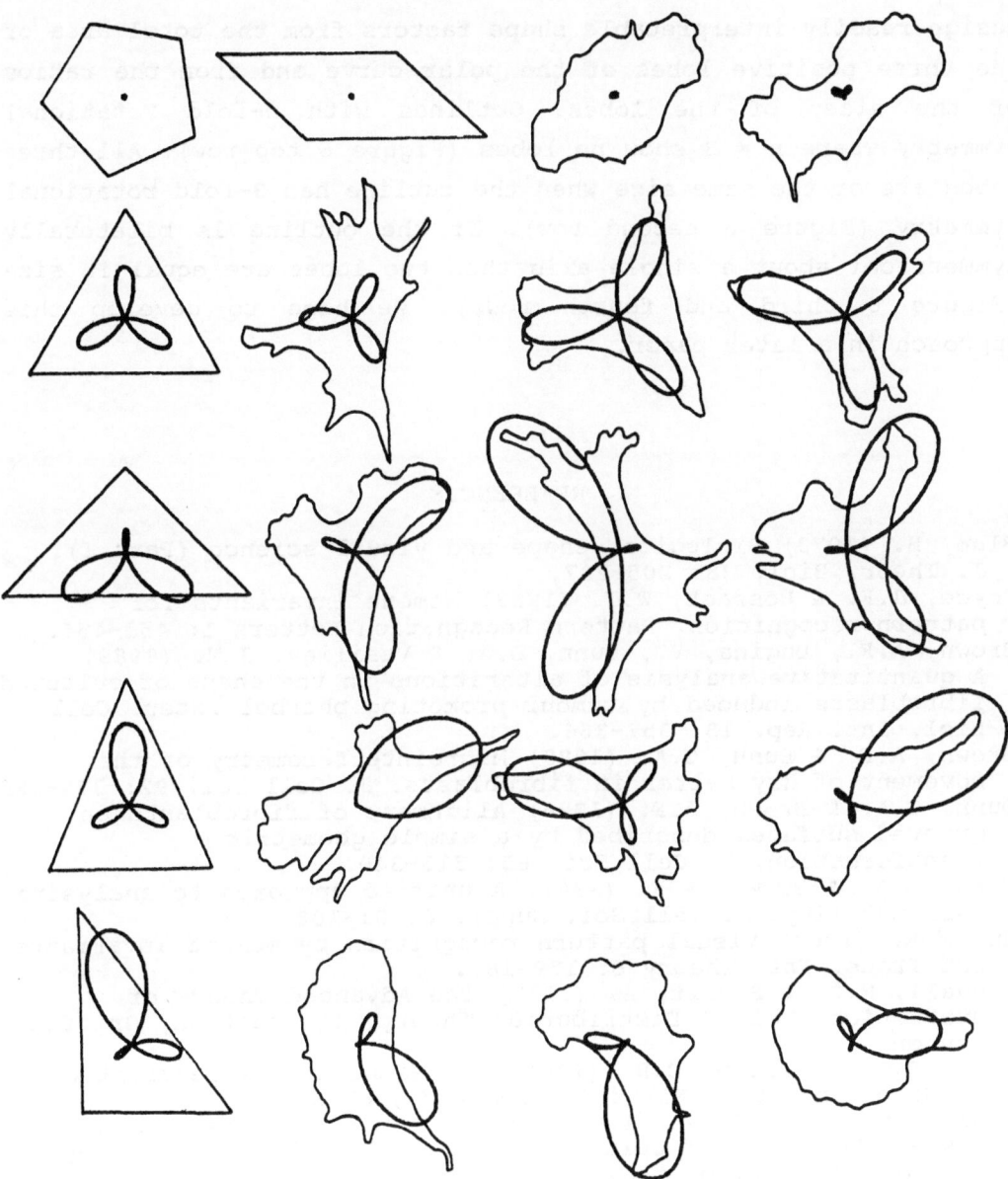

Fig.6 Polar curves of $m_{30}(\Theta)$ superimposed on geometrical outlines and outlines obtained from various fibroblasts. All outlines are centralised and normalised to unit area. The polar curves are shown at 25x the scale of the outlines.

design readily interpretable shape factors from the total area of
the three positive lobes of the polar curve and from the ratios
of the sizes of the lobes. Outlines with n-fold rotational
symmetry where $n \neq 3$ show no lobes (Figure 6 top row). All three
lobes are of the same size when the outline has 3-fold rotational
symmetry (Figure 6 second row). If the outline is bilaterally
symmetrical about a single axis then two lobes are equal in size
(Figure 6 third and fourth rows). We hope to develop this
approach in a later paper.

REFERENCES

Blum, H. (1973) Biological shape and visual science (Part I).
  J. Theor. Biol. 38: 205-287.
Boyce, J.F. & Hossack, W.J. (1983) Moment invariants for
  pattern recognition. Pattern Recognition Letters 1: 451-456.
Brown, A.F., Dugina, V., Dunn, G.A. & Vasiliev, J.M. (1989)
  A quantitative analysis of alterations in the shape of cultured
  fibroblasts induced by tumour-promoting phorbol ester. Cell
  Biol. Int. Rep. 13: 357-366.
Brown, A.F. & Dunn, G.A. (1989) Microinterferometry of the
  movement of dry matter in fibroblasts. J. Cell Sci. 92: 379-389
Dunn, G.A. & Brown, A.F. (1986) Alignment of fibroblasts on
  grooved surfaces described by a simple geometric
  transformation. J. Cell Sci. 83: 313-340.
Dunn, G.A. & Brown, A.F. (1987) A unified approach to analysing
  cell motility. J. Cell Sci. Suppl. 8: 81-102.
Hu, M-K. (1962) Visual pattern recognition by moment invariants.
  IRE Trans. Inf. Theory 8: 179-187.
Kendall, M.G. & Stuart, A. (1977) The Advanced Theory of
  Statistics. Vol. I Distribution Theory. 4th edition, Griffin,
  London.
Lord, E.A. & Wilson, C.B. (1984) The Mathematical Description
  of Shape and Form. Ellis Horwood Ltd., Chichester.
Middleton, C.A., Brown, A.F., Brown, R.M. & Roberts, D.J.H.
  (1988) The shape of cultured epithelial cells does not depend
  on the integrity of their microtubules. J. Cell Sci. 91: 337-
  346.
Middleton, C.A., Brown, A.F., Brown, R.M., Karavanova, I.D.,
  Roberts, D.J.H. & Vasiliev, J.M. (1989) The polarization of
  fibroblasts in early primary cultures is independent of
  microtubule integrity. J. Cell Sci. 94: 25-32.
Routh, E.J. (1905) Dynamics of a System of Rigid Bodies:
  Elementary Part. Constable, London (also Dover edition, New
  York 1960).
Teague, M.R. (1980) Image analysis via the general theory of
  moments. J. Opt. Soc. Am. 70: 920-930.

ANALYSIS OF LEUCOCYTE SHAPE CHANGES

John M. Lackie
Department of Cell Biology
University of Glasgow
Glasgow G12 8QQ
Scotland / U.K.

Hansuli Keller
Siegfried Eisele and
Arthur Zimmermann
Institute of Pathology
University of Bern
3010 Bern/Switzerland

Abstract

The shape changes undergone by human blood lymphocytes in suspension were analysed by using morphological criteria and by calculating shape parameters of cell outlines according to the method proposed by Dunn and Brown (1986). The biological significance of the shape changes and the utility of the different methods are discussed.

1. Introduction

1.1 Functional significance of shape changes

Shape changes of leucocytes reflect, to some extent, the functional state of the cell. Unstimulated blood leucocytes are spherical, but in order to leave the circulation or to engage in phagocytosis, the motor machinery of the cells must be activated. The activation of the internal motor machinery seems to be correlated with alteration in shape and with functional activity; non-polar cells with small surface projections show little locomotory activity, although they are active in pinocytosis, whereas polarised cells will move well. Different stimuli such as chemotactic peptides, protein kinase C activators (PMA, diacylglycerols) or inhibitors (H-7), and microtubule perturbing agents (colchicine, nocodazole, vinblastine) or heavy water, all bring about distinct shape changes (Lewis, 1934; Keller et al., 1984, 1989; Roos et al., 1987; Zimmermann et al., 1988). Although the shift from spherical to non-spherical is easily scored, it is clear even from brief inspection that diverse morphologies are elicited by various stimuli, even though it is more difficult to characterise the differences between activated cells. We have attempted to characterise the shapes induced by various agents.

Leucocytes are an excellent model system in which to study shape
changes since they are readily obtained in non-stimulated form and a
range of agonists with which to activate them is available. Since a
variety of shape changes can be demonstrated to occur in leucocytes,
and since these changes are related to function and to the nature of
the stimulus, it is important to develop efficient and reliable
methods for classifying shapes. Morphological methods and the
application of the shape parameters described by Dunn and Brown (1986)
and Box 1 are discussed below.

1.2 Morphological classification of cell shape

The shape adopted by many cells in culture is determined in part by
the locomotory machinery of the cell interacting with the substratum
through contacts which support translocation. However, many cells
will show shape changes in suspension, and these changes are
independent of interactions with the substratum and resemble those
shown on substrata of low adhesiveness. Shape change in suspension
therefore provides a more straightforward assay of the activity of the
motor system. Shape categories which have been established using
direct visual analysis are based upon a complex pattern of
information. The microscope image is not simply two-dimensional, and
by optical sectioning the three-dimensional structure can be assessed.
Other structural information is also available to the observer, for
example the presence of vacuoles, the position of the nucleus, the
granularity of the cytoplasm. All these features can contribute to
the classification. Merely considering the projected outline of the
cell may grossly oversimplify the morphological structure, particularly the nature of projections from the cell. By direct examination
it is possible to distinguish pseudopod shapes, which may vary from
thin filopodial projections, through broad flattened lamellipodia to
blunt bleb-like protrusions. Differential interference contrast
microscopy (Nomarski-type optics) is particularly suited to this sort
of morphological examination. The advantage of the direct method is
that a classification can be based on several factors, the problem is
that it requires some experience of interpretation of visual images.
Although for experienced microscopists the reproducibility of the
classification is good (Roos et al., 1987), the criticism of
subjectivity in the method is often made. A more serious problem is
that in some situations (with H-7 as a stimulus for example) there are
transitional forms and it may be difficult to assign cells to a

category; as with many biological systems there is in reality a continuum of structures which is being arbitrarily divided. If the kinetics of transition from one shape to another are altered, then more cells will be of intermediate morphology. More precise and technically simpler methods would be attractive.

1.3 Determination of mathematical parameters of shape

There are at least two reasons for wishing to derive numerical descriptions of shape, one is to automate the classification of shape using computer-assisted image analysis, the second is to try to put the classification onto a more objective footing. The major problem is that the analysis is likely to be based upon a reduced amount of information, since many of the visual cues used by an experienced observer are very difficult to incorporate into an image-processing system.

Measures of shape used so far, do not take into account some of the information used in direct visual classification, in particular the details of surface morphology. It is usually possible to determine by direct visual observation whether a narrow protrusion is a lamellipod or a tail, but this may be impossible from an outline drawing of the cell. On the other hand some information can be obtained from numerical shape analysis which is not obvious using direct observation. For example, when neutrophil leucocytes are stimulated with chemotactic peptides the proportion of cells polarised increases, but there are also differences in the shapes of the polarised cells, with maximal elongation being produced at concentrations of peptide which are optimal for stimulating locomotion (Keller, 1983). Thus there is potentially more information available than simply the proportion of polarised cells, or the proportion adopting a particular type of shape - though the proportion of non-spherical cells is a useful first parameter to measure. The numerical data is particularly valuable when comparing changes brought about by different doses of agonists. Both direct observation and numerical image analysis must be used to extract the maximum amount of information.

2. Analysis of shape changes in human blood lymphocytes

The shape parameters proposed by Dunn & Brown (1986) represent a promising new approach to cell shape analysis, and have the advantage that they are independent of the projected area of the cell (for details see Dunn and Brown, this section I/1). Recently, these shape factors have been applied to the analysis of shape changes of human blood lymphocytes (Wilkinson et al., 1988) and to the shape of cultured epithelial cells (Middleton et al., 1988). With lymphocytes the dispersion parameter (a measure of the irregularity of contour) discriminates efficiently between spherical cells (dispersion <0.02) and non-spherical cells (dispersion >0.02). The elongation parameter was found to be the most useful to describe the shapes of the non-spherical cells (the extension parameter includes both elongation and dispersion).

It was our aim to try to subdivide the shapes of non-+spherical cells, which subjectively seem to differ according to the agonist used, on the basis of these shape measures. Thus we hoped to distinguish those cells which had a distinct front-tail polarity from cells with non-polar projections, and to provide a more complete description of the shapes of stimulated cells.

The shapes of unstimulated cells and cells treated with 10^{-5} M colchicine or 10^{-8} M PMA, were assessed using visual classification (Fig. 1) and using the calculated shape parameters based on outlines. At least 100 cells were used in each category. As found by Wilkinson et al. (1988), the dispersion parameter discriminated well between spherical and non-spherical cells, but the elongation parameter for

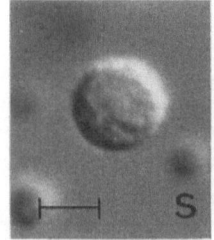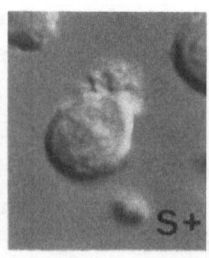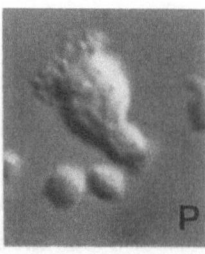

Figure 1. Typical examples of different shapes of human lymphocytes. S: spherical cell; S+: spherical cell with unifocal projections; P: cell showing front-tail polarity; NP: non-polar cell with surface projections. Differential interference contrast optics. Scale bar: 10

Table 1

Shape changes of human lymphocytes stimulated with colchicine or phorbol-myristate-acetate (PMA)

Stimulus		calculated measures of shape (median)			Visual classif.
		Extension	Dispersion	Elongation	
none	A	0.171	0.005	0.168	s=97%; s+=3%
colchicine	B	0.139	0.004	0.134	s=86%; s+=2%; p=10%
colchicine	B_1	0.958	0.125	0.773	p=100%
PMA	C	0.465	0.079	0.371	s=4%; s+=10%; np=88%

Mann-Whitney U-statistic for comparisons of treatments

	Area	Extn	Disp	Elong
C vs A	12.1	10.2	12.6	8.1
C vs B	11.0	9.5	11.2	8.1
C vs B1	2.1	7.9	4.5	7.6
A vs B	4.7	1.7	0.9	1.7
A vs B1	12.5	12.4	12.7	11.8
B vs B1	11.9	11.6	11.9	11.3

Area C vs B1 $p < 0.05$ but > 0.01. A vs B all except area not significant. All the rest $p \ll 0.001$.

Human lymphocytes were isolated, stimulated and fixed as previously described (Keller, Niggli, Zimmermann, 1989). Fixed cells were recorded on videotape, the outline of at least 100 cells was drawn and the morphology determined by a) visual classification (Keller, Niggli, Zimmermann, 1989) of the recorded cells and b) calculated measures of shape derived from the drawings (Dunn and Brown, 1986). Samples for the shape analysis have been obtained as follows: Lymphocytes were incubated for 30 minutes in either plain medium (A), medium with 10^{-5}M colchicine (B and B_1) or 10^{-8}M PMA. B and B_1 differ in that the lymphocytes in B were randomly selected, whereas cells in B_1 consisted in a selection of polarized cells only within the same preparation. Lymphocytes in visual assays were classified as spherical (s), spherical with unifocal projection (s+), polarized (p) or non-polar with surface projections (np).

non-spherical cells varied according to the treatment.   If only a
small proportion of the population changes shape, then the median
values of the parameters will reflect those of the spherical cells,
and it is therefore necessary to subdivide the population, and compare
the non-spherical cells.   For this reason non-spherical cells were
selected on the basis of direct observation in order to accumulate a
sufficiently large sample.   (The shape parameters of randomly selected
non-spherical cells did not differ from those of the selected cells,
and since it is straightforward to distinguish spherical from non-
spherical cells we did not feel that this procedure would introduce
any bias.)

Although the shapes adopted by colchicine and PMA treated cells are
clearly different, and this revealed by the median values of the
elongation parameter (Table 1), we are still trying to improve the method.   By using linear discrimination analysis we are attempting to
determine which parameter or combination of parameters will provide
the most informative description of individual cells.

Acknowledgements

Supported by the Swiss National Science Foundation. We thank Miss Ch.
Schmidhalter for technical assistance.

REFERENCES

Dunn G.A. and Brown A.F. (1986) Alignment of fibroblasts on grooved
    surfaces described by a simple geometric transformation. J. Cell
    Sci. 83: 316-340
Keller H.U. (1983) Motility, cell shape, and locomotion of neutrophil
    granulocytes. Cell Motility 3: 47-60
Keller H.U., Naef A. and Zimmermann A. (1984) Effects of colchicine,
    Vinblastine and nocodazole on polarity, motility, chemotaxis and
    cAMP levels of human polymorphonuclear leucocytes. Exp. Cell Res.
    153: 173-185
Keller H.U., Niggli V. and Zimmerman A. (1989) Diacylglycerols and PMA
    induce actin polymerization and distinct shape changes in lym-
    phocytes: relation to fluid pinocytosis and locomotion. J. Cell Sci.
    93: 457-465
Lewis W.H. (1934) On the locomotion of the polymorphonuclear neutro-
    phils of the rat in autoplasma cultures. Bull. Johns Hosp. 55: 273-
    279
Middleton C.A., Brown A.F., Brown R.M. and Roberts D.J.H. (1988) The
    shape of cultured epithelial cells does not depend upon integrity of
    their microtubules. J. Cell Sci. 91: 337-346

Roos F.J., Zimmermann A. and Keller H.U. (1987) Effect of phorbol myristate acetate and the chemotactic peptide fNLPNTL on shape and movement of human neutrophils. J. Cell Sci. 88: 399-406

Wilkinson P.C., Lackie J.M., Haston Wendy S. and Islam Laila N. (1988) Effects of phorbol esters on shape and locomotion of human blood lymphocytes. J. Cell Sci. 90: 645-655

Zimmermann A., Keller H.U. and Cottier H. (1988) Heavy water (D$_2$O)-induced shape changes, movements and F-actin redistribution in human neutrophil granulocytes. Europ. J. Cell Biol. 47: 320-326

IMAGES OF CELLS CHANGING SHAPE:
PSEUDOPODS, SKELETONS AND MOTILE BEHAVIOUR

Peter B. Noble
Department of Oral Biology
Faculty of Dentistry, McGill University
Montreal, Quebec, Canada H3A 2B2.

Abstract

Cell locomotion is of importance in embryological development, wound healing and in the invasiveness and metastasis of tumour cells. The advent of microcomputers and image processing systems has greatly facilitated the study of cell locomotion. This report comments on some of the methods that have been developed to capture and process images of cells in both two and three-dimensional environments and presents ways of quantifying and analysing the locomotory behaviours of cells.

1. Introduction

Intuitively, the simplest way of studying cell locomotory behaviour is to use a time-lapse recording, either film or video, of the locomoting cells. The positions of the cells could then be traced at specified time intervals and the cell trajectory determined. However, to do this manually for one cell is quite time consuming. To do it for many cells requires the patience of Job. Fortunately, the advent of affordable micro-computers with image processing capabilities has meant that the tremendous amount of data contained within time-lapse recordings can be readily processed and analysed. In fact, over the past few years several reports have been presented applying computer techniques to the study of cell locomotion (Berns, Berns, 1982; Coates, Harman, McGuire, 1985; Noble, Levine, 1986; Thurston, Jaggi, Palcic, 1986; Dow, Lackie, Crocket, 1987). The problems which will be discussed in this report concern methods of obtaining the images of cells, how these images can be processed and cell behavioural characteristics quantified. The cell behavioural characteristics can be broadly sub-divided into two entities; dynamic shape changes occurring within the cell and measures of cell locomotory performance. These two entities are not mutually exclusive for changes in cell shape generally reflect active locomotory mechanisms taking place within the cell which, in turn, translate to the global locomotory characteristics of the cells.

2. Cell Images

Images of cells are usually obtained using a microscope of some sort. For cells observed in tissue culture, an inverted microscope is more convenient. A video camera is attached to the microscope and the images of the cells can be either recorded on a time-lapse video recorder and /or the signal can be fed to a frame-grabber installed in the computer. This latter method will allow real-time image processing of cell behaviour whereas the former technique allows the data to be analysed at will. There are advantages and disadvantages to both methods. The primary advantages of using recorded data are that it allows reviewing of the events and selected sections of interest can then be processed. It also allows the experimenter to get a "feel" for what is going on which can be quite invaluable for interpreting the results of analysis later on. The "real-time" method obviously does not allow this luxury. The major advantage of the "real-time" method is the fact that it leads to fully automated cell analysis systems and is simpler in that it removes the added step of getting the stored images on video tape into the frame- grabber for digitization. With regard to the type of microscope optics used we have found that bright-field objectives, while perhaps not producing as clear and detailed an image to the human eye as phase and Hoffman modulating contrast objectives, do make image processing easier due to the lack of pronounced "flare" seen in certain stages of locomoting cells, especially with phase contrast.

The first step in processing the digitized image involves algorithms to detect the cell from the background. Numerous algorithms exist for this process and the reader is referred to the following references for more information about these techniques (Levine, Youssef, Noble, Boyarsky, 1980; Rosenfeld, Kak, 1976; Inoué, 1987). The determination of the shape of a cell at a given moment in time and the determination of dynamic shape profiles along with position determination are somewhat more complex and will be dealt with in more detail as these topics arise.

3. Shape Determination

As cells locomote, they go through a series of shape changes. These dynamic shape changes are different, generally speaking, from one cell type to another. For example, the polymorphonuclear leucocyte (PMN) expresses a different dynamic shape profile compared to the fibroblast (Dunn, Brown; Keller, Eisele, Zimmermann, Lackie; this section). These are rather pronounced differences in shape behaviour which are readily discernible and it remains to be seen whether more subtle differences in dynamic shape profiles can be detected within the subsets of fibroblast populations. Increasing evidence is being presented that extensive heterogeneity exists within fibroblast-like cell populations which otherwise are indistinguishable at the light microscope level (Hassell, Stanek, 1983; Schor, Schor, 1987; Conrad, Hart, Chen, 1977) . Thus, to be able to quantify the dynamic shape profiles of locomoting cells might aid in distinguishing subsets within a cell population. However, there is another reason why the study of dynamic shape profiles could be useful. Polymorphonuclear cells (PMN) are white blood cells that play an important role in host defense mechanisms. These cells respond to the presence of invading bacterial cells by actively migrating towards the foci of bacterial infection (chemotaxis). The PMN have receptors on the surface of their membranes which selectively bind to certain metabolic products (formyl peptides) produced by the bacteria. By this means, the PMN can detect and orient their locomotory behaviour up the concentration gradient of bacterial products. This chemotactic process is the efficiency step which brings PMN to the bacteria where the PMN isolate and ingest (phagocytosis) the invading organisms. Of particular interest is the fact that as the PMN responds to the chemotactic gradient it produces extensions of the cell, called pseudopods, which signal the direction of locomotion of the cell. In random movement, pseudopods form anywhere around the cell until one becomes dominant and the cell moves off in that direction. The study of dynamic cell shape profiles therefore, could be useful in investigating cell locomotion under random (no chemotactic gradient) and chemotactic conditions.

3.1 Pseudopod characterisation.

In order to study the pseudopodial kinetics of PMN migrating in the presence of bacteria, a 16 mm film record was made. The film was analysed by digitizing each frame and using the digitized image to determine the boundary of the cell and then splitting the cell into subparts and defining which of these are pseudopods or potential pseudopods (Figure 1).

The intent was to develop a knowledge based system for visual motion understanding based on the computer vision framework proposed by Levine (1978). The first step in the analysis is the segmentation of those pixels that correspond to cells from the background. Unfortunately, in locomoting cells the contrast between cell and background is

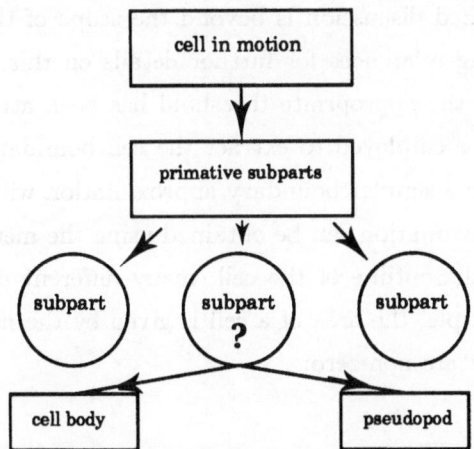

Figure 1:   Cell decomposition.

always changing due to the three-dimensional nature of the cell and it's changing shape. The use of bright field optics rather than the usual phase contrast objectives can reduce the degree of flare which complicates the image analysis process.

The cell detection protocol outline contains several stages; window adjustment, thresholding, filtering, etc. We shall be concerned here with only the first two stages; the cell detection protocol has been presented in detail (Noble, Levine, 1986). In order to track the cell a window is used of such a dimension that the displacement of the cell between consecutive frames cannot result in the cell being outside the window in the next frame. In this manner a window is created in the next frame and the cell is searched for within it. Once the cell centroid is located (see below), a new window is placed around the centroid location and the process repeated from frame to frame. If the maximum possible displacement of a cell between frames is Δ pixels and ω_α and ω_β are the dimensions of the window placed around centroid (x_k, y_k) in frame $(k + 1)$, then we set:

(1) $$\omega_\alpha = 2\Delta + \alpha_k,$$

(2) $$\omega_\beta = 2\Delta + \beta_k.$$

where α and β are the maximum cell extensions in x and y. The next step is thresholding, where the most appropriate pixel value is estimated to produce the best segmentation of the cell from background as a binary image. There are many algorithms for doing this and a detailed discussion is beyond the scope of this paper. The reader is referred to the following references for further details on this point (Noble, Levine, 1986; Inoué, 1987). Once the appropriate threshold has been attained a simple border following algorithm can be employed to extract the cell boundary and code it using a chain code. In many cases a simpler boundary approximation will suffice. For example, polygonal boundary approximation can be obtained using the method of Ramer (1972).

Having determined the outline of the cell, many different morphological features can be deduced. For example, the area of a cell is given by the number of pixels in the binary image $B(x, y)$ that are non-zero:

$$(3) \qquad\qquad F_1 = \sum_x \sum_y B(x, y)$$

where $B(i, j) = 1$ for the interior of the cell and 0 elsewhere. Several other morphological features can also be calculated such as the perimeter, minimum containing rectangle, elongation, angle regularity, etc (Levine, Noble, Youssef, 1983). Many of these features are of little value on their own due to their impreciseness in defining a specific shape but collectively can be of value as shape descriptors (Dunn, Brown. this section). The determination of the cell centroid is a useful feature for tracking cells during locomotion and is given by:

$$(4) \qquad\qquad X_c = \frac{1}{F_1} \sum \sum x B(x, y),$$

$$(5) \qquad\qquad Y_c = \frac{1}{F_1} \sum \sum y B(x, y).$$

If we wish to study the pseudopodial kinetics of a PMN locomoting in the presence of bacteria then, in order to study pseudopod behaviour, it is necessary to decompose the cell area into its constituent parts. Thus the original cell shape may consist of several subsets, some of which may be pseudopods. It is in defining what is a pseudopod that problems arise. In reviewing time-lapse recordings of locomoting PMN it is easy to see the formation of a pseudopod and, if a dominant one, to observe the intracellular contents flowing into it. The task of quantifying this procedure is made more difficult by being unable to determine exactly when a pseudopod begins and also when it ends. In our initial attempts to quantify pseudopod behaviour we considered the internal angles

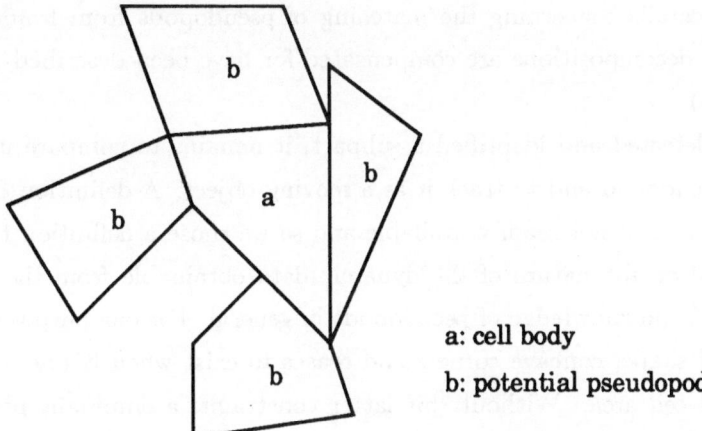

a: cell body

b: potential pseudopod

Figure 2: Cell decomposition into cell body and pseudopods.

of the polygonal approximation described earlier and noted the positions of the concave and convex angles (figure 2).

Lines drawn between concavities, interspersed in convex strings, select regions that are candidates for pseudopods. To make a dynamic statement about these potential pseudopods requires that each pseudopod be matched in the next frame $j+1$. That is, to associate each subpart v in frame $j-1$ with a subpart u in frame j. The association, from frame to frame, is based on a measure of similarity S_{uv}:

$$(6) \qquad S_{uv} = 1 - S'_{uv},$$

$$(7) \qquad S'_{uv} = \frac{1}{K} \sum_{k=1}^{k} \frac{F_{uk}(j) - F_{vk}(j+1)}{F_{uk}(j) + F_{vk}(j+1)} W_k.$$

where $F_{uk}(j)$ and $F_{vk}(j+1)$ are the k-th feature for the subparts u and v respectively. W_k is a weighting factor which allows for selected emphasis of a particular feature. Several features have been used in this matching process and include area, perimeter, angle regularity, average bending energy, centroid baseline and orientation (Levine, Noble, Youssef, 1983). The subparts u^* and v^* correspond if:

$$(8) \qquad S'_{uv} = S'_{vu}$$

Further details concerning the matching of pseudopods from frame to frame and how irregular decompositions are compensated for have been described (Levine, Noble, Youssef, 1983).

Having detected and identified a subpart, it remains to compare it with our definition of a pseudopod and to track it as a moving object. A definition of a pseudopod, in dynamic terms, is not readily available and so we chose a definition that was a compromise based on the nature of the dynamic data obtainable from the image analysis tempered with the knowledge of pseudopods in general. For our purposes, a pseudopod has elongated shape, concave corners and ceases to exist when it's area is equal to one half the total cell area. Without this latter constraint, a dominant pseudopod would eventually become the whole cell.

It is now possible to investigate the dynamic properties of pseudopods as they grow and collapse and to relate these dynamic behaviours, over a period of time, to the trajectory of the cell as a whole. The initial intention was to see which pseudopods contributed to the actual displacement of the cell; i.e. the degree of dominance as defined by the extent of pseudopod locomotion to the overall cell locomotion. If vector V_p represents the displacement of the pseudopod, determined by computing the centroid, then V_p reflects the growth pattern of the pseudopod. Likewise, if vector T defines the resultant direction of cell locomotion, pseudopod dominance d_p can be computed in the following manner; let β_p define the angle of pseudopod locomotion and β_c likewise for cell direction (figure 3).

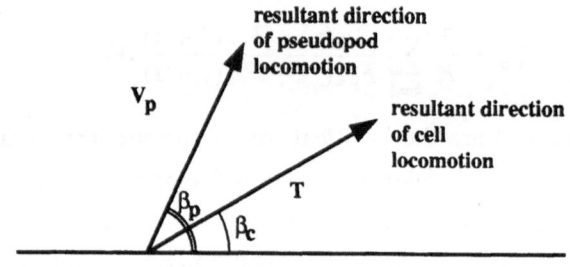

Figure 3:   Pseudopod dominance

Then

(9)
$$d_p = \frac{[180 - |\beta_c - \beta_p|]}{180}$$

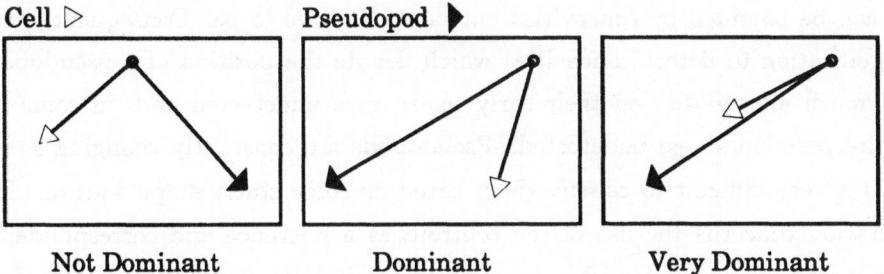

Figure 4:   Degrees of pseudopod dominance

Figure 4 shows some of the results obtained using the above equation.

Thus, by these means, the pseudopods which contribute to the overall locomotory displacement of the cell can be identified and plotted in relation to the cell trajectory as shown in Figure 5.

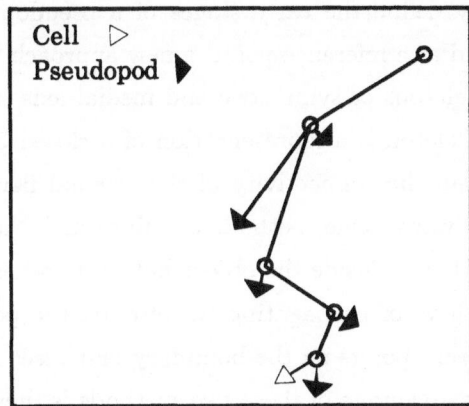

Figure 5:   Superimposition of pseudopod vectors on cell path

The changes in physical parameters of the pseudopods i.e., area, perimeter, etc, can be readily computed over time. Thus it is possible to compare the locomotory and physical features of pseudopods with varying degrees of dominancy and consequently the contribution pseudopod kinetics can be related, in a quantitative manner, to cell locomotion.

In reality, there are certain limitations to the above method of quantifying pseudopod kinetics. First, pseudopods can arise as small extensions of an already convex part of the cell boundary without any apparent deformations of the boundary. A potential pseudopod can be bounded by concavities but does not have to be. Consequently, the use of polygonization to detect concavities which denote the position of a pseudopod, means that much information on their early phase goes undetected and, in some instances, entire pseudopods go undetected. Pseudopods are constantly changing shape which makes it very difficult to classify them based on some static shape feature. The second limitation concerns the use of the centroid as a reference and correspondence point between frames. The centroid is notoriously sensitive to changes in shape. Even the slightest new protrusion at the cell boundary will displace the centroid, markedly altering reference measurements. For example, the distance from the centroid of the cell body to the centroid of a pseudopod can be altered by a protrusion forming at the opposite side of the cell even though the pseudopod has not changed it's scalar and vector quantities.

3.2 Skeleton determination.

To more accurately define the early stages of a pseudopod and to eliminate the vagueness of the centroid as a reference point, a new approach was initiated using skeletons. The concept of skeletons or symmetric and medial axis transforms was developed by Blum (1973). The skeleton is a representation of a closed boundary shape by idealized thin lines that retain the connectivity of the original figure. Skeletons have been defined in a number of ways. One method, by Blum and Nagel (1978), uses a set of maximal circular discs that fit inside the object but in no other disks within the object. In an other, using the laws of propagating wavefronts, the points at which the wavefront originated at different points on the boundary first meet form the skeleton (Blum, 1964). One common disadvantage of these two methods is their noise sensitivity. Small perturbations of the boundary translate into distortions of the corresponding skeleton branches often resulting in new branches. The net result is that a slightly altered figure can lead to a completely different skeleton.

These difficulties were overcome to some degree by the use of a new algorithm which computes the skeleton of the cell shape at multiple resolutions (Dill, Levine, Noble, 1984). Even though insensitivity to noise is a desirable feature, it is not a problem due to the fact that spurious branches that are formed in the skeleton are detected in later motion analysis. Thus, any branch classified as noise would only persist for a short period of time. However, small perturbations that eventually represent a pseudopod would persist for longer time periods and therefore would not be missed. By computing skeletons at different resolutions, a filtered version can be produced without violation of the constraints imposed by the intrinsic knowledge of PMN pseudopod behaviour.

The filtered version incorporates all the significant membrane changes that occurs at the different resolutions. Skeleton branches that persist over several resolutions arise from convexities that are both locally and globally significant.

The technique used for computing the skeleton is based on an algorithm first presented by Arcelli (1981). The resulting skeleton represents both the internal structure of the shape and specific boundary configurations. Given that the boundary of the cell is represented as a set of pixels O, the skeleton is computed as a connected set of pixels tracing its 8-connected contour C (figure 6).

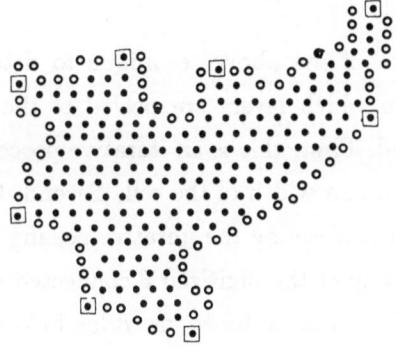

Figure 6: First iteration of skeletonization algorithm

After every iteration, contour pixels that are not multiple are assigned to a set of pixels R that are removed from set O. As we are interested only in the initial cell shape, convexities were measured at the initial iteration only. A pixel is considered to be multiple if it is traversed more than once during contour tracing, has no neighbours in the interior I of O, or if it has at least one D- neighbour which belongs to the contour but is not one of the two direct neighbours along the contour. The interior I set of O refers to the set of pixels $I = O - C$, and a D-neighbour is a horizontally or vertically displaced neighbour of a pixel. Before the next iteration, O is assigned $O = O - R$ and the algorithm stops when no more pixels can be assigned to the set R. At this stage, it remains to thin the skeleton so that the branches are all of width 1 pixel. The evaluation of noisy pixels is based on 1-code values and the reader is referred to Dill, Levine, Noble, (1987) for further information on this point. Figure 7 shows the results of the process of computing the skeleton for an image taken from a time-lapse sequence of a PMN locomoting in the presence of bacteria.

Various distances can be computed from the skeleton derived for each frame of a time-lapse record of a locomoting PMN. For example, the branches of the skeleton represent potential pseudopods and their distance from the long axis of the skeleton of

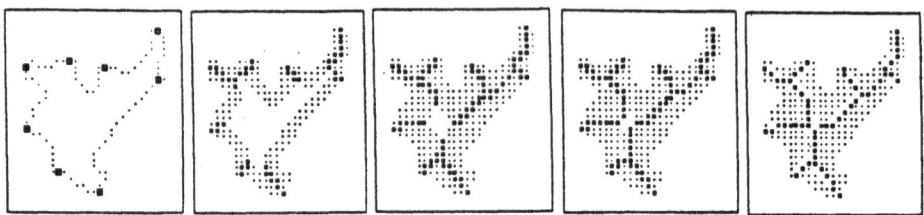

Figure 7:   The derivation of the skeleton

the cell can be readily measured. The long axis of the skeleton is much less susceptible
to small changes in cell outline than the centroid determined by the method discussed
earlier.

The essential feature of the above technique to determine the skeleton is based
on a thinning concept in which layers are peeled of the cell outline. There are two
drawbacks to this method. First, this is an iterative process whose computational time
is dependent on the maximum width of the cell. Second, the accuracy is limited by the
flow of information when performing the iterative peeling process due to the bias called
by the intrinsic connectivity of the digitized 8-connected grid.

Another class of algorithm is based on ridge-following techniques derived from
Distance Transforms applied to the cell outline. Many distance transforms are available
(Rosenfeld, Pfaltz, 1966; Arcelli, Sanniti di Baja, 1987; Arcelli, Sanniti di Baja, 1985,
Bogesfors, Sanniti di Baja, 1988; Klein, Kubler, 1987; Leymarie, Levine, 1989) however,
Euclidean Distance Transforms, which rely on ridge-following on the surface of the
distance maps to obtain the skeleton, produce more accurate and smoother results.
A new algorithm for computing the skeleton of a cell shape has been developed by
Leymarie and Levine (1989). This new method of computing skeletons combines the
advantages of *Euclidian Distance Transforms* with other attributes which ensures that
a multiscale description is immediately available. The process is based on the *grass
fire* algorithm. If a fire is started simultaneously around the periphery of a shape and
burns inward at a constant rate, then where the fire-edges meet represents the location
of the skeleton of that shape. The initiation of the grass fire is achieved by initiating
an active contour model, *Snakes* (Kass, 1987), which consist of elastic curves that tend
to minimize their energy by sensing the potential surface on which they are located. If
the grass fire propagation is considered as a function of time then a space-time graph
is obtained. If now time is replaced by distance, a *distance surface* is obtained. The
potential surface is taken as the negative of this distance surface:

(10) $$E_{pot} = -f_2 \leq 0$$

so that the snake reduces its energy by moving down the potential surface. Figure 8 shows an example of this process for a given shape representing a cell.

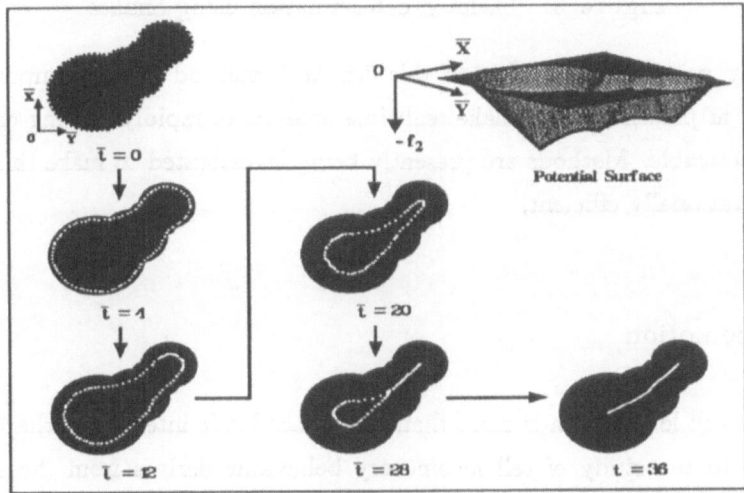

Figure 8:   An example of grass fire propagation

An attractive feature of the approach taken by Leymarie and Levine (1989) is that boundary information can be incorporated into the process such that the snake can be held at points extrema on the cell boundary. In this sense the points extrema are synonymous to the convexities used in the "peeling" algorithms mentioned earlier with the exception that points extrema are determined only on the initial boundary of the cell.

The net effect of fixing the snake at points extrema is that the snake collapses at these points to give the branches of the skeleton which reflect to points of origin of potential pseudopods (Figure 9).

It is interesting to compare this approach with the earlier attempts of polygonization to detect pseudopodial activity. The latter used concavities as pseudopod detection parameters whereas the former uses convexities which more accurately detect the growing tips of potential pseudopods. This newer technique certainly has the potential to detect and quantify pseudopod kinetics at a much earlier stage in the natural history of a pseudopod.

Figure 9:   Skeleton determination using *Snakes*

The only potential limitation to this "snakes" method is the computation time taken which, at present, would make real-time analysis of rapidly moving cells, such as PMN, impracticable. Methods are presently being investigated to make this procedure more computationally efficient.

4. Cell Locomotion

The study of cell locomotion is more than just of academic interest. Perhaps the greatest stimulus to the study of cell locomotory behaviour derives from three important biological phenomena; the migrations of cells during embryological development, the chemotactic locomotion of leucocytes towards invading organisms as the first line of host defense and the migrations of tumour cells during invasiveness and metastasis. By far the greatest amount of knowledge available regarding cell locomotory capabilities has been derived from use of the Boyden chamber in it's many forms (Wilkinson, 1982). The Boyden chamber is relatively cheap and easy to use. The major drawback to it's use is that one cannot determine the actual paths taken by the cells. Indeed, the cell can only migrate through the filter via pre-set pores. While a great deal of information can and has been obtained about cell locomotory behaviour under varying conditions from the distance that cells migrate into the filter in a unit period of time (Vicker, section II/3), much more information could be obtained if details of the actual cell trajectories were known.

Time-lapse recordings of cells locomoting on a two-dimensional surface have been available for many years. The information necessary to quantify cell locomotory trajectories is present in these recordings however the major difficulty has been how to extract and analyse this data. In fact, the amount of information available in time-lapse recordings of cell locomotory behaviour tends to overwhelming. For this reason and for many years time-lapse recordings were used merely for anecdotal purposes. With the advent of microcomputers and relatively cheap and reliable frame-grabbers, many techniques are available for tracking and hence determining cell trajectories (Berns, Berns, 1982; Coates, Harman, McGuire, 1985; Noble, Levine, 1986; Thurston, Jaggi, Palcic, 1986; Dow, Lackie, Crocket, 1987).

Whether or not the original cell data is obtained in real-time (digitization) or stored on a time-lapse recording, a decision has to be made on the sampling rate at which the cell positions will be recorded. If, for example,

a long sampling interval is used then information on cell turns in particular would be lost. On the other hand, if the sampling interval is too short virtually no difference would be observed from frame to frame. Ideally, a compromise is usually achieved where the sampling interval is such that differences in cell position are noted between frames yet cell trajectory information closely reflects the actual cell path. Obviously, the sampling rate will depend on the cell type and experimental conditions being used and can be estimated by viewing a time-lapse recording of the behaviour of the cell type in question. We have found that a sampling rate of 20 seconds is suitable for the PMN whereas a 2 minute sampling time might be more appropriate for the slower locomoting fibroblast.

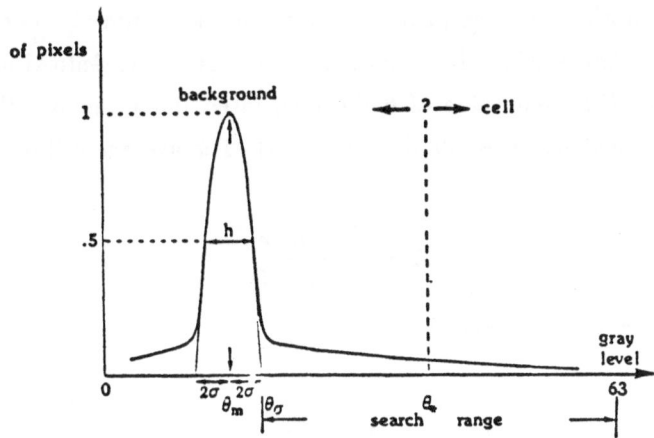

Figure 10: Searching the bright field histogram for approximate threshold

The next step is to locate the cell(s) in each frame. In this section, only the automated techniques will be discussed; for information on the more time consuming interactive techniques the reader is referred to the following reference (Noble, Levine, 1986). The only interactive step likely in the automated techniques is in the initial selection of the cells to be tracked. One technique is based on the histogram analysis of image pixel intensities. Knowing the centroid coordinates in one frame requires that they be computed in the next frame. From these two sets of coordinates and subsequent ones, the cell trajectory can be deduced. In the histogram method, the gray level image is segmented into a binary image and here the trick is in the selection of the threshold level at which the cell can be separated from the background. Figure 10 shows a typical histogram for a cell using bright-field optics. The peak represents mainly background pixels whereas the tail consists mainly of cells plus some background data. The average diameter of the cells under study can be readily estimated and the threshold θ can be selected which, after segmentation, gives a binary image in which the cell diameter D is closely matched. Referring to figure 10, the search range is restricted so that:

$$(11) \qquad\qquad\qquad\qquad \theta_\sigma \leq \theta \leq 63$$

where

$$(12) \qquad\qquad\qquad\qquad \theta_\sigma = \theta_m + 1.77h$$

based on the assumption that the peak $\theta = \theta_m$ is the maximum at a curve that approximates a Gaussian distribution. Note that the parameter h is defined as the full width at half maximum. For each value of θ the associated binary image $B(i,j)$ is labeled using connectivity analysis (Rosenfeld, Kak, 1976). The average cell area A is given by:

$$(13) \qquad\qquad\qquad\qquad A = \frac{\sum cell\,area}{no.\,of\,cells}$$

and the approximate average diameter is given by:

$$(14) \qquad\qquad\qquad\qquad D = \sqrt{A}$$

The net result of all this is that a curve can be plotted of D vs. threshold and, since D is known, θ_* can be obtained. Thus the areas pertaining to the cells can be identified and it is then an easy computation to calculate the centroid of each cell. The linkage between each identified cell, from frame to frame is accomplished by a *nearest neighbour* analysis (Noble, Levine, 1986). From these centroid locations the trajectory of the cell can be determined and various locomotory parameters computed. One other problem that has to be kept in mind is the computation of the angle between a change in cell direction. The minimum angle that can be detected is dependent on such parameters as the size of the cell (magnification used), pixel dimensions in x and y and the mean speed of the cell. Knowing these parameters, it is an easy task to compute the minimum angle detectable. In our two-dimensional systems the calculated minimum angle is about 10 degrees.

A second method for the determination of cell centroid is based on the *converging squares* algorithm proposed by O'Gorman and Sanderson (1984). In essence, this algorithm selects a brightness (or darkness) peak within an image space. There are disadvantages and advantages to this technique. The major disadvantage is that there is no attempt to ensure trajectory integrity and this does impose a constraint upon the experimental system as shall become evident. A second disadvantage, albeit minor, is that the estimation of the centroid may vary due to minor shape changes. In our experience this variation has been small compared to the overall dimensions of the cell and certainly within the acceptable limits of experimental error for cell centroid determination. The primary advantage of the *converging squares* algorithm is that it is computationally very efficient, especially compared to the *nearest neighbour* technique, and because no image segmentation is required a much higher degree of robustness with regard to noise and varying image contrast is attained. This is an important feature when the changing image quality of a moving cell is considered.

In order to initiate the process the cell has to be identified by an interactive process. The cells to be tracked are selected in our system using a light-pen on the cell images on a monitor. The process of selection places a search window around each cell and it is within this region that brightness peak is deduced. We have previously stated that a cell can move Δ pixels between frames and this parameter is dependent upon cell size, sampling rate and the average cell speed.

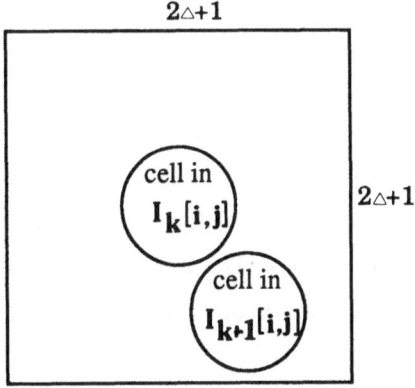

Figure 11:   The search window in $I_{k+1}(i,j)$

Referring to Figure 11, for each cell in frame $I_k(i,j)$ a window is created $(2\Delta + 1) * (2\Delta + 1)$. The brightness peak of the cell in frame $I_{k+1}(i,j)$, and hence the new centroid location, is determined by recursively reducing the size of the window as follows. At each stage of the search, the window is subdivided into four overlapping squares containing one less row and column and the brightness peak is calculated. The square with the maximum value is retained and the process is repeated until the square has the dimension of the cell diameter at which time the process stops. The dimension of the final square, i.e. cell diameter, is a programmable feature. This procedure is outlined in figure 12.

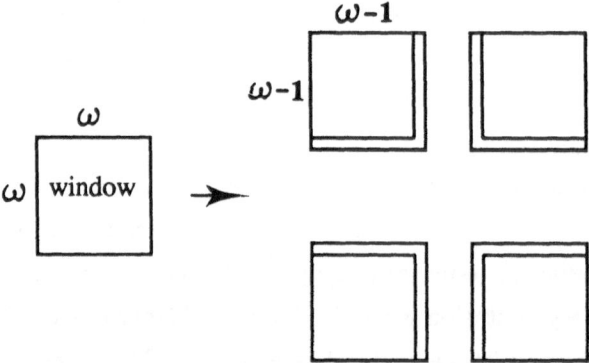

Figure 12:   Subdivision of the tracking window

Since there is a large common centre region of the square at each step of the search, the computational process is greatly enhanced. If cells divide or collide then, in either case, only the brightest image (cell) will continue to be tracked and one will be lost from the record. This drawback necessitates that cell numbers be kept low so as to keep this occurrence to an absolute minimum. Even with these limitations, the *converging squares* algorithm does a creditable job of determining the centroid coordinates of cells over time.

From these centroid coordinates the trajectory of the cell can be determined. Locomotory parameters such as speed, persistence and the angle distribution of turns can be computed. It is often necessary to be able to make a statement as to the ultimate direction that the cells will take, i.e. are the cells moving randomly or do they show a proclivity to a given direction (taxis). The directional capabilities of PMN has received particular attention due to the importance of locomotion, especially directed (chemotaxis), in the role these cells play in host defense mechanisms. We have characterized the locomotory capabilities of human PMN under neutral and chemotactic conditions by applying a continuous-time Markov probability theory to the locomotory trajectories

(Boyarsky and Noble, 1977). The Markov property assumes that future movements of a cell depend only on where it is at present and not on how it got there. The Markov chain states are angles in the cell's trajectory with respect to a given plane. Important parameters in the Markov process are the average time the cell spends in a given state, heading direction i and the transition probabilities of going from state (direction) i to state (direction) j. These parameters can be estimated from the cell trajectories and the limiting probability measure of the continuous-time Markov can be computed indicating the probabilities of direction that the cells will ultimately take. Figure 13(a) depicts how the Markov states are defined by the four quadrants of a Cartesian plane. Any number of states can be defined in the Markov and need not be of equal displacement, but in our two-dimensional experiments we have found four directional and one stationary state to be sufficient.

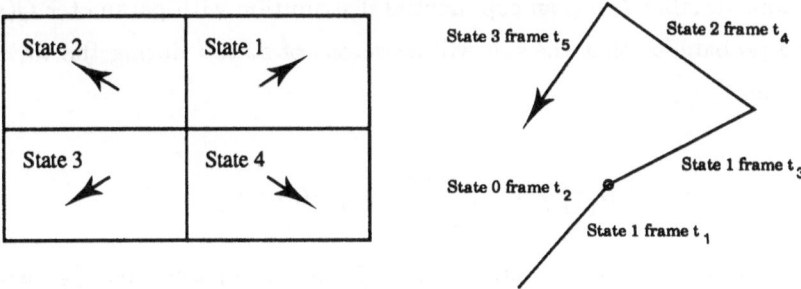

Figure 13: (a) States of the Markov; (b) Markov state assignment to a cell path

An example of the assignment of a section of a cell trajectory to the Markov states is shown in figure 13(b). State 0, which can occur anywhere on the trajectory segment, is governed by a distance threshold ϵ which we have set at one half cell diameter (Peterson, Noble, 1972).

The theoretical and practical aspects of the Markov method have been presented (Boyarsky, 1975; Boyarsky, Noble. 1977). A stochastic process $X_t : t \geq 0$ with a finite state space is consisdered a continuous-time Markov process if, at all times, $0 \leq t_1 < t_2 \cdots < t_n < t, \tau \geq 0$.

$$P\{X_{t+\tau} = j | X_t = i, X_{t_n} = i_n, \ldots X_{t_1} = i_1\}$$

$$(15) \qquad\qquad\qquad = P\{X_{t+\tau} = j | X_t = i\}$$

for states $i, j, i_1, \ldots, i_n$. As previously stated, the probability of being in state j at time $t + \tau$ depends only on being at state i at time t. If for all $t \geq 0$ and fixed $\tau \geq 0$,

$$(16) \qquad\qquad\qquad P\{X_{t+\tau} = j | X_t = I\} = p_\tau(j/i).$$

This latter equation just states that the probability of going from one state to another is independent of t.

We have already stated that, in our experiments, we have found it convenient to use four directional states and one stationary state for the Markov analysis. As the angles between vectors are the *states* of the Markov, let us in general terms divide 2π into n parts, not necessarily equal, where angles between θ_i and θ_{i-1} constitute state i. If, for example, the cell starts locomoting in state i and T_i is the first time that the cell leaves state i, then T_i is called the *waiting-time* in state i. Breiman, in 1969, showed that the random variable T_i has an exponential distribution with parameter $Q(i)$, where $Q(i)\Delta t$ is the probability that the cell will move out of state i during the time interval Δt. Thus, we have

$$(17) \qquad\qquad\qquad P\{T_i > t\} = \int_t^{+\infty} Q(i) e^{-Q(i)\tau} d\tau.$$

We wish to know which direction (state) the cells will ultimately take, i.e. what is the limiting distribution of the process. If $\overline{p}(j)$, for $j = 1, \ldots, n$, is the limiting distribution of the continuous-time Markov chain, irrespective of what the starting state i is, then

$$(18) \qquad\qquad\qquad p_t(j/i) \to \overline{p}(j)$$

as $t \to \infty$. Thus, in our four state model, if $p(4) = 0.9$ then the cells have a very high proclivity of migrating in the direction specified by state 4. If the cell trajectory angles pass from one state to another, then a unique limiting, stationary steady-state, $\{\overline{p}(j), j = 1, \ldots, n\}$ exists and is the unique solution of a set of equations (Breiman, 1967):

$$(19) \qquad Q(j)\overline{p}(j) = \sum_{\substack{l=1 \\ l \neq j}}^{n} p(j/l)Q(l)\overline{p}(l), j = 1.\ldots,n$$

where the summation is over all states except j and $p(j/l)$ is the probability of going from state l to state j. The parameters $Q(j), p(j/l)$, can be readily estimated from the experimental data and the equations can be solved for $\{\overline{p}(j); j = 1,\ldots,n\}$ by gaussian elimination (Boyarsky, 1975; Boyarsky and Noble, 1977; Noble and Levine, 1986).

We have applied the continuous-time markov method in a study investigating the locomotory behaviour *in-vitro* of patient's lymphocytes in the presence of their own lesions (Noble and Lewis, 1979; Noble and Bentley, 1980). The rational behind these studies was that certain lymphocytes have the potential to kill tumour cells and to do this required cell-cell contact. For this process to be efficient, such close cell contact would necessitate active locomotion of the cells towards the tumour cells. Time-lapse recordings were made of the patient's lymphocytes locomoting in the presence of a small biopsy of their lesion. The locomotory trajectories were then analysed using the continuous-time Markov method outlined above. Lymphocytes in the presence of pre-malignant lesions exhibited either positive or random locomotion whereas locomotory behaviour in the presence of malignant lesions was invariably negative chemotaxis (Noble and Bentley, 1980). These results suggested that in the progression in the natural history of the lesion from pre-malignant to malignant, a factor(s) was released which prevented the lymphocytes from approaching the tumour cells thus abrogating host defense mechanisms.

5. Cell Behaviour in Three-Dimensional Matrices

Much of our knowledge of cell locomotory behaviour has been derived from studies of cells in two-dimensional systems. This is for obvious reasons; ease of tissue culture on glass and plastic surfaces and the clarity of microscopical images using these systems. However, cells *in-vivo* exist within a three-dimensional environment. With the fairly recent introduction of reconstituted three-dimensional collagen gels, the means now exists with which to study cell behaviour in a environment which more closely reflects *in-vivo* reality.

Notable differences have already been described between cell behaviour observed in three-dimensional environments compared to two-dimensional. The morphology of fibroblasts in a three-dimensional collagen gel is virtually the same as has been observed in histological sections of connective tissue (Elsdale, Bard, 1972; Allen, Schor, Schor, 1984). Similarly, the rate of cell division is comparable to fibroblasts *in-vivo* (Yoshigato, Taira, Yamamoto, 1985). However, the study of cell locomotory behaviour in three-dimensional environments does pose some degree of difficulty if cell locomotory trajectories are desired. Most studies of cell behaviour in three-dimensional gels have relied on the measure of the rate of invasiveness of cells into the gel after layering on the gel surface (Schor, Schor, Rushton, Smith, 1985; Haston, Shields, Wilkinson, 1982) In many ways the approach is similar to the Boyden chamber technique and, like the latter, it does not provide information on the locomotory characteristics used by the cells both individually or collectively. Without this information it is more difficult to understand the mechanisms behind the locomotory behaviour observed.

In order to track cells within a three-dimensional environment, a computer assisted cell tracking unit has been designed to provide the x, y and z coordinates of the cells at specified time intervals as they migrate through the collagen matrix. The methodology has been reported elsewhere (Noble, 1987). Briefly, the tracking unit consists of a programmable interface which regulates the operation of a stepper-motor attached to the fine focusing control of an inverted microscope. The interface also communicates with a computer complete with frame-grabber. The number of optical slices (a section) to be taken through the gel, the inter-slice distance, the number of sections to be taken and the inter-section delay time are all programmable via the interface. The interface instructs the computer to digitize the camera image at each slice level. The cells to be tracked are initially selected using a light-pen from the slices in the first section, after which the process is automatic for the number of sections required. During the inter-section delay time, all the stored digitized image slices are processed to provide only those pixels corresponding to the x, y and z coordinates of the cells. Thus the computer memory is cleared ready to receive the next section of digitized slices. A schematic drawing of the tracking system is shown in figure 14.

The cells are tracked and the three-dimensional coordinates obtained using a three-dimensional version of the *converging squares* algorithm previously described. However, instead of a square the tracking window is now a cube which is recursively reduced over six faces and twelve sides until a cube of size approximating the cell diameter is obtained. From this smaller cube, the centroid is calculated and a new tracking cube placed around these coordinates. The determination of the dimensions of the tracking cube is the same in principle as already described for two-dimensional cell tracking.

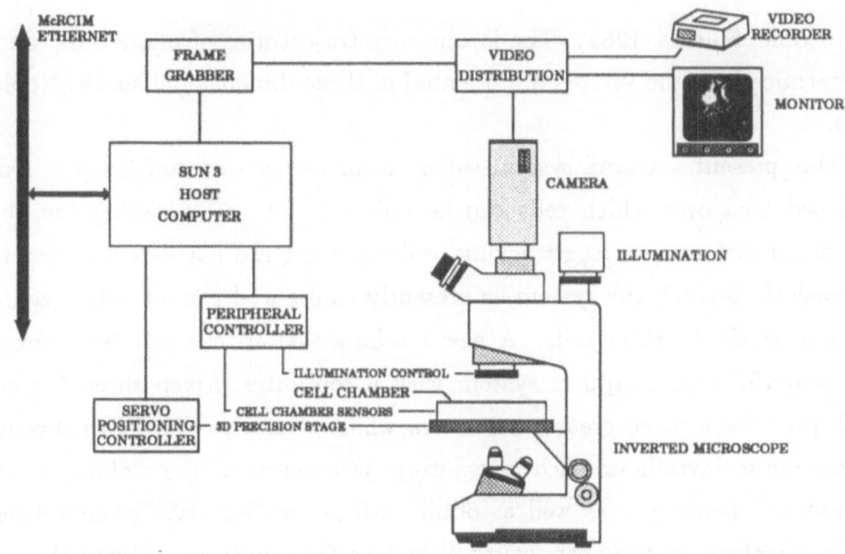

Figure 14: Three-dimensional cell tracking unit

Data analysis of the x, y and z coordinates over time is performed using a three-dimensional version of the Markov method. In this case, instead of four directional states, we have used eight directional states and one stationary state as outlined in figure 15.

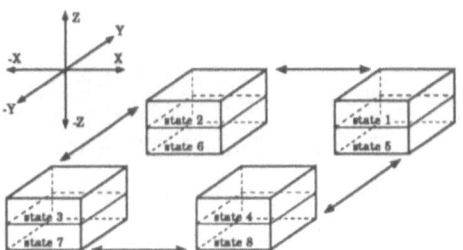

Figure 15: The three-dimensional Markov states

Again, there is no fixed number of states and we have found eight to be adequate for our needs. As shown in figure 16, any difference in coordinates from one section to another which results in a $+x, +y, +z$ is assigned to state 1; likewise, $-x, +y, +z$ would be assigned to state 2, and so on. From these coordinates, the cell trajectory can be deduced and various parameters computed. Factor analysis of the angles between the vectors of changing cell direction has suggested that, in the population of fibroblasts studied, a measure of heterogeneity in locomotory parameters exists (Shields, Noble,

1987; Noble, Shields, 1989). The locomotory trajectories of many cells can be analysed to determine how the vectors are oriented in three-dimensional space (Noble, Boyarsky, 1988).

This present tracking system suffers from two major deficiencies. One being the restricted area over which cells can be followed, i.e. cells leaving the objective field of view are lost and, during tracking, cells entering are not detected and therefore are not tracked. Second, the system as presently configured cannot detect cell division and therefore track daughter cells. A new tracking system has just been developed based on a powerful Sun computer system with a computer driven three-dimensional stage which provides a much greater volume in which to track cells. With this system it will be possible to investigate such locomotory parameters as the stability of daughter cell locomotory phenotypes as well as obtain cell generation data to complement locomotory parameters. In the near future, it will be feasible to investigate three-dimensional shape changes in cells and correlate these with locomotory parameters. And in the not to distant future, it should be feasible to study real-time cell-cell interactive behaviour in the three-dimensional milieu and to relate these changes to overall cell locomotory behaviour. To do this will require an intelligent tracking system which uses dynamic shape descriptions and profiles in conjunction with characteristic locomotory behaviour. At this stage we shall be very close to studying dynamic cell behaviour under conditions which accurately reflect *in-vivo* reality.

Acknowledgements

The support of the Medical Research Council of Canada is gratefully acknowledged. The author is indebted to the following members of the McGill Research Centre for Intelligent Machines for discussions, collaborations and access to their research : M.D.Levine, Y. Youssef, A. Dill, F. Leymarie and D. Gauthier.

REFERENCES

Allen T. D., Schor S. L., Schor A. M. (1984) An ultra-structural review of collagen gels: A model system for cell matrix, cell basement membrane and cell-cell interaction. Scan. Elect. Micros. 1: 375-390

Arcelli C. (1981) Pattern thinning by contour tracing. Computer Graphics and Image Processing 17: 130-144.

Arcelli C., Sanniti di Baja G. (1987) A one pass two-operation process to detect skeletal pixels on a 4-distance transform. PAMI 1987

Arcelli C., Sanniti di Baja G. (1985) A width dependent fast thinning algorithm. IEEE Trans. Pattern and Machine Intelligence, PAMI-7: 463-474.

Berns G. S., Berns M. W. (1982) Computer based tracking of living cells. Exp, Cell Res. 142: 103-109

Blum H. (1964) A transformation for extracting new descriptors of shape. Models for the perception of speech and visual form. Symposium, Boston, 1964. Ed. W. Wathen-Dunn, M.I.T. Press, Cambridge, Mass, 1967, pp. 362-380.

Blum H. (1973) Biological shape and visual science. J. Theoret. Biol. 38: 205-287.

Blum H., Nagel R. (1978) Shape description using weighted symmetric axis features. Pattern Recognition 10: 167-180.

Bogefors G., Sanniti di Baja G. (1988) Skeletonizing the distance transform on the hexagonal grid. Proc. Natl. Conf. Pattern Recognition 1: 504-507.

Boyarsky A. (1975) A Markov chain model for human granulocyte movement. J. Math. Biol. 2: 69-78.

Boyarsky A., Noble P. B. (1977) A Markov chain charaterization of human neutrophil locomotion under neutral and chemotactic conditions. Canad. J. Physiol. Pharm. 55: 1-6.

Breiman L. (1967) Probability and Stochastic Processes. Addison-Wesley, Mass.

Coates T. D., Harman J. T. McGuire, W. A. (1985) A computer based program for video analysis of chemotaxis under agarose. Comp. Methods and Programs in Biomed. 21: 195-212.

Conrad G. W., Hart G. W., Chen Y. (1977) Differences *in-vitro* between fibroblast-like cells from cornea, heart and skin of embryonic chicks. J. Cell Sci. 26: 119-137.

Dill A. R., Levine M. D., Noble P. B. (1987) Multiple resolution skeletons. IEEE Trans. Pattern Anal. Machine Intelligence PAMI-9, 495-504.

Dow J. A. T., Lackie J. M., Crocket K. V. (1987) A simple microcomputer system for real-time analysis of cell behaviour. J. Cell Sci. 87: 171-182.

Dunn G. A., Brown A. F. (1986) Alignment of fibroblasts on grooved surfaces described by a simple geometric transformation. J. Cell Sci. 83: 313-341.

Elsdale T., Bard J. (1972) Collagen substrata for studies on cell behaviour. J. Cell Biol. 54: 626-637.

Hassell T. M., Stanck E. J. (1983) Evidence that healthy human gingiva contains functionally heterogeneous fibroblast subpopulations. Arch. Oral Biol. 28: 617-625.

Haston W. S., Shields J. M., Wilkinson P. C. (1982) Lymphocyte locomotion and attachment on two-dimensional surfaces and in three-dimensional matrices. J. Cell Biol. 92: 747-752.

Inoué S. (1987) Video Microscopy. Plennum Press. New York.

Kass M., Witkin A., Terzopoulos D. (1987) Snakes: active contour models. IEEE Proc. Ist. Int. Conf. Computer Vision. 259-268.

Klein F., Kubler D. (1987) Euclidean distance transformation and model- guided image interpretation. Pattern Recog. Letters 5: 19-30.

Levine M. D. (1978) A knowledge-based computer vision system. In Computer Vision Systems. Eds. Hanson, A., Riseman, E., Academic Press, New York. pp. 335

Levine M. D., Youssef Y. M., Noble P. B., Boyarsky A. (1980) The quantification of blood cell motion by a method of automatic digital picture processing. IEEE Trans, Patter Anal. Machine Intelligence PAMI-2: 444-450.

Levine M. D., Noble P. B., Youssef Y. M. (1983) NATO ANSI series F2. Image sequence processing and dynamic scene analysis. Ed. Huang, T.S., Springer-verlag. Berlin.

Levine M. D., Noble P. B., Youssef Y. M. (1983) Understanding blood cell motion. Comp. Vision, Graphics and Image Processing 21: 58-84.

Leymarie F., Levine M. D. (1989) Snakes and skeletons. Technical report TR-CIM -89-3, McGill Research Centre for Intelligent Machines.

Noble P. B., Lewis M. G. (1979) Lymphocyte migration and infiltration in melanoma. Pigment Cell 5: 174-181.

Noble P. B., Bentley K. C. (1980) An in-vitro study of lymphocyte migration in the presence of pre-malignant and malignant oral lesions. Int. J. Oral Surgery 9: 148-153.

Noble P. B., Levine M. D. (1986) Computer-assisted analyses of cell locomotion and chemotaxis. CRC Press, Boca Raton, Florida. pp 150

Noble P. B. (1987) Extracellular matrix and cell migration: Locomotory characteristics of Mos-11 cells within a three-dimensional hydrated collagen lattice. J. Cell Sci. 87: 241-248

Noble P. B., Boyarsky A. (1988) Analysis of cell three-dimensional locomotory vectors. Exptl. Cell Biol. 56: 289-296.

Noble P. B., Shields E. D. (1989) Time-based changes in fibroblast three-dimensional locomotory characteristics and phenotypes. Expl. Cell Biol. 57: 238-245

O'Gorman L., Sanderson A. C. (1984) The converging squares algorithm: An efficient method for locating peaks in multidimensions. IEEE Trans. Pattern Anal. Machine Intelligence PAMI-6: 280-288.

Peterson S. C., Noble P. B. (1972) A two-dimensional random walk analysis of human granulocyte movement. Biophys. J. 12: 1048-1055.

Ramer U. (1972) An iterative procedure for polygonal approximation of plane curves. Compt. Graphics Image Process. 1: 244

Rosenfeld A., Pfaltz J. L. (1966) Sequential operations in digital picture processing. J. ACM. 13: 471-494.

Rosenfeld A., Kak A. C. (1976) Digital picture processing. Academic Press, New York.

Schor S. L., Schor A. M., Rushton G., Smith L. (1985) Adult, foetal and transformed fibroblasts display different migratory phenotypes on collagen gels: Evidence for isoformic transformation during foetal development. J. Cell Sci. 73: 221-234.

Schor S. L., Schor A. M. (1987) Clonal heterogeneity in fibroblast phenotype: Implications for the control of epithelial-mesenchymal interactions. BioEssays 7: 200-204.

Shields E. D., Noble P. B. (1987) Methodology for detection of heterogeneity of cell locomotory phenotypes in three-dimensional gels. Exptl. Cell Biol. 55: 250-256.

Thurston G., Jaggi B., Palcic B. (1986) Cell motility measurements with an automatic microscope system. Exp. Cell Res. 165: 380-390.

Wilkinson P. C. (1982) Chemotaxis and Inflammation. 2nd. Edition, Churchill Livingstone, Edinburgh.

Yoshigato K., Taira T., Yamamoto J. B. (1985) Growth inhibition of human fibroblasts by reconstituted collagen fibrils. Biomed. Res. 6: 81-77.

CONTINUITY OF MOVEMENT
AND
PRESERVATION OF ARCHITECTURE DURING CELL LOCOMOTION

Jürgen Bereiter-Hahn, Norbert Braun, Monika Vöth
Cinematic Cell Research Group.
J. W. Goethe-Universität Frankfurt a. M.
Senckenberganlage 27
6000 Frankfurt / M, F.R.G.

Abstract

Continuous fluxes of cytoplasm and membrane material are required to sustain cell locomotion. In some examples, e.g. epidermal cells of various vertebrates very small changes in the overall morphology of the cells take place during locomotion, in others e.g. fibroblasts or lymphocytes, at least the general appearance remains unaltered. Thus any model on cell locomotion has to include the continuous organization and disorganization of cell architecture under steady state conditions. Such a model is presented, based on intracellular pressure differences providing the source for the motive force and on the distribution of cytoskeletal elements providing the structural basis for force generation and cell shape. A detailed description of very small changes in cell surface topography presents the basis on which the models of cell locomotion and the control of this event can be discussed appropriately. Cytosolic calcium controls force generation and the direction of locomotion. Ca^{2+} concentration is highest in the lamella/cell body transition region and at the leading front, as has been revealed by scanning fluorometry.

1. Introduction

Maintenance of individuality and characteristics of shape despite of continuous movement is one of the most fascinating properties of living entities and can be observed on all levels of organization. To understand these phenomena we have to consider living systems as being self organizing, thus emphasizing the process character of life. In the present contribution we will concentrate on cell motility, in particular on locomotion of vertebrate cells in culture.

The morphology of stationary cells differs from locomoting ones by reduction or lack of focal contacts (Rees et al. 1982) in locomoting cells and by the development of one or more lamellar or cylindrical protrusions into the direction of movement. The extent of such extensions determines cell polarity, a prerequisite for locomotion. Cells without lamellae or those which are totally surrounded by a lamella ("fried egg shaped") do not move unless polarity is induced by an external stimulus.

Isolated vertebrate keratocytes glide relatively fast (Brown and Middleton 1985) over a substratum. During this locomotion they do not change shape considerably, the arrangement of cytoskeletal elements is preserved. Thus this behaviour is an excellent example for a self organizing process on the single cell level. Any hypothesis on the mechanism of these motions has to take into account the overall stability of the system while it is rebuilding itself continuously. In the following we describe some aspects of this behaviour of tadpole skin keratinocytes in culture.

2. Hypotheses on the locomotion of cells.

The question "how does a cell move?" is still unsolved. Several mechanisms have been proposed. Some of the more recent ideas concerning the generation of the motive force are

    actin assembly at the tip of a pseudopodium or leading
    edge (Condeelis et al. 1988; Hall et al. 1988; Tilney
    1980)

    sliding of filaments along each other in the direction of
    locomotion (Huxley 1976),

    contraction of gel like cytoplasm with solation on the op-
    posite side (Taylor et al. 1976; Taylor and Fechheimer
    1982)

    extension of the plasma membrane (Bretscher 1988; Kupfer
    et al. 1987)

    hydrostatic pressure in the cell causing bulging or cyto-
    plasmic outflow at any site of mechanical weakness (Harris
    1973; Komnick et al. 1972; DiPasquale 1975; Bereiter-Hahn
    et al. 1981).

These factors are by no means exclusive for each other and
some of them may be acting cooperatively.
One of the most important factors involved in control of cell
motility is calcium (Gail et al., 1973; Wilkinson, 1975; Moore
and Pastan, 1979; Strohmeier and Bereiter-Hahn, 1984; Cooper
and Schliwa, 1985; Mittal and Bereiter-Hahn, 1985). Control of
Ca^{2+} fluxes offers the opportunity to inhibit or to stimulate
locomotion of cells in culture. The concommittant changes in
cell shape and distribution of actin, myosin and tubulin (the
proteins most probably engaged in production of motive force)
should provide indications on the site of motive force production. Reflection interference microscopy (RIC) allows to follow very small thickness changes in cells and thus to reconstruct the surface topography (Gingell 1981; Gingell and Todd
1979; Bereiter-Hahn et al. 1983).

3. Results

3.1 Shape changes of cells in the presence of lanthanum.

Moving amphibian or fish keratocytes exhibit a large flat lamella comprising approximately half of the area adhering to
glass, and a distinct cell body region containing all the membrane bound organelles and masses of tonofibrils. The overall
shape of these cells is approximately semi-circular or semi-
elliptic. First order interference fringes in the lamella seen
with RIC at an illuminating numerical aperture (I.N.A.) of
0.17 and a wavelength of 578 nm correspond to a layer thickness around 125 nm (dark areas) and 250 nm (bright areas):
Higher orders represent multiples of these values. In normally
moving cells the leading front of the lamella exhibits a row
of swellings (microcolliculi, arrowheads in Fig. 8), between
250 and 375 nm thick, while the central part of the lamella is
around 125 nm (dark grey or black), then in the lamella/cell
body transition region thickness may increase suddenly so that
fringes are no longer recognized because of scattering and
obliqueness of the surface (Fig.1).

On addition of 1 mM La^{3+} in Ca-free Ringer solution, locomotion ceases immediately or in a few minutes. In the very elongated cell shown in Fig. 1 the middle part of the cell body
advances towards the edge of the lamella which does not extend
as it would do under normal conditions. The thickness of the
marginal parts of the lamella increases by flattening of the
slope between lamella and cell body (Fig. 1: 1 to 10 min). During 20 min in La^{3+}, actin and myosin distribution change considerably (comp. Figs. 2a, 3a, 4 and 2b, 3b, 5). F-actin has
been revealed by staining with TRITC-phalloidin. In controls
it is highly concentrated at the very periphery of the leading
edge and in the cell body region. In some cells a further accumulation may be found at the lamella/cell body transition
region. Regarding that the tickness of the lamella is less
than 1/30th that of the cell body, the high concentration of

F-actin in this zone becomes obvious. Myosin is found primarily in the cell body, in lamella/cell body transition region
and in approximately 2/3 of the proximal lamella. The very end
of the lamella is totally devoid of myosin in moving cells as
well as in the lamella of marginal cells anchored to a sheet
of keratinocytes (Fig. 3).

In La^{3+}-treated cultures both these cytoskeletal substances
become homogenously distributed throughout the cell (Fig. 2b,
5). In many cells a ring of F-actin is seen close to the cell
margin. This ring represents the actin accumulation from the
lamella/cell body transition region in the controls which has
been dislocated to the periphery under the influence of La^{3+}
(Fig. 1).

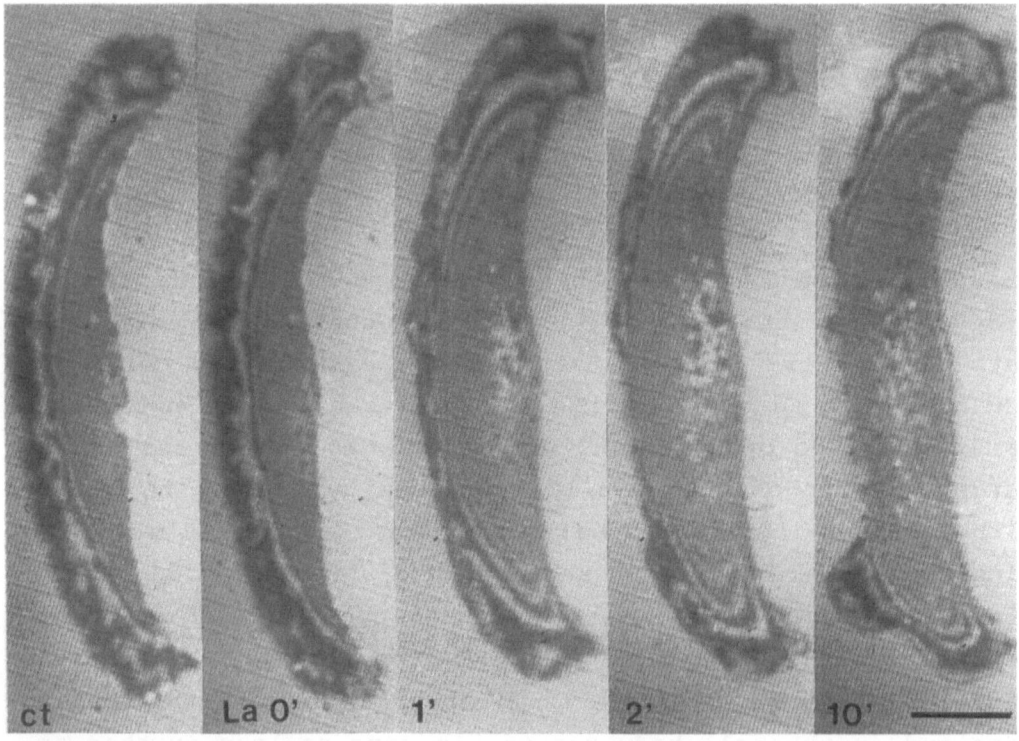

Figure 1. Reaction of a single Xenopus tadpole keratinocyte to
La^{3+}. ct: control in Ca-Ringer, the numbers indicate the time
interval in min. passed since addition of 1 mM La^{3+} in Ca-free
saline. Bar: 10 μm.

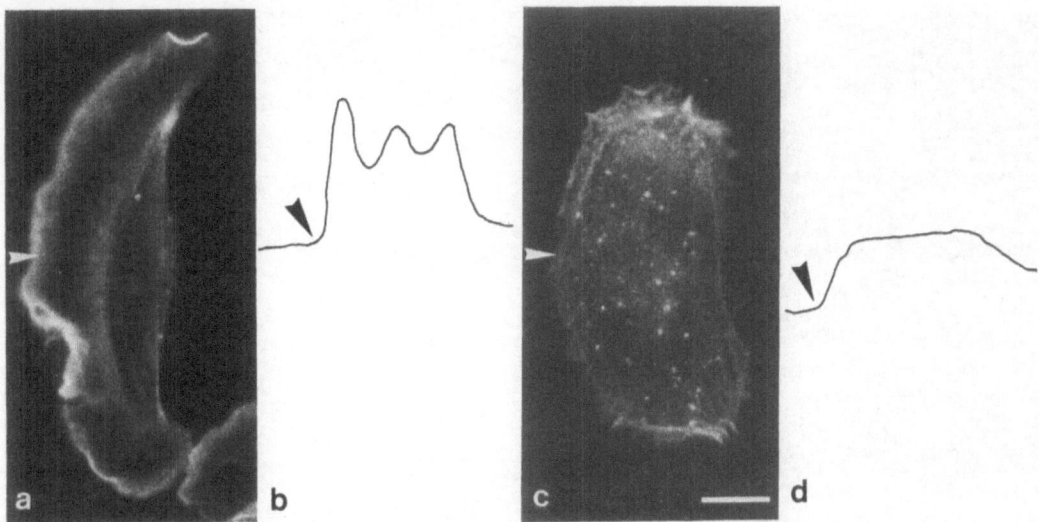

Figure 2. Actin distribution (TRITC-phalloidin staining) in single keratocytes before (a, b) and 20 min after exposure to La^{3+} (c, d). Fluorescence intensity has been plotted (b, d) along the lines indicated by arrows in a and c. In the control cell accumulations of F-actin are obvious at the front of the leading edge, the lamella/cell body transition region and at the rear end of the cell (a, b). Treatment of the cultures with La^{3+} destroys this compartmentation (c, d).

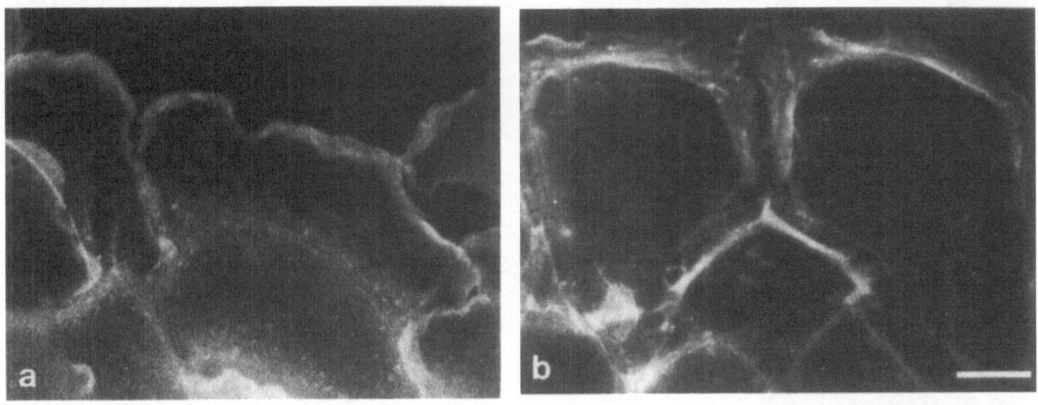

Figure 3. Margin of a sheet of cells, TRITC-phalloidin staining. In the control culture (a) the marginal cells develop broad lamellae, the actin distribution corresponds to that of moving cells (comp. Fig 2b). In La^{3+} most of the lamella area is lost due to an extension of the lamella/cell body transition region towards the lamella. Bar: 10 μm

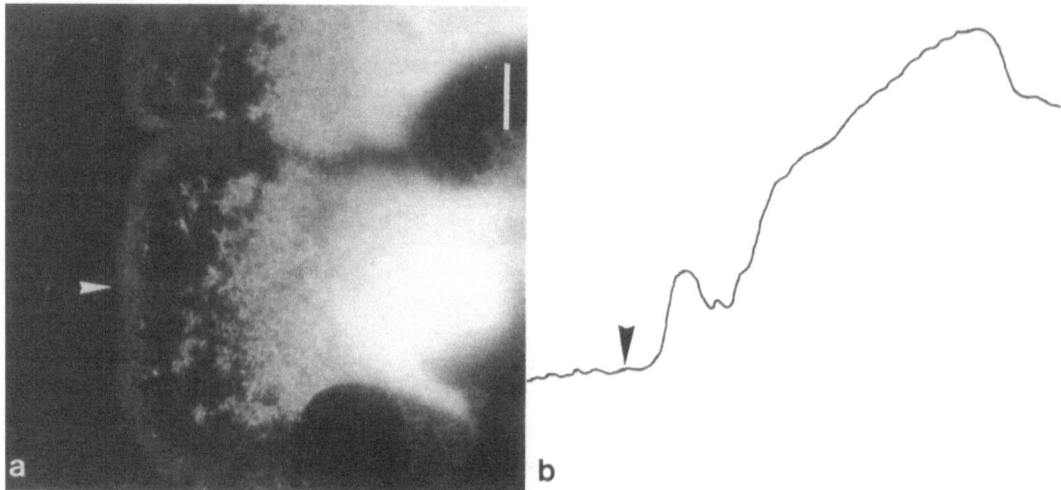

<u>Figure 4</u>. Visualization of non-muscle cell myosin by indirect immunofluorescence (a). The intensity distribution along the line indicated by arrows in (a) is shown in (b). Just before the margin (which is devoid of myosin and thus not to be seen) an accumulation of myosin takes place, in the lamella/cell body transition region a further strong increase is noted and with increasing cell thickness towards the rear end myosin content also increases.

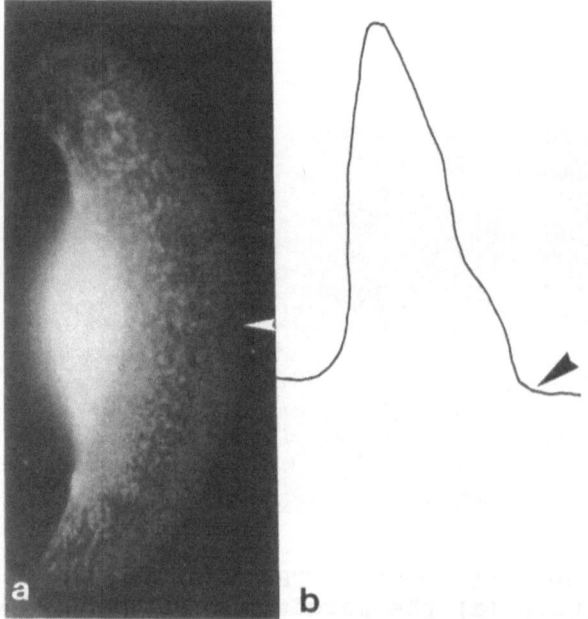

<u>Figure 5.</u> Myosin distribution in a cell treated with La^{3+} for 20 min. Most of the irregularities seen in Fig. 4 disappeared, myosin now is found up to the very edge of the lamella.

3.2 Shape changes and locomotion induced by ionomycine.

Treatment of cells with the Ca^{2+} specific ionophore ionomycine is supposed to lead to an influx of calcium from the culture medium (Ca^{2+} concentration: 1.7 mM) and thus should oppose the effects observed in presence of lanthanum. Two conditions have been chosen to test this hypothesis: 1μM ionomycine in Ringer's saline and 0.5 μM ionomycine in amphibian culture medium. The reactions have been followed using time-lapse video to record the images obtained with a RIC microscope, which visualizes cell shape and its adhesion to the glass surface. The parameters for the measurements were the adhesion area of the cells and their locomotion.

Locomotion was quantified by determination of the area covered by the moving cell in one minute. This value is measured in percentage of the adhesion area of the cell. This procedure assures that curved movements and polarized spreading of the lamella without displacement of the whole cell are represented properly (Mittal and Bereiter-Hahn 1985). Regarding large differences of the size of the primary cells in culture percentage measurements seem to be more appropriate than giving absolute values. The mean displacement of untreated cells lies in the range between 5 and 30 μm per min.

1 μM ionomycine in saline: A typical experiment is shown in Fig. 6. In ACM the adhesion area is relatively constant and small (25-35 min) while locomotion varies considerably. On addition of serum free saline the cell spreads continuously and its locomotory activity decreases (35 - 48 min). Addition of ionomycine accelerates locomotion, a maximum is reached in 7 min (48 - 55 min). During this time no further spreading of the adhesion area takes place. After about 12 min (at 60 min in Fig. 6) locomotion declines suddenly because the cell loses polarity by becoming fried egg shape. This is accompanied by a further increase in the adhesion area which then remains more or less constant.

Influence of 1 µM Ionomycin on
velocity and adhesion area

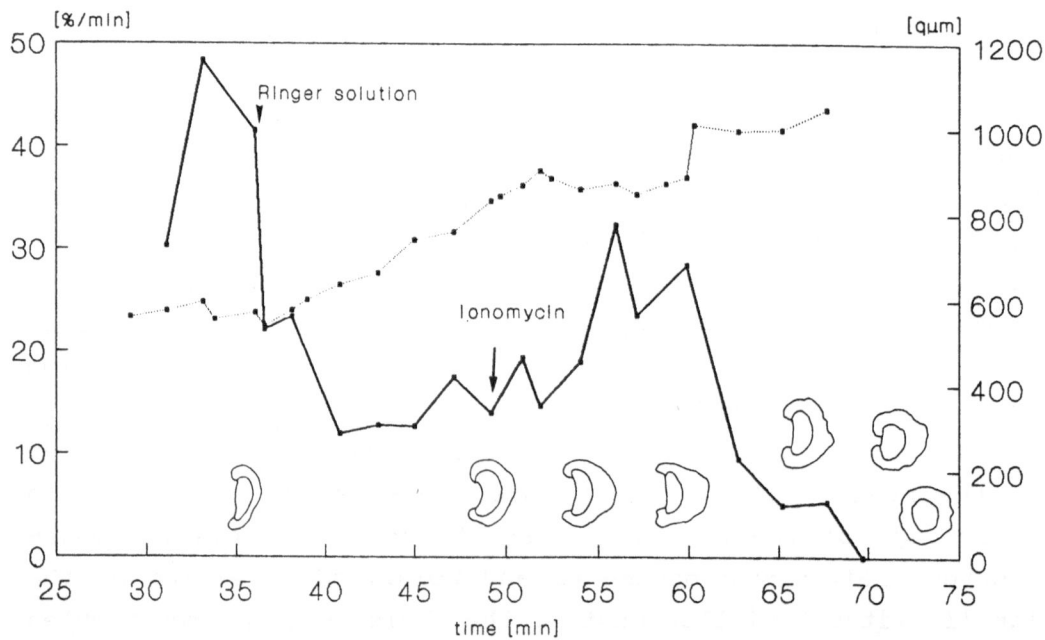

--•-- adhesion area   —•— velocity

Figure 6. Locomotion (full line, left ordinate) and adhesion area (dotted line, right ordinate) of a Xenopus tadpole keratocyte in ACM, Ringer's saline and 1µM ionomycine in Ringer's solution. The shape of the cell and of the cell body are outlined by the inserted drawings. [The cells have been isolated from tadpole epidermis [Nieuwkoop and Faber (1956) stage 47-52] by mild trypsinization (5 min at room temperature) and resuspension in ACM (amphibian culture medium according to Wolf and Quimby 1962, obtained by Gibco, Glasgow, U.K.)]. Abscissa: time in min.

For interpretation of these observation a more detailed knowledge of shape changes of the cell as revealed by RIC is helpful:

Immediately before addition of ionomycine the cell shown in Fig. 6) had a well developed lamella with small microcolliculi at the leading edge. Thickness of the lamella was in the range between 125 to 280 nm (as seen from the interferences in RIC), the cell body was clearly delineated by a dark second order interference fringe. About 4 min after addition of ionomycine the cell body adhesion area began to decrease indicating a

culi around 150 nm in thickness and also broadening, the central part of the lamella thickened by about 250 nm and the
relative amount of the adhesion area occupied by the lamella
increased by approximately the amount the cell body area was
diminished (see insets in Fig. 6). This morphology was maintained during the period of fast movement (until 60 min in
Fig. 6). Locomotion ceased because the lamella was flowing
around the cell body. Also in the final immobile stage, the
periphery of the lamella showed prominent microcolliculi which
were in steady change.

0.5 µM ionomycine in ACM: In this condition the serum content
of the medium is supposed to reduce the activity of the
ionomycine, cell physiology, however, should be promoted by
the presence of serum. The treatment results in a transient
peak of locomotory activity which lasts about 3 min, lateron
the cell seems to be able to control its cytosolic calcium
levels despite of the presence of some ionomycine (Fig. 7).

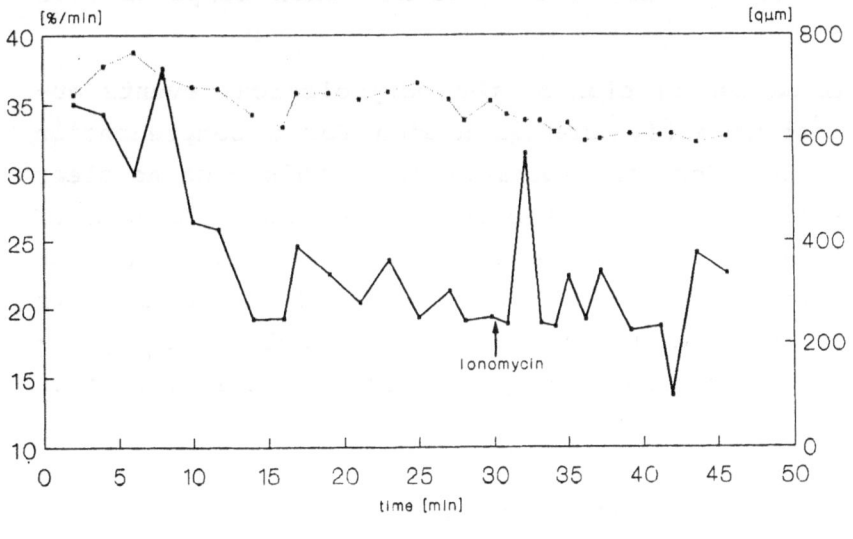

Influence of 0,5 µM Ionomycin on velocity and adhesion area

Figure 7. Locomotion (full line, left ordinate) and adhesion
area (dotted line, right ordinate) of a Xenopus tadpole
keratocyte in Ringer's saline and 0.5 µM ionomycine in
Ringer's solution. Abscissa: time in min.

29:35   **30:48**   **31:38**   **32:32**   **35:49**

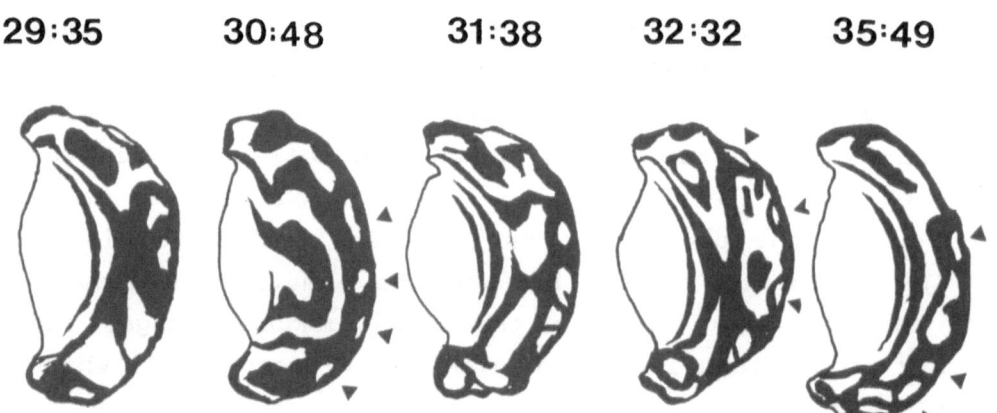

Figure 8. Changes in surface topography revealed by RIC; same cell as that measured for Fig. 7. For better understanding the grey level distribution has been simplified by drawing minima and maxima of the interferences only. The numbers indicate the time (min:sec) marked on the abscissa in Fig. 7. At 29:35 the lamella exhibits first order dark and bright zones with one 2nd order dark zone (arrow). 30:48: A 2nd order fringe extends in the centre of the lamella (ionomycine had been added at 29:40), microcolliculi (arrowheads) at the margin become prominent, cell body area is reduced. During the next minute the cell tries to reestablish the original shape, at 32:32 maximal speed is reached, the thickness of the lamella periphery is increased considerably (2nd order dark). 35:49: Cell body is extended again, lamella thickness is increased slightly (1st order dark to 1st order bright) when compared with 29:35.

Again, a detailed description of the morphological events during this transient activation is needed for a comprehension of the mechanisms underlying locomotion. In this case no clear change in the size of the cell body occurs, the thickness of the lamella starts to increase 1 min after addition of the ionomycine. Thickening proceeds from proximal to distal with the development of a few prominent microcolliculi. These processes last for four minutes only, then the original situation is reestablished (Fig. 8).

3.3 Distrubution of cytosolic free Ca^{2+}

The distribution of cytosolic free Ca^{2+} has been followed in an untreated normally locomoting keratinocyte using the calcium specific fluorochrome in Fig. 9. The site with the highest Ca^{2+} concentration is obviously the lamella/cell body transition region. The margin of the lamella also shows an increased Ca^{2+} concentration. These are the regions reacting most sensitive to the treatments described above.

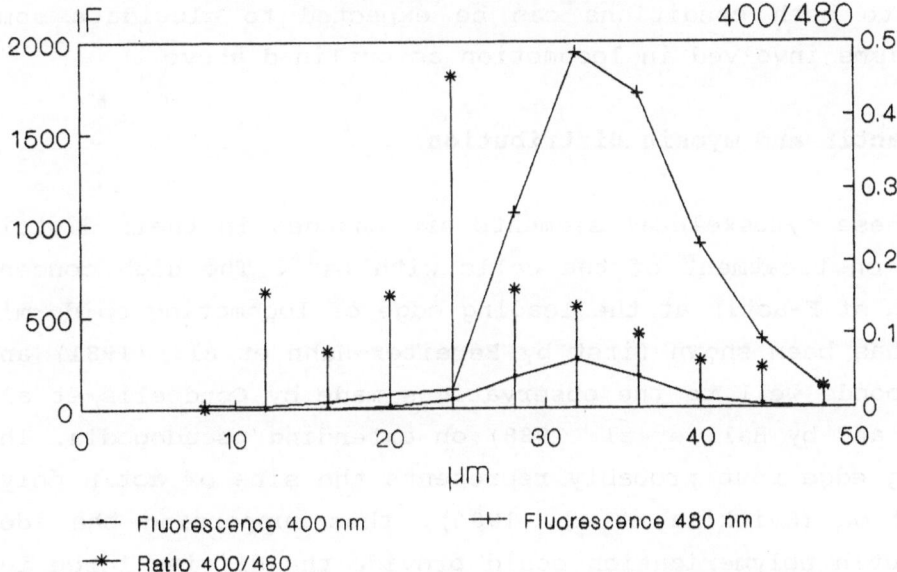

—•— Fluorescence 400 nm   —+— Fluorescence 480 nm

—*— Ratio 400/480

<u>Figure 9</u>. Distribution of Ca^{2+} in a single keratocyte along a scanning line (abscissa, units in μm) corresponding that in Fig. 2 a. The cells had been loaded with Indo-1/AM (2.5 μg/ml) for 8 min, then they were kept for 40 min in Indo-1-free saline. Fluorescence intensity has been measured at 400 nm and at 480 nm, excitation was 340-370 nm (with a HBO 50 high pressure mercury arc lamp). Scanning steps were 4 μm, the width of the measuring aperture was 4x4 μm. The ratio of the intensity measured at 400 (lower trace) and at 480 nm (upper trace: +)gives a measure of the cytosolic free Ca^{2+} concentration (not calibrated). Using esters, an exact determination of Ca^{2+} concentration is very uncertain because of an unknown degree of ester hydrolysis (Lückhoff, 1986). Therefore only the 400/480 ratio (bars with *) is given.
The values on the abscissa start on the left with the leading edge (first value) and end with the rear margin of the cell on the right.

4. Discussion

Two experimental setups of opposite effect on keratinocyte locomotion have been presented. Lanthanum is thought to block calcium channels in the plasmamembrane (Caswell, 1979; Mela, 1968) and may reduce the negative surface charge of the

membrane; ionomycine is known to make it permeable for
calcium. Ionomycine induced at least transient acceleration of
locomotion while lanthanum inhibited it. Thus cellular reactions to both conditions can be expected to elucidate some
mechanisms involved in locomotion as outlined above.

4.1 F-actin and myosin distribution

Both these cytoskeletal elements are changed in their distribution on treatment of the cells with La^{3+}. The high concentration of F-actin at the leading edge of locomoting epidermis
cells has been shown first by Bereiter-Hahn et al. (1981) and
corresponds well to the observations made by Condeelis et al.
(1988) and by Hall et al. 1988) on extending pseudopodia. The
leading edge most probably represents the site of actin polymerization (Svitkina et al. 1986), thus supporting the idea
that actin polymerization could provide the driving force for
the movements. In this case the cell body and the lamella/cell
body transition region would have a passive role only. Observations on the motility of dissected cytoplasm, like the
actinoplasts (Vasiliev 1987; Cooper and Schliwa 1985) support
this idea. However, actinoplasts are not able to move in one
direction without an external gradient which is not required
for normal cell locomotion. Thus the direction of locomotion
seems to be determined by cell polarity based on the relation
of the lamella (or lamellae) and the rest of the cell. In
addition, G-actin has to be transported from the site of actin
disassembly to the lamella front. The existence of such a bulk
flow is proven by direct observation with phase contrast cinemicrography (Bereiter-Hahn, 1967), it is indicated by the
curved course of microtubules in locomoting and their high
number and straightness in stationary cells (Strohmeier and
Bereiter-Hahn, 1984) and by the distribution of myosin in stationary and moving cells. Myosin is supposed to be present in
oligomeric form, this complex being much larger than actin
oligomeres. Therefore myosin is facing larger resistance while
percolating through the fibrillar meshwork filling the lamella

than smaller molecules or aggregates. The distrubution of myosin would thus result from a continuous cytoplasmic flow, it is a dissipative structure. At the moment of cessation, a homogenous distribution should be achieved by diffusion, as is found in La^{3+} treated cells.

4.2 Shape changes induced by La^{3+} and ionomycine

Characteristics which have to be preserved during locomotion are the flat lamella and the thick cell body and the polarity of the cell. This distinction is lost by La^{3+} treatment, the cell body extends into the lamella. The alternative way of lamella retraction can be excluded by RIC cinemicrography (Fig. 1). Extension of the cell body region is preceeded by a slight decrease in lamella thickness. On the other hand acceleration of movements by ionomycine is related to a decrease of cell body area and an increase in lamella thickness. These observations are easily explained by the assumption of contractile forces developed in the lamella/cell body transition region and the cortex of the cell body. These forces produce a hydraulic pressure in the cell which drives the bulk flow of cytoplasm towards the zone with the least resistance, i.e. the lamella. The other role of contraction is to maintain the slope between lamella and cell body. An outward directed flow of cytoplasm alone is insufficient to preserve cell shape and to sustain continuous movement. For this purpose a reverse flow mechanism has to be postulated. The movement of membrane components from the periphery to the lamella/cell body transition region which has been described many times for moving cells indicates the existence of such a reverse mechanism. The relative high concentration of actin and myosin in this area and the formation of fibrillar strands radiating from this area towards the periphery of the lamella are a strong argument for actomyosin based contraction pulling the cell body towards the lamella and thereby allowing the re-uptake of fibrillar material into the cell body where it may be disassembled again and recirculated into the lamella.

The contractions are supposed to result from actomyosin interactions because of its calcium sensitivity and the localizi-
tion of myosin and Ca^{2+} distribution. Weakening of cytoplasm
could be due to solation of the actin meshwork at the leading
front controlled by a relatively high Ca^{2+}-concentration
(Fig.9). This solation requires less Ca^{2+} than contraction
(e.g. Taylor et al., 1973, 1976; Taylor and Fechheimer 1982).
A comparable behaviour of Acanthamoeba extract, described by
Pollard (1977), has been discussed by Dembo et al. (1986) and
has been included in the mechanochemical cytogel model proposed by Oster (1984) and Oster and Odell (1984).

Whether locomotion could be driven without myosin (obviously
without any of the different isoforms of myosin), as described
for Dictyostelium amebae, remains unclear. Injection with antibodies directed against non muscle cell myosin also did not
inhibit motility, however it also did not allow directed locomotion (Höhner et al. 1988; Höner and Jockusch 1988). Thus po-
lymerisation and depolymerisation alone might be considered to
drive motions without myosin being involved. Formation of an
ordered fibrillar pattern and directed locomotion, however,
seem to require actin and myosin interactions.

On the basis of the observations presented here the following
working hypothesis is stated about factors determining cell
shape:
Influx of Ca^{2+} causes the cell body to contract, this contraction extends the lamella in width and in thickness. Prolonged
increase of cytosolic Ca^{2+} evoked the formation of a fried egg
shaped cell. This is thought to result from loss of mechanical
resistance in the lamella periphery ("swollen" margin with
many microcolliculi). Decrease of cytosolic Ca^{2+} causes relaxation from contractions, the cell body flattens thus the slope
of the lamella/cell body transition region flattens into the
lamella which is no more extended by cytoplasmic flow.

The evidence provided here is indirect, measurements of cytosolic Ca^{2+} concentrations under various conditions are required.

REFERENCES

Bereiter-Hahn J. (1967): Dissoziation und Reaggregation von Epidermiszellen der Larven von Xenopus laevis (Daudin) in vitro. Z.Zellf. 79: 118-156

Bereiter-Hahn J., Strohmeier R., Beck K. (1983): Determination of the thickness profile of cells with the reflection contrast microscope. Scient. Techn. Inf. VIII: 125-150

Bretscher M.S. (1988): Fibroblasts on the move. J. Cell Biol. 106: 235-237

Brown R.M., Middleton C.A. (1985): Morphology and locomotion of individual epithelial cells in culture. J. Cell Sci. 78: 105-115

Condeelis J., Hall A., Bresnick A., Warren V., Hock R., Bennett H., Ogihara S. (1988): Actin polymerization and pseudopod extension during amoeboid chemotaxis. Cell Motil. Cytoskel. 10: 77-90

Cooper M.S., Schliwa M. (1985): Electrical and ionic controls of tissue cell locomotion in DC electric fields. J. Neuroscience Res. 13: 223-244

Dembo M., Maltrud M., Harlow F. (1986): Numerical studies of unreactive contractile networks. Biophys. J. 50: 123-137

DiPasquale A. (1975): Locomotory activity of ephithelial cells in culture. Exp.Cell Res. 94: 191-215

Gingell D. (1981): The interpretation of interference reflexion images of spread cells: Significant contributions from thin peripheral cytoplasm. J. Cell Sci. 49: 237-247

Gingell D., Todd I. (1979): Interference reflection microscopy: a quantitative theory for image inter pretation and its application to cell-substratum separation measurement. Biophys. J. 26: 507-526

Hall A.L., Schlein A., Condeelis J. (1988): Relationship of pseudopod extension to chemotactic hormone-induced actin polymerization in amoeboid cells. J. Cell. Biochem. 37: 285-299

Harris A.K. (1973): Cell surface movements related to cell locomotion. In: Ciba Found.Symp. Locomotion of Tissue Cells. pp.3-26, Elsevier North-Holland.

Huxley H.E. (1976): Introductory remarks: The relevance of studies on muscle to problems of cell motility. In: R. Goldman, T. Pollard, J. Rosenbaum (eds.) Cell Motility, Book A, Motility, of muscle and non-muscle cells. pp.115-126, Cold Spring Harbor Conf. on Cell Prolif. Vol 3

Höner B., Citi S., Kendrick-Jones J., Jockusch B.M. (1988): Modulation of cellular morphology and locomotory activity by antibodies against myosin. J. Cell Biol. 107: 2181-2189

Höner B., Jockusch B.M. (1988): Stress fiber dynamics as probed by antibodies against myosin. Europ. J. Cell Biol. 47: 14-21

Komnick H., Stockem W., Wohlfarth-Bottermann K.E. (1972): Ursachen, Begleitphänomene und Steuerung zellulärer Bewegungserscheinungen. Fortschritte d. Zoologie 21: 3-60

Kukulies J., Brix K., Stockem W. (1985): Fluorescent analog cytochemistry of the actin system and cell surface morphology in Physarum microplasmodia. Europ.J. Cell Biol. 39: 62-69

Kupfer A., Kronebusch P.J., Rose J.K., Singer S.J. (1987): A critical role for the polarization of membrane recycling in cell motility. Cell Motility and Cytoskeleton 8: 182-189

Lückhoff A. (1986): Measuring cytosolic free calcium concentration in endothelial cells with Indo-1: The pitfall of using the ratio of two fluorescence in tensities recorded at different wavelengths. Cell Calcium 7: 233-248

Mittal A.K., Bereiter-Hahn J. (1985): Ionic control of locomotion and shape of epithelial cells: I. Role of calcium influx. Cell Motility 5: 123-136

Mela L. (1968): Interaction of La^{3+} and local anesthetic drugs with mitochondrial Ca^{2+} and Mn^{2+} uptake. Arch Biochem. Biophys. 123: 286-293

Moore L., Pastan J. (1979): A calcium requirement for movement of cultured cells. J. Cell Physiol. 101: 101-108

Nieuwkoop P.D., Faber J. (1956): Normal tables of Xenopus laevis (DAUDIN) North Holland Publ.Amsterdam

Oster G.F. (1984): On the crawling of cells. J. Embryol. Exp. Morphol. 83: 329-364

Oster G.F., Odell G.M. (1984): The mechanochemistry of cytogels. Physica 12D: 333-350

Pollard T.D. (1977): Cytoplasmic contractile proteins. In: B.R.Brinkley, K.R.Porter (ed.) International Cell Biology 1976-1977 pp.378-387

Rees D.A., Couchman J.R., Smith C.G., Woods A., Wilson G. (1982): Cell substratum interactions in the adhesion and locomotion of fibroblasts. Phil.Trans. R. Soc. Lond. B299: 169-176

Svitkina T., Neyfakh A.A. jr. Bershadsky A. (1986): Actin cytoskeleton of spread fibroblasts appears to assemble at the cell edges. J. Cell Sci. 82: 235-248

Taylor D.L., Condeelis J.S., Moore P.L., Allen R.D. (1973): The contractile basis of amoeboid movement. I. The chemical control of motility in isolated cytoplasm. J. Cell Biology 59: 378-394

Taylor D.L., Fechheimer M. (1982): Cytoplasmic structure and contractility: the solation-contraction coupling hypothesis. Philos. Trans. R. Soc. Lond. (Biol.) 299: 185-197

Taylor D.L., Moore P.L., Condeelis J.S., Allen R.D. (1976): The mechanochemical basis of amoeboid movement. I. Ionic Requirments for maintaining viscoeleasticity and contractility of amoeba cytoplasm. Exp. Cell Res. 101: 127-133

Tilney L.G. (1980): Polymerization of actin. V. A new organelle, the actomere, that initiates the assembly of actin filaments in Thyone sperm. J. Cell Biol. 77: 551-564

Wilkinson P.C. (1975): Leucocyte locomotion and chemotaxis. The influence of divalent cations and cation ionophores. Exp. Cell Res. 93: 420-426.

PERIODIC CONTRACTION WAVES IN CYTOPLASMIC EXTRACTS

Thomas Pohl

Abtlg. Theoretische Biologie, Botanisches Institut der Universität Bonn

Kirschallee 1, 53 BONN 1, FRG

Abstract

The motility of most eucaryotic cells is based on constitution and contraction of an actomyosin network. In some in—vitro experiments numerous consecutive waves of gel assembly and gel contraction have been observed when cell free cytoplasmic extracts were incubated. The corresponding model considers the cytoplasmic matrix (F—actin, myosin, G—actin, ATP, ...) as a highly viscous two component mixture. The components are the solution and the interpenetrating fibroid network phase. Application of mass and momentum conservation laws of fluid mechanics to both components leads to a system of partial differential equations of hyperbolic—elliptic type. The presented numerical solutions of these equations reflect the experimentally observed autonomous recurrent contraction waves.

1. Introduction

The mechanical interaction of actin and myosin performing movement may be described by a <u>sliding filament mechanism</u>, cp. e.g. Lackie (1986), Dembo et al. (1984), Alt (1986). In this model G—actin assembles within the cytomatrix to F—actin. Some actin binding proteins (ABP) may crosslink several actin filaments constructing so—called actin nodes (see Fig.1). Also myosin molecules are interconnected at the long tail regions to thick bipolar filaments. When the heads of

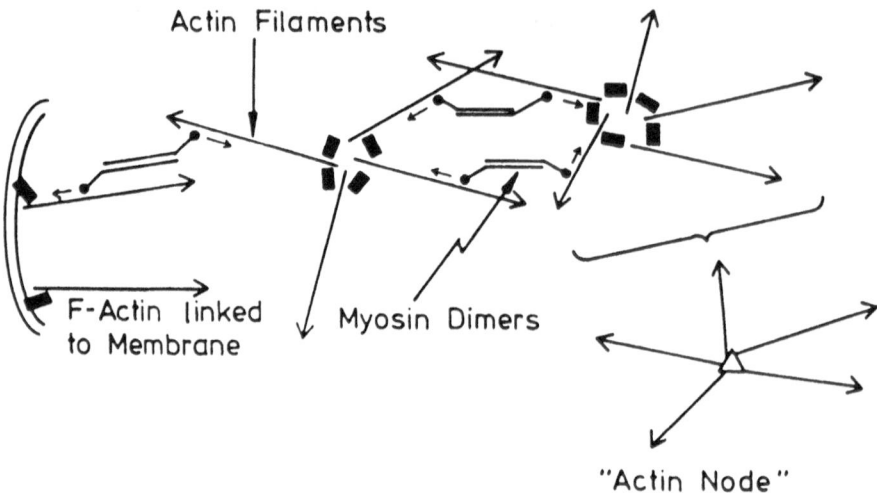

Figure 1. Illustration of the sliding filament mechanism (from Alt, 1986)

these dimers bind to actin filaments chemical energy is transformed into mechanical work by hydrolising ATP. As a result interconnected actin nodes are contracted. The process of assembly and disassembly of molecules to filaments and
the crosslinking mechanism of all components to a contractile gel is thought to
work continuously hand in hand. The sum of those conformational changes constitutes the often observed contractile behaviour of cells and cytogels. The briefly
described model above is used here to simulate the subsequently described experiment.

In order to examine the regulation of actin fibrillation and contraction Ezzell et
al. (1983) used cell free cytoplasmic extracts of Xenopus eggs. When these extracts
(prepared with ATPγS) are incubated actomyosin based recurrent waves of
assembly and contraction are observed. Fig.2 shows a series of photographs illus-
trating the waves. The cuvette contains G—actin, myosin, ATP, some other
proteins (e.g. ABP) and also, in low concentration, F—actin. During the process
more and more proteins are polymerized and the actomyosin complex starts to
contract. The contracted gel constitutes a contraction center (Fig.2,A—C). Because
of some trapped air bubbles due to outgassing from incubating the extract the
highly contracted gel moves to the free surface. This first contraction wave is
followed by a period of gelation and another contraction wave (Fig.2,D—F), and

again by a third cycle (Fig.2,G—I). The gel formed during each period occupies
the entire volume in the cuvette. Remarkable is that the newly performed gel
always aims at the initially constituted center. (Arrows in Fig.2 indicate the edges
of contracting gels.) Each wave appears to be autonomous and so the process
continues; —probably until all of the available actin is incorporated into the
contraction center. Ezzell et al. report on about 30 consecutive waves within a
time period of 100 minutes.

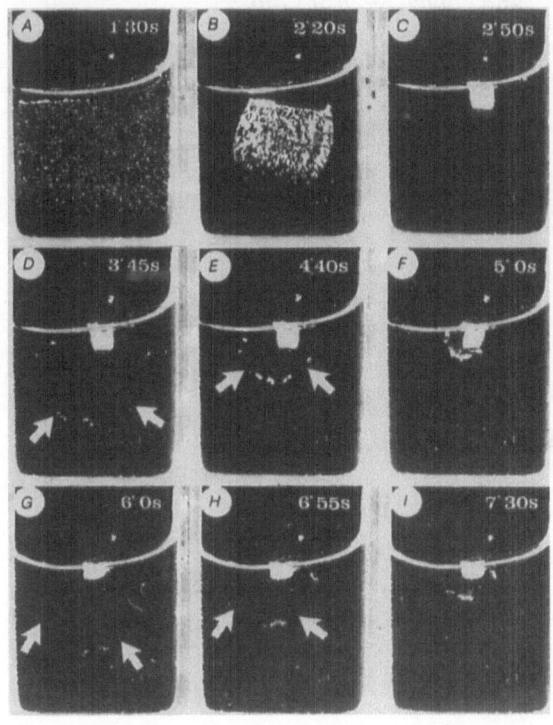

Figure 2. Photo series showing recurrent contraction waves (from Ezzell et al.
(1983), with kind permission by the author)

2. Mathematical Formulation of the Problem

The mathematical formulation of the model is based on the reactive flow model
of Dembo and Harlow (1986). Fibroid network phase and aqueous solution are

viewed as homogeneous Newtonian fluids of incompressible, highly viscous materials. Thus the physical behaviour of the suspension is completely described by
distributions of volume fractions (θ, θ_s) and velocity fields (v, w). Ignorance of the
molecular details of configuration and motion of network and solution leads to the
application of the one–dimensional mass and momentum conservation equations as
commonly known in fluid mechanics. Thereby the viscosity of the network phase
material is simply assumed to increase with an inverse proportional power law of
volume fraction of solution. Thus the following functions may be applied:

$$\mu_n(\theta) \;=\; \mu_{no}\theta/(1-\theta)^m \quad \text{and} \quad \mu_s(\theta) \;=\; \mu_{so}(1-\theta)$$

where μ_{no}, m and μ_{so} are constant parameters. The following system of partial
differential equations in non–dimensionless quantities results (for a detailed
derivation see Pohl, 1989):

(1)
$$\partial_t\theta + \partial_x(\theta v) \;=\; \Pi_{Str}(\theta_{eq}-\theta)$$

(2) $\Pi_{Re_n}\left[\partial_t(\theta v) + \left[\frac{1-2\theta}{1-\theta}\right]\partial_x(\theta v^2) - \theta^2 v^2\partial_x\left[\frac{\theta}{1-\theta}\right]\right] \;=\; -\,\Pi_\phi\theta v + \Pi_{Ar}\theta(1-\theta)$

$$+ (1-\theta)\partial_x\left[\left[\frac{\theta}{(1-\theta)^m}\right]\partial_x v + \tfrac{1}{2}\Pi_\psi\theta^2 + \Pi_\sigma\ln(1-\theta)\right] \;+\; \frac{\Pi_{Re_n}}{\Pi_{Re_s}}\,\theta\,\partial_x\left[(1-\theta)\partial_x\left[\frac{\theta v}{1-\theta}\right]\right]$$

The effect of the second phase is implicitly incorporated in the equations above.
Velocity field and volume distribution of the solution may be explicitly calculated
from:

(3)
$$w = -\frac{\theta}{1-\theta}\,v \;; \quad \theta_s = 1-\theta$$

In a mathematical sense the similarity numbers (denoted by Π) are just the parameters of the system, but, moreover, they describe the effect of the concerned
physical force in comparison to others. So the Strouhal number Π_{Str} compares the
physical time of locomotion with the chemical relaxation time. The Reynolds
numbers Π_{Re} relate the viscous forces to the inertial effects. Drag resistance
number Π_ϕ (Dembo et al., 1986) and Archimedes number Π_{Ar} (Gebhart and Pera,
1971) reflect the ratios between friction forces exhibited at the interface,
respectively buoyancy forces and inner friction. The parameters describing
contraction Π_ψ and swelling Π_σ are also related per definitionem to the friction
forces caused by viscosity of the network.

The orders of magnitudes of most terms are dominated by these similarity numbers. Other factors obtain the order $O(1)$. Bereiter–Hahn (1987) points out that for cytoplasm the Reynolds number varies in the range of 10^{-4}–10^{-5} which is generally confirmed by Lackie (1984). Janmey et al.(1988) and Zaner et al.(1983 and 1986) stated that the viscosity of fibroid network and aqueous solution are very high. So it can be predicted that both Reynolds numbers are small and that the Reynolds number of network is much smaller then the Reynolds number for the aqueous phase:

$$\Pi_{Re_n} \ll \Pi_{Re_s} \ll 1.$$

Consequently inertial effects —described by the terms on the left of the equality sign in eq.(2)— are negligible. This prediction is experimentally confirmed by Oster and Odell (1984) who found that motion within cytoplasm immediately ceases if the driving force vanishes: inertial overshoots did not occur. Since the ratio Π_{Re_n}/Π_{Re_s} is also predicted to be much smaller than one, the term dominated by this ratio is neglected too. Based on these facts the second partial differential equation reduces to a force balancing equation:

$$(4) \qquad \partial_x \left[\frac{\theta}{(1-\theta)^m} \partial_x v + \frac{1}{2}\Pi_\psi \theta^2 + \Pi_\sigma \ln(1-\theta) \right] - \Pi_\phi \frac{\theta}{(1-\theta)} v + \Pi_{Ar}\theta = 0$$

3. Boundary Conditions and Initial Distribution

With respect to the hyperbolic type of the mass balancing equation it is needed to define, besides boundary conditions, also initial data for the network volume fraction. According to the experiment the cytoplasma added to the vessel is prepared in state of its chemical equilibrium indicated by θ_{eq}:

$$t=0, \quad 0 \leq x \leq 1: \quad \theta = \theta_{eq}$$

The boundary conditions arise from the physics of the system. Since the fluid does not interact with the surrounding neither influx nor efflux of mass are approved:

$$(5) \qquad\qquad t: \quad x=0 \text{ and } x=1: \quad v\theta=0$$

The force balancing equation is elliptic. Then it is just needed to specify conditions at the edges of the bounded domain. Physically is unalterable required to
restrict solutions appropriately in order to obtain inward directed flow only; just
then the volume of the suspension remains constant. In addition the natural
boundary conditions are provided when the underlying energy functional is considered (for details see Pohl, 1989): a Neumann condition results as indicated
below:

$$(6) \qquad \partial_x v = \frac{\Pi_\sigma |\ln(1-\theta)| - \frac{1}{2}\Pi_\psi \theta^2}{\theta/(1-\theta)^m}$$

Then the combination of the mathematical and physical requirements leads to
application of special, inhomogeneous boundary conditions as follows:

(7)

<u>x=0:</u>   IF $v|_{x=0} \geq 0$ results with the natural (Neumann) boundary condition

        THEN this solution is accepted,

        ELSE the corresponding Dirichlet condition $v|_{x=0}=0$ is applied.

<u>x=1:</u>   IF $v|_{x=1} \leq 0$ results with the natural (Neumann) boundary condition

        THEN this solution is accepted,

        ELSE the corresponding Dirichlet condition $v|_{x=1}= 0$ is applied.

These IF–ELSE conditions mean first to apply the Neumann condition (eq.6) by
trial and error for each time step. IF the obtained results verify the conditions
x=1:v≤0, and(!) x=0:v≥0 respectively, calculations are proceeded with the next
time step. Otherwise (ELSE) the solution is rejected and calculations are repeated
by application of the corresponding Dirichlet condition at both or even one boundary. The here presented conditions at the boundaries are called "stress–free" (B.
Tranquillo, –personal communication) since no extra potential of energy is exhibited. It is remarked that for vanishing buoyancy forces a contraction formation in
the center position of the vessel is obtained. If Π_{Ar} is chosen to be non–zero then
the spatially centered contracting gel is forced to move in a certain direction: for
$\Pi_{Ar} < 0$ the contraction center moves in negative x–direction and conversely.
Numerical results obtained by choice of a negative Archimedes number are subsequently presented and discussed.

4. Numerical Results

With regard to the hyperbolic type of eq.(1) an explicit predictor–corrector scheme is applied to evaluate a succeeding time–step. By upwind differencing the spatial discretization is yield. The truncation error of this method is of the order $O(\Delta t^2, \Delta x)$. Centered in space finite differencing over half–mesh increments $\Delta x/2$ of the corresponding force balancing equation (4) results in an implicit tridiagonal block matrix which is solved by the Gaussian algorithm. The total number of compartments is chosen to be $N=100$. Then Δx equals $1/100$ and the width of each time step is defined by $\Delta t = 0.95\Delta x / \max|v_i|$, $1 \leq i \leq N$. Fig.3 perspectively

represents the calculated distributions of network volume fraction and velocity. Each line indicates the data of the variable versus x at a certain time t. For better graphical demonstration the time axis starts in the back directed to the forth. So the initial distributions for θ and v are drawn in the back. It is shown

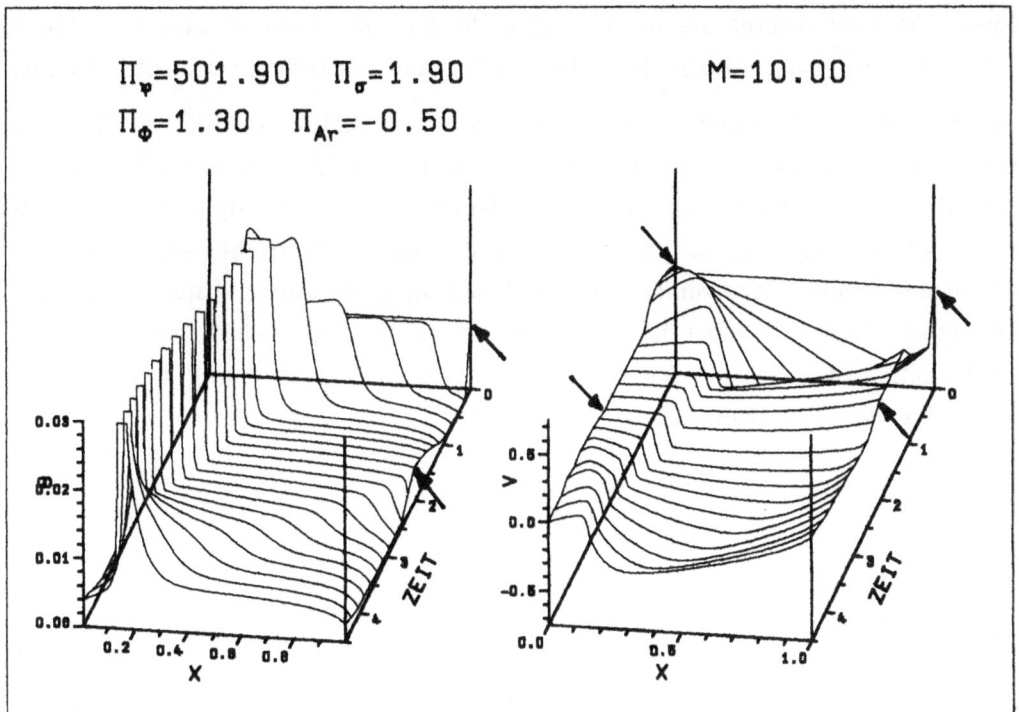

Figure 3. Representation of a numerical simulation of autonoumous recurrent contraction waves. It is shown that contraction waves disrupt at both boundaries (indicated by arrows) which are moving inward constituting a contraction formation.

that contraction waves disrupt at both boundaries moving inward and constituting a contraction center with its maximum at the spatial position x≈0.25. By effects of the buoyancy forces the highly contracted gel moves to a position at x=0.1. The contraction center is sharply edged and covers about 5% of the total height (length) of the cuvette. Over its width the velocity linearly decreases and changes its sign. Moreover, additional waves are induced during the course of time, at the right hand side as well as at the left handed boundary. The initiation of a new oscillation is indicated each time by the first non–zero velocity at the considered boundary. New waves are finally absorbed by the already existing contraction center. Before this happens the new local maxima are trying to pull the highly contracted gel towards itself; waves starting at the right are attracting the contraction center to the right and vice–versa. So the contraction center is not a static phenomenon but sways a little around the basic position.

Concentrating on the time behaviour it can be constituted that the oscillations induced at both boundaries are perfectly periodic. Moreover, the periodicity of waves moving in positive x–direction exactly equals the periodicity of the other waves. The wave lengths are determined to be 3.1 time units (1 time unit equals 1 min for $\Pi_{Str}=1$), which fits the time scale of the experiment. Another characteristic of this result is that the point of disruption is different for both boundaries. An initially transient period is evident until perfect oscillations are retained, but even afterwards the waves detaching from the right start a little earlier when compared to results obtained from calculations with vanishing Archimedes number. Accordingly, waves detaching from the left are disrupting a little later. Thus this effect is due to the buoyancy forces which are supporting the disruption of waves moving in the same direction of the acting force and are suppressing the other ones.

5. Concluding Remarks

The present paper applies fundamental laws of fluid mechanics with regard to some external forces to simulate an in–vitro experiment showing numerous consecutive actin–myosin based waves of gel assembly and gel contraction (cp. Fig.2). The corresponding model, which is based on the sliding filament mechanism and the reactive flow model considers the cytoplasmic matrix as a two component mixture, neglecting microtubules and other stuctural materials. The fibroid network phase and the aqueous solution are viewed as incompressible, highly viscous materials. A comparitive study of the effect of single forces show that inertial forces and inner friction of the soluted phase are negligible. These predictions lead to a set of two partial differential equations describing assimilation and movement of the network phase. The effect of the other phase is implicitly incorporated in it. The equations are numerically analyzed with respect to constant initial data (chemical equilibrium) and natural inhomogeneous boundary conditions.

The results strongly depend on an appropriate choice of the parameters which allows simulations reflecting the experiment. The parameters are basically determined on biological data, but within certain ranges such that, 1. a wave length of about 3 min is exhibited, 2. the very first contraction wave realizes "visible" distributions ($\theta(x) \geq \theta_{eq}$, $0 \leq x \leq 1$) of network volume fraction allmost all over the domain (cp.Fig.2B), 3. local contraction centers arise which are finally completely absorbed by the already existing ones, 4. the highly contracted gel is supposed to be placed close to the left handed boundary (x=0.), and, 5. the width of the contraction center should cover about 10% of the total height of the cuvette.

As a main result perfectly periodic waves of gelation and contraction within a suitable period are illustrated. The physical and biological behaviour of the cytoplasm is reflected, but the constituted contraction center appears to be too small and actually could be constructed closer to the free surface when compared to the experiment. Therefore it might be recommended to include an additional force regarding the surface tension at the interface between cytoplasm and surrounding. Also the irreversible transition of G—actin into the network should be regarded which finally ceases the occurrence of even more waves after consumption of G—actin.

Acknowledgements

Gratefully acknowleged are many valuable discussions with Wolfgang Alt, the author is indebted to him for introduction and cooperative guidance in modelling intra—cellular activities. For providing video tapes and stimulating discussions I want to thank R.M. Ezzell, R.T. Tranquillo and M. Dembo. The work reported above is sponsored by the German research Society in the Collaborative Research Programme (Sonderforschungsbereich) 256 "Nichtlineare partielle Differential-gleichungen".

REFERENCES

Alt W. (1985) Contraction and oscillation in a simple model for cell plasma motion. In: L.Rensing and W.J.Jaeger (ed.) Temporal Order: 163–174. Springer Verlag, Berlin

Alt W. (1986) Mathematical models in actin–myosin interaction. In: K. E. Wohl-farth–Bottermann (ed.) Nature and function of cytoskeletal proteins in motility and transport: 219–230. Gustav Fischer Verlag, Stuttgart

Alt W. (1988) Models of cytoplasmic motion. In: M.Markus, S.C.Müller and G. Nicholis (ed.) From chemical to biological organization: 235–247. Springer Verlag, Berlin

Bereiter–Hahn J. (1987) Mechanical principles of architecture of eucaryotic cells. In: J.Bereiter–Hahn, O.R.Anderson and W.E.Reif (ed.) Cytomechanics —the mechanical basis of cell form and structure—: 3–30. Springer Verlag, Berlin

Dembo M., Harlow F.H. and Alt W. (1984) The biophysics of cell surface moti-lity. In: A.S.Perelson, C.DeLisi and F.W.Wiegel (ed.) Cell surface dynamics —concepts and models—: 495–542. Marcel Dekker, Inc., New York

Dembo M. and Harlow F. (1986) Cell motion, contractile networks, and the physics of interpenetrating reactive flow. Biophys. J., 50: 109–121

Dembo M., Maltrud M. and Harlow F. (1986) Numerical studies of unreactive contractile networks. Biophys. J., 50: 123–137

Ezzell R.M., Brothers A.J. and Cande W. (1983) Phosphorylation dependent contraction of actomyosin gels from amphibian eggs. Nature, 306: 620–622

Gebhart B. and Pera L. (1971) The nature of vertical natural flows resulting from the combined buoyancy effects of thermal and mass diffusion. Int. Journal of Heat and Mass Transfer, 14: 2025–2049

Janmey P.A., Hvidt S., Peetermans J., Lamb J., Ferry J.D. and Stossel T.P. (1988) Viscoelasticity of F–actin and F–actin/gelsolin complexes. In Press.

Lackie J.M (1986) Cell movement and cell behaviour. Allen & Unwin, London

Oster G.F.and Odell G.M. (1984) Mechanics of cytogels I: oscillations in physarum. Cell Motility, 4: 469–503

Pohl T. (1989) Fluid dynamical approach in modeling cyclic movements of acto-myosin gels. Preprint no.94, SFB 256, Universität Bonn, Wegelerstr.6, 5300 BONN 1, FRG

Zaner K.S. and Hartwig J.H. (1986) The effect of filament shortening on the mech-anical properties of gel–filtered actin. J. of Biological Chemistry, 28: 7615–7620

Zaner K.S. and Stossel P.T.(1983) Physical basis of the rheologic properties of F–actin. J. of Biological Chemistry, 25: 11004–11011

MOTION ANALYSIS OF INTRACELLULAR OBJECTS:

TRAJECTORIES WITH AND WITHOUT VISIBLE TRACKS

Dieter G. Weiss, Günther Galfe,
Josef Gulden, Dieter Seitz-Tutter,

Institut für Zoologie
Technische Universität München
Lichtenbergstr. 4
D-8046 Garching, Fed. Rep. Germany

George M. Langford,

Department of Physiology
School of Medicine
University of North Carolina
Chapel Hill, NC 27599, USA

Albrecht Struppler and Adolf Weindl

Neurologische Klinik und Poliklinik der
Technischen Universität München, Möhlstr. 28
D-8000 München 80, Fed. Rep. Germany

Abstract

The study of the motion of intracellular organelles which are far smaller than the limit of resolution of light microscopy has become possible only since the advent of video microscopy, that is the coupling of fast image processors to state-of-the-art light microscopes. The techniques required to make such objects visible are summarized and procedures to extract and analyse positional data of moving organelles are explained. Examples of our work on organelle motion along free, native microtubules from squid giant axons and on movements in human nerve fibers are presented.

Such studies have revealed that in animal cells microtubules serve as tracks for organelle movement. Thus, the questions studied at present deal with the analysis of motion along tracks, and especially with the possible existence of regular, high frequency components (steps), or low frequency velocity fluctuations in such movements. The question of whether this movement is stochastic or deterministic is also treated.

The present status of our analysis points to the existence of a molecular motor producing constant motion with only little variations of stochastic rather than deterministic nature. An additional stochastic component is especially pronounced in intact cells but almost absent from the cell-free motile system studied. It is therefore ascribed to cytoskeletal impediment and not to the motor itself.

1. Introduction

1.1. Active transport of organelles

Our report deals with what are probably the smallest biological objects which can be made visible by light microscopy and subjected to motion analysis, namely the cell organelles. Some of them reach the size of bacteria but most of them are much smaller. The various classes of organelles are responsible for the different basic cellular functions such as membrane synthesis (endoplasmic reticulum and Golgi apparatus), transport of membrane material (vesicles), providing the energy for metabolism and motion (mitochondria), and macromolecule degradation (lysosomes).

The cytoplasm of all animal cells is a highly structured gel. It consists of a fluid phase (cytosol) and the cytoskeleton made up from several types of proteinaceous filaments. Embedded in this matrix which has in animal cells usually a high viscosity (see Gross and Weiss, 1982 for review) are the various kinds of membrane-bounded organelles ranging from 30 nm vesicles to the mitochondria which are up to several micrometers long. From the size of many cells ranging from several tens to hundreds of micrometers it is understandable that the redistribution of newly synthesized organelles and of those to be disposed of cannot be achieved by Brownian motion alone. Instead these cells have developed ATP-consuming mechanisms to move actively and redistribute the organelles. The cytoskeletal filaments serve as tracks, as part of the force-generating mechanism, and they organize the distribution of organelles to their places of destination (Schliwa 1984, Weiss 1985).

All objects subjected to this kind of motion are structurally well characterized by electron microscopy in the fixed and dehydrated state. The study of the motion itself in the living cell was, however, impossible for almost all but the largest organelles due to the limited resolution of light microscopy. When using visible light, this limit is around 200 nm thus making the observation of most organelles impossible such as secretory vesicles, synaptic vesicles, peroxisomes, small lysosomes and of essentially all cytoskeletal filaments such as actin filaments (6 nm diameter), intermediate filaments (10 nm), and microtubules (25 nm).

Considerable progress in the field had to await the advent of video microscopy and especially of Allen video-enhanced contrast-differential interference contrast (AVEC-DIC) microscopy (Allen et al. 1981a,b, Allen 1985) which finally allows us by coupling high-speed digital image processing to the microscope to make visible all membrane-bounded organelles and the microtubules, whose motion and interaction we want to study. The new technique was quickly used to study the most prominent system of intracellular organelle motion, namely the transport in nerve cell processes (axons) called axoplasmic transport (see Grafstein and Forman 1980, Weiss 1982a, 1986b for review). From these and subsequent studies (Hayden and Allen 1984, Allen et al. 1985, Vale et al. 1985a, Weiss 1986b) it became clear that organelle motion in animal cells proceeds exclusively along microtubules. In plant cells this role is taken over by actin filament bundles (e.g. Sheetz and Spudich 1983, Schliwa 1984, Lichtscheidl and Weiss 1988).

Based on these studies many more breakthroughs followed, i.e. the finding that microtubules themselves can move actively (Allen and Weiss 1985, Allen et al. 1985), that on one microtubule organelles move bidirectionally (Hayden and Allen 1984, Allen et al. 1985, Schnapp et al. 1985), and that proteins could be isolated which are involved in bringing about these microtubule–based movements (Brady 1985, Scholey et al. 1985, Vale et al. 1985c, Paschal et al. 1987). All biochemical studies had to employ video–microscopic motion analysis as the only means to detect and assay the force–generating proteins. While it is sufficient for this purpose to detect merely the existence of motion, we chose to perform a detailed quantitative analysis of the occurring motion to better characterize the events of intracellular motility and to better understand the underlying molecular mechanism.

Studying organelle motion in nerve cell axons has two advantages over the use of other cells. (i) The movement along curved microtubules which occur in round–shaped cells yields curved trajectories which make the quantitative analysis difficult. In axons the microtubules form parallel bundles and the motion of organelles is therefore more or less straight, however bidirectional (see Weiss and Gross 1983, Weiss 1986b, for reviews). (ii) In most living cells including axons, microtubules cannot be directly observed microscopically, even with AVEC–DIC microscopy due to their close packing. However, the giant axon of the squid offers the unique opportunity that the cytoplasm can be extruded and dispersed. In this situation individual microtubules can be made visible by video microscopy (Brady et al. 1982, Brady 1985, Weiss 1986a, Weiss et al. 1990).

In this report we give examples of our analyses of the motion of organelles in two systems: (i) Human nerve biopsy material serves as an example for organelles moving on invisible tracks. (ii) On free native microtubules obtained from squid giant axons organelle motion and the underlying tracks can be observed at the same time.

2. Models for the motile behavior of organelles

The analysis of organelle motion parallels the biochemical studies and is meant to cross–fertilize the interdisciplinary research aiming at understanding the molecular mechanism of cellular force generation.

In biochemical studies using purified enzymes and organelles we can study only the basic enzymatic mechanism. Much more information is needed, however, to understand how these events take place in the cell, how their many components are integrated in good order to form the living cytoplasm and how the structural and functional integrity of this transport system is maintained and regulated in the cell.

Objects may move freely in three dimensions or be constrained in movement. They may translocate actively by themselves, by a driving force or simply by diffusion. Therefore, before describing our approach to analyse the motion of intracellular objects we will briefly outline the basic circumstances and assumptions on organelle motion. As microscopists we are restricted to a more or less two dimensional motion analysis, because at high magnification the depth of focus is very small (0.2–0.3 µm) so that only trajectories of objects moving in the plane of focus can be obtained.

A few hypotheses were put forward to define more clearly the basic questions to be asked about the organization of organelle motion.

Hypothesis 1: Organelle motion is not Brownian but directed.

When analysing the motion of intracellular objects it is of prime interest to know whether the object is undergoing random or directed movement. A good way to distinguish between these modes is to examine the mean square displacement MSD of the observed object, which is related to the diffusion coefficent D and the velocity v by

$$(1) \qquad\qquad MSD(t) = \langle(x(t)-x(t_0))^2\rangle = 4Dt + (vt)^2$$

where x is the position of the particle. Free diffusion would result in a linearly rising MSD, constant motion without diffusion in a quadratic rising MSD. A recent application of this method to the problem of the movement of proteins in the plasma membrane has been demonstrated by Sheetz et al. (1989). For intracellular transport, especially in nerve cell processes (axons), diffusion had however been ruled out by a different approach, namely observing for several hours the redistribution of radioactive tracers which take part in this transport (e.g. Weiss 1982b, Schmid et al. 1983, Weiss and Gross 1983, Weiss 1986b). Further analyses of the MSD and its derived descriptional parameters for the motile behavior, such as directional persistence index, persistence time and motility coefficient, as described by Alt (1988, see also Section II/1), have not yet been applied to organelle motion although this would most probably be very beneficial.

Hypothesis 2: Organelles move on tracks, although such tracks may often be invisible.

Next, we should consider whether the objects move freely or are constrained in their motion. Here again experimental work and especially the application of AVEC-DIC light microscopy has provided direct insight. It is now commonly accepted that organelles cannot move freely by themselves, but need an underlying track, the microtubule, for directed translocation (Hayden and Allen 1984, Allen et al. 1985, Schnapp et al. 1985, Vale et al. 1985a, Weiss et al. 1986a). Studies on free microtubules clearly showed that the motion changes qualitatively if organelles reach the end of a microtubule and detach (Allen and Weiss 1985, Allen et al. 1985). In this case the typical behavior of particles in

Brownian motion can be observed which is virtually absent from intact animal cells, but may be observed in some plant cells.

Therefore, when analyzing organelle motion, we have to take into consideration that probably all translocations occur along preformed tracks.

Hypothesis 3: The organelles are not streaming along the microtubules but they are moved interactingly by a stepping or ratchet type motor.

Two classes of hypotheses have been put forward to describe the generation of the driving force between microtubule and organelles. One type proposed a continous streaming type of motion exerted by electroosmotic or repulsive forces which were suggested to arise from ATP hydrolysis (Gross 1975, Odell 1976, Weiss and Gross 1982, Morel and Bachouchi 1988).

The other type of hypothetical force generation was borrowed from muscle contraction. Here it was suggested that the organelle interacts with the microtubule mechanically by going through cycles in which formation of a cross bridge, a change of conformation in the cross bridge, and ATP hydrolysis alternate in such a way that the organelle steps along the microtubule (Schmitt 1968, Ochs 1972, see Weiss and Gross 1983 for review).

Direct observation by AVEC-DIC microscopy combined with the use of non-hydrolyzable ATP analogs showed that the organelles are arrested in the state of the cross bridge cycle in which they are attached to the microtubules ('rigor') (Lasek and Brady 1985). This led to the general acceptance of an interacting, stepping mechanism (Vale et al. 1985b, Weiss and Allen 1985, Langford et al. 1987).

Hypothesis 4a: Velocity variations are of stochastic nature;
or Hypothesis 4b: Velocity variations obey a deterministic function.

Possible deterministic functions of the velocity could give insight into the molecular mechanism. High-frequency events could reveal the step size of the enzymatic motor, lower-frequency rhythmicity could indicate spiralling of the organelle around the microtubule while progressing or other features of the mechanism. Also knowledge whether the characteristic pauses, which gave this motion the name "saltatory organelle movement" (Rebhun 1959, 1972; Schliwa 1984) occur at fixed or random intervals would be important for understanding organelle motion. It was also important to look whether the findings of Koles, Smith and collaborators (Koles et al. 1982 a,b, Smith and Forman 1988) could be verified for other model systems. These authors had pioneered the field of organelle motion analysis and reported on a wave-like velocity function in organelles moving in *Xenopus* frog axons. Thus, their results were speaking for hypothesis 4b.

In order to answer such questions and to distinguish between hypotheses 4a and 4b, we presently analyse the time series of positional data of translocating organelles according to the following scheme:

The position x of a particle in the time course can be described in a recursive way by

(2) $x(t+1) = x(t) + \Delta x(t+1)$

(3) $\Delta x(t+1) = v(t+1) * \Delta t$

where Δx is the displacement during a sampling interval Δt and v is the instantaneous velocity. We are interested to analyse whether the particle is transported at a constant velocity with some random deviations or at a time-dependent velocity due to some underlying mechanism. So we have to distinguish between

(4a) $v(t) = v_c + e$

where v_c is the constant velocity and e a normally distributed random variable with mean 0 and variance σ^2, that is $N(0, \sigma^2)$, and

(4b) $v(t) = f(t)$

where f(t) denotes an unknown deterministic function of the instantaneous velocity.

Rewriting (2) and using (3) and either (4a) or (4b) yields

(5a) $(x(t+1) - x(t))/\Delta t = v_c + e$

(5b) $(x(t+1) - x(t))/\Delta t = f(t).$

Looking at (5a) and (5b) reveals that, when analysing the time series given by the left side, we should be able to distinguish between either a random process with mean v_c (the mean velocity) (5a), or in the case 5b the unknown function f(t).

3. Materials and methods

3.1. Cells used

Squid (*Loligo pealei*) were obtained at the Marine Biological Laboratory (Woods Hole, MA), the cytoplasm of the giant axons was extruded and free microtubules were obtained by the techniques described elsewhere (Allen et al. 1985, Brady et al. 1985, Weiss et al. 1990).

Segments (1–2 cm) of human nerves (mainly N. suralis) from diagnostic biopsies and transplantations or from tumor operations were obtained from the Neurology and the Otorhinolaryngology Clinic of the Technical University of Munich. Bundles of 20 – 100 axons were dissected and subjected to microscopy in Ringer's solution.

3.2. Visualization of objects at and below the limit of resolution

The squid axoplasm was observed on a Zeiss Axiomat microscope with an internally corrected 100x planapochromatic oil immersion objective (N.A. 1.30) and a 50 W mercury arc lamp. The human material was inspected on a Polyvar-Met microscope from Reichert/Cambridge Instruments using a 200W mercury arc lamp, asymmetric illumination and a 100x planapochromatic objective (N.A. 1.32). In both cases additional magnification changers had to be used to reach the necessary magnification on the TV monitor (screen width 25 cm) of up to x10,000.

The motile organelles were made visible by Allen video-enhanced contrast differential interference contrast (AVEC-DIC) microscopy (Allen et al. 1981a, b, Allen 1985, Weiss 1986a, Weiss et al. 1990) in the case of the free microtubules, and by video-enhanced contrast asymmetric illumination contrast (VEC-AIC) in the case of the myelinated human axons. Real time analog contrast enhancement and digital image processing with the Hamamatsu C 1966 Photonic Microscope System (Hamamatsu Photonics, Herrsching, FRG; and Photonic Microscopy Inc., Oak Brook, IL, USA) was applied consisting of the following steps: (a) AVEC analog enhancement of the full aperture image (at an instrumental compensator setting of 1/9 of a wavelength in the AVEC-DIC case) with gain and offset adjusted from about one-third to one-half of their respective ranges, (b) digital subtraction of the fixed pattern of mottle in the optical path, (c) manipulation of the gray scale by stretching of the pixel brightness histogram, (d) in some cases, reduction of pixel noise by a real-time digital rolling average operation (recursive filtering) over two frames. All video microscopy procedures used were described in full detail elsewhere (Weiss et al. 1989, 1990, Weiss 1989).

Video-enhanced microscope images were recorded in real time by a normal speed video cassette recorder and in some cases by a time lapse recorder (Sony model TV0-9000) for playback with an acceleration of 10 times. The squid preparations were studied in parallel by electron microscopy as described before (Allen et al. 1985, Langford et al. 1987, Weiss et al. 1987, 1990).

3.3. Extraction of trajectories from video sequences

We obtain trajectories of selected particles in two ways. Particles of suitable size and contrast are tracked automatically by a X-Y tracker (X-Y Tracker C1055, Hamamatsu Photonics). This device detects objects whose brightness is above or below an adjustable threshold grey-value which are located in a relatively small, adjustable region of interest defined by a frame on the monitor. If the object moves, the center cross-hair of this frame follows automatically (Fig. 1). The position of the cross hair is output to the computer. Small particles with low contrast are tracked by using the X-Y tracker in its manual mode. With both modes the x-y coordinates are transferred via a digital interface to an IBM compatible microcomputer.

In automatic mode the coordinates are updated at video frequency (1/50 or 1/60 sec depending on the video standard used). In manual mode usually the coordinates of every 10th full frame are determined, but other frequencies can be selected as well. The x- and y-values for the coordinates range from 0 to 511, and are measured with an accuracy which was determined empirically to be ±2 pixels or video lines in automatic mode and ±1 in manual mode. These coordinates mark the position of the particle in the video frame. They have to be transformed into physical x-y coordinates according to the magnification factor of the microscope and the aspect ratio of the video frame.

With the method described above trajectories of particles have to be determined one by one. To overcome this restriction we are now developing a system which will be capable of multiple object tracking. It is based on a digital image processing system (MaxVideo, Datacube, Inc., Peabody, Mass.) and an area parameter accelerator (APA512-MX, Vision Systems, Adelaide, Australia). The latter describes a two dimensional binary image in the form of a set of descriptive parameters (e.g. number of pixels, perimeter, minimum and maximum x and y coordinates) for connected regions of pixels of the same range of grey levels, usually called 'blobs'. The set of parameters for up to 255 blobs (blob pattern) is generated for every video frame. From this sequence of blob patterns the multiple trajectories can be derived.

3.4. Motion Analysis

The subsequent analysis is performed interactively with a software package developed in our laboratory (Keller 1986, Weiss et al. 1986, 1988) which is similar to the procedures developed by Koles et al. (1982a, b).

3.4.1. Estimating underlying tracks

In the cell the moving object and the underlying track cannot be seen simultaneously. Hence we have to estimate the track by fitting a curve to the observed x,y-positions of the object. We have, however, to be aware that the commonly used regression model to fit a curve is not appropriate in this case. Just recall, that it is based on the assumption that the y-value is measured with random errors, while the x-value is measured without error. The estimated regression curve minimizes the sum of the squares of the distances of the measured y-value and the y-value of the curve. In our case, both, the x- and the y-values are subjected to measurement errors. Therefore the sum of squares of the perpendicular distances of the x,y-positions to the line (track) has to be minimized.

Visualization of microtubules revealed that the majority forms straight tracks and that their diameter is small compared to the size of the organelles being transported (Langford et al. 1987, Weiss et al. 1988). This holds true for microtubules in axons, as well as for free microtubules. Additionally, we exclude from the analysis trajectories which grossly deviate from straight tracks. Estimating the track is hence reduced to fitting a straight line to the observed positional data of the particle. The slope a and offset b of the line $g(x) = \hat{y} = a x + b$ are given by equations

(6) $a^2 S_{xy} + a(S^2_x - S^2_y) - S_{xy} = 0$ and

(7) $\bar{y} = a\bar{x} + b$

where S^2_x, S^2_y and S_{xy} are the variances and the covariance, and $\bar{y}$ and $\bar{x}$ are the means of the x- and y-values. For a more detailed explanation and confidence intervals for the estimated parameters refer e.g. to Madansky (1959), Barnett (1967) or Wong (1989).

Solving the above equations we obtain an estimate for the track. For any observed position of the particle the point of intersection of the perpendicular with the line yields its estimated position on the track. If we transform the coordinate system in a way, that the origin is translocated to the estimated starting position of the particle on the track and the x-axis is rotated onto the estimated line (track), we obtain the desired representation for further analysis. Then the equations for the translocation and rotation are

(8) $x' = x - (x_0 + (y_0 - \hat{y}_0) * \sin(\alpha) * \cos(\alpha))$

(9) $y' = y - (y_0 - (y_0 - \hat{y}_0) * \cos(\alpha) * \cos(\alpha))$

where $\alpha = \arctan(a)$ und $\hat{y}_0 = g(x_0)$, and

(10) $\begin{pmatrix} x'' \\ y'' \end{pmatrix} = \begin{pmatrix} \cos(\alpha) & \sin(\alpha) \\ -\sin(\alpha) & \cos(\alpha) \end{pmatrix} \begin{pmatrix} x' \\ y' \end{pmatrix}$.

The motion of the particle is now described by the time series of x", y"-coordinates, where x" denotes the position on the track, and y", its perpendicular deviation. This description is used in the further analysis.

3.4.2. Derivation of motion parameters

The most often used parameter to describe the motion of an organelle in more detail and to compare the status of different cells is the average velocity of particles in the main direction of movement. It is determined in different ways, such as the slope of the line fitted to the positional data over time, the average of frame to frame velocities, or the quotient of distance and time for the tracked path. The latter is used in our work.

In studying the molecular mechanism of the intracellular transport the analysis of time- and track-dependent variations of the velocity found in a single trajectory are in the focus of view. In contrary to the average velocity this is in the literature often called 'instantaneous' velocity. The methods to obtain the estimate of this velocity vary (Koles et al. 1982a, Breuer et al. 1987), so that we describe our approach in more detail.

As mentioned above the coordinates are likely to be affected with measurement errors. Thus some difficulties will emerge when estimating the velocity from the positional data. The velocity of a particle is obtained by differentiation of the time series of x positions. First-order differencing according to

(11) $v(t) = (x(t) - x(t-1))/\Delta t$

where Δt is the time between successive x positions, is for good reasons considered to be a rather gross approximation to differentiation. However, we decided to use this method, for two reasons.

(i) Due to our sampling frequencies (10 Hz or 25 Hz) the maximum resolvable frequencies of the positional variation of a particle are either 5 Hz or 12.5 Hz. In nearly all observed spectra more than 99% of the power of the signal is well below 1 Hz. Obviously the signal is 'oversampled' and we therefore obtain, with respect to the signal, a 'small' Δt and hence a better approximation to differentiation. Noise components at higher frequencies, induced by measurement errors, would, however, still degrade the estimate severely, because high frequency components are 'amplified' by differentiation.

(ii) Therefore, before differencing the time series according to (11), we remove critical noise components at higher frequencies, in the following way. First, we transform the trend-removed time series of the x positions into the frequency domain, using the Cooley-Tukey (1965) FFT algorithm. Second, a cut-off frequency for low-pass filtering is determined. With the criterion that all significant peaks of the signal have to be within the pass band, we usually obtain cut-off frequencies of about 0.5 – 0.7 Hz. Third, the trend-removed time series is digitally filtered using a low-pass filter (second order Butterworth seemed appropriate) with the respective cut-off frequency. Fourth, the trend is re-added to the filtered time series. After differencing, according to (11), we finally obtain the instantaneous velocity of the particle.

With the methods described above we obtain the desired variables to describe quantitatively and further analyse the motion of organelles on underlying tracks.

4. Results

Figure 1 demonstrates that organelles can be detected similarly in preparations of extruded squid axoplasm and in intact cells by AVEC-DIC microscopy. It should be mentioned that the larger organelles (0.2 – 0.5 μm) are depicted in real size, i.e. optically resolved, while all organelles and the microtubules which are smaller than the limit of resolution (around 0.2 μm) appear inflated by diffraction to a diameter of about 0.2 μm (cf. Inoué 1986, Weiss et al. 1989). While free microtubules can be clearly visualized (Fig. 1A) it is obvious that in intact cells they are too closely packed to be distinguishable (Fig. 1B).

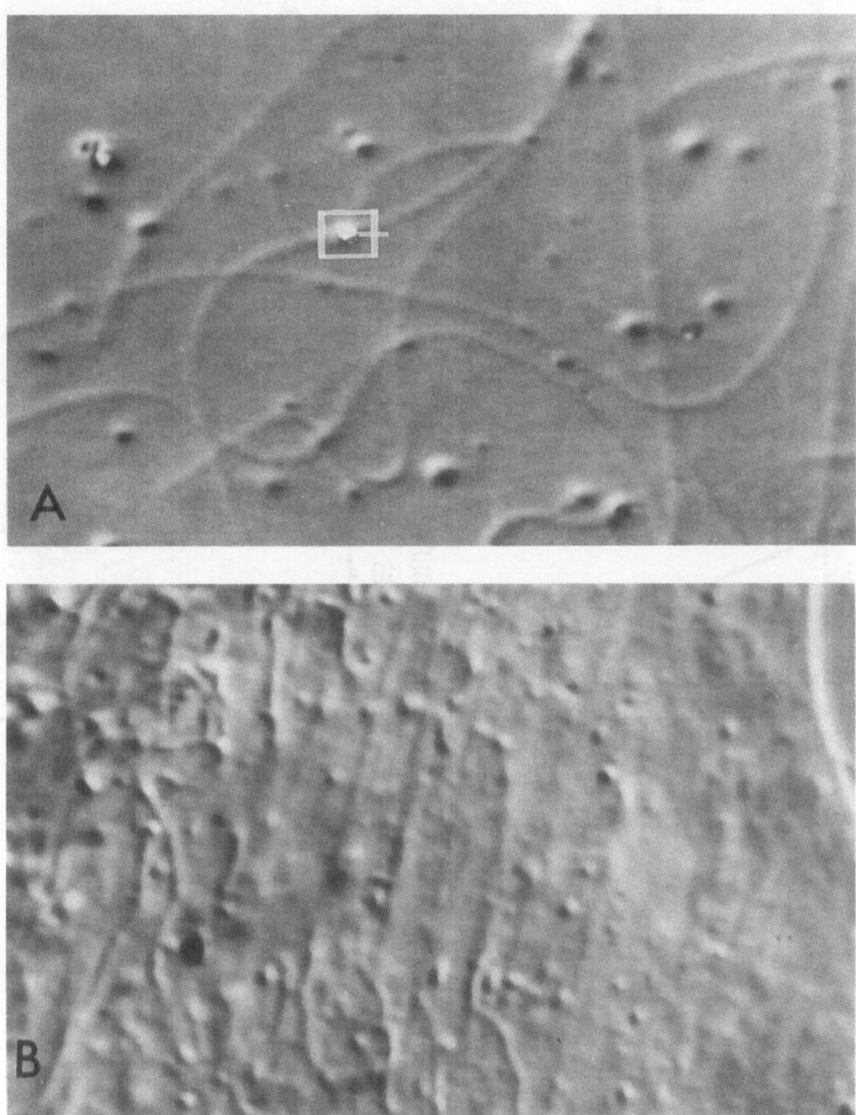

Figure 1: Video-microscopic imaging of organelles and microtubules. A: Extruded and dispersed cytoplasm from a squid giant axon. Microtubules and most organelles (axoplasmic vesicles) appear inflated by diffraction to about 250 nm. The size of the frame generated by the x, y–tracker and the pointer defining the center of a detected particle are schematically indicated. Frame width 18 μm.
B: Region near the cell body of a neuroblastoma cell. Linear elements consisting of cytoskeletal filament bundles are barely visible. The curved, elongated structures are mitochondria. The larger round objects are the prelysosomal organelles studied here in more detail. Frame width 30 μm.

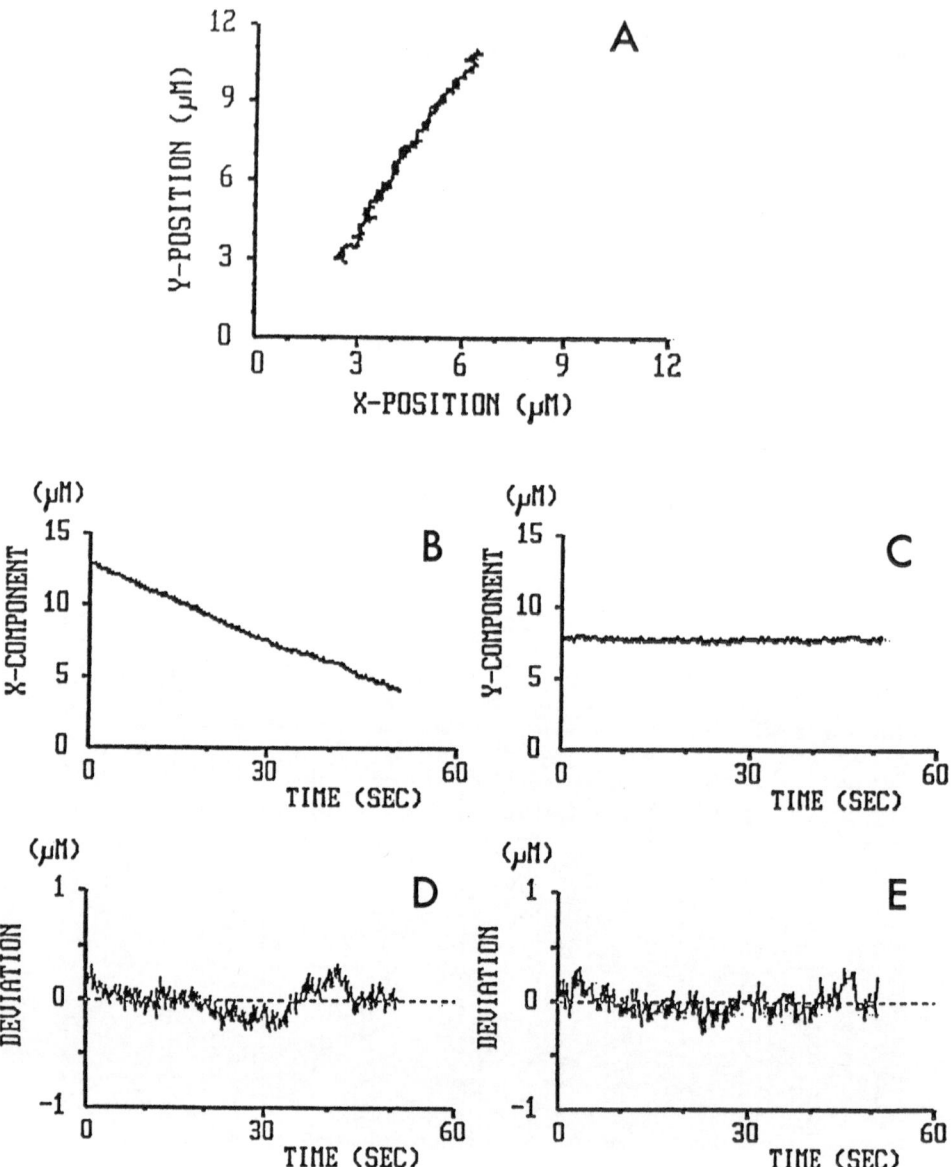

Figure 2: Quantitative analysis of organelle movement along isolated native microtubules from extruded squid axoplasm. A: Trajectory of a representative large organelle which translocated along a microtubule as seen on the TV-screen. The mean velocity is 0.18 µm/s. For further analysis the track was rotated that the main direction of movement was horizontal. B: X-component plotted versus time. The movement is relatively constant and does not show reversals in direction or pauses. C: The y-component, i.e. the displacement perpendicular to the microtubule is much less pronounced. D: Deviations from the estimated positions in the main direction of movement plotted versus time. E. Same plot for motion in the direction perpendicular to the track plotted versus time.

4.1. Organelles moving along free, visible microtubules

The trajectories of 15 round organelles of about 0.2 - 0.5 μm diameter on free native microtubules attached to the cover glass surface were analysed. Curved microtubules or moving ones were excluded from this study. Electron micrographs that were obtained from parallel preparations revealed that these organelles are smooth-surfaced and most probably of the prelysosomal type (Langford et al. 1987, Weiss et al. 1988). The average velocity was 0.48 ± 0.21 μm/s. It is well known that the majority of these organelles move retrogradely, i.e. toward the microtubule's (-) end, even when on dissociated microtubules (Allen et al. 1985), but in this study no attempt was made to determine the direction relative to the polarity of the microtubule.

The trajectories and their x- and the y-components are shown for one representative particle in Fig. 2A-E. The plots of position versus time (where the slopes represent the mean velocities) are seen in Fig. 2B and C. The movement in the main (x-) direction, i.e. along the microtubule, is relatively constant and does not show reversals of direction (Fig. 2B). The movement perpendicular to the microtubule is much less pronounced and does not result in net displacement from the microtubule (Fig. 2C).

When the linear regression lines were subtracted (trend removal) the fluctuations of the respective directional component resulted (Fig. 2D and E). Figures 3A-D show power spectra of the trend removed positional function of the particle's x- and y-components. There is no distinct frequency component visible. The shape of the spectra implies that stochastic events dominated the trend-free movement of the organelles, because prominent peaks at distinct frequencies were absent.

When we analyse particles observable for various durations (5-50 sec.) and plot the frequencies of the first peaks of the power spectra versus the sampling duration, i.e. the length of the data set, we obtain a hyperbolic relationship (Weiss et al. 1988). This also implies that the first peak does not contain useful spectral information on the potential presence of a low frequency component such as an oscillation. From these findings we conclude there are no distinct high frequency components in this kind of movement between 1 Hz and 5 Hz. The velocity fluctuation is shown for the axial direction in time (Fig. 3E) and along the length of the microtubule (Fig. 3F).

4.2. Organelles moving on invisible tracks in intact axons

The positional data for medium-sized, i.e. lysosome-like organelles moving in myelinated human axons were processed in essentially the same way as described above (Figs. 4 and 5). The most striking difference was the much less constant movement with pronounced changes of velocity and even direction (Fig. 4). The power spectra of the trend-free positional deviations of these movements (Fig. 5A-D) had an even smoother shape than in the case of the free microtubules. Nevertheless, no distinct frequential component could be detected; instead, the frequencies of the peak also reflected the sample length.

As Fig. 5C and D show, the spectra above 0.2 Hz were interpreted to be caused by noise. Therefore, the positional data were low–pass filtered at this frequency and subsequently the instantaneous velocities for each 0.1 sec interval calculated. Fluctuations in the axial direction of the instantaneous velocity with time were much more pronounced than those of the organelles moving on free

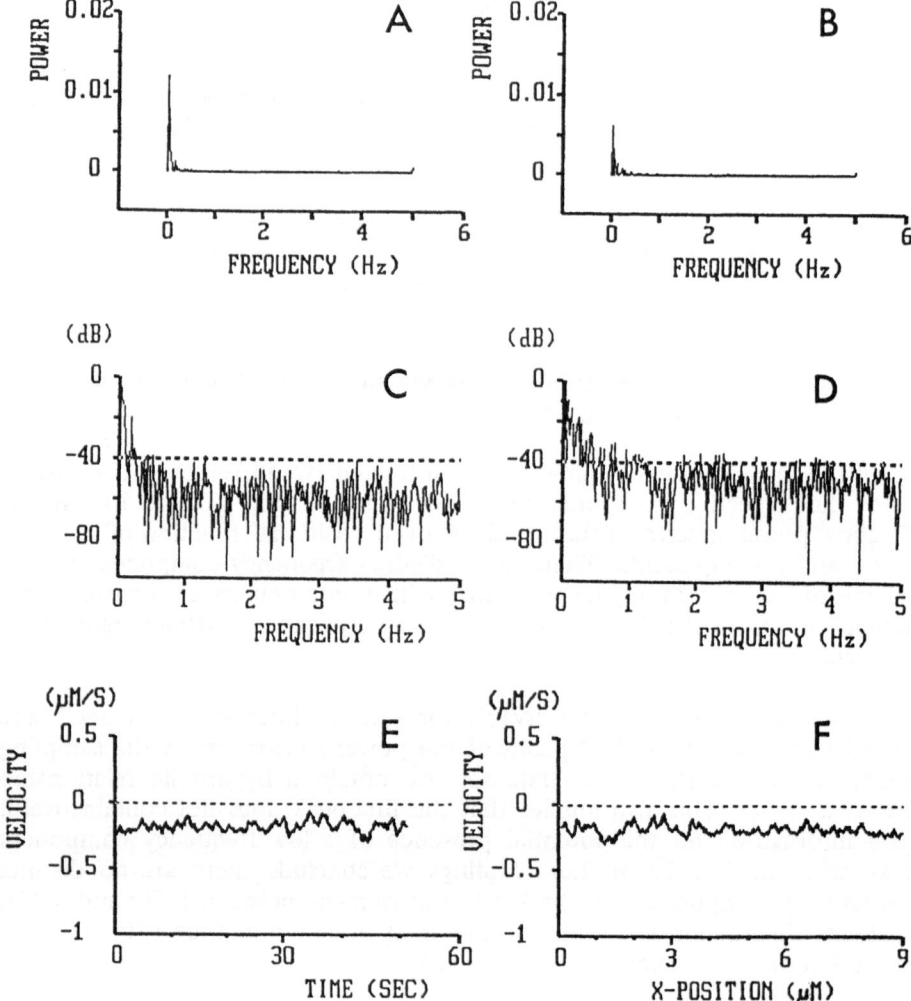

Figure 3: Spectral analysis of the translocation of particles along native microtubules. A,B: The power spectra of the deviation from the estimated positions in either direction show no distinct frequency components above approximately 0.5 Hz but the amplitudes decrease rapidly from a maximum value. The amplitude of the spectrum of the component axial to the microtubule (A) is somewhat higher than that of the component perpendicular to it (B). C,D: The logarithmic representation of the power spectra shows no significant peak, i.e. frequency component, at lower amplitudes (–40 dB = 1/1000 of maximal amplitude). The instantaneous velocity was plotted versus time (E) and along the length of the track (F).

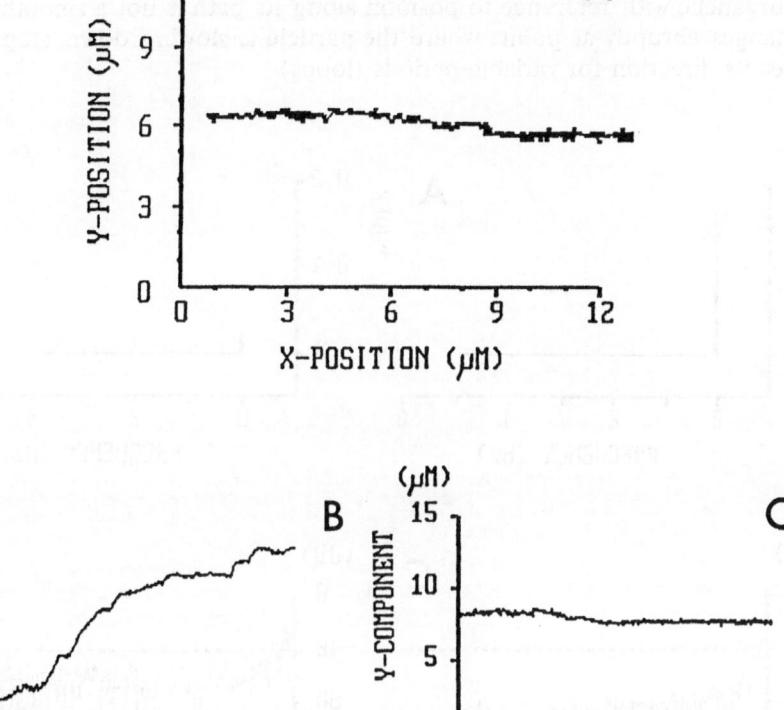

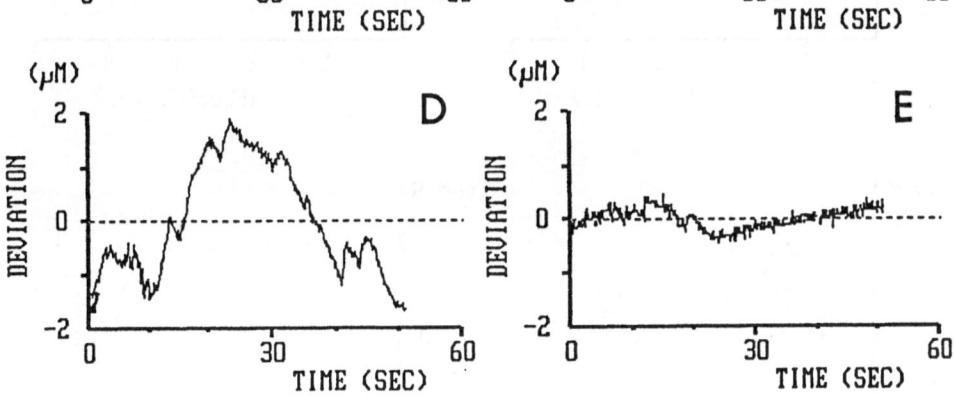

Figure 4: Example for a similar analysis as in Fig. 2 of the movement of a representative lysosome–like organelle in a human axon. A. Path of the organelle. The position of the organelle is plotted versus time for the main direction of movement (B) and perpenicular to it (C). D,E: Positional deviations from the hypothetical constant movement (regression line).

microtubules (Fig. 5E). Also Fig. 5F shows that instantaneous velocity of the same organelle with reference to position along its path is not a smooth function but changes abruptly at points where the particle is slowing down, stops or even reverses its direction for variable periods (loops).

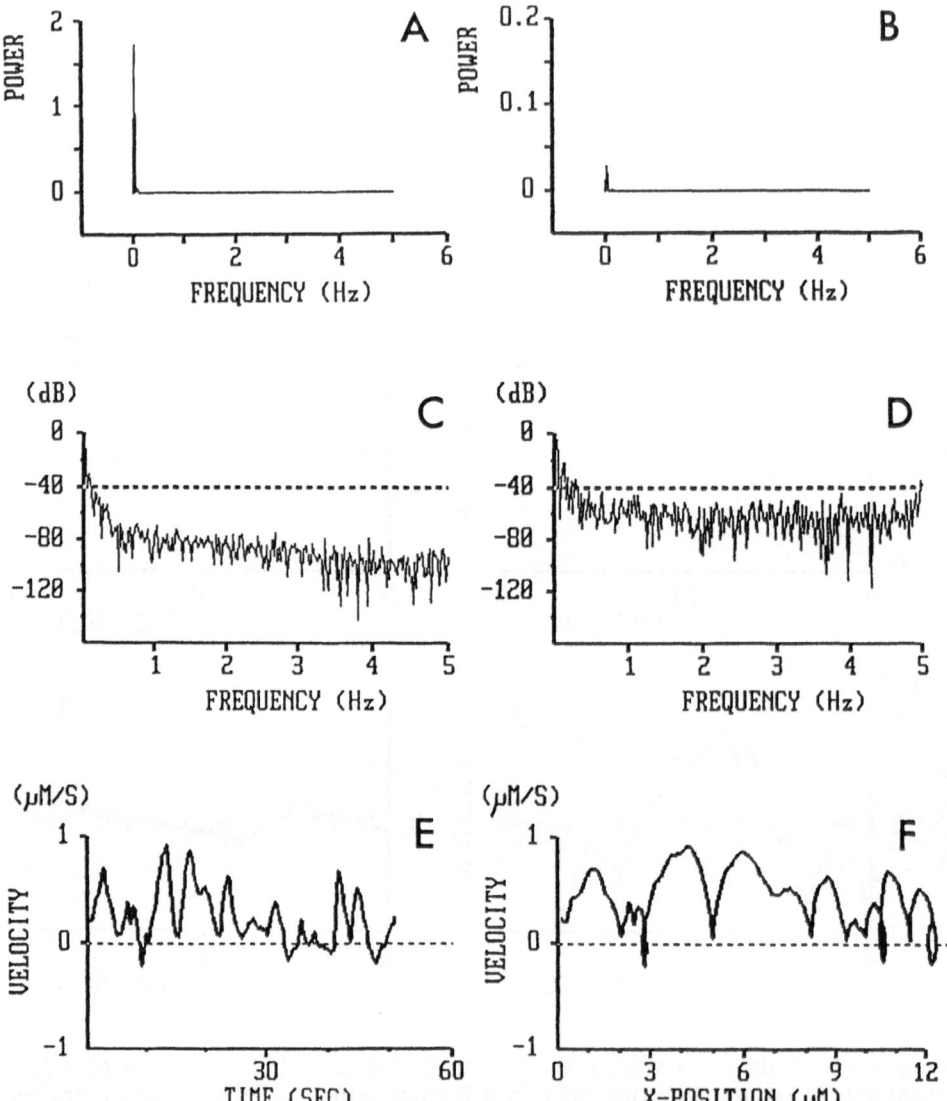

Figure 5: Similar analysis as in Fig. 3 of the movement of the same organelle as in Fig. 4. The respective power spectra for the two perpendicular directions (A,B) show again very little evidence for frequential components. The logarithmic representation (C,D) yields the same result as in Fig. 3 C,D. The instantaneous velocity plotted versus time (C) and along the length of the track (D) shows considerable fluctuations including changes of direction.

5. Discussion

Some of the questions raised above can already be discussed in the light of our organelle motion analyses. A few results and further potential applications are also summarized.

5.1. Tracks

As there is no evidence for directed motion in animal cells except along microtubules, several authors have used motion analysis to gain insight into the arrangement of the underlying but invisible microtubule tracks (Koles et al. 1982a, Lynn et al. 1986, Gulden and Weiss 1988). In these studies mainly information on the stability of local tracks with time and on the distance between neighboring tracks could be obtained (Lynn et al. 1986, Gulden and Weiss 1988).

Also information on the molecular mechanism can be obtained from motion analyses. We described in detail the qualitative properties of the behavior of organelles which attach to or move along microtubules and its implications (Allen et al. 1985, Weiss 1986a, 1987). Some quantitative results will be discussed below.

From the motion analysis point of view interesting results can be expected from the analysis of organelles moving on curved tracks and especially on intersecting microtubules, where they very readily switch tracks and undergo abrupt changes of direction.

5.2. Absence of low frequency oscillations

The movement of prelysosomal organelles and mitochondria in certain frog and lobster axons had been reported to be oscillatory in nature (Koles et al. 1982a,b, Kendal et al. 1983, Smith and Forman 1988). However, the low frequency component (first peak) present in our amplitude and power spectra (Fig. 3A-D) did not indicate a true oscillatory component but reflected the sampling time. The absence of regular oscillations was also found in studies on axoplasmic vesicles (Gulden and Weiss 1988), and on large organelles (unpublished) of unmyelinated crayfish motoneurons, as well as in myelinated human axons (see Fig. 5A-D and Gulden et al. 1988). We conclude, therefore, that these velocity fluctuations are stochastic, although they seem to differ from a purely random process, so that at present we would prefer Hypothesis 4a over 4b.

5.3. Regularity of the pattern of organelle-microtubule interactions

The power spectra (Fig. 3A-D, 5A-D) implied that mainly a stochastic component is present. This more or less random motion was so pronounced that also no regular component in the range between 1 and 5 Hz such as stepping along or circling around the microtubule could be detected. Because of the resolution of the system with a pixel size of about 50 nm, we could not have detected very small step lengths of the crossbridge cycle. We showed that at velocities of about 0.5 µm/s, regular interactions between the microtubule and the organelle which influence the speed of the translocation, such as stepping, were not detectable in the range covered by our spectral analysis, i.e. between 50 and 300 nm step size.

Gelles et al. (1988) reported the existence of 4 nm steps for artificial beads moving very slowly along microtubules in a very low ATP and elevated viscosity situation. This distance corresponds with the size of the tubulin subunits but it would be a very small and energy-consuming step size for the cross bridge cycle. The motor proteins kinesin and dynein have a length of 70 - 80 nm (Hirokawa et al. 1989) and can be expected to progress under physiological conditions with larger steps, as does also the myosin head in muscle contraction (13 nm steps; Huxley and Kress 1985).

5.4. Components of organelle motion

We conlude from our analysis that the organelles' translocation behavior consists of three different components: (1) an active transport process whose velocity is remarkably constant; (2) a small, high-frequency component superimposed on the active motion (probably sampling noise), and (3) stochastic, low-frequency, retarding events. The latter were absent in the study of translocations on free microtubules but very prominent in the intact axon (compare Figs. 2B and 4B) (Gulden et al. 1988). We conclude, therefore, that these are impediments due to cytoskeletal structures in the intact cytoplasm which obstruct the particles path and cause local decelerations and even pauses.

5.5. Classification

Our motion analyses were used to classify organelle motion in various cellular systems (Weiss et al. 1986). According to this preliminary scheme, we had suggested that the motion of the small axoplasmic vesicles (Allen et al. 1982) should be considered as continuous motion, i.e. with a relatively constant velocity, and without reversals and pauses. The larger prelysosomal organelles in intact axons were classified as interrupted motion because they often pause without reversal of direction (Smith 1982, Weiss et al. 1986, Gulden et al. 1988). This analysis, however, shows that the motion of the larger organelles, translocating along free microtubules, is also continuous. We conclude, therefore, that the interrupted type of motion (formerly called saltatory motion) observed for the same class of organelles in intact cytoplasm is due to transient blockage of the progress of large organelles by the cytomatrix.

5.6. Future trends

Organelle motion is a vital and very basic property of eukaryotic cells. Therefore, a quantitative characterization of the motile events inside a given cell type is one means to describe the cell's physiological situation. Once this is accomplished the motile situation of the cell can be used as a very sensitive and multifacetted indicator in cytopharmacological and cytotoxicological studies. Also the question can be tested whether defined exogenous test particles, such as colloidal gold (Geerts et al. 1987), show the same behavior as the endogenous organelles. Once we have quantitative data on organelle movement in human nerve biopsies, for example, then we may be able to base a diagnostic test system for certain neuropathies on a quantitative determination of changes in organelle motion. Such efforts are presently made in our laboratory as well as in others (e.g. Breuer et al. 1987).

Acknowledgments

Parts of this work were supported by Deutsche Forschungsgemeinschaft (We 790/11-2), NSF Grant DCB 851 7983, Wilhelm-Sander-Stiftung (Grant 83.010), Bayer. Staatsministerien für Landesentwicklung und Umweltfragen und für Wissenschaft und Kunst, and The President of the Technical University Munich.

REFERENCES

Allen, R.D. (1985) New observations on cell architecture and dynamics by video-enhanced contrast optical microscopy. Annu. Rev. Biophys. Biophys. Chem. 14: 256-290

Allen, R.D. and Weiss, D.G. (1985) An experimental analysis of the mechanisms of fast axonal transport in the squid giant axon. In Ishikawa, H., Hatano, S., and Sato, H. (eds): "Cell Motility: Mechanism and Regulation", 10. Yamada Conference, Sept. 1984, Tokyo: University of Tokyo Press, pp 327-333

Allen, R.D., Allen, N.S. and Travis, J.L. (1981a) Video-enhanced contrast, differential interference contrast (AVEC-DIC) microscopy: A new method capable of analyzing microtubule-related motility in the reticulopodial network of Allogromia laticollaris. Cell Motil. 1: 291-302

Allen, R.D., Metuzals, J., Tasaki, I., Brady, S.T. and Gilbert, S.P. (1982) Fast axonal transport in squid giant axon. Science 218: 1127-1129

Allen, R.D., Travis, J.L., Allen, N.S. and Yilmaz, H. (1981b) Video-enhanced contrast polarization (AVEC-POL) microscopy: A new method applied to the detection of birefringence in the motile reticulopodial network of Allogromia laticollaris. Cell Motil. 1: 275-289

Allen, R.D., Weiss, D.G., Hayden, J.H., Brown, D.T., Fujiwake, H. and Simpson, M. (1985) Gliding movement of and bidirectional transport along native microtubules from squid axoplasm: Evidence for an active role of microtubules in cytoplasmic transport. J. Cell Biol. 100: 1736-1752

Alt W. (1988) Modelling of Motility in Biological Systems. In: McKenna J., Teman R. (Edits.) ICIAM'87: Proceedings of the First International Conference on Industrial and Applied Mathematics. pp. 15-30 Philadelphia: SIAM.

Barnett V.D. (1967) A note on linear structural relationships with both residual variances are known. Biometrika 54: 670-672

Brady, S.T. (1985) A novel brain ATPase with properties expected for the fast axonal transport motor. Nature (Lond.) 317: 73-75

Brady, S.T., Lasek, R.J. and Allen, R.D. (1982) Fast axonal transport in extruded axoplasm from squid giant axon. Science 218: 1129-1131

Brady, S.T., Lasek, R.J. and Allen, R.D. (1985) Video microscopy of fast axonal transport in extruded axoplasm: A new model for study of molecular mechanisms. Cell Motil. 5: 81-101

Breuer A.C., Lynn M.P., Atkinson M.B., Chou S.M., Wilbourn A.J., Marks K.E., Culver J.E. and Fleegler E.J. (1987) Fast axonal transport in amyotrophic lateral sclerosis: An intra-axonal organelle traffic analysis. Neurology 37: 738-748

Cooley, J.W. and Tukey, J.W. (1965) An algorithm for the machine calculation of complex Fourier series. Math. of Comp. 19: 297-301

Geerts, H., De Brabander, M., Nuydens, R., Geuens, S., Moeremans, M., De Mey, J. and Hollenbeck, P. (1987) Nanovid tracking: a new automatic method for the study of motility in living cells based on colloidal gold and video microscopy. Biophys. J. 52: 775-792

Gelles, J., Schnapp, B.J. and Sheetz, M.P. (1988) Tracking kinesin-driven movements with nanometre-scale precision. Nature 331: 450-453

Grafstein, B. and Forman, D.S. (1980) Intracellular transport in neurons. Physiol. Rev. 60: 1167-1283

Gross, G.W. (1975) The microstream concept of axoplasmic and dendritic transport. Adv. Neurol. 12: 283-296

Gross, G.W. and Weiss, D.G. (1982) Theoretical considerations on rapid transport in low viscosity axonal regions. In: Axoplasmic Transport, D.G. Weiss ed., Springer Verlag, Berlin, pp. 330-341

Gulden, J. and Weiss, D.G. (1988) Axonal transport of small particles (vesicles) visualized in crayfish axons by AVEC-DIC microscopy. Cell Motil. Cytoskel. 10: 343

Gulden, J., Weiss, D.G., and Clasen, B. (1988) The velocity fluctuations of organelles transported in crustacean and human axons are random. Eur. J. Cell Biol. 46 Suppl. 22: 24

Hayden, J.H. and Allen, R.D. (1984) Detection of single microtubules in living cells: Particle transport can occur in both directions along the same microtubule. J. Cell Biol. 99: 1785-1793

Hirokawa, N., Pfister, K.K., Yorifuji, H., Wagner, M.C., Brady, S.T. and Bloom, G.S. (1989) Submolecular domains of bovine brain kinesin identified by electron microscopy and monoclonal antibody decoration. Cell 56: 867-878

Huxley, H.E. and Kress, M. (1985) Crossbridge behaviour during muscle contraction. J. Muscle Res. Cell Motil. 6: 153-161

Inoué, S. (1986) Video Microscopy. New York and London, Plenum Press

Keller, F. (1986) Computergestützte Analyse von Organellenbewegungen in Zellen dargestellt an den Axopodien von Raphidiophrys ambigua und Echinosphaerium nucleofilum. Thesis, Faculty of Biology, University of Munich, FRG.

Kendal, W.S., Koles, Z.J. and Smith, R.S. (1983) Oscillatory motion of intra-axonal organelles of Xenopus laevis following inhibition of their rapid transport. J. Physiol. (Lond.) 345: 501-513

Koles, Z.J., McLeod, K.D. and Smith, R.S. (1982a) The determination of the instantaneous velocity of axonally transported organelles from filmed records of their motion. Can. J. Physiol. Pharmacol. 60: 670-679

Koles, Z.J., McLeod, K.D. and Smith, R.S. (1982b) A study of the motion of organelles which undergo retrograde and anterograde rapid axonal transport in Xenopus. J. Physiol. (Lond.) 328: 469-484

Langford, G.M., Allen, R.D. and Weiss, D.G. (1987) Substructure of sidearms on squid axoplasmic vesicles and microtubules visualized by negative contrast electron microscopy. Cell Motil. Cytoskel. 7: 20-30

Lasek, R.J. and Brady, S.T. (1985) Attachment of transported vesicles to microtubules in axoplasm is facilitated by AMP-PNP. Nature (Lond.) 316: 645-647

Lichtscheidl, I.K. and Weiss, D.G. (1988) Visualization of submicroscopic structures in the cytoplasm of Allium cepa inner epidermal cells by video-enhanced contrast light microscopy. Eur. J. Cell Biol. 46: 376-382

Lynn, M.P., Atkinson, M.B. and Breuer, A.C. (1986) Influence of translocation track on the motion of intra-axonally transported organelles in human nerve. Cell Motil. Cytoskel. 6: 339-346

Madansky A. (1959) The fitting of straight line when both variables are subject to error. J. AM Statist. Assoc. 54: 173-205

Morel, J.E. and Bachouchi, N. (1988) Muscle contraction and movement of cellular organelles: Are there two different types of mechanisms for their generation? J. theor. Biol. 132: 83-96

Ochs, S. (1972) Fast transport of materials in mammalian nerve fibers. Science 176: 252-260

Odell, G.M. (1976) A new mathematical continuum theory of axoplasmic transport. J. theor. Biol. 60: 223-237

Paschal, B.M., Shpetner, H.S. and Vallee, R.B. (1987): MAP 1C is a microtubule-activated ATPase which translocates microtubules in vitro and has dynein-like properties. J. Cell Biol. 105: 1273-1283

Rebhun, L.I. (1959) Studies of early cleavage in the surf clam, *Spisula solidissima*, using methylene blue and toluidine blue as vital stains. Biol. Bull. 117: 518-545

Rebhun, L.I. (1972) Polarized intracellular particle transport: saltatory movements and cytoplasmic streaming. Int. Rev. Cytol. 32: 93-137

Schliwa, M. (1984) Mechanisms of intracellular organelle transport. In: Shay, J.W. (ed.) Cell and Muscle Motility, Vol. 5. New York, Plenum, pp. 1-84

Schmid, G., Wagner, L. and Weiss, D.G. (1983) Rapid axoplasmic transport of free leucine. J. Neurobiol. 14: 133-144

Schmitt, F.O. (1968) Fibrous proteins - neuronal organelles. Proc. Natl. Acad. Sci. USA 60: 1092-1101

Schnapp, B.J., Vale, R.D., Sheetz, M.P. and Reese, T.S. (1985) Single microtubules from squid axoplasm support bidirectional movements of organelles. Cell 40: 455-462

Scholey, J.M., Porter, M.E., Grissom, P.M. and McIntosh, J.R. (1985) Identification of kinesin in sea urchin eggs and evidence for its localization in the mitotic spindle. Nature (Lond.) 318: 483-486

Sheetz M.P., Turney S., Qian H. and Elson E.L. (1989) Nanometre-level analysis demonstrates that lipid flow does not drive membrane glycoprotein movements. Nature 340: 284-288

Sheetz, M.P. and Spudich J.A. (1983) Movement of myosin-coated fluorescent beads on actin cables in vitro. Nature 303: 31-35

Smith, R.S. (1982) Axonal transport of optically detectable particulate organelles. In: Weiss, D.G. (ed): "Axoplasmic Transport", Berlin: Springer-Verlag, pp. 181-192

Smith, R.S. and Forman, D.S. (1988) Organelle dynamics in lobster axons: anterograde and retrograde particulate organelles. Brain Res. 446: 226-236

Vale, R.D., Reese, T.S. and Sheetz, M.P. (1985c) Identification of a novel, force-generating protein, kinesin, involved in microtubule-based motility. Cell 42: 39-50

Vale, R.D., Schnapp, B.J., Mitchison, T., Steuer, E., Reese, T.S. and Sheetz, M.P. (1985b) Different axoplasmic proteins generate movement in opposite directions along microtubules in vitro. Cell 43: 623-632

Vale, R.D., Schnapp, B.J., Reese, T.S. and Sheetz, M.P. (1985a) Organelle, bead and microtubule translocations promoted by soluble factors from the squid giant axon. Cell 40: 559-569

Weiss, D.G. (Ed.) (1982a) Axoplasmic Transport. Springer Verlag, Berlin, 477 pp

Weiss, D.G. (1982b) 3-0-methyl-D-glucose and ß-alanine. Rapid axoplasmic transport of metabolically inert low molecular weight substances. Neurosci. Letters 31: 241-246

Weiss, D.G. (1985) Dynamics and cooperativity in the organization of cytoplasmic structures and flows. In: Complex Systems - Operational Approaches in Neurobiology, Physics, and Computers, H. Haken ed., Springer Verlag, Berlin, pp. 179-191

Weiss, D.G. (1986a) Visualization of the living cytoskeleton by video-enhanced microscopy and digital image processing. J. Cell Sci. Suppl. 5: 1-15

Weiss, D.G. (1986b) The mechanism of axoplasmic transport. In: Iqbal, Z. (ed): "Axoplasmic Transport", Boca Raton FL:CRC Press, pp. 275-307

Weiss, D.G. (1987) Visualization of microtubule gliding and organelle transport along microtubules from squid giant axons. In: Nature and Function of Cytoskeletal Proteins in Motility and Transport, K.-E. Wohlfarth-Bottermann ed., Progress in Zoology, Vol. 34, pp. 133-144

Weiss, D.G. (1989) Video-microscopic measurents in living cells: Dynamic determination of multiple endpoints for in vitro toxicology. Molec. Toxicol. 1: 465-488

Weiss, D.G. and Allen, R.D. (1985) The organization of force generation in microtubule-based motility. In: De Brabander, M., and De Mey, J. (eds): "Microtubules and Microtubule Inhibitors 1985," Amsterdam: Elsevier, pp 232-240.

Weiss, D.G. and Gross, G.W. (1982) The microstream hypothesis: Characteristics, predictions, and compatibility with data. In: Axoplasmic Transport, D.G. Weiss ed., Springer Verlag, Berlin, pp. 362-383

Weiss, D.G. and Gross, G.W. (1983) Intracellular transport in axonal microtubular domains. I.Theoretical considerations on the essential properties of a force-generating mechanism. Protoplasma 114: 179-197

Weiss, D.G., Keller, F., Gulden, J. and Maile, W. (1986) Towards a new classification of intracellular particle movements based on quantitative analyses. Cell Motil Cytoskel 6: 128-135

Weiss, D.G., Langford, G.M., Seitz-Tutter, D. and Keller, F. (1988) Dynamic instability and motile events of native microtubules from squid axoplasm. Cell Motil. Cytoskel. 10: 285-296

Weiss, D.G., Maile, W. and Wick, R.A. (1989) Video microscopy. (Chapter 8) In: Light Microscopy in Biology. A Practical Approach. A.J. Lacey ed., IRL Press Oxford, pp. 221-278

Weiss, D.G., Meyer, M.A. and Langford, G.M. (1990) Studying axoplasmic transport using video microscopy and the squid giant axon. In: Squid as Experimental Animals, D.L. Gilbert, W.J. Adelman, J. Arnold eds., Plenum Press, New York, pp. 303-321

Weiss, D.G., Seitz-Tutter, D., Langford, G.M. and Allen, R.D. (1987) The native microtubule as the engine for bidirectional organelle movement. In Smith, R.S., and Bisby, M.A. (eds): "Axonal Transport". New York: Alan R. Liss, pp. 91-111

Wong M.Y. (1989) Likelihood estimation of a simple linear regression model when both variables have error. Biometrika 76: 141-148

CENTRIN-MEDIATED CELL MOTILITY IN EUKARIOTIC CELLS

Michael Melkonian

Universität zu Köln

Botanisches Institut, Lehrstuhl 1

Gyrhofstraße 15, D-5000 Köln 41, Germany

Abstract

In eukaryotic cells special sets of motor proteins (myosins, flagellar and cytoplasmatic dyneins, kinesins) interact with protein filaments (actin filaments and microtubules) to produce biological motion. These movements have in common that they depend on ATP hydrolysis and the sliding of a protein filament against another filament or against a cell organelle or vesicle. Although it has been known for some time that rapid cell movements exist that differ mechanistically from the common sliding-based motility systems (Hoffmann-Berling 1985; Amos 1971), interest in this type of cell motility system has only increased relatively recently. The identification of the major protein involved in this type of cell motility and the availability of polyclonal and monoclonal antibodys raised against this protein has greatly facilitated further research (Salisbury et al. 1984).

The phenomenollogical characteristics of the novel cell motility system are: 1) a rapid contraction of a filamentous structure within less than 20 msec to usually less than 50% of its original length, 2) a much slower re-extension of the filamentous structure to its original length (in the range of a few seconds up to about one hour). Because of these characteristics this cell motility system has been called a "shock-motility system" (Höhfeld et al. 1988). Contraction of filaments is based on supercoiling, not on sliding, is initiated by the binding of Ca^{++} to the major filament protein and is independent of ATP-hydrolysis. Binding of Ca^{++} to the major protein presumably leads to large conformational changes of the protein that are responsible for filament contraction. The Ca^{++}-modulated contractile protein belongs to a family of low molecular weight (16 - 22 kDa) acidic phosphoproteins that have been termed spasmins, centrins or caltractins (Amos et al. 1975, Salisbury et al. 1984, Salisbury et al. 1988, Huang et al. 1988) and which show considerable sequence homology to calmodulin (45 - 48 % identity) and other EF-hand Ca^{++}-binding proteins. It is clear that re-extension of the contracted filaments requires removal of Ca^{++} from the protein (Salisbury et al. 1984, Salisbury et al. 1987) and phosphorylation/dephosphorylation of the protein is necessary for repeated cycles of contraction/re-extension of the filaments (Salisbury et al. 1987).

Many aspects of the molecular function of centrin-based cell motility are still unknown including the organization of the centrin filaments (involvement of other proteins ?), the nature of the conformational changes of the protein upon Ca^{++}-binding, the regulation of the phosphorylation/dephosphorylation cycle of the protein and the mathematical modelling of this new system of extremely fast biological motions.

REFERENCES

Amos W.B. (1971) Reversible mechanochemical cycle in the contraction of Vorticella. Nature 229: 127-128

Amos W.B., Routledge L.M. and Yew F.F. (1975) Calcium-binding proteins in a vorticellid contractile organelle. J. Cell Sci. 19: 203-213

Höhfeld I., Otten J. and Melkonian M. (1988) Contractile eukaryotic flagella: centrin is involved. Protoplasma 147: 16-24

Hoffmann-Berling H. (1958) Der Mechanismus eines neuen, von der Muskelkontraktion verschiedenen Kontraktionszyklus. Biochim. Biophys. Acta 27: 247-255

Huang B., Mengerson A. and Lee V.D. (1988) Molecular cloning of cDNA for caltractin, a basal body-associated Ca^{++}-binding protein: homology in its protein sequence with calmodulin and the yeast CDC31 gene product. J. Cell Biol. 107: 133-140

Salisbury J.L., Baron A., Surek B. and Melkonian M. (1984) Striated flagellar roots: isolation and partial characterization of a calcium-modulated contractile organelle. J. Cell Biol. 99: 962-970

Salisbury J.L., Sanders M.A. and Harpst L. (1987) Flagellar root contraction and nuclear movement during flagellar regeneration in *Chlamydomonas reinhardtii*. J. Cell Biol. 105: 1799-1850

Salisbury J.L., Baron A.T. and Sanders M.A. (1988) The centrin-based cytoskeleton of *Chlamydomonas reinhardtii*: distribution in interphase and mitotic cells. J. Cell Biol. 107: 635-642

Conclusion

One problem that is peculiar to the study of crawling motility in tissue cells is the great variability in their rapidly changing shape and structure. Special methods are needed even to define the position and orientation of a motile cell with sufficient consistency and objectivity for detailed motion analysis. The apparently simple problem of classifying cellular shape in a biologically meaningful way and relating this to motility gave rise to much discussion in the working group and two radically different approaches were presented in the first three papers. The method of moment invariants (see also Hu, 1962) gives a concise numerical description of many different aspects of the shape but the individual moments become harder to interpret as the moment order is increased. On the other hand, the median axis transform (see also Blum, 1973) is easy to interpret as the 'skeleton' of the cell shape, and can be used for counting the processes or protrusions of the cell, but it does not give a simple numerical description of other aspects of the shape.

As with Kepler's laws of planetary motion, the crowning achievement of a quantitative description of cellular motion would be that it inspired new insight into natural processes and mechanisms. I cannot pretend, however, that we are now on the verge of discovering by numerical methods the fundamental laws of cell motility, even if such generalities exist, and I do not believe that a satisfactory account of cell motility will ever be found at any single level of description. But the first three contributions together with Weiss and Langford's demonstrate an immediate and important advantage of numerical description: large databases of cell motion can be collected in a form suitable for systematic exploration and subsequent numerical analysis. Such techniques can be far more informative than mere verbal description of motility and will eventually prove essential for characterising the effects of molecular modification or genetic manipulation of the motile machinery.

A general theme that emerged in the working group is that studying the dynamics of intracellular motion (including changes in shape and molecular structure) is an essential step towards understanding the mechanism of cell translocation. The molecular biology of the motile machinery is now possibly well enough understood to begin to propose tentative but realistic models of the intracellular dynamics as exemplified by the last four papers. Thomas Pohl's approach shows that a remarkably realistic theoretical model of cytoplasmic contraction can be constructed from the basic principles of fluid mechanics. Weiss and Langford demonstrate the considerable progress that can be made using the alternative approach of detailed analysis of the dynamics of intracellular motility. Their stochastic path analysis of particles along microtubules has much in common with the methods for analysing the paths of whole organisms (see Part II, Section 1). Bereiter-Hahn, Braun and Voth propose an ingenious hypothesis of how intracellular contraction might be converted into translocative motion. Finally, Melkonian's article raises the question of how many molecular transducers of the motile machinery remain undiscovered.

It is clear from the articles in this Section, as from the Workshop as a whole, that integrating the various levels of organisation in order to obtain a global picture of cellular motion will require the close cooperation of specialists from practically all major fields of scientific endeavour.

References and Further Reading

Akkas, N. (ed.) (1990) 'Biomechanics of active movement and deformation of cells' Conference Proceedings Istanbul 1989, NATO ASI Ser. H, Vol. 42, Springer, Berlin & Heidelberg.

Blum, H. (1973) Biological shape and visual science (Part I). J. Theor. Biol. 38: 205-287.

Dembo M.(1989) Mechanics and control of the cytoskeleton in Amoeba proteus. Biophys.J. 55: 1053-1080

Dembo M.(1989) Field theories of the cytoplasm. Comments Theor.Biol. 1: 159-177.

Hu, M-K. (1962) Visual pattern recognition by moment invariants. IRE Trans. Inf. Theory 8: 179-187.

Lord, E.A. & Wilson, C.B. (1984) The Mathematical Description of Shape and Form. Ellis Horwood Ltd., Chichester.

Oster G.F. and Perelson A.S.(1987) The physics of cell motility. J.Cell Sci. Suppl. 8: 35-44.

CILIA AND FLAGELLA

(Coordinator: Hans Machemer)

Introduction

Microtubules are now seen as universal cytoskeletal elements suited for various polar motion of organelles in eucaryote cells (see Section I/1, in particular Weiss). Cilia and flagella generate cyclic motion based on a sliding-microtubule principle. Nine doublets of microtubules are arranged to form a cylinder enclosing a core of two single microtubules. Transient and permanent links, radial as well as tangential, establish the axoneme. Structural cues (enantiomorphy of the 9+2 pattern) and unidirectionality in force generation of the dynein arms suggest that a "closed" microtubule sliding unit represents a unique function: rotary sliding transfer and hence generation of oscillatory bending motion.

Principles of the ciliary sliding machine have been unraveling during the last 30 years, but the detailed motor functions and controls are still difficult to isolate for several reasons: 1. The rapid response of the diminutive axoneme (200 nm in diameter) is modulated temporally as well as spatially. 2. Recording of ciliary beating meets technical and interpretation problems (framing rates, illumination, two-dimensionality of image). 3. Because external stimuli can alter the ciliary motor response *via* modulation of the Ca membrane conductance and intraciliary ionic composition, experiments designed to characterize the intrinsic signalling chain include electrophysiological controls of the live cell, or monitoring of the components of aqueous solutions applied to demembranated ciliary "models".

*Basal body and ciliary shaft are structurally and functionally
different entities*

The basal body consists of nine triplets of microtubules and is
the primordium for ciliary growth and regeneration (Fig. 1B).
Fibrous interconnections prohibit displacements between the triplets. The structural integrity of the basal body is a basis of
the sliding-microtubule hypothesis of axonemal bending. The basal body is, in addition, an organelle for ciliary anchoring. Cortical microtubular bands and striated roots connect to the basal
body so that it does not rotate or tilt. Kinetodesmata which originate from basal bodies in some ciliates do not contract and
are hence unlike root structures of phytoflagellates which may
include a contractile protein (see Melkonian on centrin-based
motility, Section I/1).

The basal body is part of the cell soma. Transverse plate
structures distal to the basal body, and a peculiar ring of
membrane-integrated particles ("ciliary necklace") at the transition between soma and ciliary shaft suggest that the cell soma
and the cilium may be separate compartments of ionic regulation. Electrophysiological data are in support of this view: In
ciliates, the soma membrane has a substantial leakage conductance for Ca^{2+} and K^+, so that the membrane resting potential
is intermediate between the equilibrium potentials of these ions.
The leakage conductance of the resting cilium is minimal, that
is, below experimental resolution; removal of the cilia (in *Paramecium*: equivalent to 50% reduction of the total membrane area)
did not affect the resting conductance and the resting potential
(Machemer and Ogura, 1979; Pape and Machemer, 1986).

The axoneme consists of nine microtubular doublets and one
central pair of microtubules. In addition to the tubulin-based
longitudinal polarity of a microtubule, each doublet displays a
transverse polarity in that only one microtubule ("subfiber a")
is a complete cylinder and gives rise to dynein arms and spokes. The neighbouring microtubule ("subfiber b") shares part of
the wall of subfiber a. The transverse polarity of the nine doublets generates an enantiomorphic axonemal structure which
suggests a function of circular directionality (Fig. 1A). An apparent bilaterality of the axoneme might be derived, at first

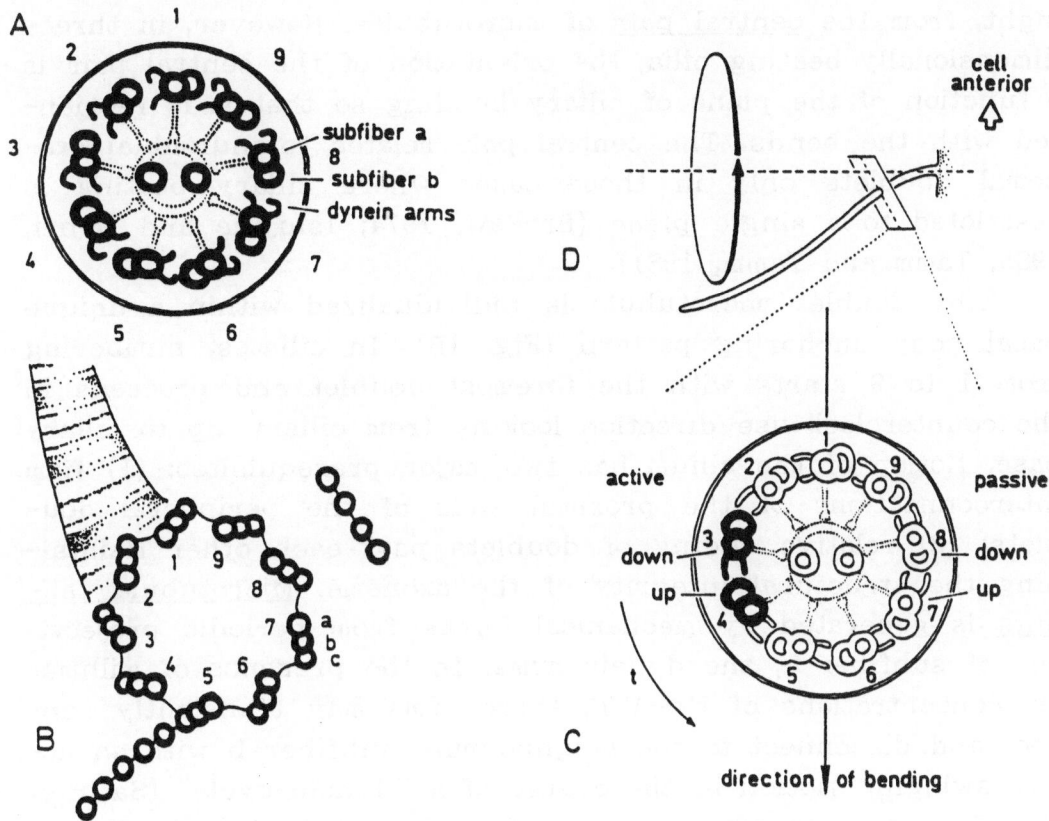

Figure 1. Ciliary structure, sliding of microtubules and resulting gyration of the cilium. **A.** Cross-section of the cilium as viewed tip-to-base. Numbering starts with the most anterior doublet of microtubules and proceeds in the counterclockwise direction. Dotted lines indicate radial spokes and central sheath (nexin links between doublets not shown). Note the covering of the axoneme by the cell membrane. **B.** Cross-section of basal body. A third microtubule (subfiber c) joins the doublets. Striated roots, tangential and postciliary microtubular bands help to anchor the basal body in the cell cortex. Not shown are fibrillar interconnections between the triplets. **C.** Active sliding, such as between the doublets 3 and 4, causes the cilium to bend (straight arrow); at the same time, the quasi-antipodes of the active pair 3-4 (that is, 7-8, 8-9) are passive backsliders. The counterclockwise translocation of active sliding (bent arrow) causes ciliary bending to rotate in the same direction. **D.** Repetitive unidirectional sliding translocation generates a counterclockwise gyration of the cilium. A, C, D from Machemer (1977); B, modified after Pitelka (1968).

sight, from the <u>central pair</u> of microtubules. However, in three-dimensionally beating cilia, the orientation of the central pair is a function of the plane of ciliary bending so that it is reoriented with the bends. The central pair relates to individual axonemal doublets only in those cases where ciliary beating is restricted to a single plane (Brokaw, 1974; Ishijima and Mohri, 1985; Tamm and Tamm, 1981).

Any doublet microtubule is individualized within a unique basal body anchoring pattern (Fig. 1B). In ciliates, numbering from 1 to 9 starts with the foremost doublet and proceeds in the counterclockwise direction looking from ciliary tip to ciliary base. Motion of the cilium has two major prerequisites: (1) firm interconnections of the proximal ends of the peripheral doublets; (2) relative sliding of doublets past each other maintaining the structural integrity of the axoneme. <u>Microtubular sliding</u> is generated by mechanical forces from periodic projections of subfiber a, the dynein arms. In the presence of millimolar concentrations of Mg-ATP, these arms may transiently connect and disconnect to the neighbouring subfiber b with an active swinging motion in the course of a "dynein-cycle" (Satir et al., 1981). As a result, the arms of doublet n "walk" along doublet n+1. Walking is unidirectional (= downward to the ciliary base) so that net shear forces occur between an active pair of doublets, which generates positive sliding (doublet n+1 upward with respect to doublet n) (Sale and Satir, 1977). Positive sliding of a pair of doublets is limited and is succeeded by passive, negative sliding (or backsliding) which may proceed beyond the neutral state. Negative sliding of a pair of doublets is caused by, and coincides with, positive sliding of the active "antipode" pair(s) of doublets on the other side of the axoneme. The local and temporal controls of this sliding mechanism are so far unknown.

In many cases the ciliary cycle is three-dimensional, in particular in protozoan cilia. In accordance with the structural enantiomorphy, it has been assumed that neighbouring doublets slide actively in temporal sequence (n/n+1 6 n+1/n+2 6 n+2/n+3 6 ...), that is, <u>active sliding travels around</u> the ring of doublets in the course of the ciliary cycle (Machemer, 1977). Active sliding does not include one entire pair of doublets at one time, but proceeds from base to tip of the cilium (Sugino and Naitoh

1982). Along an observed longitudinal bending wave, segments
of active (and positive) sliding alternate with segments of passive (and negative sliding). Thus, in protistan cilia, areas of
active sliding occur, at one time, in a leftwinding helix around
the axoneme from the ciliary base to the tip. It is important to
note that a bending wave, as observed at a particular plane,
includes components of bending that result from shear vectors
in various planes. In the case of two-dimensional ciliary beating, plausible arguing suggests that mechanical constraints and
a corresponding compliance in sliding can restrict vectors from
nine active pairs to subvectors in a single plane (see **Brokaw**
for planar flagellar motility, this Section I/2).

*Active microtubular sliding and ciliary bending are geometrically
corresponding*

The planes of the developing shear force between an <u>active pair
of doublets</u> at one time, and of the resulting ciliary bend, seem
to correspond to each other because walking of the arms is
unidirectional, and the axoneme does not develop a net twist
(Hines and Blum, 1985; Omoto and Brokaw, 1983). If bending is
determined by sliding, the geometry of the ciliary cycle may
reveal sliding properties of identifiable components of the
axoneme. This hypothesis was applied by Sugino and Naitoh
(1982) for ciliary movement, using three-dimensional wire models
which approximated ciliary images in *Paramecium* (Machemer,
1972a, b). In recording from live cilia, it was expected that an
axial view would supply undistorted images of ciliary reorientation with respect to the major axis of the organelle. High-speed
film data from the frontal cirri of *Stylonychia* taken at axial
view revealed oblique side views of the cycle because the beat
cone was tilted (Machemer and Sugino, 1986). This bend at the
proximal shaft ("inclination") was steady for a given membrane
potential. Exploiting geometric cues, the angle of inclination was
determined, and an axial view of the "beat cone" was reconstructed using computer assistance (Sugino and Machemer,
1987). We call a cycle "normalized", after it was corrected for

inclination. Figure 2 summarizes the spatial relationships between a normalized cycle and axonemal function.

Regression from the 3D-t record of the ciliary cycle to the level of microtubular sliding (see contribution by **Sugino**, this Section I/2) suggests that two functional parameters of the axoneme, sliding rate and rate of translocation of sliding, can generate the cyclic motion of the cilium. Application of these functions to the nine pairs of doublets and along any of these pairs is equivalent to modelling ciliary motion.

Modelling of mechanisms and controls

The planar cycle of the sea urchin spermatozoon includes bends of sometimes different curvature (principle bend, reverse bend) travelling in the proximo-distal direction. Mechanisms of two-dimensional wave propagation along the axoneme and extraction of function from the asymmetric wave form have been major objectives of the analysis of spermatozoan motility. **Brokaw**'s contribution emphasizes the problems and promises of quantitative modelling of flagellar activity for the assessment of the underlying active sliding process. In their contribution, **Baba and Mogami** show that digital image analysis of spermatozoan flagella reveals discrete curvature changes. Some temporo-spatial discreteness of the ciliary cycle would be expected from the postulated alternation of active sliding and interdoublet translocation of sliding (see contribution by **Sugino**). The contribution by **Machemer** summarizes current lines of electrophysiological research in the ciliate protozoa. Properties of three-dimensional ciliary activity are continuous functions of the membrane potential from hyperpolarization to depolarization, which controls intraciliary concentrations of Ca^{2+}. A reassessment of published cell reactivation studies suggests a regulation of ciliary function by binding of both Ca^{2+} and Mg^{2+} to axonemal proteins. In their contribution, **Mogami and Machemer** present a quantitative model predicting observed hyperpolarization-induced and depolarization-induced parameters of the ciliary cycle from competitive Ca^{2+}-Mg^{2+} binding.

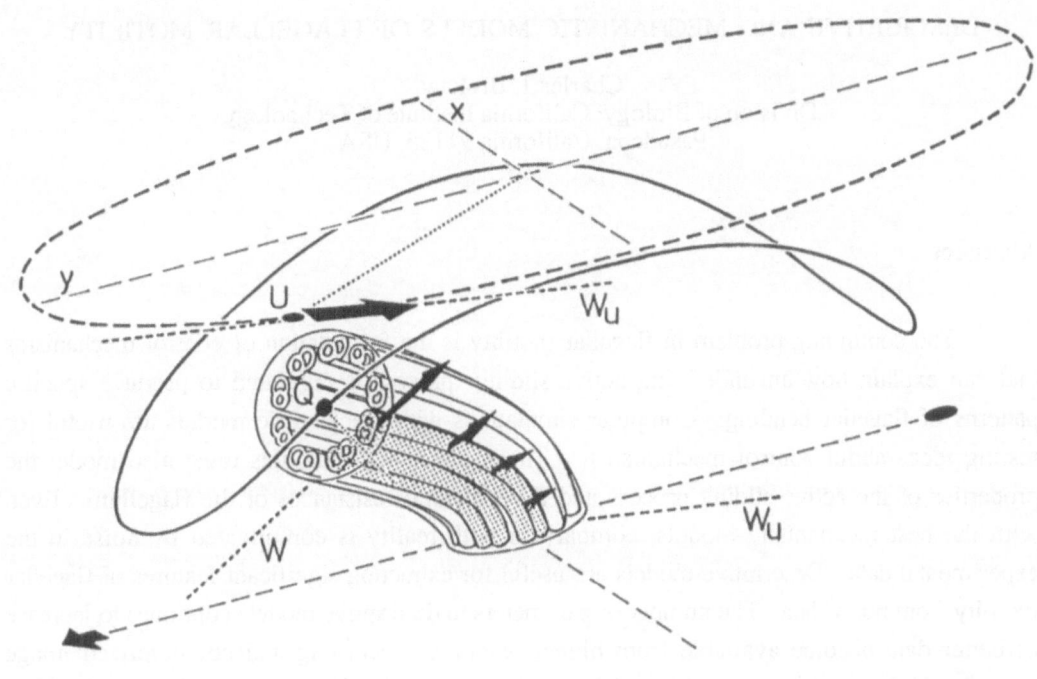

Figure 2. Spatial relations between spherial movement of a ciliary segment (continuous line) and the actively sliding doublets generating this movement (shaded). Procedures for planification of the track transform point **Q** to **U**, and the direction **W** to **Wu**. U (and Q) are defined by their connection to the point of ciliary origin. W corresponds to the common plane, and the direction of bending, of the active doublets, Wu to ciliary direction in the x-y plane. Heavy dot and short arrow mark the beginning and end of the power stroke which determines the major (y-) axis of the cycle. Note that angular velocities (arrows) of a segment differ from those in the planified projection. From Sugino and Machemer (1990).

DESCRIPTIVE AND MECHANISTIC MODELS OF FLAGELLAR MOTILITY

Charles J. Brokaw
Division of Biology, California Institute of Technology
Pasadena, California 91125, USA

Abstract

The continuing problem in flagellar motility is the formulation of control mechanisms that can explain how an underlying active sliding process is regulated to produce specific patterns of flagellar bending. Computer simulations with mechanistic models are useful for testing ideas about control mechanisms. These mechanistic models must also model the properties of the active sliding process and the structural resistances of the flagellum. Even with the best mechanistic models, comparison with reality is complicated by noise in the experimental data. Descriptive models are useful for extracting significant features of flagellar motility from noisy data. The number of parameters in descriptive models continues to increase as better data become available from higher resolution recording and computerized image analysis. Ultimately, the parameters of descriptive models should be the significant variables in mechanistic models, but that goal is still a long way off.

1. Introduction

Flagellar motility is exemplified by the photograph of a sea urchin spermatozoon in Fig. 1. The photograph shows that the sperm flagellum is oscillating and generating new bends at a frequency corresponding to about 12 images per beat, or 42 beats per second. This is only one example of a great many different kinds of bending patterns generated by flagella and cilia, with frequencies ranging from less than 1 Hz to over 100 Hz. Many of these bending patterns are significantly three-dimensional, but the sea urchin sperm flagellum has a nearly two-dimensional bending wave, and all of this discussion will be limited to such two-dimensional patterns, which can be photographed with high resolution.

For the last 20 years or so, we have realized that the bending of flagella and cilia is generated by active sliding between the outer doublet microtubules of the flagellar axoneme. (cf. introduction to this section) The continuing question is what are the control mechanisms that regulate this active sliding process to produce oscillation and bending wave propagation? I'm not going to answer that question here, but instead simply describe some of the ways in which modeling can be used to investigate this problem.

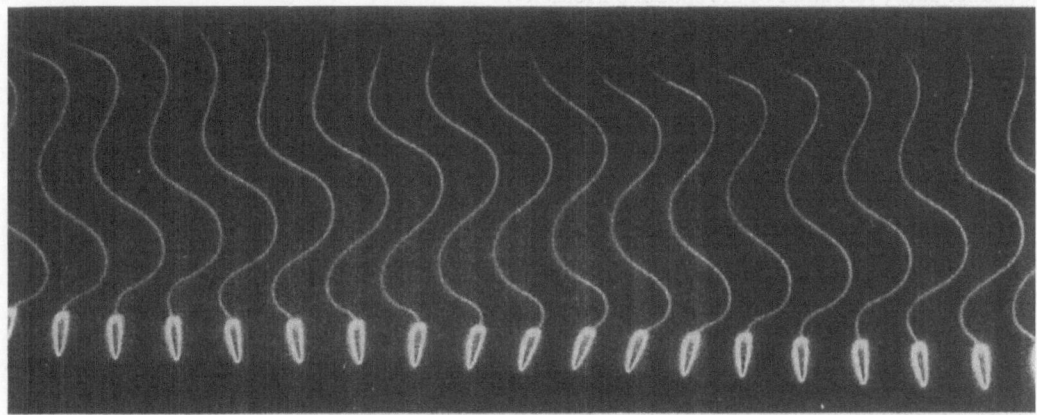

Figure 1. A sea urchin spermatozoon photographed with strobe flashes at 500 Hz, using 35 mm film moving at 1 m/sec in an oscilloscope camera. The tail, or flagellum, of this spermatozoon is about 45 μm in length.

2. Analysis of flagellar bending patterns with descriptive models.

2.1 Descriptive models for symmetric flagellar bending waves.

Descriptive modeling of flagellar movement probably began when Sir James Gray (1955) used a sinusoidal wave model, defined by the parameters of <u>frequency, amplitude,</u> and <u>wavelength,</u> to describe the movement of sea urchin sperm flagella. The most important component of this model is the idea that the movement has a regular temporal periodicity. Anyone who has used stroboscopic illumination to "stop" the movements of rapidly beating flagella has tested this model and found it to be extremely valid. Actually, this is somewhat surprising. A sea urchin sperm flagellum generates its bending waves using about 30,000 dynein arms -- individual molecular units. Different parts of the flagellum are not operating in synchrony but are doing different things at different times. Nevertheless, the movement appears to be regular rather than stochastic. The regularity is of the sort that we typically associate with macroscopic systems such as vibrations in an elastic beam. It leads us to suspect that the elastic properties of a flagellum are major determinants of its behavior.

The frequency of oscillation is obviously a natural parameter to use to describe flagellar bending waves. It is a good example of the type of measure that needs to be extracted from photographic data such as Fig. 1. The use of a wavelength parameter to describe the bending wave implies spatial periodicity, or propagation of wave phase points with constant velocity. All flagellar bending waves deviate from this ideal in the region near the base of the flagellum where bends are developing, and few flagella display a sufficient number of complete waves to allow adequate evaluation of the spatial periodicity.

A sinusoidal wave is not a good model for flagellar bending waves, because it is defined relative to an x axis that is the axis of symmetry of the wave. Such an axis has little relevance to the internal mechanisms in a flagellum. Much better are models that are defined relative to distance measured along the centerline of the flagellum, by specifying either sliding displacement or curvature as a function of position along the length of the flagellum. Sliding displacement is measured by the tangent angle at any point on the flagellar centerline, relative to the basal end of the flagellum where sliding between doublets is restricted (Satir, 1968; Warner and Satir, 1974).

The first such alternative to sinusoidal bending waves was the arc-line model of Brokaw and Wright (1963) and Brokaw (1965), which defined the wave as the result of a generating function containing regions of zero curvature (producing straight lines) and regions of constant, non-zero curvature (producing circular arcs). This model was originally developed with reference to "contractile" models for producing flagellar bending, in which bends were considered to be regions in which contractile elements on one side of the flagellum were activated. In these models, sharp transitions between regions of contraction (bends) or no contraction (straight regions) were propagated along the flagellum (Brokaw, 1966a). With subsequent recognition of the generation of flagellar bending by active sliding, it became clear that sharp transitions between regions of different curvature are unlikely, because such transitions would require extreme concentrations of active shear force to overcome the elastic bending resistance of the flagellum. This led to the "trapezoid-generated" model (Brokaw, Goldstein and Miller, 1970) or "constant curvature" model (Brokaw, 1983) , in which regions of constant curvature are connected by transition regions with a linear change in curvature. With both the arc-line model and the constant curvature model, a fourth descriptive parameter, in addition to frequency, wavelength, and amplitude, is required to describe the relative lengths of the constant curvature regions and the transition regions.

At about the same time, it was realized that sine-generated bending waves, in which curvature and shear angle are sinusoidal functions of length, give good approximations to some flagellar bending waves [Brokaw et al., 1970; Silvester and Holwill, 1972]. These bending waves are interesting because they are a close approximation to the equilibrium shape for an elastic filament that is bent by applying forces (not moments) that prevent it from being fully extended. In the simplest case, a sine-generated bending wave is described by only three parameters, and is essentially indistinguishable from a constant curvature wave having transition regions that are 0.34 times the wavelength (Eshel and Brokaw, 1988).

Consideration of such models suggests that there are two problems in interpreting data from photographs of moving flagella. One is determining whether a particular model is a sufficiently accurate representation of the true shape of the flagellum, and the second is determining the quantitative parameters of a particular descriptive model to obtain the best fit to the photographic data.

2.2 Digitization of flagellar images

Now that powerful microcomputers are readily available, analysis of photographic images of flagella, such as those in Fig. 1, should begin by getting data from the image into a computer in useful form. A simple way to do this is to use a digitizer to manually trace the image of the sperm tail, so that a series of points approximating the center line of the flagellum is entered into the computer. Although slow and tedious, this method allows the operator to ignore or compensate for imperfections in the images, so that useful data can often be obtained from low quality images. My laboratory has used this method extensively for parameter extraction from sequences of low resolution photographic images (Brokaw, 1984). Where higher resolution photographs are analyzed, manual digitization is less appropriate because of its susceptibility to distortion if the manual tracing of the flagellar curve is influenced by the operator's model of what the waveform should be like. This is the worst form of modeling, because the model is undefined!

Silvester and Johnston (1976) developed an automatic system using photodiodes to track the photographic image of a flagellum.

Alternatively, a digitizing camera, or a video camera and a digitizing frame grabber, can be used to transfer an entire image frame into a computer for analysis by software. Baba and Mogami (1985) pioneered the development of methods for this analysis. In their method, the starting point at the base of the flagellum was manually located, and then slices of pixels crossing the flagellum were examined. In each slice, pixels with intensity greater than a threshold were selected, and the centroid of these pixels then yielded a point for the centerline of the flagellar image. This process was repeated in stepwize manner along the length of the flagellar image, to give a series of points at irregular intervals, just as with manual digitization. I have obtained similar results with algorithms that fit the image intensity data in each slice with either a Gaussian or trapezoidal model of the image intensity profile in order to obtain points on the centerline. These methods, which look at only a small region of the flagellum at any time, are advantageous because they assume no model for the shape of the flagellar bends. However, for the same reason, they lack the intelligence that allows a human operator to compensate for imperfections in the photographic images, so they are not particularly suited for off-line digitization without operator supervision unless the images are all of very high quality.

These digitizing methods, whether manual or automatic, end up with the same type of representation of the data as a set of closely-spaced points approximating the centerline of the flagellum. Further processing is required to convert these data to a useful form, such as values of tangent angle or curvature as functions of length along the flagellum. Processes based on overlapping straight lines drawn between every nth point and processes based on fitting of circles to a group of points have been described (Johnston, Silvester and Holwill, 1979; Brokaw, Luck and Huang, 1982; Brokaw, 1984; Hiramoto and Baba, 1978; Baba and Mogami, 1985; etc.). A new approach will be described here.

2.3 Modeling flagellar images with spline functions

2.31 Introduction

Spline functions are polynomial functions that are used to approximate or interpolate sets of data points (cf. de Boor, 1978). Most of the established methods for their use involve single-valued functions, y(x), defined on the real line x. For an nth order spline, the nth derivative, $d^n y/dx^n$, is a step function, with the steps occurring at x values referred to as "breakpoints". Between the breakpoints, the spline function is a polynomial, with all (n–1)th order derivatives continuous at the breakpoints. For work with flagellar images, large numbers of closely spaced data points are available, so it is desirable to use a smaller number of breakpoints and seek an approximating spline that falls close to the data points rather than passing through each data point.

For flagella, the spline function should be defined in terms of derivatives with respect to length, s, along the curve of the flagellum. This is both desirable because it is more directly relatable to the internal mechanisms of the flagellum and necessary because some flagellar waveforms are reentrant and cannot be expressed as single-valued y(x) functions for any x axis.

A first order spline fit will be one in which the slope of the curve, ϕ = arc tan (dy/dx), is a step function of s. Even a cursory examination of flagellar photographs provides evidence that the slope is not a step function unless the steps are extremely small. Nevertheless, a first order spline fit with uniformly spaced breakpoints can be a very useful approximation, and will be described more fully below.

A second order spline fit is one in which the curvature, κ = $d\phi/ds$, is a step function. If a minimum number of breakpoints is used, the second order spline is equivalent to fitting with "circular arcs and straight lines" (Brokaw, 1965), except that the sections between the circular bends are not necessarily straight lines.

A third order spline fit, known as a cubic spline fit, is the most widely used member of this family. Recent studies on flagellar waveforms have presented evidence that the third derivative, $d\phi^2/ds^2$, is actually a step function (Baba and Mogami, 1985; Eshel and Brokaw, 1988). Third order spline fits will not be appropriate for testing this proposition, but if it is accepted, they will be the most appropriate type of spline fit for determining parameters of a particular shape.

Higher order splines, such as quintic splines, have been used for some purposes (Wood and Jennings, 1979) and might be most appropriate for testing the true shape of flagellar bending waves.

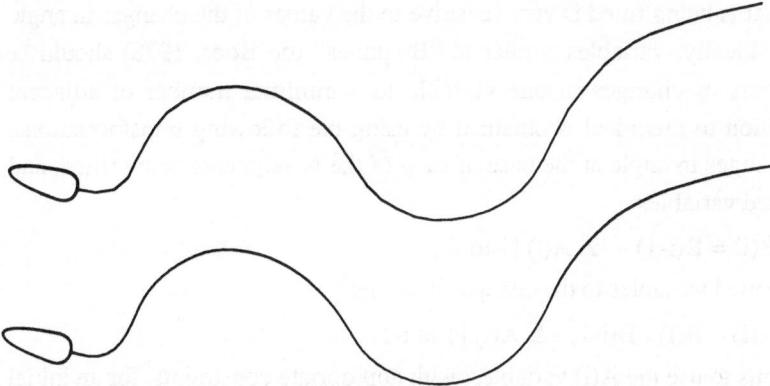

Figure 2. Results of automatic analyses of a digitized image frame containing one image of a sea urchin spermatozoon. The upper curve was obtained using 1 μm segments and the lower curve was obtained using 0.5 μm segments.

2.32 First order spline fitting

Fig. 2 demonstrates results from first order spline fitting of a flagellar image. For these photographs, a 512 x 1024 pixel array containing one sperm image was scanned, at a magnification corresponding to about 25 pixels/μm. The first task was then accurately locating the sperm head. This was done by taking advantage of the bright outline of the sperm head image (Fig. 1). A three-parameter model of this outline was created, and the best fit to the image was then found. This was done by moving the model around in the field, looking up the image intensity values for points on the contour of the model, and by trial and error finding the position that gave the minimum value for the image intensity along this contour. The Simplex method for minimization was used.

Once the head is located, it provides a starting point for tracing the flagellum. The flagellum is modeled as a connected series of straight segments. For the upper curve in Fig. 2, each segment was 1 μm in length. Starting at the head, three 1 μm segments were fitted to the image data. After the best fit for these three segments was obtained, the procedure took a 1 μm step and then repeated the fitting of three segments. This was repeated until the end of the flagellum was reached. The three values for the change in angle at the base of each segment (equivalent to a measure of the curvature of the flagellum) can be used as the variables that are optimized to give the best fit to the data. This choice of variables is advantageous because it can easily accommodate constraints on the magnitude of the changes in angle at each segment. For 1 μm segments, constraining these changes in angle to ± 1 radian, or even less, helps to ensure successful tracking of the flagellar image. However, with these variables, the position of the distal segment in the set that is being fitted is very sensitive to the values of the changes in angle for the earlier segments. Ideally, variables similar to "B-splines" (de Boor, 1978) should be used that isolate the effects of changes in one variable to a minimal number of adjacent segments. An approximation to this ideal is obtained by using the following transformations, where the A(i)s are the changes in angle at the base of each of the N segments being fitted, and the B(i)s are the transformed variables:

(1) $B(1) = A(1)$ and $B(i) = B(i-1) + \Sigma A(i)$ [1 to i] .

To return from the transformed variables to the changes in angle:

(2) $A(1) = B(1)$ and $A(i) = B(i) - B(i-1) - \Sigma A(i)$ [1 to i-1] .

I have found it advantageous to use the A(i) variables, with appropriate constraints, for an initial optimization at each step, and the B(i) variables, without constraints, for subsequent optimizations at each step.

I have examined three methods for comparing the model with image data. Method 1, used for Fig. 2, calculates an integral of image intensity along the segments that are being optimized. Using points at 0.1 μm intervals, the closest image pixel to each point is found, and the intensities of this pixel and the 8 surrounding pixels are summed for each point. For three 1 μm segments, this sum is obtained for 30 points, and is then divided by 270 to obtain an average intensity along these three segments. Depending on the sense of the image data, this intensity is either minimized or maximized to find the best fit. I have used this method successfully with up to 10 segments.

Method 2 has been used with sets of center line points obtained by manual digitization or digitization with image slices, as in the method of Baba and Mogami (1985). In this case, the sum of the squares of distances from the center line points to the model segments is minimized.

Method 3 calculates a more complete model of the image in a rectangular region containing three segments of the flagellar model. The flagellar image cross-section is approximated by a Gaussian function of fixed width. Two additional variables, for the height of the Gaussian function and the background intensity, are added to the optimization. The minimum least squares difference between the model and the image data is sought. This method requires much more computation, but it uses the image data more completely, and it is easily augmented to include other objects in the model, such as beads that are used as indicators of microtubule sliding (Brokaw, 1989a).

As seen in Fig. 2 (upper curve), first order spline fitting works well, but its segmental character is evident in regions of high curvature near the sperm head. A smoother appearing fit can be obtained by using shorter segments. The lower curve in Fig. 2 was obtained using 0.5 μm segments, and fitting five of these segments at each step. Figs. 3 and 4 show additional results using this procedure, including plots of curvature vs. length.

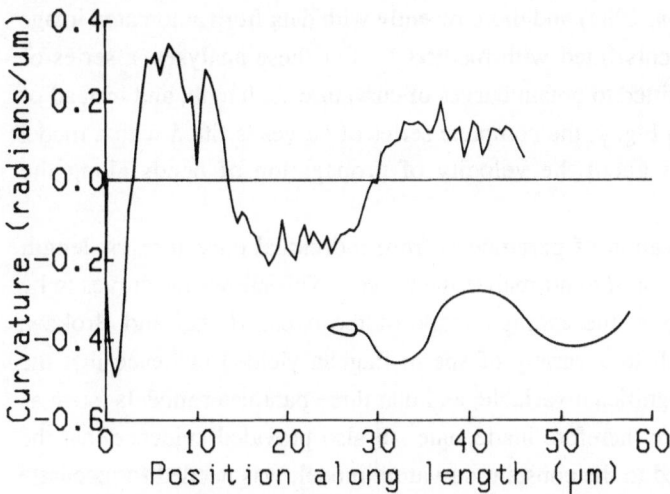

Figure 3. Curvature vs. length plot of results of fitting another image from the same series as Fig. 2, using 0.5 μm segments.

2.33 Higher order spline fitting

An alternative to the use of a large number of short, straight segments of equal length, as in the lower curve of Fig. 2, might be the use of, for example, 1 or 2 μm <u>curved</u> segments. However, I have not found this to be a more successful alternative.

Another, more successful, use of curved segments, with a minimal number of breakpoints, is shown by the heavy lines in Fig. 4. In this case, the gradient curvature model has been fitted to the plot of curvature vs. length. The proper number of breakpoints was first determined by counting the zero crossings using a heavily filtered version of the data curve. The locations of the breakpoints, and the values of curvature at the breakpoints, were then obtained by least squares fitting of the model to unfiltered curvature data, using the Simplex algorithm.

In principle, it should be possible to integrate a gradient curvature model, such as the one shown in Fig. 4, to obtain the curve in the x,y plane, and then to optimize the parameters of the gradient curvature model to obtain the best fit of the integrated curve to the image data. I have not yet had success with this method. There are at least two difficulties. One is finding a set of variables similar to B-splines that adequately localizes the effects of changes in each variable. Another is the fact that the gradient curvature model is probably not an exact description of the waveform.

A more successful result of fitting these types of models to the curvature vs. length plots is the derivation of descriptive parameters such as wavelength and bend angle. I have made extensive use of the constant curvature model for this type of parameter estimation, originally with manually digitized data (Brokaw, 1984) and more recently with data from automated image analysis, using 1 μm straight segments fitted with method 1. For these analyses, a series of images of each sperm flagellum is fitted to obtain curves of curvature vs. length, and instead of fitting each curve individually, as in Fig. 4, the complete series of curves is fitted with a model that also incorporates assumptions about the velocity of propagation of bends along the flagellum (Brokaw, 1984).

Another application is derivation of parameters from individual curvature vs. length curves, so that the parameters can be used to normalize the curves. This allows the curves to be averaged, to obtain information about the average shape of the bends (Eshel and Brokaw, 1988). Application of this approach to a variety of sperm flagella yielded evidence that the length of the transition region is a significant variable, and that three parameter models (such as the sine-generated bending wave) are therefore inadequate. It also provided evidence that the gradient curvature model, as opposed to the constant curvature model, was needed for accurate fitting.

2.4   Smoothing, filtering, and averaging

Although the results representing the configuration of the flagellum in x,y space, shown in the lower curve of Fig. 2, appear smooth, differentiation to obtain curves for angle and curvature as functions of length gives results that are significantly noisy, as shown in Fig. 3. Some of this noise may be a true representation of the stochastic, molecular events underlying the bending of the flagellum. Nevertheless, the search for simple patterns of regulated sliding activity is confused by the presence of this noise and may be facilitated by its elimination.

2.41   Filtering

One approach is filtering of the results after fitting to the data. One of the simplest filters generates a new set of data points, $y(i)$, from the original set, $x(i)$, by using

(3) $y(i) = 0.25\, x(i-1) + 0.5\, x(i) + 0.25\, x(i+1)$.

Although this filter effectively removes high frequency noise, if applied repeatedly it will ultimately eliminate the low frequency bends on the flagellum also. I have obtained better results using the following algorithm (Brokaw, 1984):

(4)   Let $g = x(i) - (x(i+1) + x(i-1))/2$.

Let $h = x(i+1) - (x(i+2) + x(i))/2$.

If g,h have opposite signs, then $y(i) = x(i) - (g-h)/2$ and $y(i+1) = x(i+1) + (g-h)/2$.

With a series of N values, this algorithm is applied for i from 2 to N-2 and then repeated in the opposite direction for i from N-1 to 3. This filter can be applied repeatedly to a curve resulting from integration of a step function, with no effect on the curve. It has relatively little effect if applied to the second integral of a step function. However, if applied to a step function, it will eventually convert it to a curve that is the integral of a step function. Therefore, the use of this filter requires making an assumption about the level at which the data curve may be assumed to be a step function. When applied to a noisy curve, it successfully recovers the curve generated by the original step function, but may pass the results of spurious steps resulting from noise. Either of these filters will eliminate small, but possibly significant, features of the curves such as the two points near 0 curvature at the beginning of the last bend on the flagellum in Fig. 3, which are indicative of a "straight region" between these bends.

I have obtained good results by using the filtering algorithm in (4) after each step to smooth the variables that are transferred to the next step, with the final fits at each step recorded without filtering. This procedure was used to obtain the results shown in Figs. 2 and 3.

2.42 Smoothing

Smoothing terms are commonly added to spline fitting procedures. For example, "energy-minimizing spline approximations" usually include in the function that is minimized a weighted sum of the squares of the curvatures at points along a model. A first order spline fit using method 1 with an energy-minimizing term is very similar to the method of snakes (Kass, Witkin and Terzopoulos,1987), except that the shortening term has been omitted. In general, a weighting factor for the energy-minimizing term needs to be determined empirically (de Boor, 1978). If it is too large, the fitted waveform can be unnaturally forced towards a "sine-generated" waveform. To avoid this distortion, I have used a smoothing algorithm derived from the

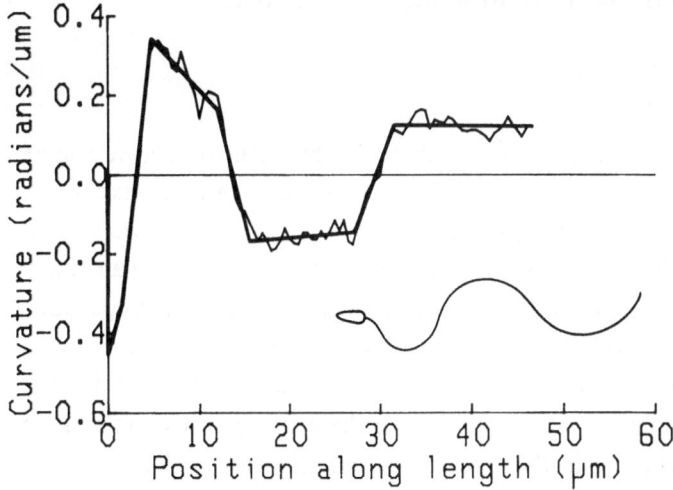

Figure 4. Averaged results from 3 independent fittings of the same image used for Fig. 3 are shown by the curve. The heavy lines are the result of fitting the gradient curvature model to this data.

filtering algorithm given in section 2.41(Equation (4)). In this case, if g and h have opposite signs, then the weighted square of (g-h) is added to the function that is minimized. When multiple fittings are carried out at each step, the initial fitting with constrained variables is usually performed without adding smoothing terms, and the fully weighted smoothing terms are added to the final fitting at each step.

2.43   Averaging

Repeated digitization of the same photograph, and repeated fitting of a linear spline model to the same digitized image, can provide a basis for averaging that will eliminate noise arising during the digitizing and fitting processes. Fig. 4 shows the results of averaging 3 independent spline fits to the same digitized image data used for Fig. 3. The noise reduction, compared to Fig. 3, is an indication that the fitting procedure is not yet optimal.

A major source of noise, the graininess of the photographic negative, is not eliminated by these forms of averaging. One method that could be used to reduce the effects of photographic noise would be to take advantage of the regular temporal periodicity of flagellar movement, by taking a series of photographs with strobe illumination at the same frequency as the beat frequency, so that each image should be identical. Digitization, fitting, and averaging of these images should then reduce the noise in the imaging and analysis steps, as well as the possibly more interesting noise resulting from the behavior of the flagellum itself.

3. Synthesis of flagellar bending patterns with mechanistic models

3.1 Introduction

As planar bending waves propagate along a flagellum, the directions of relative sliding of microtubules must periodically alternate. Observations on disintegrating axonemes (Summers and Gibbons, 1971) and on gliding microtubules (Weiss, this volume, section I/1) indicate that the flagellar active sliding mechanism only works in one direction (Sale and Satir, 1977). Therefore, the alternation in direction of sliding associated with bend propagation presumably involves an alternation between active sliding in one direction and passive sliding in the other direction. At any point on the flagellum, the dynein arms on outer doublet microtubules on one side of the flagellum can be in their active phase, while the dynein arms on the other side of the flagellum must be in their passive phase (Brokaw, 1972a; see also introduction to this section, Fig. 1).

In principle, any pattern of sliding and bending could be generated by regulating the timing and velocity of active sliding at every point on the flagellum. However, the experimental evidence indicates that the velocity of sliding is not regulated in this manner. In particular, mechanical influences, such as the viscosity of the surrounding fluid, significantly alter sliding velocities (e.g., Brokaw, 1966b, 1975a). These observations are more easily explained by thinking of the active sliding mechanism as a regulated force generator, with the velocities resulting from a balance between active force and various resistances.

3.2 Solving the moment balance equation

To test this idea, computer simulations have been carried out using programs that solve the equation for the balance of active and resistive bending moments on a model flagellum (Box 1). These simulations are now feasible on readily available microcomputer workstations. Numerical methods for solving the moment balance equation to simulate the movement of a flagellum have been described by Hines and Blum (1978) and in my own work (Brokaw, 1972b, 1985, 1988). Only the latter will be summarized here. The flagellum is modeled as a connected series of straight segments of equal length (usually 1 μm in length), with moment balance equilibrium at each intersegmental joint. The configuration of the model at any time is described in terms of curvature, κ, represented by the angle between each segment. The behavior of the model is described by the time rate of change of each of these values of curvature; these are the unknowns to be obtained by solving the moment balance equation at each time point and then used for the forward integration in time.

Exact integrations of the viscous moments are used for each segment, rather than the usual finite difference expressions, so that the length of the segments does not affect the accuracy of the viscous terms. However, because of non-linearities, my methods are restricted to explicit consideration of the viscous terms, as functions of $\kappa(t)$. To obtain stability, the elastic terms are included in implicit form, as functions of $\kappa(t+\Delta t)$ (Brokaw, 1985). These resistance terms introduce into the model a minimum of three parameters: two viscous drag coefficients, C_N and C_L, and an elastic bending resistance, E_B. Reasonable estimates for all three of these parameters are available. Parameters for elastic shear resistance can be added, but less information about their appropriate values is available.

The first successful models of this type used the simple assumption that the active sliding system generated a constant force (that is, independent of sliding velocity) with direction and magnitude controlled as a function of the curvature of the flagellum (Brokaw, 1972b). This type of force generator is unrealistic, because it has been known for a long time that force generation in muscle decreases with sliding velocity, and force-velocity data are now also available for microtubule sliding in disintegrating axonemes (Oiwa and Takahashi, 1988). It is more likely that the flagellar control mechanisms regulate the degree of activity of a force generator that produces a force that is a function of steady state sliding velocity and also has significant transient behavior.

Much more realistic models for force generation have been constructed in terms of detailed kinetics of dynein cross-bridge attachment and detachment (Hines and Blum, 1979; Brokaw, 1982, etc.). The details of these models are speculative and not yet readily testable.

BOX 1

MOMENT BALANCE EQUATIONS FOR PLANAR FLAGELLAR BENDING

Charles J. Brokaw

Planar flagellar bending can be modeled by a partial differential equation for the motion of a massless elastic filament in a viscous medium, as studied in early work by Machin and by Rikmenspoel:

(1) $$\partial^2 M_A/\partial x^2 + E_B \partial^4 y/\partial x^4 + C_N \partial y/\partial t = 0.$$

M_A represents the active bending moments that are generating the movement, E_B represents the elastic bending resistance of the flagellum, and C_N represents the normal viscous drag coefficient. With appropriate boundary conditions, this equation can be used to obtain analytic expressions for the active bending moments required for sinusoidal wave solutions. Equation(1) is valid only for small amplitude motions. For numerical analysis of the large amplitude bending typical of real flagella, a more generalized form of the equation is:

(2) $$M_A(s,t) + M_E(s,t) + M_V(s,t) = 0,$$

with M_E representing bending moments resulting from elastic resistances of the flagellum and M_V representing bending moments resulting from viscous resistances to movement of the flagellum. For large amplitude bending, it is better to use a coordinate, s, measuring distance along the flagellum, rather than an external x axis, and it is convenient to express the shape of the flagellum in terms of its curvature, $\kappa(s,t)$. For large amplitude bending, M_V must be expanded to include forces resulting from longitudinal movement of the flagellum, depending on a longitudinal drag coefficient, C_L. M_V, obtained by a quadruple integration along the flagellum, becomes a non-linear function of κ and $\partial\kappa/\partial t$. The use of these drag coefficients represents a very useful approximation of flagellar hydrodynamics, first introduced by Gray and Hancock. M_V could also include internal viscous resistances. For a free swimming flagellum, the boundary conditions needed for integration of M_V are 0 force and moment at the ends of the flagellum. For a sliding microtubule model of a flagellum, M_E must include not only the elastic bending resistance of outer doublet microtubules, etc., but also elastic shear resistance (probably non-linear) of connections between the microtubules. M_A is obtained by integrating the shear forces generated by the active sliding mechanism, with the boundary condition of 0 shear force at the distal end of the flagellum. The basal end of the flagellum is assumed to contain a high shear resistance that can absorb the integrated shear force with negligible strain, so that M_A at the base is not constrained by a boundary condition. (For details, see Brokaw 1972b, 1985; Hines and Blum, 1978.)

I have found useful an intermediate level of modeling of the force generator that avoids the complexities of the cross-bridge models but provides some of their more realistic properties (Brokaw, 1985,1988). In this model, the cross-bridges are assumed to be force generators that behave instantaneously as elastic elements, but recover to their isometric force with first-order kinetics. When sliding at constant velocity, the force is a linearly decreasing function of increasing sliding velocity. Although this steady state behavior could be obtained more simply just by adding an internal shear viscosity to a constant force, the model is more useful because it also shows semi-realistic transient behavior in non-steady-state situations. This model for the force generator contains three independent parameters, constrained by our knowledge of maximum microtubule sliding velocities seen in disintegrating axonemes, by direct force measurements, and by the assumption that the energy available from one ATP dephosphorylation cycle is used efficiently by one cross-bridge attachment-detachment cycle. An additional parameter, determined by the geometry of the axoneme, is required to relate the intensity of sliding force to the local intensity of active moment, m_A. The active bending moment, M_A, is then determined by integration of $dM_A/ds = -m_A$.

Although these elements of the model represent a great simplification of the actual situation, they provide a model that is defined by a relatively small number of reasonably well-known parameters, so that attention can be focused on the unknown mechanisms that control the active force generator.

3.3 Control mechanisms

Stimulated by the work of Machin (1958,1963), my mechanistic flagellar models have incorporated simple feedback relationships between curvature and active sliding that make the models self-oscillatory (Brokaw, 1971, 1972b). Possibly the simplest such relationship is one that locally changes the direction of operation of the active force generator when the curvature of the flagellum at that location reaches a critical value. This introduces only one new parameter, the critical curvature. Computations with this model have demonstrated that it can oscillate and propagate bending waves, even in the absence of external viscosity, with the only resistance being that provided by an elastic bending resistance (Brokaw, 1985). This control specification is not sufficient to determine both the frequency and the wavelength of the movement, and the operating point can be varied over a wide range by small changes in boundary conditions. A major conclusion is that the model is underdetermined, and does not contain a sufficiently detailed control mechanism to match the behavior of the model to the behavior of real flagella (Brokaw, 1985).

Fig. 5 shows an example of the output of such a model, including a plot of curvature vs. length. There is no strong resemblance to the data curves shown in Figs. 3 and 4.

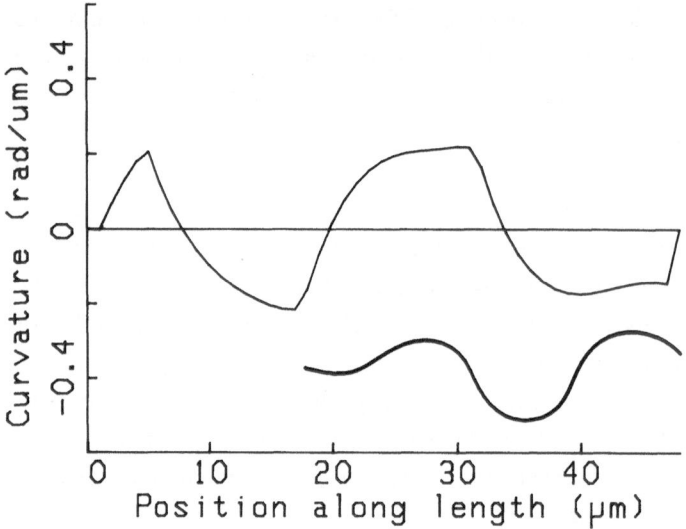

Figure 5. Output from a computer simulation of the mechanistic flagellar model described by Brokaw (1985). Results are shown at one time point from a stable, periodic motion.

Even simpler models, lacking curvature control of the active sliding process, can be constructed by making the active sliding process self-oscillatory, so that sliding in one direction continues until sufficient resistance is met to start it back in the opposite direction (Brokaw, 1975b). These models are essentially a series of independent shear oscillators, strongly coupled by the continuity of the microtubules. They oscillate and propagate bending waves at zero external viscosity, but at normal viscosities they become unstable and fail to propagate realistic bending waves (Brokaw, 1989b, and unpublished observations).

3.4 Conclusions from mechanistic modeling

The general conclusion from this work with mechanistic models is that computer simulation methods are available for examining models containing appropriate control mechanisms and for making specific predictions about the behavior of flagella containing these control mechanisms. None of the control mechanisms examined so far is very successful in producing behavior that matches the wide array of experimental data available even for simple flagella such as sea urchin sperm flagella.

We can imagine that we might simulate the movement of a particular mechanistic model with a particular set of parameters, and then compare the results directly with photographic images of flagellar motility. Our computer could then adjust the parameters of the mechanistic model in order to optimize the fit between the model bending patterns and the photographic data. This is probably not a very good approach, first because it is very costly in terms of computational resources, but more importantly because it lumps together differences between the model and the data resulting from inadequacies of the model and differences resulting from noise in the image data and possibly in the behavior of the real flagella. It is probably better to separate these differences explicitly by seeking descriptive models that attempt to eliminate the noise from the data, so that these descriptive models can be compared with the results from computer simulations of mechanistic models.

As our ability to analyze the data improves, we find that neither the simple descriptive models such as the constant curvature model nor the simple mechanistic models that we have tested are adequate. More complicated descriptive models, and mechanistic models with more complicated control mechanisms, appear necessary. If the mechanistic models are valid, they should generate movement that can be fitted well with the same descriptive models that work well for real flagellar images. That goal has not yet been achieved.

Acknowledgements

My work has been generously supported by the U. S. National Institutes of Health by grant GM 18711.

REFERENCES

Baba, S. and Mogami Y. (1985) An approach to digital image analysis of bending shapes of eukaryotic flagella and cilia. Cell Mot. 5: 475-489.

de Boor C. (1978) A Practical Guide to Splines. New York: Springer Verlag.

Brokaw C. J. (1965) Non-sinusoidal bending waves of sperm flagella. J. Exp. Biol. 43: 155-169.

Brokaw C. J. (1966a) Bend propagation along flagella. Nature 209: 161-163.

Brokaw C. J. (1966b) Effects of increased viscosity on the movements of some invertebrate sperm flagella. J. Exp. Biol. 45: 113-139.

Brokaw C. J. (1971) Bend propagation by a sliding filament model for flagella. J. Exp. Biol. 55: 289-304.

Brokaw C. J. (1972a) Flagellar movement: A sliding filament model. Science 178: 455-462.

Brokaw C. J. (1972b) Computer simulation of flagellar movement I. Demonstration of stable bend propagation and bend initiation by the sliding filament model. Biophys. J. 12: 564-586.

Brokaw C. J. (1975a) Effects of viscosity and ATP concentration on the movement of reactivated sea-urchin sperm flagella. J. Exp. Biol. 62: 701-719.

Brokaw C. J. (1975b) Molecular mechanism for oscillation in flagella and muscle. Proc. Natl. Acad. Sci. USA 72: 3102-3106.

Brokaw C. J. (1982) Models for oscillation and bend propagation by flagella. Symp. Soc. Exp. Biol. 35: 313-338.

Brokaw C. J. (1984) Automated methods for estimation of sperm flagellar bending parameters. Cell Mot. 4: 417-430.

Brokaw C. J. (1985) Computer simulation of flagellar movement. VI. Simple curvature-controlled models are incompletely specified. Biophys. J. 48: 633-642.

Brokaw C. J. (1988) Bending wave propagation by microtubules and flagella. Math. Biosciences 90: 247-263.

Brokaw C. J. (1989a) Direct measurements of sliding between outer doublet microtubules in swimming sperm flagella. Science 243: 1593-1596.

Brokaw C. J. (1989b) Operation and regulation of the flagellar oscillator, pp.267-279. In: F. D. Warner, I. R. Gibbons, and P. Satir (Eds.) Cell Movement, Vol. 1: The Dynein ATPases. New York: A. R. Liss, Inc.

Brokaw C. J., Goldstein S. F. and Miller R. L. (1970) Recent studies on the motility of spermatozoa from some marine invertebrates. pp. 475-485. In: B. Baccetti (Ed.) Comparative Spermatology. New York: Academic Press.

Brokaw C. J., Luck D. J. L. and Huang B. (1982) Analysis of the movement of Chlamydomonas flagella: The function of the radial-spoke system is revealed by comparison of wild-type and mutant flagella. J. Cell Biol. 92: 722-732.

Brokaw C. J. and Wright L. C. (1963) Bending waves of the posterior flagellum of Ceratium . Science 142: 1169-1170.

Eshel D. and Brokaw C. J. (1988) Determination of the average shape of flagellar bends: A gradient curvature model. Cell Mot. Cytoskel. 9: 312-324.

Gray, J. (1955) The movement of sea-urchin spermatozoa. J. Exp. Biol. 32: 775-801.

Gray J. and Hancock G. J. (1955) The propulsion of sea urchin spermatozoa. J. Exp. Biol. 32: 802-814.

Hines M. and Blum J. J. (1978) Bend propagation in flagella. I. Derivation of equations of motion and their simulation. Biophys. J. 23: 41-57.

Hines M. and Blum J. J. (1979) Bend propagation in flagella. II. Incorporation of dynein cross-bridge kinetics into the equations of motion. Biophys. J. 25: 421-442.

Hiramoto Y. and Baba S. (1978) A quantitative analysis of flagellar movement in echinoderm spermatozoa. J. Exp. Biol. 76: 85-104.

Johnston D. N., Silvester N. R. and Holwill M. E. J. (1979) An analysis of the shape and propagation of waves on the flagellum of Crithidia oncopelti. J. Exp. Biol. 80: 299-315.

Kass M., Witkin A. and Terzopoulos D. (1987) Snakes: active contour models. Internat. J. Computer Vision 1: 321-331.

Machin K. E. (1958) Wave propagation along flagella. J. Exp. Biol. 35: 796-806.

Machin K. E. (1963) The control and synchronization of flagellar movement. Proc. Roy. Soc. Lond. B 158: 88-104.

Oiwa K. and Takahashi K. (1988) The force-velocity relationship for microtubule sliding in demembranated sperm flagella of the sea urchin. Cell Struct. Funct. 13: 193-205.

Sale W. S. and Satir P. (1977) Direction of active sliding of microtubules in Tetrahymena cilia. Proc. Nat. Acad. Sci. USA 74: 2045-2049.

Satir P. (1968) Studies on cilia. III. Further studies of the cilium tip and a "sliding filament" model of ciliary motility. J. Cell Biol. 39: 77-94.

Silvester N. R. and Holwill M. E. J. (1972) An analysis of hypothetical flagellar waveforms. J. theor. Biol. 35: 505-523.

Silvester N. R. and Johnston D. N. (1976) An electro-optical curve follower with analog control. J. Phys. (E: Sci. Instrument.) 9: 990-995.

Summers, K. E. and Gibbons I. R. (1971) Adenosine triphosphate-induced sliding of tubules in trypsin-treated flagella of sea urchin sperm. Proc. Natl. Acad. Sci. USA 68: 3092-3096.

Warner F. D. and Satir P. (1974) The structural basis of ciliary bend formation. J. Cell Biol. 63: 35-63.

Wood G. A. and Jennings L. S. (1979) On the use of spline functions for data smoothing. J. Biomech. 12: 477-479.

DISCRETE NATURE OF FLAGELLAR BENDING DETECTED BY DIGITAL IMAGE ANALYSIS

Shoji A. Baba, Yoshihiro Mogami
Department of Biology
Ochanomizu University
Otsuka, Tokyo 112

Koichi Nonaka
Department Hygiene
Teikyo University School of Medicine
Itabashi–ku, Tokyo 173

Abstract

Bending shapes of live and demembranated sea–urchin sperm flagella have been studied at high resolution by means of digital image analysis. The curvature of a flagellum tended to change abruptly from one value to another along the flagellum, whereas it remained rather constant within a segment of variable length. The length of the segment, in which the curvature is kept constant, varied in a beat cycle. The transition points between these segments of constant curvature were not fixed at particular distances from the flagellar base. These dynamic aspects of discrete bend steps may reflect discreteness of active bend generation in flagella.

1. Introduction

Baba and Mogami (1985) have examined the curvature of sea urchin sperm flagella and embryo cilia by means of digital image analysis, and reported abrupt changes in the slope of the curvature plotted against the distance from the base. It has been demonstrated from an analysis by high–speed cinematography that, at a given distance from the base, the angular direction of a cilium shows alternating rapid and slow phases when plotted as a function of time (Baba, 1979; Baba and Mogami, 1987).

The presumed discrete nature of flagellar and ciliary bends would reflect some of so far unknown characteristics of the mechanism of bend generation. It is therefore of crucial importance to analyze the discreteness of flagellar bends more accurately. In this paper, we will report recent advances, which have enabled us to present a constant–curvature wave model for flagellar bends on the basis of the sliding microtubule hypothesis.

2. Digital image analysis

The bending shapes of live and demembranated sperm flagella from the sea urchins, *Pseudocentrotus depressus*, *Astriclypeus manni* and *Hemicentrotus pulcherrimus*, were recorded on 35-mm Kodak Plus-X or Panatomic-X film using an oscilloscope camera at up to 80 frames per s (fps) and a xenon flash tube (Chadwick-Helmuth STROBEX 35S). Flagellar images were digitized from enlarged prints or directly from the negatives viewed through a Hamamatsu C1000 TV camera at a final magnification of about 10 pixels/μm on an image processor NEXUS 6400 interfaced with a minicomputer OKITAC 50/10 or a main frame computer HITAC M-240D (Fig. 1). Errors accompanying digitization were reduced by digital picture integration and filtration performed on the image processor.

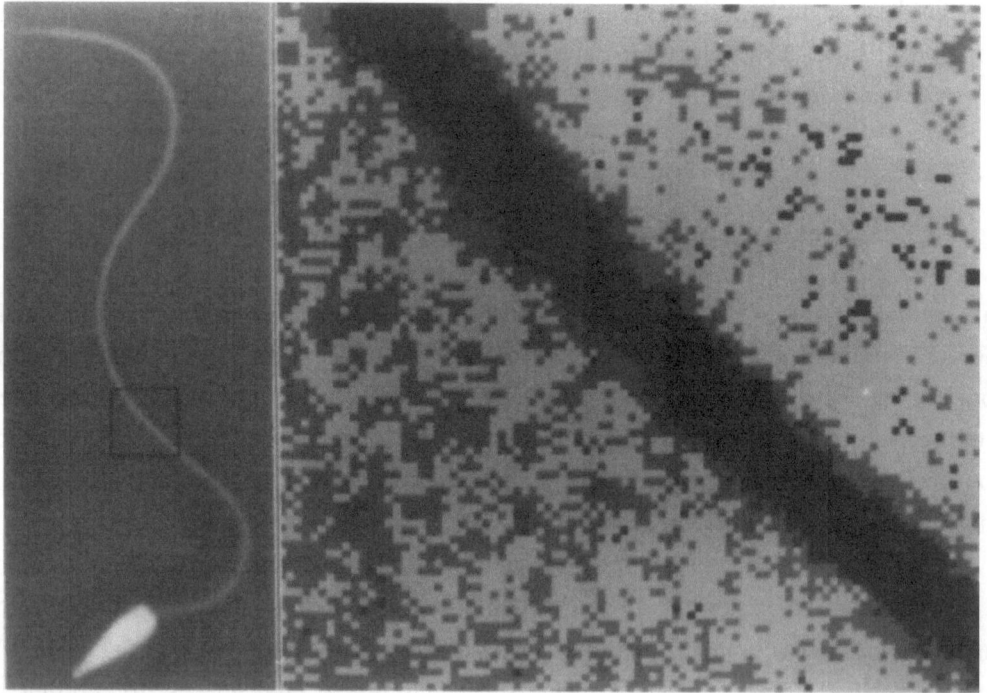

Figure 1. Phase-contrast micrograph of a sperm flagellum of *Pseudocentrotus depressus* (left) and its digital picture. Magnification: 27 pixels/μm.

Our image processor has 512 x 480 pixels/frame. For higher resolution, we developed the following techniques:

(1) Rotation method. The image of a flagellum on the same photograph is analyzed repeatedly using several repeatedly digitized images each obtained through the TV camera rotated around its optic axis. The results of repeated analyses are averaged to reduce digitization errors (Fig. 2, left).

(2) Local–magnification method. Several portions of the image of a flagellum are analyzed independently at high magnification, and the resultants are connected by computer for analysis of the whole image (Fig. 2, right).

(3) 1024–mode method. Digitized images of 512 x 480 pixel resolution are expanded to 1024 x 960 pixel resolution by digital interpolation. These techniques have been combined to achieve a high quality analysis.

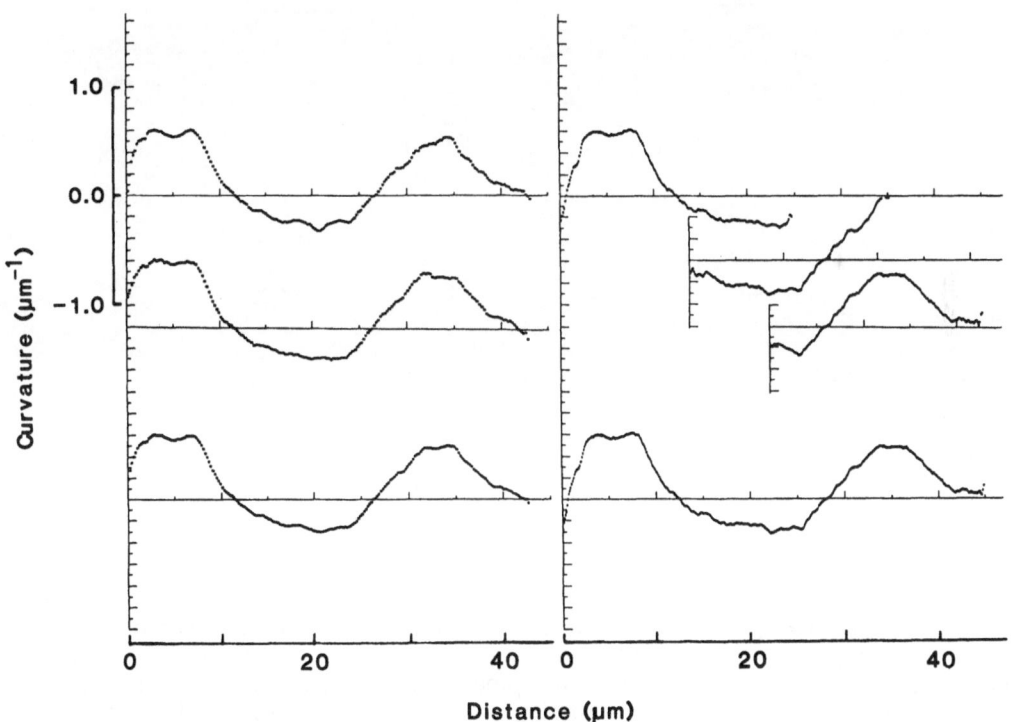

Figure 2. Digital image analyses of one picture of a live sperm flagellum of *Pseudocentrotus depressus*. Left, rotation method (45°) in the 512–mode; resolution: 0.08 μm/pixel. The bottom curve is the average of upper two curves. Right, local–magnification method in the 512–mode; resolution: 0.04 μm/pixel. See text.

3. Discrete steps in curvature plots

By recording and analyzing some 20 images per sperm, we examined most bending shapes in one beat cycle of individual sperm (Fig. 3). There were prominent steps and flattened portions in the outline of superimposed curvature plots of the bending shapes. These steps and flattened portions are not resulting from superimposition. In fact, individual wave shapes often possessed corresponding steps and flattened portions as shown in Fig. 3.

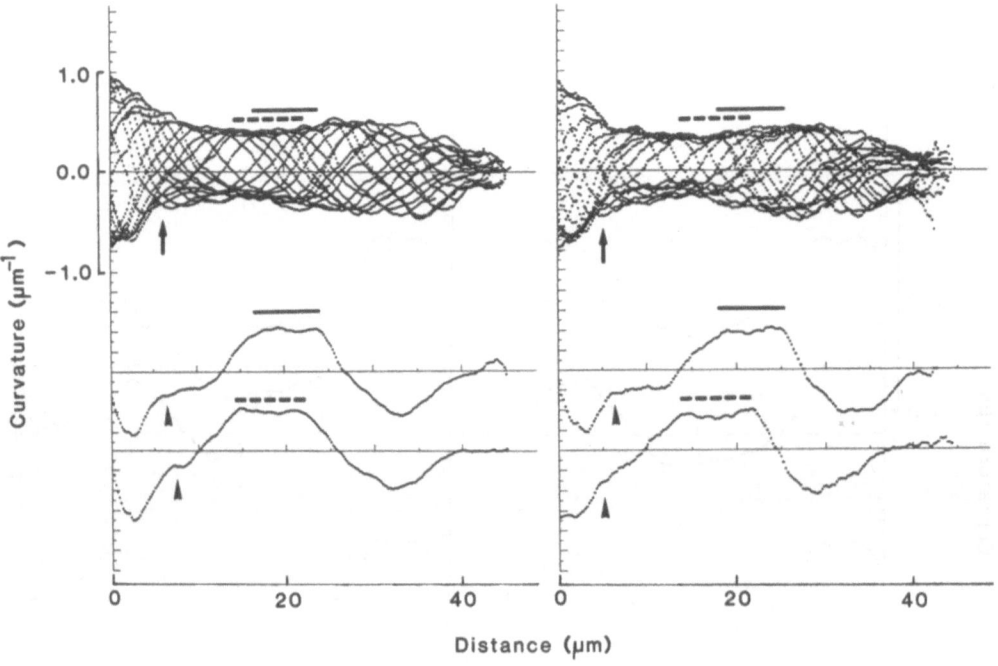

Figure 3. Digital image analyses of live sperm flagella of *Pseudocentrotus depressus*. Upper, superimposed curvature plots of flagella beating at 44 Hz (left) and 38 Hz (right) at 20 °C. Lower, selected curvature plots showing discrete curvature steps, whose correspondence to those seen in superimposed plots are indicated by arrows and horizontal lines. Left, local−magnification method in the 512−mode; 0.04 μm/pixel. Right, 512−mode; resolution: 0.08 μm/pixel. Framing rate: 80 fps.

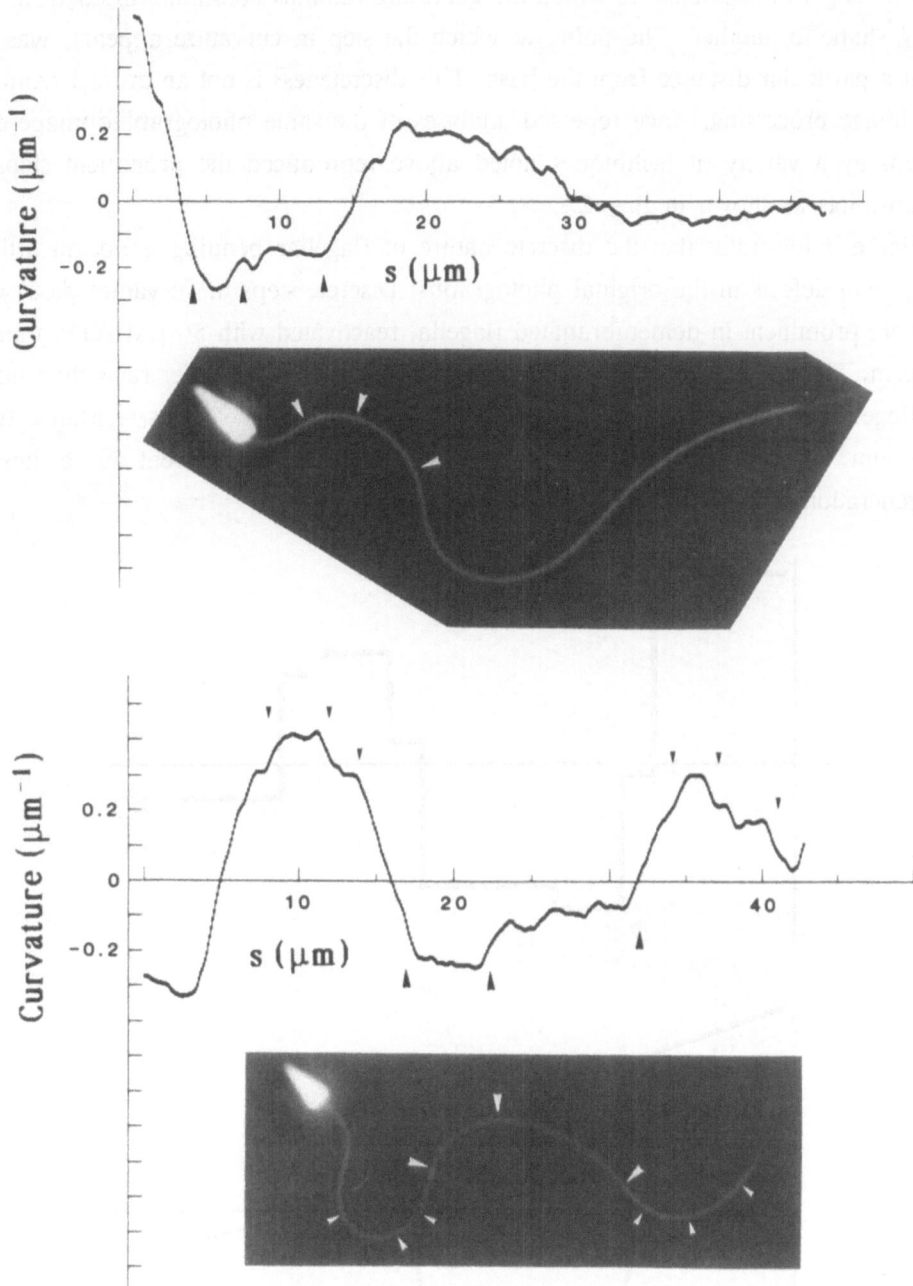

Figure 4. Stepwise changes in curvature detected by digital image analysis of sperm flagella. Rotation method (10 repeats by 5°) in the 1024–mode; resolution, 0.08 μm/pixel. Arrowheads indicate discrete points in the curvature plots, white arrowheads the corresponding points in the original photographs. Upper, live sperm of *Astriclypeus manni*. Beat frequency 48 Hz at 28 °C. Lower, *Hemicentrotus pulcherrimus* sperm reactivated with 50 μM MgATP under potentially asymmetric conditions (for details of the method, see Sato, et al., 1988). Beat frequency, 12.4 Hz. 23 °C.

The length of segments, in which the curvature remains constant, varied from one bending shape to another. The point, at which the step in curvature appears, was not fixed at a particular distance from the base. This discreteness is not an artifact from the digital image processing, since repeated analyses of the same photographic image of a flagellum by a variety of techniques noted above reproduced the prominent steps in curvature plots as shown in Fig. 2.

Figure 4 illustrates that the discrete nature of flagellar bending is not an artifact coming from defects in the original photographs. Discrete steps in curvature plots were even more prominent in demembranated flagella, reactivated with 50 µM ATP, than in live sperm. The demembranated flagella are thinner and beat at lower rates than active intact flagella, and hence would be supposed to meet a smaller viscous resistance from the medium. Therefore, the discrete bend steps may reflect an inherent discreteness in bend generation of the flagella.

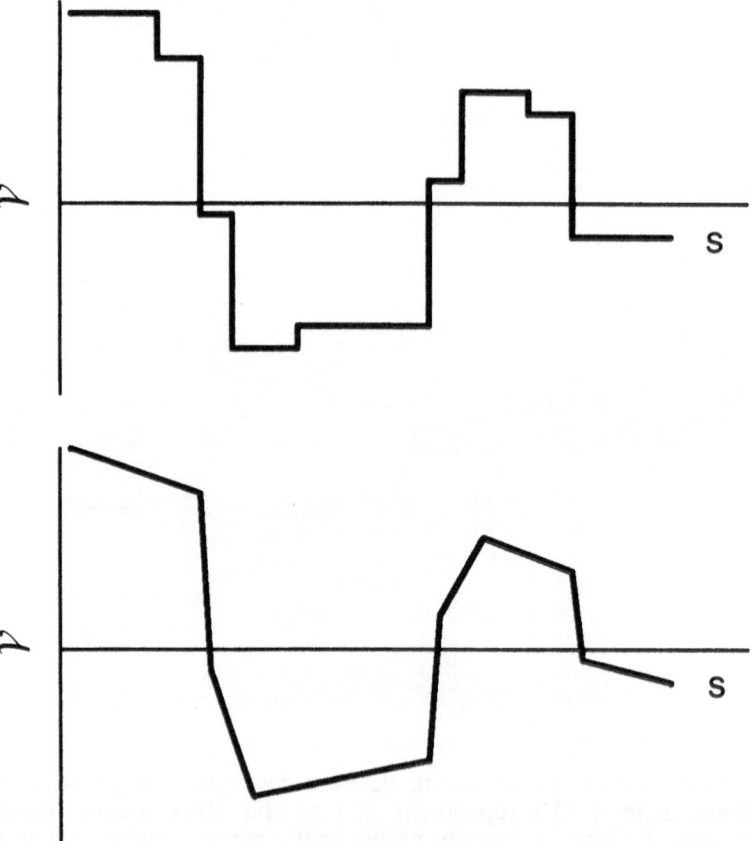

Figure 5. Models for flagellar bending shapes. Upper, constant–curvature model. Lower, constant–curvature–slope model. γ is the curvature and s the distance along the flagellum.

4. Models for flagellar bending shapes

We propose two idealized models for the shape of flagellar bends (Fig. 5):

(1) The constant–curvature bending wave model. The curvature measured along a flagellum tends to remain constant within a region of substantial length and changes abruptly to a different stable value upon transition to a neighboring region.

(2) The constant–curvature–slope bending wave model. The rate of change in curvature is kept constant within regions of variable length.

The constant–curvature model includes the arc–straight line wave that was originally proposed by Brokaw (1965) and supported by means of Fourier analysis (Johnston et al., 1979). The constant–curvature–slope model was proposed by Baba and Mogami (1985) and the gradient curvature model of Eshel and Brokaw (1988).

In our previous study (Baba and Mogami, 1985), intrinsic curvature steps as shown in the upper graph of Fig. 5, may have been smoothed out due to reduction of noise, resulting in linear slope of curvature as shown in the lower graph. However, our present analysis does not completely exclude the possibility that (1) the curvature of a flagellum changes with a constant rate over a distance along the flagellum and (2) the rate changes abruptly.

The constant–curvature wave may indicate that the axoneme has a curvature–specific mechanical stability. Alternatively, this wave may result from discreteness in the dynamic properties of bend generation. When the bending wave of a flagellum is in a steady–state, the angular direction $\phi(s, t)$ has a form of

(1) $$\phi(s - vt) \equiv \phi(x),$$

where s is the distance along the flagellum, v the wave velocity and t the time. Therefore,

(2) $$d\phi/dt = -vd\phi/ds,$$

if v is kept constant along the flagellum. Since the curvature $\gamma(s, t)$ is $d\phi/ds$ by definition, the fact that γ is a stepwise function of s and hence of x indicates that $d\phi/dt$ is also a stepwise function of s and t. This may imply that the sliding rate du/dt is a similar stepwise function of s and t, since u has been related proportionately to ϕ on the basis of reasonable assumptions on the axonemal structure and tubule sliding (Baba, 1979). The legitimacy of the relationship is to be tested by direct measurements (see Brokaw, 1989). The sliding rate would be a function of active sliding between pairs of doublets and resistances due to the axonemal structure and the surrounding medium. Therefore, the constant–curvature wave may suggest that the activation of sliding is discrete.

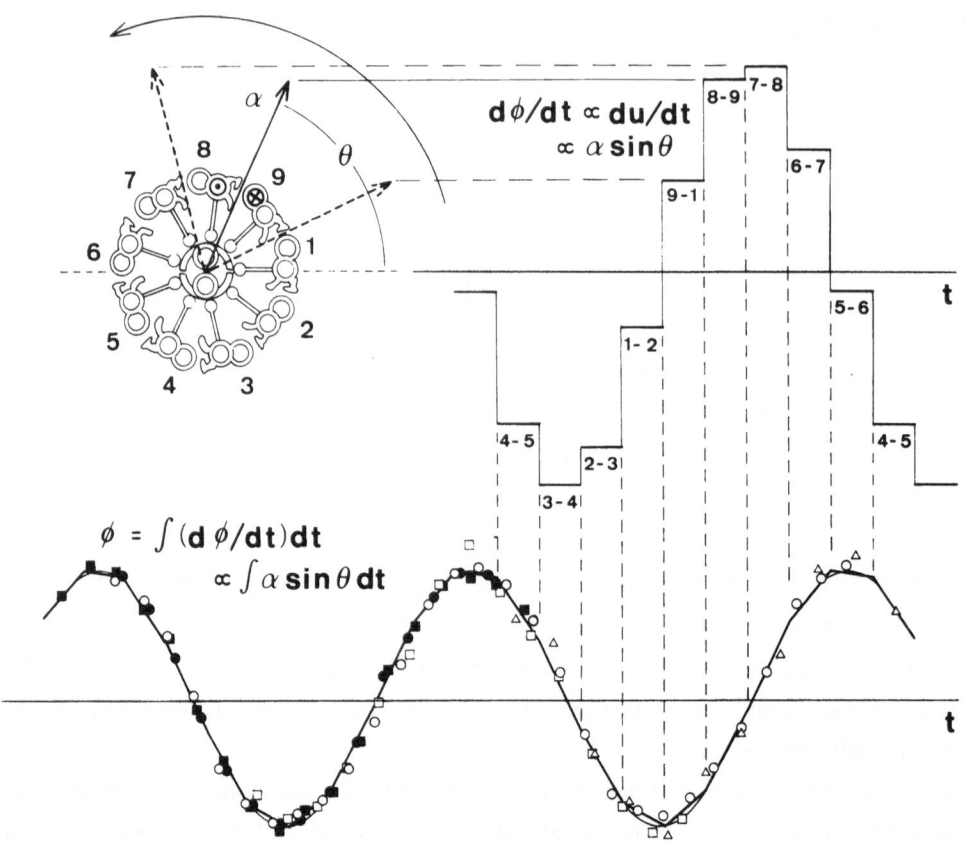

Figure 6. A model for flagellar bend generation. The curved arrow indicates the direction of activation transfer of sliding. The sliding transfer is supposed to be counterclockwise when viewed from base for sea urchin sperm (Hiramoto and Baba, 1978). The solid arrow indicates activated sliding vector couple of the amplitude α; the dashed arrows are vectors inactivated at the same time. Five different symbols are for $\phi(s=30 \ \mu m, t)$, measured with respect to the head axis and normalized for amplitude and beat frequency, of live starfish sperm, data from five specimens including those from Hiramoto and Baba (1978). See text for details.

Hiramoto and Baba (1978) have obtained an empirical equation, in which ϕ is described as a sine function of t, that is, the bending wave is sine−generated with respect to time. Taking account of the sine−generated wave, the unipolar sliding (Sale and Satir, 1977) and the right−handed helical bending, they have also proposed a model for bend generation, in which activation of sliding is transferred among doublet pairs in the counterclockwise direction when viewed from the base. We propose here a model for bend generation, which is based on Hiramoto and Baba's proposal and can explain the constant−curvature wave (Fig. 6). In this model, sliding activation is limited to a particular pair of doublets at one time, of the all−or−none type with full activation value α, and transferred to the neighboring pair in the counterclockwise direction at a constant speed. The sliding rate is postulated to be proportional to $\alpha \sin\theta$, where θ is the angle made by the sliding vector couple and the major bend plane, and $d\phi/dt$ to the sliding rate as noted above. The form of ϕ obtained from this model by integration of $d\phi/dt$, is so close to that of a sine function of t as shown in Fig. 6 (lower) that it is hard to determine which of these model functions matches the observation.

The implication of the constant–curvature–slope wave is rather complicated. Since the internal shear force S is given by

(3) $\qquad S = - E_b(d\gamma/ds) - F_n,$

where E_b is the elastic bending resistance and F_n the normal component of the external viscous force (Blum and Hines, 1979), the constancy and abrupt changes in $d\gamma/ds$ may indicate discreteness in generation of S.

In conclusion, extensive analyses of the discreteness of flagellar bends using data of high temporal and spatial resolution and simulation based on proposed models will be useful tools in elucidation of the mechanism of flagellar bend generation.

Acknowledgements

We are most grateful to Prof. Teiji Miura for the use of the image analysis system of Teikyo University School of Medicine. This work was supported by Grants–in–Aid for Scientific Research from the Ministry of Education, Science, and Culture to S. A. Baba.

REFERENCES

Baba, S.A. (1979) Regular steps in bending cilia during the effective stroke. Nature 282: 717–720.

Baba, S.A. and Mogami, Y. (1985) An approach to digital image analysis of bending shapes of eukaryotic flagella and cilia. Cell Motil. 5: 475–489.

Baba, S.A. and Mogami, Y. (1987) High time–resolution analysis of transient bending patterns during ciliary responses following electric stimulation in sea urchin embryos. Cell Motil. Cytoskel. 7: 198–208.

Blum, J.J. and Hines, M. (1979) Biophysics of flagellar motility. Q. Rev. Biophys. 12: 103–180.

Brokaw, C.J. (1965) Non–sinusoidal bending waves of sperm flagella. J. Exp. Biol. 43: 155–169.

Brokaw, C.J. (1989) Direct measurements of sliding between outer doublet microtubules in swimming sperm flagella. Science 243: 1593–1596.

Eshel, D. and Brokaw, C.J. (1988) Determination of the average shape of flagellar bends: A gradient curvature model. Cell Motil. Cytoskel. 9: 312–324.

Hiramoto, Y. and Baba, S.A. (1978) A quantitative analysis of flagellar movement in echinoderm spermatozoa. J. Exp. Biol. 76: 85–104.

Johnston, D.N., Silvester, N.R. and Holwill, M.E.J. (1979) An analysis of the shape and propagation of waves on the flagellum of Crithidia oncopelti. J. Exp. Biol. 80: 299–315.

Sale, W.S. and Satir, P. (1977) Direction of active sliding of microtubules in Tetrahymena cilia. Proc. Natl. Acad. Sci. USA 74: 2045–2049.

Sato, F., Mogami, Y. and Baba, S.A. (1988) Flagellar quiescence and transience of inactivation induced by rapid pH drop. Cell Motil. Cytoskel. 10:374–379.

AN APPROACH TO QUANTITATIVE ANALYSIS AND MODELLING OF THREE-DIMENSIONAL CILIARY MOTION

Kazuyuki Sugino

Institute of Biological Sciences, The University of Tsukuba
Tennodai 1-1, Tsukuba, 305 Japan

Abstract

Computer simulation of three-dimensional ciliary motion suggests that mechanisms controlling the activity of ciliary beating are localized at the basal region of cilia. An analysis of photographic data of the proximal shaft of a beating cilium, which had been recorded under membrane potential control, allows us to quantitatively describe motion at the level of axonemal functions. There are three parameters of ciliary beating: sliding velocity of outer doublets, transfer rate of sliding activity around the axoneme, and the steady inclination of the axis of beating.

1. Introduction

Cilia and flagella are dynein-microtubule based sliding systems (Satir, 1965) consisting of a minimal number of substructure elements and generating more complex movements (Machemer, 1972a) than the actin-myosin sliding system of muscle. Basic mechanisms of generation of two- or three-dimensional movements of flagella and cilia, respectively, have been investigated theoretically using, for the most part, simulations by computer (Brokaw, 1971, 1972, 1976a, b; Sugino and Naitoh, 1982, 1983). Computer simulations of the ciliary movement in *Paramecium* re-

vealed that (1) sliding activities initiate at the basal end of cilium, (2) an active sliding region propagates toward the ciliary
tip with constant velocity, (3) active doublet pairs change continuously in a beat cycle and propagate counterclockwise about
the axoneme as observed from the ciliary tip. These properties
suggest that an axonemal bending is generated in most cases at
the proximal region and propagates automatically. Previously, an
expedient method has been established for the analysis of
three-dimensional ciliary movement. Our method uses a series of
two-dimensional images of a cilium observed parallel to the ciliary axis at its base, or perpendicular to the membrane (Sugino
and Machemer, 1987). The method is summarized in Figure 1.

2. Theoretical background of the analysis

Most of the active force is transmitted to the surrounding medium in directions normal to the ciliary axis. Ciliary bending is
thought to occur between the proximal end of an active sliding
region and a fixed part, such as the ciliary base or the distal

Figure 1. Flow of process to isolate functional parameters of the
ciliary beat cycle. The route from A to D through B shows the
conversion of orientation-time data. The second route from A to
D through C shows the translation of angular data into the
track of a given point on the ciliary axis. One example is indicated as a direction marked with δ (before normalization) or D
(after normalization) to clarify the data flow. **A.** Outlines of a
cilium (or a cirrus) are recorded at fixed time intervals, and
proximal orientations are determined. **B.** The time course (a) of
observed ciliary orientations is plotted and recalculated after
normalization (b). **C.** Extremes of orientation and projected ciliary length determine the major and minor diameter and the inclination (θ) of the beat cone (a and a'). Compensation for the
inclination angle transforms an observed orientation of point p
into an axial view of the circular movement displaying orientation D in the x-y plane (b and b'). A proximal ciliary segment
moves on a spherical pathway (dashed line) along the surface of
the beat cone (c). This pathway is transformed into planified
courses to represent either the correct beating amplitude
(movement of point U (d)) or the direction of movement (tangent
W in U (e)). **D.** Polar diagrams of the ciliary angular velocity
(ω; a and a'), active sliding rates (V) of identified pairs of doublets (b and b') and their rates of translocation (W') around the
axoneme (c and c') result from the combination of temporal (B)
and spatial (C) data. Polarograms Da to c are from a single cycle, a' to c' after averaging over sequential cycles. From Sugino
and Machemer, 1988, with permission.

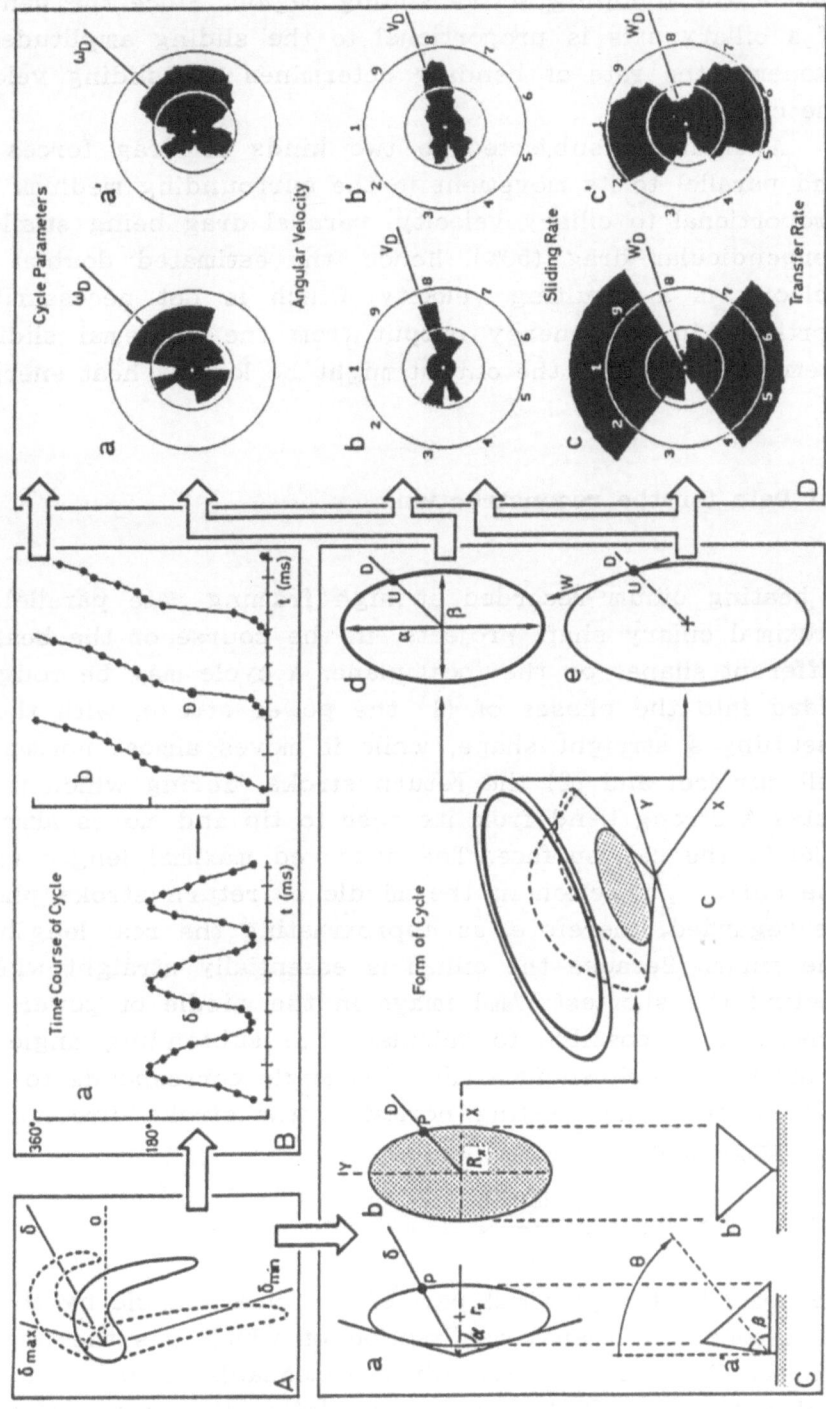

end of the following active sliding region. Since the bend angle
of a ciliary axis is proportional to the sliding amplitude of the
axoneme, the rate of bending determines the sliding velocity of
the ciliary shaft.

A cilium is subjected to two kinds of drag forces normal
and parallel to its movement in the surrounding medium. Drag is
proportional to ciliary velocity, parallel drag being smaller than
perpendicular drag (50%). Hence, the estimated doublet sliding
velocity is a resulting velocity, which is not necessarily proportional to the energy output from the axonemal sliding system. Some part of the output might be lost as heat energy.

2.1 Data for the reconstruction

A beating cilium recorded at high framing rate parallel to the
proximal ciliary shaft projects, in the course of the beat cycle,
different shapes on the focal plane. A cycle may be roughly divided into the phases of (1) the power stroke, with the cilium
assuming a straight shape, while it moves almost normal to the
cell surface, and (2) the return stroke, during which it propagates a strong bend from its base to tip and moves almost parallel to the cell surface. The observed maximal length (l_{max}) of
the ciliary projection in the middle of return stroke phase can
be regarded, therefore, as approximating the real length (L) of
the cilium. Because the cilium is essentially straight when projecting the shortest (l_{min}) image in the middle of power stroke
phase, it is possible to calculate the subtending angle of the
shaft to the cell surface (β). This angle corresponds to the amplitude of ciliary beating normal to the stroke direction (= minor amplitude):

(1) $\beta = \arccos \dfrac{l_{min}}{L}$.

We describe the proximal part of the ciliary cycle by the mantle
of a cone, the transverse section of which has the form of an
ellipse. This simplifying, albeit reasonable assumption (Sugino
and Machemer, 1988) allows us to define the course of the proximal shaft using the swing amplitude (2α) of the projected cili-

ary path and the minor amplitude (β). The swing amplitude is determined as the angle between two extreme directions (δ_{max} and δ_{min} respectively) of the projected ciliary axis:

(2) $2\alpha = \delta_{max} - \delta_{min}$.

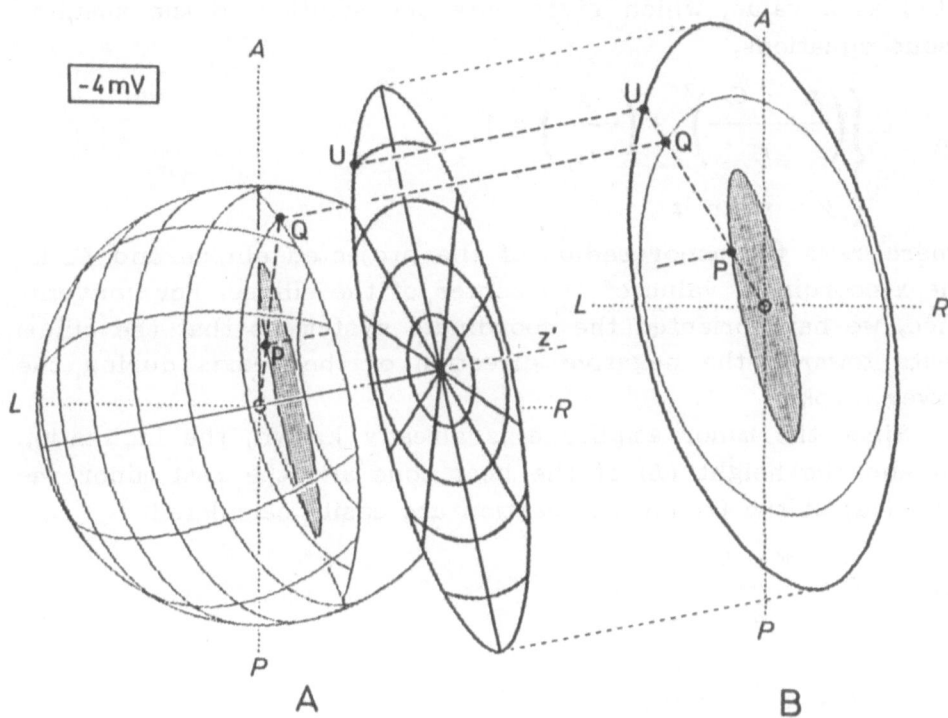

Figure 2. Geometry of the ciliary cycle as represented by a hemisphere arising from a planar surface. The observer looks down to the cell surface (A-P, anterior-posterior or y-axis; L-R, left-right or x-axis). The ciliary *base* is thought to be aligned along the axis z (normal to x-y plane). **A.** Upon application of a stimulus (-4 mV hyperpolarization in the present example) the ciliary *shaft* starts to gyrate about the inclined polar axis (z') of an ellipsoidal beat cone (shaded area, point P in Fig. 1 Cb). For a particular segment of the cilium, the beat cone cuts the surface of the hemisphere at point Q. Planification of the pathway of Q leads to the course of point U which may be calculated for correct representation of length on longitudes or of orientational relationship between points on the hemisphere (only the former course is shown). **B.** After correction for inclination of the cycle, three different representations of the track of the cilium are seen in the plane of the cell surface (x-y plane). Only point U on the planified course can represent the track of a particular axonemal cross-section. From Sugino and Machemer, 1988, with permission.

2.2 Reconstruction of the three–dimensional cycle

So far the two extreme ciliary directions of the inclined beat cone correspond to two tangential lines to the ellipse from the ciliary origin. Hence, the size of the major radius (r_y) is calculated as a value, which gives only one solution to the simultaneous equations,

(3)
$$\begin{cases} \left(\dfrac{x - C_x}{r_x}\right)^2 + \left(\dfrac{y}{r_y}\right)^2 = 1 \quad , \\ y = x \tan \alpha \end{cases}$$

where r_x is the minor radius of the projected ellipse and C_x is the x–coordinate value of the center of the ellipse. For convenience, we have oriented the coordinate system so that the cilium beats towards the negative direction of the y–axis during the power stroke.

Since the minor amplitude is already known, the inclination (θ) and the height (h) of the beat cone and the real minor radius (R_x) of the transverse section are easily calculated:

(4)
$$\theta = \frac{\pi - \beta}{2} \quad ,$$

(5)
$$h = L \cos \frac{\beta}{2} \quad ,$$

and

(6)
$$R_x = L \sin \frac{\beta}{2} \quad ,$$

respectively.

2.3 Localization on the beat cone

In order to calculate the angular velocity and direction of bending, it is necessary to determine the spatial position of the ciliary shaft on each picture frame. In the two–dimensional image, only the direction of the proximal ciliary shaft is a reliable information, so that the spatial coordinates must be calculated from the direction on the reconstructed beat cone.

Approximating the proximal ciliary shaft with a straight line, the intersection (P) of the shaft and transverse section of beat cone (Fig. 2) has the same y-coordinate value (y_p) as that of projected figures $(p;$ see Fig 1Ca, b). y_p is, therefore, calculated as the solution of y in the following simultaneous equations:

$$(7) \quad \begin{cases} \left(\dfrac{x - C_x}{r_x}\right)^2 + \left(\dfrac{y}{r_y}\right)^2 = 1 \\[2mm] y = x \tan \delta \end{cases}$$

where δ is the angle of the projected ciliary shaft from the x-axis, which corresponds to the orientation of inclination (Fig. 1A):

$$(8) \quad o = \frac{\delta_{max} + \delta_{min}}{2} .$$

δ is the ciliary direction as determined counterclockwise from posterior $= 0°$ (Fig. 1A). The x-value (x_p) of the inclination in space is then calculated as the corresponding solution of x in the equation:

$$(9) \quad \left(\frac{x}{R_x}\right)^2 + \left(\frac{y_p}{R_y}\right)^2 = 1 ,$$

where $R_y = r_y$. The beat cone is now normalized, i.e., rotated so that the proximal ciliary axis stands perpendicularly to the cell surface. The corresponding angle D (Fig. 1Bb, Cb) is given by

$$(10) \quad D = \text{arc tan } \frac{y_p}{x_p} .$$

In order to trace the path (Q) of a certain cross-section of a cilium on the mantle of the reconstructed ellipsoidal beat cone (see Fig. 2 of Introduction to Section I/2), the distance from its origin must be adjusted according to the following equation:

$$(11) \quad Q = \frac{\epsilon \cdot P}{\sqrt{x_p^2 + y_p^2 + z_p^2}} , \quad P = (x_p, y_p, z_p) ,$$

where $\epsilon = 1$ at the level of the transverse section of the beat cone $(z_p$).

2.4 Sliding parameters

Ciliary beating is the result of active sliding between outer doublets. The bending moment and the bending angle are functions of the sliding velocity of outer doublets. The bending direction is determined by the plane and direction of active sliding (Fig. 3).

Localization of a tracing point on the ciliary shaft at a certain phase of the cycle allows us to estimate the spatial angular velocity (ω) and the sliding velocity (V) of the outer doublets:

$$(12) \qquad \omega = \frac{d \text{ arc } Q}{dt} = \lim_{t \to 0} \frac{\text{arc } (Q(t + \Delta t) - Q(t))}{\Delta t},$$

$$(13) \qquad V = a \cdot \omega,$$

where a is the distance between the centers of two neighbouring doublets.

In order to calculate the rate of shift of the active sliding center around the axoneme (translocation rate), the direction of bending is to be adapted to axonemal geometry. When the curved beat path of point Q in Figure 2 is translated into a path on a plane using the conformal projection method (point U), the tangent to U corresponds to the bending direction at this moment. With this translation, the rate of translocation (W') at time t is:

$$(14) \qquad W'(t) = \frac{d \text{ arc } W(t)}{dt},$$

and the distance (l) between the center of the projected ellipsoid and a given point on the path is changed according to:

$$(15) \qquad l = 2 \tan \frac{\varphi}{2},$$

where φ is the angle between the center and the point.

Figure 3 summarizes the relationship between a normalized cycle and axonemal function. The planified track (**A**, point U) represents the form of the cycle, as referred to the topography of the cell surface, and including a major and minor amplitude. The tangent (W) to the track in U corresponds to the ciliary

bending plane and is parallel to the plane of active sliding between an identified pair (or pairs) of doublets. The angular velocity along W is proportional to the sliding velocity of the active pair. The rate of bending reorientation corresponds to the
rate, at which active sliding translocates around the ring of
doublets.

2.5 Inclination parameters

In addition to the form and frequency of the cycle, the direction and amplitude of the inclination depend on the membrane
potential. The direction of inclination (o) is the bisector between

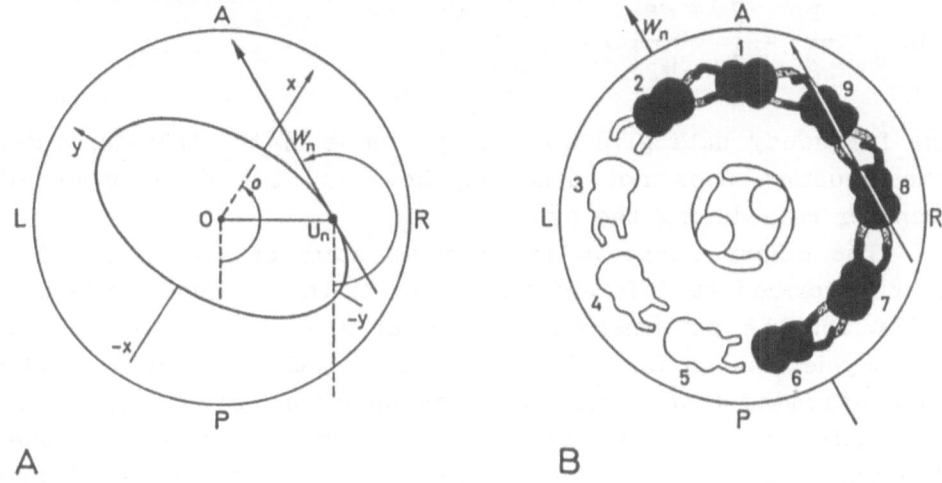

Figure 3. Inference of active sliding from the normalized track
of a ciliary segment. **A.** Ellipsoidal planified course of a ciliary
segment as referred to coordinates of the cell surface (A, anterior; R, right; P, posterior; L, left) and of the cycle (x, minor
amplitude and direction of inclination (o); y, major amplitude).
At t_n, the cilium points toward the right and moves in the direction of W_n which is the tangent to the track in U_n. **B.** Ciliary cross-section as oriented in U_n at time t_n. In order to
move the cilium in the direction W_n, the doublet pair 8-9 (and
potentially the neighbours 6-7, 7-8, 9-1, 1-2) slide actively
(dark). With the time advancing, the center of active sliding will
shift from 8-9 to 9-1. Because of the noncircular beat form, an
angle of ciliary reorientation during the cycle ($U_n \rightarrow U_{n+1}$) does
not necessarily correspond to the same angle of sliding translocation. From Sugino and Machemer (1987).

the two extreme angles (δ_{min} and δ_{max}). The amplitude of inclination is defined as the angle (θ) subtended between the beat
cone axis (z') and axis of the ciliary base (z) (Fig. 1Ca', Fig. 2).

3. Analytical description of the cycle

A ciliary cycle has been usually described using a temporal and
a spatial feature, i.e., beat frequency and shape. The frequency
(f) of *Paramecium* cilia increases with the strength of a voltage
stimulus (Machemer and Eckert, 1975). Since the duration ($1/f$)
of a beat cycle corresponds to the integral of the inverse of
the transfer rate (W'),

$$(16) \qquad \frac{1}{f} = \int_0^{2\pi} \frac{d\theta}{W'} \, ,$$

the frequency data give an average value of the transfer rate.
The equation does not describe, however, the distribution of
transfer rate during the cycle.

The ciliary shape in the resting state of *Paramecium* has
been described at different viscosities of the medium (Machemer,
1972a, b). The potential-dependence of spatial polarities of the
beat cycle was so far related to the direction of power stroke
only, assuming that a three-dimensional ciliary cycle consists of
a largely planar power stroke and a curving return stroke.
However, the form of a beat cycle under depolarized membrane
condition is very different from that in the hyperpolarized or
resting membrane condition. The rough shape of the cycle will
be determined by the ratio of active sliding transfer rate to
sliding velocity of active outer doublets (Fig. 4).

4. Application of ciliary activity under voltage-clamp

For the ultimate goal of precise description of orientational
distribution of beat parameters, the so-called ciliary "beat

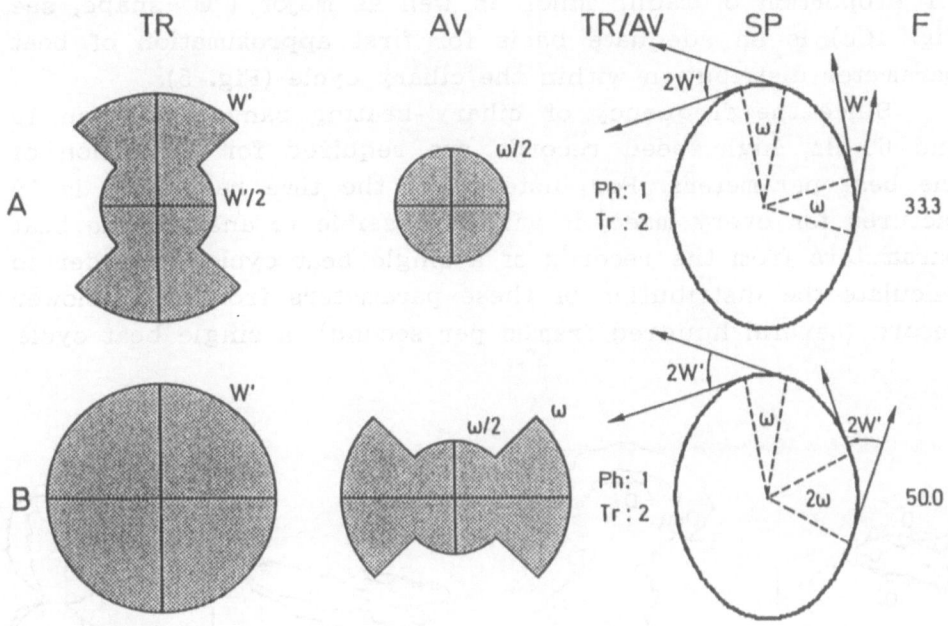

Figure 4. A thought model for the relationship between the translocation (or transfer) rate (TR) and the angular velocity (AV) to determine a beat form (SP; spatial polarity). TR and AV are represented by polar diagrams. **A.** The case of constant angular velocity. Transfer rate during transitions between power and return stroke is doubled as compared to that during the stroke phases. **B.** The case of constant transfer rate. Angular velocity during transitions is half of that during the stroke phases. Since the distribution of the ratio of transfer rate over angular velocity (TR/AV) is the same in both cases, the beat form is identical. The important difference is the beat frequency (F), which is a function of the transfer rate (frequency in B being 150% of frequency in A).

form", that is, the transverse section of axonemal pathway, should be determined prior to 3D-application of angular data of the ciliary cycle. In the present method, two independent definitions of the transverse section are available in principle: (1) the proportion of the major radii of an ellipse, which determines the curvature of the pathway during transitions between power and return strokes; (2) the proportion of the minor radii determining the curvature in the course of the power and return strokes. In modelling our data with these hypothetical alternatives, we have convinced ourselves that the simple assumption of

1:1 proportion of radii, minor as well as major ("O"-shape, see
Fig. 1Cc) is an adequate basis for first approximation of beat
parameter distribution within the ciliary cycle (Fig. 5).

Since the frequency of ciliary beating ranges between 10
and 60 Hz, high-speed records are required for extraction of
the beat parameters. For instance, if the time resolution is 10
pictures for every msec, it will be possible to analyze the beat
parameters from the records of a single beat cycle. In order to
calculate the distribution of these parameters from much slower
record (several hundred frames per second), a single beat cycle

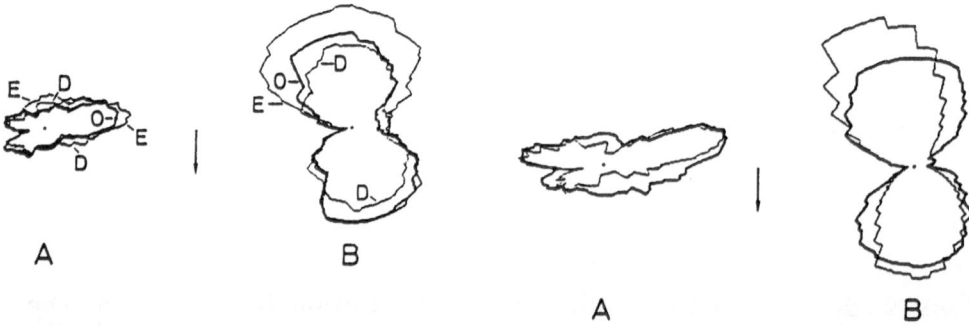

Figure 5. Comparison of beat
parameters (**A**, angular velocity, **B**, transfer rate) under
hyperpolarized condition
(–4 mV) using the same method for three different assumptions on the beat
shape. The "O"-shape polarograms are obtained with
the assumption that a beat
cone shows an elliptical
transverse section. "D"-sha-
pes are on the assumption
that ciliary pathway is more
planar in power stroke and
more round in return
stroke. The assumption for
"egg"-shape (E) is that the
pathway during transition
from return to power stroke
is more round than that during transition from power
to return stroke. All po-
larograms show similar characteristics. The arrow
points to the cell posterior.
From Sugino and Machemer,
1988, with permission.

Figure 6. Comparison of calculated beat parameters (**A**,
angular velocity, **B**, transfer
rate) in the hyperpolarized
condition (–10 mV) using
different averaging methods.
In the present analysis
(heavy-line polarograms),
the parameters for each of
tens of cycles were calculated before averaging. Thin
lines indicate the parameters
calculated after averaging
over all cycle, i.e. from a
generalized beat cycle. Both
plots differ only in some
details from each other, e.g.,
a small rotation of the polarograms, caused by the selection of phases to match
the cycles. An arrow points
to the cell posterior. From
Sugino and Machemer, 1988,
with permission.

is insufficient. Therefore, in order to obtain statistically reliable data, we employed the voltage–clamp technique to record tens of cycles under fixed membrane potential.

Even under constant voltage condition, a cilium shows slight changes in beat frequency, direction or form (Fig. 1). It is necessary, therefore, to average the orientations and to calculate a generalized cycle from all recorded cycles. As an example, we generated a cycle from recorded sequences of cycles with a matching transition from return to power stroke (Fig. 6). This generalized cycle still includes some uncertainties such as direction and timing of power stroke, and the duration of the return stroke.

The membrane potential dependency of ciliary activities in *Stylonychia* (Sugino and Machemer, 1988, 1990) is apparent in several parameters: the transfer rate of sliding activity around the axoneme increased with the depolarization during both the power and the return stroke phase. The sliding velocity of outer doublets rose with the degree of depolarization only during the power stroke, and was relatively constant during the return stroke. While the sliding velocity of doublets was constant in extracted and reactivated axonemes (Kamimura and Takahashi, 1981; Mogami and Takahashi, 1983), our analysis shows that it changed.

Possible mechanisms to explain these changes are: (1) the dyneins might change their activity relating to the phase of the cycle, or (2) resistive forces during the cycle might disturb the relationship between the sliding velocity and the number of doublets being active at one time.

Acknowledgements

This work was done at the Arbeitsgruppe Zelluläre Erregungsphysiologie, Ruhr-Universität Bochum, and supported by the Deutsche Forschungsgemeinschaft, Sonderforschungsbereich 114, TP A5, and the Forschergruppe Konzell 3.

REFERENCES

Brokaw C.J. (1971) Bend propagation by a sliding filament model
for flagella. J. Exp. Biol. 55: 289–304

Brokaw C.J. (1972) Computer simulation of flagellar movement. I.
Demonstration of stable bend propagation and bend initiation
by the sliding filament model. Biophys. J. 12: 564–586

Brokaw C.J. (1976a) Computer simulation of movement-generating
cross-bridge. Biophys. J. 16: 1013–1027

Brokaw C.J. (1976b) Computer simulation of flagellar movement.
IV. Properties of an oscillatory two-state cross-bridge model.
Biophys. J. 16: 1029–1041

Kamimura S. and Takahashi K. (1981) Direct measurement of the
force of microtubule sliding in flagella. Nature (London) 293:
566–568

Machemer H. (1972a) Properties of polarized ciliary beat in
Paramecium. In: Motile systems of cells, Int. Symp. Cracow.
Acta Protozool. 11: 295–300

Machemer H. (1972b) Ciliary activity and the origin of metachrony in *Paramecium*: effects of increased viscosity. J. Exp.
Biol. 57: 239–259

Machemer H. and Eckert R. (1975) Ciliary frequency and orientational responses to clamped voltage steps in *Paramecium*. J.
Comp. Physiol. 104: 247–260

Mogami Y. and Takahashi K. (1983) Calcium and microtubule sliding in ciliary axonemes isolated from *Paramecium caudatum*. J.
Cell Sci. 61: 107–121

Satir P. (1965) Studies on cilia II. Examination of the distal region of the ciliary shaft and the role of the filaments in motility. J. Cell Biol. 26: 805–834

Sugino K. and Machemer H. (1987) Axial-view recording: an approach to assess the third dimension of the ciliary cycle. J.
Theor. Biol. 125: 67–82

Sugino K. and Machemer H. (1988) The ciliary cycle during hyperpolarization-induced activity: an analysis of axonemal functional parameters. Cell Motil. Cytoskeleton 11: 275–290

Sugino K. and Machemer H. (1990) Depolarization-controlled
parameters of the ciliary cycle and axonemal function. Cell
Motil. Cytoskeleton 16: (in press)

Sugino K. and Naitoh Y. (1982) Simulated cross-bridge patterns
corresponding to ciliary beating in *Paramecium*. Nature
(London) 295: 606–611

Sugino K. and Naitoh Y. (1983) Computer simulation of ciliary
beating in *Paramecium*. J. Submicrosc. Cytol. 15: 37–42

BIOELECTRIC CONTROL OF THE CILIARY CYCLE

Hans Machemer

Arbeitsgruppe Zelluläre Erregungsphysiologie
Fakultät für Biologie, Ruhr-Universität, D-4630 Bochum 1
Federal Republic of Germany

Abstract

Electrophysiological research in ciliates has established the central role of ionic Ca^{2+} in membrane excitation and ciliary electromotor coupling. Ca^{2+} passes depolarization-sensitive ciliary channels and is thought to bind to axonemal proteins. During hyperpolarization, the concentration of axonemally bound Ca^{2+} is presumably reduced. The ciliary motor response - frequency and beat direction - is a monotonous function of the intensitiy of a stimulus impinging on the cell. Intermediate steps in sensory-motor coupling: potentials of either polarity, concentration of the messenger substance Ca^{2+}, and the binding of Ca^{2+} to axonemal target proteins reflect the transmission of gradedness to the ciliary motor response.

1. The cilium: an intracellular motor organelle covered by large area of excitable membrane

For a cilium of *Paramecium* (10 µm length, 0.25 µm diameter), a surface area of 8 µm² encloses a volume of less than 0.5 µm³, a fraction of which is fluid space. Assuming a resting intracellular Ca^{2+} concentration of 10^{-7} M, an average of 10 Ca^{2+} ions will be suspended in the fluid space of the cilium. It is anticipated that

activation of Ca channels in the ciliary membrane can change
the intraciliary Ca^{2+} concentration within orders of magnitude
which then determines intracellular Ca-binding to mediators and
target proteins (see contribution by Mogami and Machemer, this
Section I/2). The axoneme, commonly known as the 9+2 structure, is truly intracellular (see Introduction to I/2: Fig. 1A).
The axoneme is thereby linked to the cellular metabolism, and
disconnected from bioelectric membrane events. An intracellular
messenger, presumably the concentration of membrane-regulated
Ca^{2+}, can transmit to the axoneme those signals which have
been received from the external world and, following sensory
transduction, expressed in terms of a modulated membrane conductance.

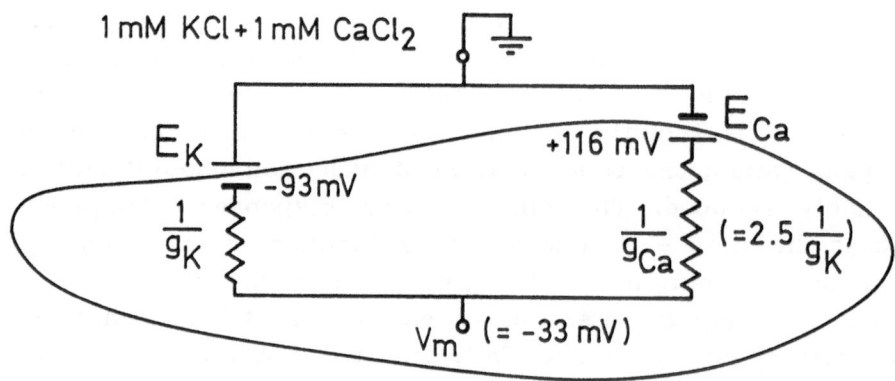

Figure 1. Simplified circuit diagram of membrane electrogenesis
in a ciliate cell. Two electrochemical batteries, as represented
by the equilibrium potentials E_K and E_{Ca}, and their internal resistances, $1/g_K$ and $1/g_{Ca}$, represent K and Ca channels. The
membrane potential, V_m, results from ohmic voltage division. V_m
depolarizes with graded increases in g_{Ca} (or decreases in g_K),
and V_m hyperpolarizes with graded increases in g_K (or decreases in g_{Ca}). From Machemer (1988).

2. Ciliary cyclic activity is a bimodal continuous function of the membrane potential

The basis of the membrane potential in marine as well as freshwater ciliates are the electrochemical batteries of Ca^{2+} and K$^+$

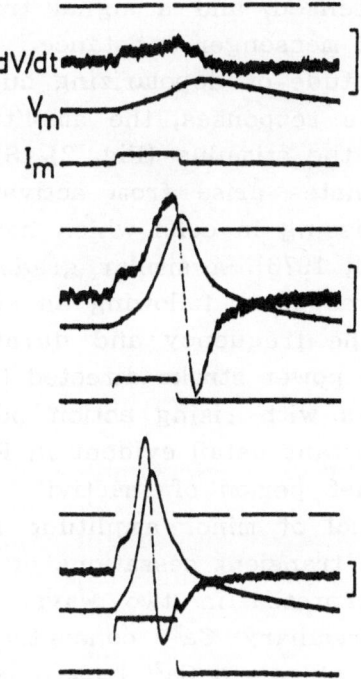

dV/dt
V_m
I_m

Figure 2. The graded action potential in *Paramecium*. Rising initial depolarizations, such as by receptor potentials, are here represented by injected 20ms-square currents from 0,3 to 1,3nA (lower of triple traces); the regenerative depolarization appears as a positive inflection of the voltage trace (middle) and the size of the time derivative of the voltage (top trace; scale: $2Vs^{-1}$). Shown action potential amplitudes rise from 9 to 40mV. Modified after Machemer and Eckert (1973).

ions (Deitmer and Machemer, 1982; De Peyer and Machemer, 1977; Naitoh et al., 1972; Pape and Machemer, 1986). These batteries establish the electric circuit of the ciliate cell (Fig. 1). Changes in the series resistances due to channel opening and closure lead to shifts of the membrane potential. An interesting type of channel is the depolarization-sensitive Ca channel located in the ciliary membrane. This channel supplies Ca^{2+} to the ciliary space. Depending on the depolarization amplitude, final Ca^{2+} concentrations may exceed 10^{-5} M. Hyperpolarizing stimuli activate rectifying K channels in the somatic membrane; hyperpolarization does not modulate the conductance of the ciliary Ca channel which is minimal at the membrane resting potential. Because cilia raise their frequency and shift the beat direction clockwise following hyperpolarization, it follows that also this type of stimulus is sensed, and a signal transmitted intracellularly by a presumed messenger substance.

Raising the amplitude of depolarizing current pulses feeds regenerative membrane responses, the amplitudes of which reflect the strength of the stimulus (Fig. 2). Stimulus-graded action potentials in ciliates arise from activation of ciliary Ca channels; they are missing in cells which have shed their cilia (Ogura and Takahashi, 1976). A similar gradedness also applies to the ciliary motor responses following an action potential. Figure 3 shows that the frequency and duration of "reversed" beating (i.e., with the power stroke directed toward the anterior end of the cell) rises with rising action potential amplitudes. There is another important detail evident in Figure 3: the cilium passes through a brief period of inactivity following reversal, and an action potential of minor amplitude may induce ciliary inactivity only. These transient cessations of cyclic ciliary activity have been interpreted in two ways: (1) as related to slightly elevated intraciliary Ca^{2+} concentrations following a progressive reduction of raised Ca^{2+} from a large action potential, or (2) following a minor initial Ca^{2+} increase from a small depolarization (Machemer, 1974; Machemer and Eckert, 1975).

The depolarization-dependent gradual reduction of ciliary activity has long been overlooked because, with a typical action potential amplitude, the cilia "switch" to high beating rates within a few milliseconds (less than the duration of a single cycle). The voltage-clamp technique allows us to predetermine the rate of rise of the membrane potential. In agreement with the predicted Ca^{2+}-dependence, ciliary inactivity was seen to precede anteriad reorientation and increase in frequency, when the rate of depolarization was sufficiently reduced (Fig. 4). In screening the range of physiologically meaningful potentials with slow linear voltage ramps, a "frequency-voltage profile" of cilia was established (Fig. 5). Each cilium can augment frequency in a graded manner with hyperpolarization *and* depolarization, and shows a frequency minimum during minor depolarization. Potential-graded transitions in beat orientation were established using optical sections of the metachronal wave system in *Paramecium* (Fig. 6). Since the lines of ciliary synchrony ("wave fronts") occur parallel to the major plane of the power stroke, changes in beat direction are visualized as chan

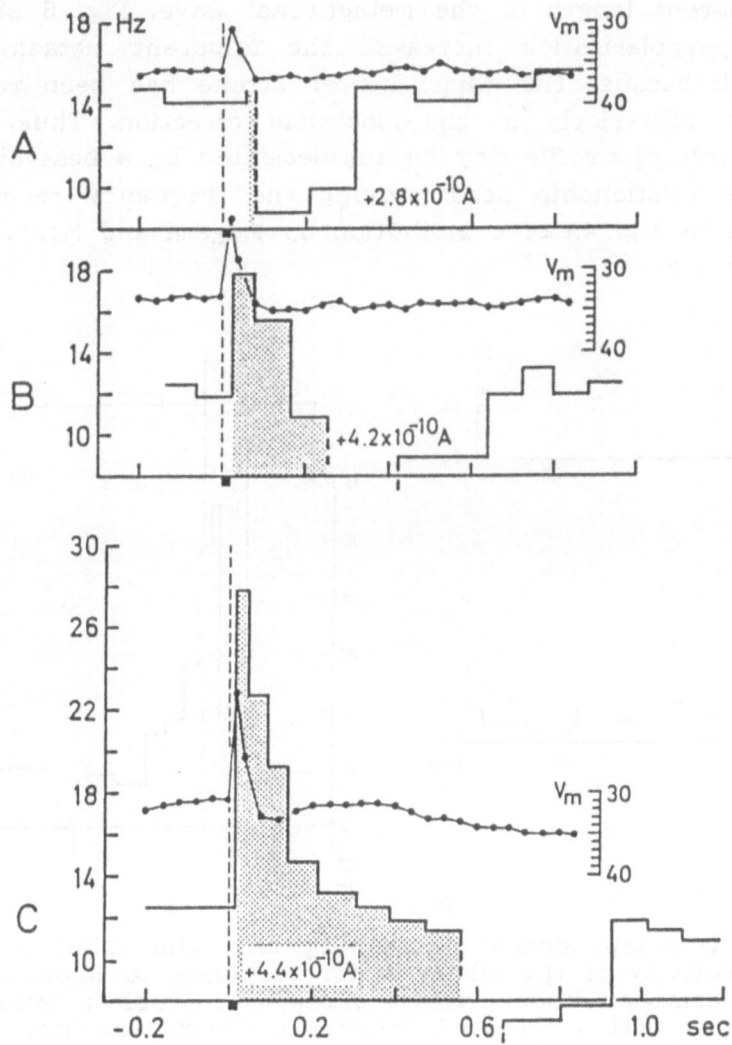

Figure 3. Graded depolarization-dependent ciliary activation (DCA) of *Paramecium*. Current stimuli of rising amplitude induce increasingly stronger anteriad ("reversed") ciliary beating responses (shaded area of frequency-time histogram). Reversed beating decreases with time towards transient inactivity of the cilium. Posteriad "normal" beating begins after the inactivation period. Note that weak depolarization (A) may induce inactivation following an abortive reversal of the cilium. Filled circles show the time course of the action potential. From Machemer (1974).

ges in apparent length of the metachronal wave. Fig. 6 shows
that a hyperpolarization increased the apparent metachronal
wave length because the ciliary power stroke had been reoriented more posteriorly in the clockwise direction. Thus, the
frequency-voltage profile may be supplemented by a beat direc-
tion-voltage relationship accompanying the frequency response
(see arrows in Fig. 4A of contribution by Mogami and Machemer,
this Section I/2).

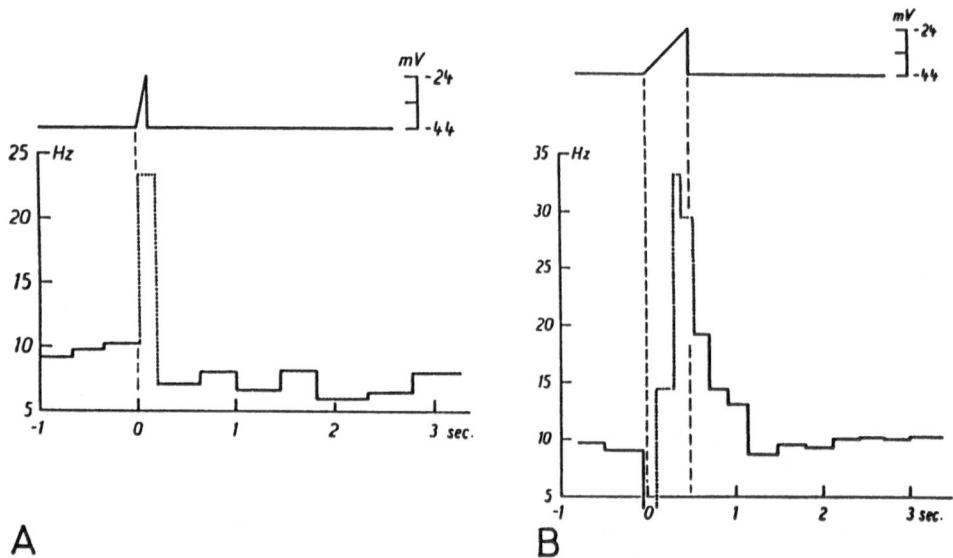

Figure 4. Rate-dependence of masking and unmasking of the
voltage-sensitivity of the ciliary motor response. **A.** Rapid membrane depolarization under voltage-clamp, comparable to rates of
rise during an action potential, induces a direct transition from
posteriad beating (continuous line) to anteriad beating of the
cilia (dotted line). **B.** Reduction in the rate of depolarization
shows that the cilia inactivate prior to transition to ciliary reversal. See text for explanation. From Machemer (1975).

3. Continuously modulated $[Ca^{2+}]_i$ is the presumed intraciliary messenger for electromotor coupling

The present contribution does not intend to fully document the
role of Ca^{2+} in ciliary electromotor coupling. Various experiments have shown that with rising depolarization intraciliary

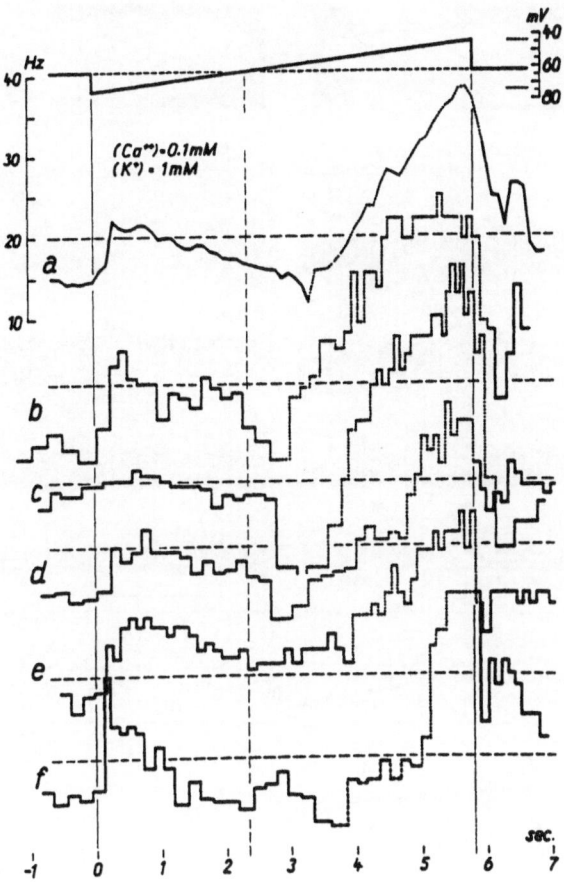

<u>Figure 5.</u> Frequency-voltage profiles of cilia in *Paramecium.*
Motor responses of five sample cilia (**b** - **f**) from one cell following exposure to a linear voltage ramp under voltage clamp.
The ramp started with a hyperpolarization of 12 mV and ended
with a depolarization of 18 mV. The cilia inactivated or reduced
frequency with minor depolarization and augmented frequency
with larger depolarization, as well as with hyperpolarization.
Anteriad beating is indicated by dotted course of the frequency
histogram. Averaging of the motor response (**a**) reveals a bimodal, linear frequency-voltage relationship. The averaged frequency minimum exceeds zero because the frequency profile varied in individual cilia. The falling (left) slope shows a continuous transition from hyperpolarization-induced to depolarization-induced ciliary activity. See text for conclusions. From Machemer (1976).

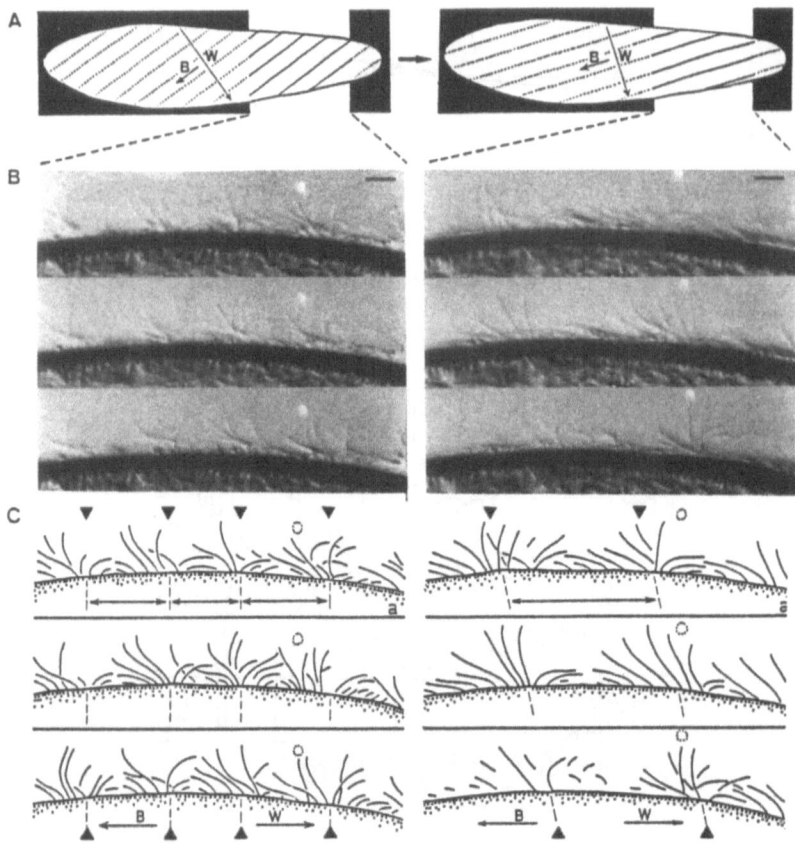

<u>Figure 6.</u> Graded reorientation of the ciliary power stroke as visualized by an increase in apparent length of the metachronal wave in *Paramecium*. Cilia in profile were taken by a high-speed 16mm camera with frames following at 4 ms intervals (middle column). Prior to stimulation, an evaluation of the ciliary images (lower triple frames of left column) reveals a posteriad orientation of the power stroke and metachronal waves travelling anteriorly (a, anterior cell end). Following membrane hyperpolarization, the frequency of the cilia rose, and the apparent wave length increased (upper triple frames). The interpretation of these images suggests a clockwise reorientation of the metachronal wave pattern and hence beat direction associated with an increase in frequency of the cilia (right column). From Machemer and Eckert (1977).

Ca^{2+} is elevated due to an increased Ca^{2+} conductance (Eckert et al., 1976). One crucial test may be given as an example: In accordance with the membrane theory predicting that extreme positive shifts of the membrane potential reduce and eventually remove the inward driving force for Ca^{2+}, a Ca^{2+}-dependent motor response was suppressed by applying these potentials. Suppression of ciliary reversal was documented in *Paramecium*, *Stylonychia* and *Didinium* (Machemer and Eckert, 1973; De Peyer and Machemer, 1982; Pernberg, unpublished).

Rising positive step potentials between 60 and 100 mV applied to *Paramecium* also established the same sequence in ciliary responses as seen with minor depolarization, that is, decrease in frequency of anteriad beating → inactivity → increase in frequency of posteriad beating, suggesting a parallel decrease in $[Ca^{2+}]_i$ with continuously reduced Ca^{2+} driving force (Machemer and Eckert, 1975). The depolarizing course of the ciliary activity-voltage curve (Fig. 5) may thus be interpreted as being regulated by rising concentrations of intraciliary Ca^{2+}. By inference, the negative-going part of the activity-voltage curve is regulated by continuously decreasing $[Ca^{2+}]_i$. Indirect support comes from the observation that, under voltage-clamp, hyperpolarizing and depolarizing stimulation in sequence lead to mutual suppression of the hyperpolarization- and depolarization-induced ciliary responses (Machemer and De Peyer, 1982). Direct evidence that hyperpolarization-induced ciliary activation is mediated by a decrease in $[Ca^{2+}]_i$ was obtained previously from cirri of *Stylonychia*. A "transverse" group of these organelles shows a noncyclic posteriad inclination upon physiological hyperpolarization, which differs from the conventional cyclic reversal response following depolarization. The hyperpolarization-induced response was also obtained when the membrane potential was clamped to positive potentials exceeding the Ca equilibrium potential (Mogami and Machemer, unpublished data).

Cyclic nucleotides have been suggested as putative ciliary activating factors during membrane hyperpolarization. Cell demembranation and nucleotide-reactivation experiments support this view (Bonini et al., 1986). However, *in-vivo* injection of cAMP and analogs into *Paramecium*, while holding the membrane potential at rest under voltage-clamp, have not confirmed expectations that cAMP can shunt the membrane (and Ca^{2+}) control of the cilia (Hennessey et al., 1985). The nucleotide-injection experiments have been expanded previously showing that in *Paramecium* the whole range of voltage-activated ciliary frequency was unaffected by cAMP and cGMP (Nakaoka and Mache-

mer, 1990; Fig. 7). The data indicate that the cyclic nucleotides
were able to slightly shift the voltage of transition from posteriad to anteriad ciliary beating. This suggests, in agreement
with findings by Izumi and Nakaoka (1987) on effects of Ca^{2+}
and cAMP on ciliary reversal, that cyclic nucleotides may ant-
agonize Ca^{2+}-calmodulin in reorienting the ciliary power stroke.

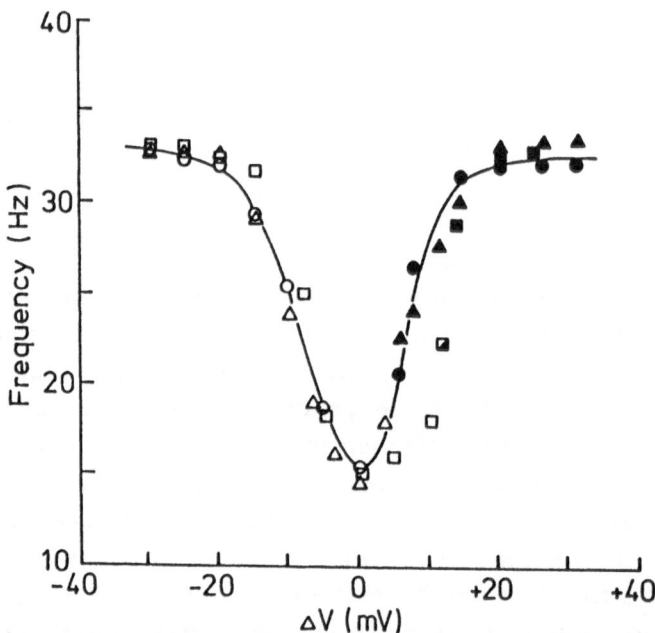

Figure 7. A test of the effect of injected cyclic nucleotides on
ciliary activity in *Paramecium*. Applying voltage-clamp steps at 5
mV intervals, the frequency-voltage curve (circles) was essentially unchanged after injection of the cyclic nucleotides
(squares, cAMP; triangles, cGMP). Transition from posteriad
beating (open symbols) to anteriad beating (closed symbols) was
slightly shifted in the positive direction of the voltage scale.
Modified after Nakaoka and Machemer (1990).

4. A perspective of calcium, axoneme, and cell behaviour

Current electrophysiological work on the ciliary motor control
strives to quantify the axonemal response following shifts in
membrane potential (see contribution by Sugino, this Section

I/2). The steady-state transmembrane potential is thought to reflect levels of intraciliary Ca^{2+} concentrations. New techniques of $[Ca^{2+}]_i$-modulated fluorescence (Ohmori, 1988) promise to document the action of Ca^{2+} *in-vivo.* A step in the hyperpolarization-induced signalling chain to the axoneme has now been established: manipulations of the polarity and amplitude of the Ca^{2+} driving force in *Didinium* and *Stylonychia* show that ciliary motor responses following membrane hyperpolarization are related to *decreases* in intraciliary Ca^{2+} (Pernberg, Mogami and Machemer, unpublished data). These data agree with previous ATP-reactivation studies in *Paramecium* (Nakaoka et al., 1984). At present, no evidence is available suggesting that ciliary resting Ca^{2+} concentrations are depressed by pumping or sequestering mechanisms following hyperpolarization. An attractive idea is that proportions of Ca^{2+} bound to the axoneme and to the membrane vary depending on the membrane potential (Fig. 8). In permeabilized and reactivated models of *Paramecium*, Mg^{2+} antagonized Ca^{2+} in regulating the frequency and the directional response of the cilium (Nakaoka and Toyotama, 1979). At millimolar intraciliary concentrations, Mg^{2+} is an unlikely candidate for cellular messenger, but this cation may compete with Ca^{2+} for binding to a divalent cation-sensitive axonemal protein. Model calculations show that the observed bimodal potential-dependence of frequency and the unimodal potential dependence of orientational responses of the cilium are predicted from differential Ca^{2+}-Mg^{2+} binding of an axonemal protein (see contribution by Mogami and Machemer, this Section I/2).

Physiological work on sensory-motor coupling of cilia suggests that transduction of a limited range of stimuli converges on a single intracellular messenger, Ca^{2+}, to control the axonemal motor. The ciliary responses, and hence locomotion in freeliving protozoans, vary as continuous functions of the membrane potential (Fig. 9). This being established, a major concern of research are the still unknown transitions from free intraciliary Ca^{2+} to a modulated cycle of the cilium.

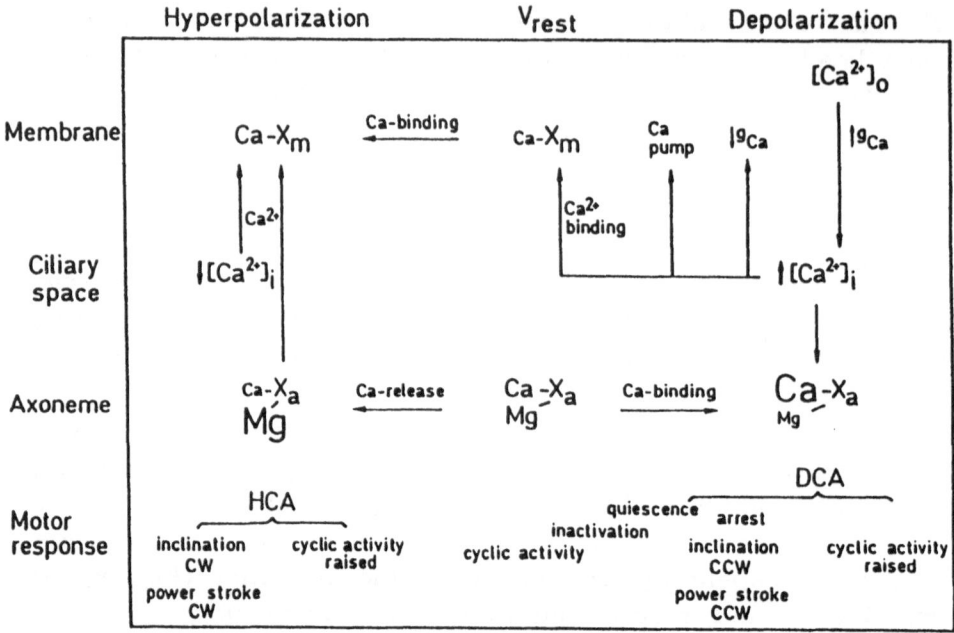

Figure 8. Summary of facts and conjectures of ciliary electro-motor coupling. Four levels of events (membrane, ciliary space, axoneme, motor response) are considered for three bioelectric states of the membrane: the resting potential (V_{rest}), hyperpolarization, and depolarization. Ionic calcium in the ciliary space is regulated by (1) the conductance for Ca^{2+} (g_{Ca}), (2) Ca^{2+} pumping, and (3) binding of Ca^{2+} to a voltage-sensitive, membrane integrated protein, (X_m), and an axonemal protein, (X_a). Intracellular Mg^{2+} competes with Ca^{2+} for binding to X_a. Because X_m absorbs Ca^{2+} at negative membrane voltages, the ratio of Ca/Mg bound to X_a is low during hyperpolarization, and increased during depolarization of the membrane. Graded Ca-Mg-induced conformational changes of X_a are the presumed signals to the dynein complex of the sliding machine which generates hyperpolarization-induced ciliary activation (HCA) or depolarization-induced ciliary activation (DCA). CW, clockwise; CCW, counterclockwise. From Machemer (1986).

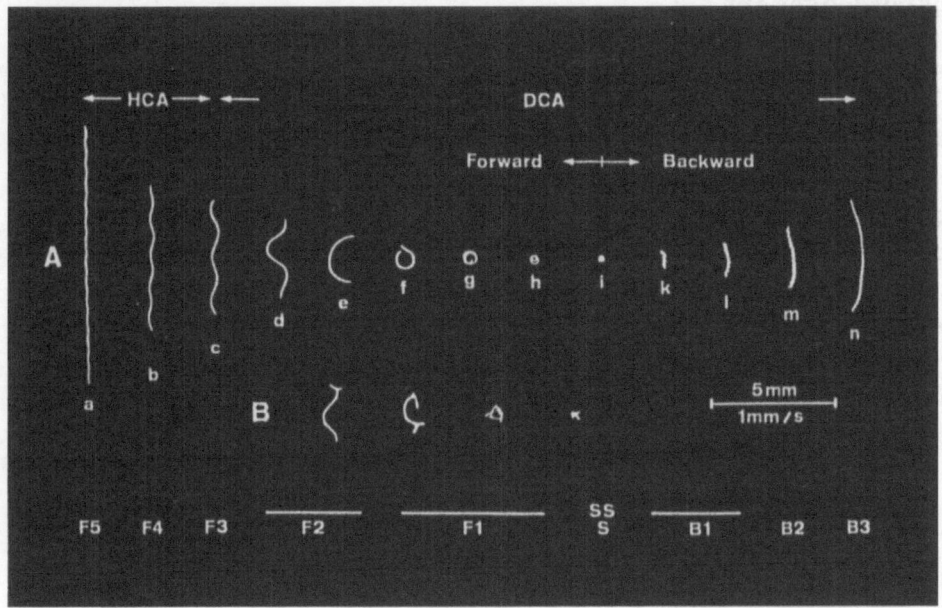

Figure 9. *Paramecium* swimming patterns reveal graded transitions from hyperpolarization-induced (HCA) to depolarization-induced activity (DCA). 5s-exposure traces of single cells under darkfield illumination were taken during stimulation in solutions of various Ca^{2+}-K^+ composition which hyperpolarized or depolarized the cell. The length of a trace represents the swimming velocity and the rate of ciliary beating, respectively. The pitch of the swimming helices grossly indicates the orientation of the ciliary power stroke. **A.** Forms of continuous forward swimming (a to h) are subdivided into five classes (F5 to F1), backward swimming into three classes (B1 to B3). F3 shows the typical forward swimming pattern of a fully adapted cell at electrophysiological rest. Note that patterns of forward and backward swimming do not coincide with HCA and DCA. **B.** Raised concentrations of Mg^{2+}, Ba^{2+} or Na^+ are favourable conditions for the generation of repetitive "reversals" which superimpose on continuous swimming (Machemer, 1989). A reversal may occur at potentials fluctuating near the threshold of the action potential. From Machemer and Sugino (1989).

Acknowledgement

I would like to thank my collegues, past and present, Dr.
Joachim W. Deitmer, Dr. Jacques E. De Peyer, Dr. Roger Eckert
(†), Dr. Todd M. Hennessey, Dr. Sigrun Machemer-Röhnisch, Dr.
Boris Martinac, Dr. Yoshihiro Mogami, Dr. Akira Murakami, Dr.
Yutaka Naitoh, Dr. Yasuo Nakaoka, Dr. Akihiko Ogura, Dr. Chri-
stian Pape, Joachim Pernberg, Dr. Kazuyuki Sugino and Martin
Weskamp, who by their patient and continuous work, have con-
tributed in establishing data and conclusions on ciliary electro-
motor coupling. This work was supported by the Deutsche For-
schungsgemeinschaft, Forschergruppe Konzell 3.

REFERENCES

Bonini N.M., Gustin M.C. and Nelson D.L. (1986) Regulation of ci-
    liary motility by membrane potential in *Paramecium*: a role for
    cyclic AMP. Cell Motility Cytoskel. 6: 256-272
Deitmer J.W. and Machemer H. (1982) Osmotic tolerance of Ca-de-
    pendent excitability in the marine ciliate *Didinium*. J. Exp.
    Biol. 97: 311-324
De Peyer J.E. and Machemer H. (1977) Membrane excitability in
    Stylonychia: properties of the two-peak regenerative Ca-re-
    sponse. J. Comp. Physiol. 121: 15-32
De Peyer J.E. and Machemer H. (1982) Electromechanical coupling
    in cilia I. Effects of depolarizing voltage steps. Cell Motility 2:
    483-496
Eckert R., Naitoh Y. and Machemer H. (1976) Calcium in the bio-
    electric and motor functions of *Paramecium*. Symp. Soc. Exp.
    Biol. 30: 233-255
Hennessey T., Machemer H. and Nelson D.L. (1985) Injected cyc-
    lic AMP increases ciliary beat frequency in conjunction with
    membrane hyperpolarization. Eur. J. Cell Biol. 36: 153-156
Izumi A. and Nakaoka Y. (1987) cAMP-mediated inhibitory effect
    of calmodulin antagonists on ciliary reversal of *Paramecium*.
    Cell Motility Cytoskel. 7: 154-159
Machemer H. (1974) Frequency and directional responses of cilia
    to membrane potential changes in *Paramecium*. J. Comp. Phy-
    siol. 92: 293-316
Machemer H. (1975) Modification of ciliary activity by the rate of
    membrane potential changes in *Paramecium*. J. Comp. Physiol.
    101: 343-356
Machemer H. (1976) Interactions of membrane potential and ca-
    tions in regulation of ciliary activity in *Paramecium*. J. Exp.
    Biol. 65: 427-448

Machemer H. (1986) Electromotor coupling in cilia. In: H.C. Lütt-gau (ed.) Membrane control of cellular activity. Fortschr. Zool. 33: 205-250

Machemer H. (1988) Electrophysiology: 185-215. In: H.D. Görtz (ed.) *Paramecium.* Springer Verlag, Berlin, Heidelberg, New York, Tokyo

Machemer H. (1989) Cellular behaviour modulated by ions: electrophysiological implications. J. Protozool. 36: 463-487

Machemer H. and De Peyer J. (1982) Analysis of ciliary beating frequency under voltage-clamp control of the membrane. Cell Motility Suppl. 1: 205-210

Machemer H. and Eckert R. (1973) Electrophysiological control of reversed ciliary beating in *Paramecium.* J. Gen. Physiol. 61: 572-587

Machemer H. and Eckert R. (1975) Ciliary frequency and orientational responses to clamped voltage steps in *Paramecium.* J. Comp. Physiol. 104: 247-260

Machemer H. and Eckert R. (1977) Electromechanical coupling of ciliary activity in *Paramecium.* Fortschr. Zool. 24: 211-215

Machemer H. and Sugino K. (1989) Electrophysiological control of ciliary beating: a basis of motile behaviour in ciliated protozoa. Comp. Biochem. Physiol 94 A: 365-374

Naitoh Y., Eckert R. and Friedman K. (1972) A regenerative calcium response in *Paramecium.* J. Exp. Biol. 56: 667-681

Nakaoka Y. and Machemer H. (1990) Effects of cyclic nucleotides and intracellular Ca^{2+} on voltage-activated ciliary beating in *Paramecium.* J. Comp. Physiol. A 166: 401-406

Nakaoka Y. and Toyotama H. (1979) Directional change of ciliary beat effected with Mg^{2+} in *Paramecium.* J. Cell Sci. 40: 207-214

Nakaoka Y., Tanaka H. and Oosawa F. (1984) Ca^{2+}-dependent regulation of beat frequency of cilia in *Paramecium.* J. Cell Sci. 65: 223-231

Ogura A. Takahashi M. (1976) Artificial deciliation causes loss of Ca-dependent responses in *Paramecium.* Nature (London) 264: 170-172

Ohmori H. (1988) Mechanical stimulation and Fura-2 fluorescence in the hair bundle of dissociated hair cells of the chick. J. Physiol. 399: 115-137

Pape H.C. and Machemer H. (1986) Electrical properties and membrane currents in the ciliate *Didinium.* J. Comp. Physiol. A 158: 111-124

Ca-Mg CONTROL OF CILIARY MOTION: A QUANTITATIVE MODEL STUDY

Yoshihiro Mogami and Hans Machemer

Arbeitsgruppe Zelluläre Erregungsphysiologie
Fakultät für Biologie, Ruhr-Universität, D-4630 Bochum 1
Federal Republic of Germany

Abstract

We have developed a generalized model for Ca-mediated control of ciliary beating using established data from the literature. According to the model, both direction and frequency of beating are controlled by membrane-regulated Ca^{2+} as the intraciliary messenger including a modulatory function of Mg^{2+} which competes with Ca^{2+} for binding to an axonemal protein.

1. Introduction

Ciliate protozoa exhibit a close coupling between the electric activity of the cell membrane and ciliary motility. A large body of data on ciliary electromotor coupling has been accumulating, but the signalling pathway from modulated membrane conductances to responses of the motile machine is not satisfactorily understood. Ionic calcium is a highly favoured candidate for a universal intraciliary messenger among several presumed mechanisms linking ciliary responses to the membrane (see e.g. Machemer, 1986). In *Paramecium* and other ciliates, it has been shown that Ca^{2+} enters the intraciliary space following an opening of depolarization-sensitive Ca channels; this causes the cilia to activate or (augment existing activity) and to specifically reorient beat direction. Although a hyperpolarization of the membrane is nonexcitatory, the ciliary frequency of ciliates such as *Paramecium* and *Stylonychia* rises as well. This motor response was - so far - not conclusively linked to Ca^{2+}. However, indirect and new direct evidence suggests that intraciliary Ca^{2+} is depressed during hyperpolarization (Machemer, this Section I/2).

The highly specific voltage-dependence of the frequency and directional responses of the cilia has been a permanent challenge to applicability of a universal Ca hypothesis of ciliary motor control. Can the concentration of a single ciliary messenger substance duplicate the established bimodal frequency-voltage relation and, at the same time, the unimodal shifts of beat orientation with decreasing membrane negativity? In addition, if Ca^{2+} is the presumed messenger, how can a simple analog signal convey messages to generate the observed complex motor responses? According to our model, Ca^{2+} can do a dual regulation because, during part of its function, it competes with a second species of intraciliary divalent cation, Mg^{2+}, for binding to an axonemal target protein. Mg^{2+} is thought to be kept at high (that is, millimolar) intracellular concentration. A Ca-Mg dependent regulation of ciliary activity was demonstrated in previous ATP-reactivation studies of *Paramecium* (Naitoh and Kaneko, 1972, 1973; Nakaoka and Toyotama, 1979; Nakaoka et al., 1984). Our model will be examined with respect to reactivation data from these authors in addition to evidence established in live cells.

2. Effects of Ca^{2+} and Mg^{2+} on ciliary activity

Demembranated models of *Paramecium* may be reactivated under minimal artificial conditions (Fig. 1). Naitoh and Kaneko (1972) discovered a pronounced modulatory effect of Ca^{2+} on the ciliary beat direction. Using a reactivation medium with the Ca^{2+} concentration kept below 10^{-6} M, the ciliary beating was toward the posterior of the cell, and the cell models swam forwards. Above concentrations of 10^{-6} M, the models reversed the direc-

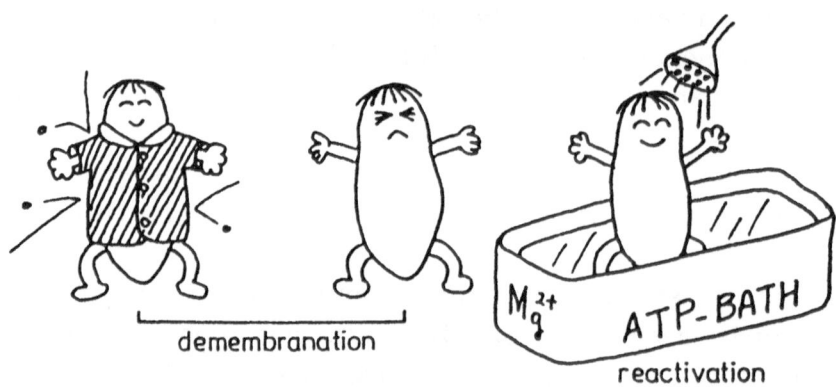

Figure 1. Principle of cell reactivation studies. The membrane (jacket) separates the internal organization from the environment. Detergent tears the membrane away from the cell which turns into a "model". Bathing of the cell model in a suitable medium reactivates motility of the cell.

tion of beating and swam backwards. From these observations, Naitoh and Kaneko (1973) derived a "two-machine hypothesis" with one Ca^{2+}-dependent machine supposed to serve beat direction, and a second Mg^{2+}-dependent machine powering cyclic ciliary activity.

Interactions between Ca^{2+} and Mg^{2+} in reactivating *Paramecium* cilia were reported by Nakaoka and collaborators. These authors obtained their results from cell models which had been prepared in the presence of the Ca^{2+}-chelating agent, EGTA. The EGTA-models behave differently as compared to the models by Naitoh and Kaneko, who employed a Ca^{2+}-Mg^{2+} chelator, EDTA (EDTA-model). The major differences in the motor behaviour of these models can be summarized as follows:

I. When EGTA-models were reactivated in low concentrations of Mg^{2+}, they swam backward even with Ca^{2+} $<10^{-7}$ M. Increases in Mg^{2+} changed the swimming direction from backward to forward.

II. EGTA-models changed the swimming direction in high Mg^{2+} depending on the Ca^{2+} concentration. Increases in Ca^{2+} changed the swimming from forward to backward.

III. The beat frequency of EGTA-models changed in a Ca^{2+}-concentration dependent manner. The Ca^{2+} concentration for minimal frequency approximated that for reorientation of beat orientation from posteriad to anteriad.

IV. A Ca^{2+}-dependent modulation of beat frequency was not observed in the EDTA-models. In these models, the Ca^{2+}-dependence of swimming direction was comparable to that of EGTA-models at millimolar Mg^{2+} concentration.

3. Model for Ca-Mg control of ciliary motility

For transmission of the Ca^{2+} signal to the cilium, Machemer (1986) introduced a hypothetical intraciliary component, X_a, which can bind Ca^{2+} and Mg^{2+} at different ratios. We here assume that X_a has a single binding site for Ca^{2+} and Mg^{2+} with association constants K_{Ca} and K_{Mg}. Based on the binding equations for these cations,

(1) $[X_a\text{-}Ca^{2+}] = [X_a][Ca^{2+}]K_{Ca}$, and
(2) $[X_a\text{-}Mg^{2+}] = [X_a][Mg^{2+}]K_{Mg}$,

the equilibrium concentrations of the free ligand and the metal-bound ligand are given as follows:

(3) $[X_{a\ free}] = [X_{a\ total}]/(1+\alpha+\beta)$,
(4) $[X_a\text{-}Ca^{2+}] = [X_{a\ total}]\beta/(1+\alpha+\beta)$,
(5) $[X_a\text{-}Mg^{2+}] = [X_{a\ total}]\alpha/(1+\alpha+\beta)$,

where $[X_{a\ total}]$ is the total concentration of X_a in free and metal-bound form, and α and β are the following abbreviations:

(6) $\alpha = K_{Mg}[Mg^{2+}]$,
(7) $\beta = K_{Ca}[Ca^{2+}]$.

We have developed equations of the beating parameters on the assumption that X_a acts as a regulatory factor in the metal-bound form. The equations for beat direction and frequency imply a proportionality to the following expressions of metal-bound X_a:

(8)   beat direction $= C \cdot A$,
(9)   beat frequency $= C' \cdot (A+B-A \cdot B)$,

where A and B are molar fractions, as follows,

(10) $A = [X_a\text{-}Ca^{2+}]/[X_{a\ total}]$ and
(11) $B = [X_a\text{-}Mg^{2+}]/[X_{a\ total}]$, and

C and C' are proportionality factors. Figure 2 shows a plot of fractional values of beat direction and frequency as a function of ß assuming different values of α.

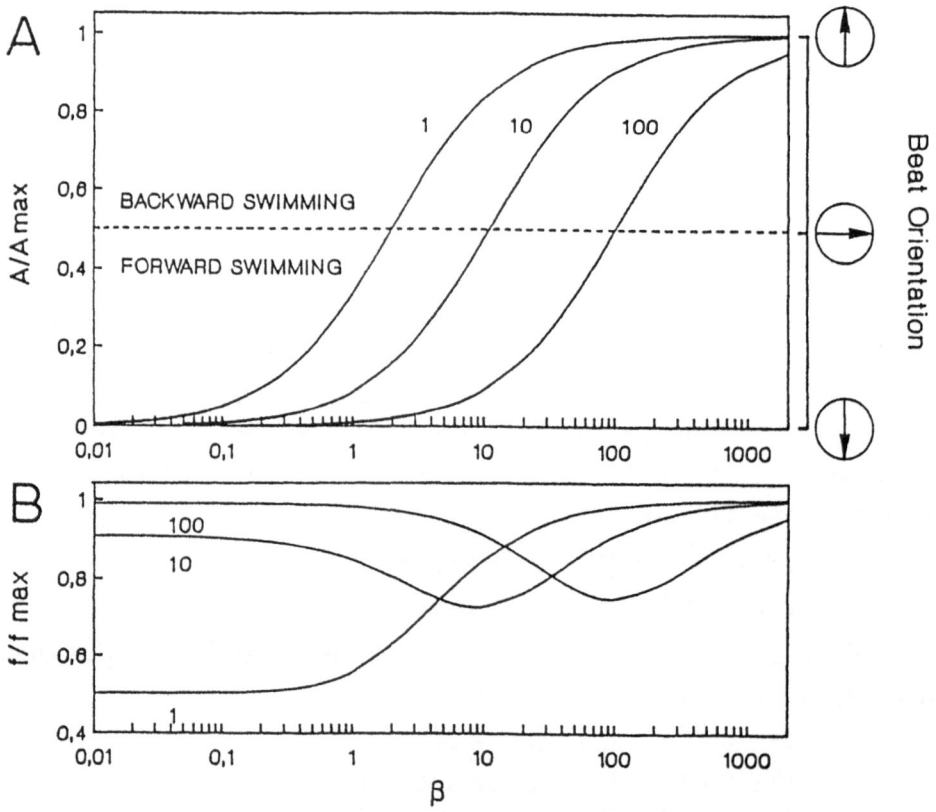

Figure 2. Calculated behaviour of the Ca-Mg control model. Fractional values of Ca^{2+} bound to the axonemal component, X_a, and of beat orientation (equation 8) and frequency (equation 9) as functions of α (1, 10 and 100 indicated beside lines) and ß (abscissa). **A.** Fraction of bound calcium (left ordinate) and beat orientation (right ordinate). **B.** Frequency.

4. Modulation of beat direction

The model assumes, for simplicity, a 180 degree range of beat directions from posteriad to anteriad with a state of transition coupled to a half-maximum value derived from equation 8. The swimming direction will be reversed from forward to backward lowing an increase in ß. Figure 2A indicates that the effective value of ß (abscissa) rises with increasing α. The figure predicts a change from backward to forward swimming, and from anteriad to posteriad beating of the cilia, with increasing value of α and fixed value of ß. The figure also suggests that, at small values of α, the direction of swimming will be backward beyond a critical level of ß. At larger values of α, and near an equivalence with ß, the swimming direction changes as a function of α. These predictions correspond to the experimental data summarized in points I and II (see above).

5. Modulation of the beating rate

The beat frequency in ciliates approximates a minimum close to the threshold value of ß for reversal of the swimming direction (Fig. 2B). This relationship is in agreement with the experimental data summarized in point III and with voltage-clamp experiments in *Paramecium* and *Stylonychia* (Machemer 1976; Machemer and De Peyer, 1982). Equation 9 predicts a suppression of the augmentation of frequency with increasingly reduced values of α. Correspondingly, a modulation of ciliary frequency in the ciliate *Didinium* is restricted to depolarizing stimulation, whereas hyperpolarization does not alter the resting rate of beating (Pernberg and Machemer, unpublished observations). An interpretation of equation 9 is that the beating rate is determined by the total amount of divalent cation-bound X_a including a molar-fraction effect due to a mutual inhibition between the two cation-bound species.

6. EDTA- and EGTA-models

It seems reasonable to assume that the values of $[X_a\ total]$, K_{Ca}, and K_{Mg} are kept constant in live cells and in models prepared with reproducible methods. Since α and ß are proportional to ionic concentrations, they may be expressed by variables having the dimension of concentration. Figure 3C shows the result of calculations using the values of 10^7 M^{-1} for K_{Ca} and 10^4 M^{-1} for K_{Mg}. These values are equivalent to a ratio, K_{Ca}/K_{Mg}, of 10^3 corresponding to ratios of other Ca^{2+}-binding proteins under physiological concentrations of Mg^{2+} (calmodulin: 10^3; troponin C: 10^4; Potter et al., 1983).

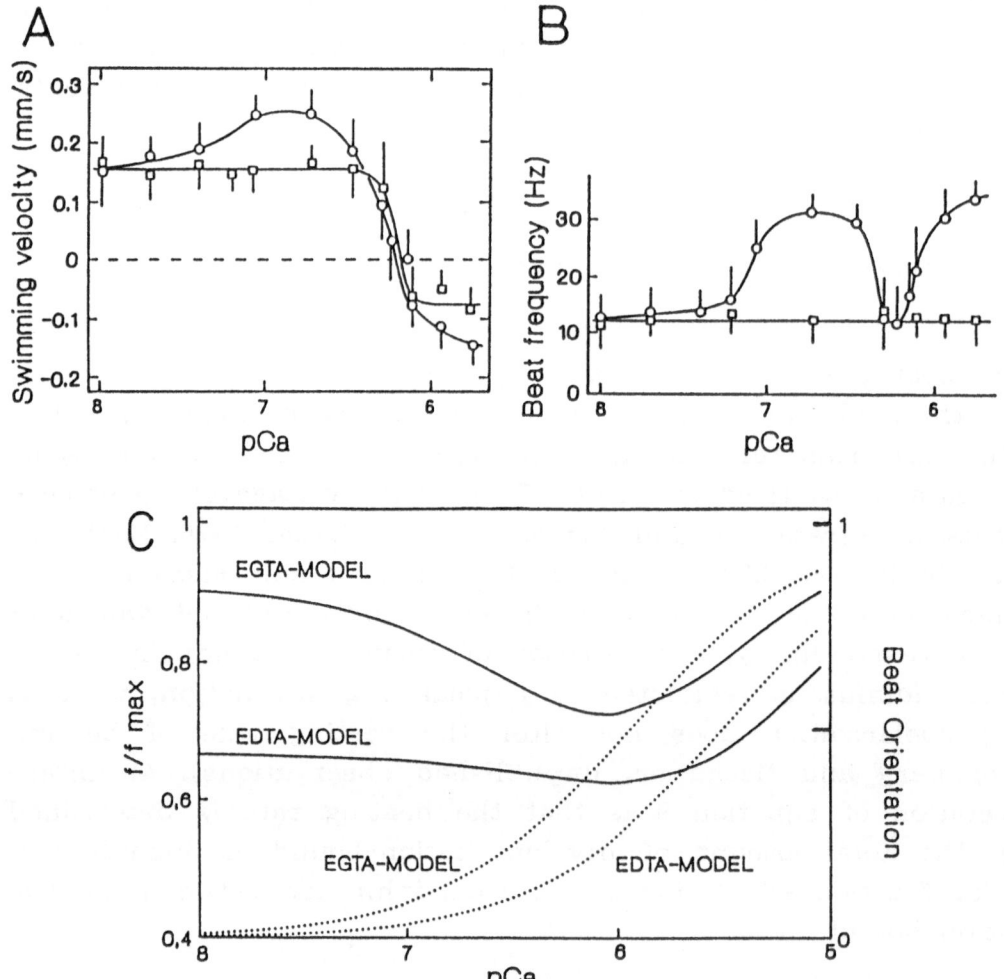

Figure 3. Modelling of published ATP-Mg reactivation experiments of *Paramecium* which used EDTA and/or EGTA during Triton extraction (A and B from Nakaoka et al., 1984). **A.** Swimming velocity of reactivated cells prepared with EGTA (circles) or EDTA (squares). **B.** Ca^{2+}-dependence of ciliary frequency. Symbols comparable to (A). **C.** Calculated frequency (solid) and directional response of the cilia (dotted) as functions of the free Ca^{2+} concentration (Mg^{2+} = 10^{-3} M throughout). The EDTA-model assumes reduced association constants for Ca^{2+} and Mg^{2+} (K_{Ca} = 10^6 M^{-1}; K_{Mg} = 2×10^3 M^{-1}) as compared to the EGTA-model (K_{Ca} = 10^7 M^{-1}; K_{Mg} = 10^4 M^{-1}).

The model allows us to derive a large variety of curves by definition of parameters. Accordingly, the functional properties of cilia has been reported to change with variations in the preparation of the models (Brokaw et al., 1974; Izumi and Nakaoka, 1987). Figure 3C exemplifies one of the various possible cases that can be obtained from different sets of parameters: reduction in association constant assumed for the EDTA-model as compared to the EGTA-model. The result shows a reduced frequency modulation of the EDTA-model, whereas the directional response remains comparable to the EGTA-model. This predicted change in responsiveness of the reactivated model cells corresponds to the experimental data (Fig. 3A and B).

7. Ca^{2+} as intracellular regulator

Our hypothesis can successfully model rather complex experimental data which would be difficult to explain assuming that Ca^{2+} is the sole regulator of frequency as well as beat orientation. Since Mg^{2+} was inferred to be kept constant at millimolar levels in the intraciliary space (see Machemer, 1986), this cation is an unlikely candidate for short-term regulation of ciliary activity. Similar conclusions apply to ATP which is thought to be supplied to the the cilia at millimolar concentration (Naitoh and Kaneko, 1973). Cyclic nucleotides have been tentatively implicated in direct ciliary regulation (Bonini and Nelson, 1988; Bonini et al., 1988; Schultz et al., 1984), but crucial experiments using voltage-clamp and nucleotide pressure injection methods show that ciliary electromotor coupling, in particular the frequency-voltage relationship, is unaffected even in the presence of millimolar intracellular concentrations of cAMP or cGMP (Hennessey et al., 1985; Nakaoka and Machemer, 1990).

In our model, Mg^{2+} serves as a ubiquitous cation at constant concentration which competes with Ca^{2+} for binding to an intervening axonemal protein. The concentration of free Ca^{2+} is thought to be an intermediate signal which is passed to the Ca^{2+}-binding molecule. The model reconciles cell reactivation experiments with the voltage-clamp data from viable cells (cf. Machemer, this Section I/2). The Ca^{2+}-dependent response of the ciliary motor response is consistent with the voltage-dependent profile of ciliary activity (Fig. 4). Our model predicts (1) the

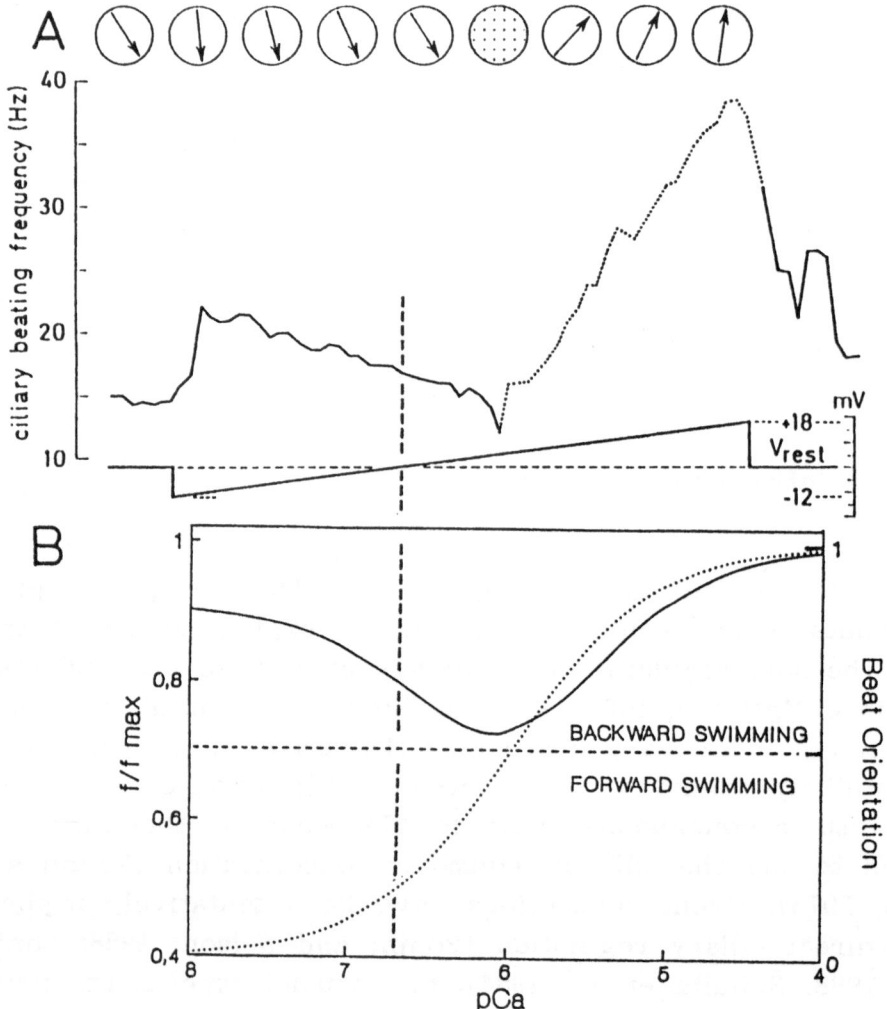

Figure 4.   Experimental data and model results of monotonous shifts of beat orientation and a bimodal regulation of frequency in *Paramecium*.   **A.** Voltage-dependence of ciliary frequency and beat orientation (posteriad: solid line; anteriad: dotted line) as obtained under voltage-clamp. The cell was hyperpolarized from the resting potential (V_{rest}) and then linearly depolarized. The arrows on top represent graded changes in beat direction (anterior end of cell up). The vertical dashed line marks the frequency and beat orientation in the resting state. From Machemer, 1976.   **B.** Modelling of beat orientation (dotted) and frequency (solid) as functions of the intraciliary concentration of ionic calcium using equations 8 and 9 ($K_{Ca} = 10^7$ M^{-1}; $K_{Mg} = 10^4$ M^{-1}; $[Mg^{2+}] = 10^{-3}$ M). The curves duplicate the voltage-dependent activity profiles of live cilia (A).

depression of the beating rate by a small increment in $[Ca^{2+}]_i$ from minor depolarization and (2) the substantial increase in frequency by an even more raised $[Ca^{2+}]_i$ from large depolarization. The model can (3) also simulate the hyperpolarization-induced motor response on the assumption that hyperpolarization reduces $[Ca^{2+}]_i$ from a level at membrane rest. Our model is in line with a previous hypothesis suggesting that electromotor coupling is due to unimodal regulation of $[Ca^{2+}]_i$ within the full range of stimulation from hyperpolarization to depolarization of the ciliate membrane (Machemer, 1974, 1976; Machemer and De Peyer, 1982).

Previously, direct evidence was obtained showing that a reduction in intraciliary Ca^{2+} parallels hyperpolarization-induced ciliary activation. Clamping the membrane potential of *Stylonychia* far positive to the Ca equilibrium potential establishes an artificial outward Ca^{2+} electromotive force across open Ca channels which acts to depress the intraciliary concentration of Ca^{2+}. Under these conditions, a ciliary organelle of *Stylonychia* performed a motor response which is commonly evoked after hyperpolarizing stimulation only (Mogami and Machemer, unpublished observations).

In its present form, our model would be too primitive to be expressed in terms of molecular components of the cilium. For example, our simple assumptions of the binding properties of the axonemal component X_a bear presumably only remote similarity to properties of Ca^{2+}-binding proteins which have, according to biochemical evidence, multiple binding sites, different affinities and cooperativity between these sites. However, it is to be expected that continuing efforts in analyzing the temporospatial properties of the ciliary responses and their controls, together with refinements in quantitative modelling, will improve our so far limited insight into one of the finest machines developed at the eucaryotic level.

Acknowledgements. We thank Dr. Yasuo Nakaoka, Osaka University, for fruitful discussions during his stay at our Bochum laboratory, and Jörg Pernberg for his help in programming the graphic display of our calculations. This work was supported by the Deutsche Forschungsgemeinschaft, Forschergruppe Konzell 3.

REFERENCES

Bonini N.M. and Nelson D.L. (1988) Differential regulation of *Paramecium* ciliary motility by cAMP and cGMP. J. Cell Biol. 106: 1615-1623

Bonini N.M., Gustin M.C. and Nelson D.L. (1986) Regulation of ciliary motility by membrane potential in *Paramecium*: a role for cyclic AMP. Cell Motility Cytoskel. 6: 256-272

Brokaw C.J., Josslin R. and Bobrow L. (1974) Calcium ion regulation of flagellar beat symmetry in reactivated sea urchin spermatozoa. Biochem. Biophys. Res. Commun. 58: 795-800

Hennessey T., Machemer H. and Nelson D.L. (1985) Injected cyclic AMP increases ciliary beat frequency in conjunction with membrane hyperpolarization. Eur. J. Cell Biol. 36: 153-156

Izumi A. and Nakaoka Y. (1987) cAMP-mediated inhibitory effect of calmodulin antagonists on ciliary reversal of *Paramecium*. Cell Motility Cytoskel. 7: 154-159

Machemer H. (1974) Frequency and directional responses of cilia to membrane potential changes in *Paramecium*. J. Comp. Physiol. 92: 293-316

Machemer H. (1976) Interactions of membrane potential and cations in regulation of ciliary activity in *Paramecium*. J. Exp. Biol. 65: 427-448

Machemer H. (1986) Electromotor coupling in cilia. In : H.C. Lüttgau (Edit) Membrane control of cellular activity. Fortschr. Zool. 33: 205-250

Machemer H. and De Peyer J.E. (1982) Analysis of ciliary beating frequency under voltage-clamp control of the membrane. Cell Motility (Suppl.) 1: 205-210

Naitoh Y. and Kaneko H. (1972) Reactivated Triton-extracted models of *Paramecium*: modification of ciliary movements by calcium ions. Science 176: 523-524

Naitoh Y. and Kaneko H. (1973) Control of ciliary activity by adenosinetriphosphate and divalent cations in Triton-extracted models of *Paramecium caudatum*. J. Exp. Biol. 58: 657-676

Nakaoka Y. and Machemer H. (1990) Effects of cyclic nucleotides and intracellular Ca^{2+} on voltage-activated ciliary beating in *Paramecium* J. Comp. Physiol. A 166: 401-406

Nakaoka Y. and Toyotama H. (1979) Directional change of ciliary beat effected with Mg^{2+} in *Paramecium*. J. Cell Sci. 40: 207-214

Nakaoka Y., Tanaka, H. and Oosawa F. (1984) Ca^{2+}-dependent regulation of beat frequency of cilia in *Paramecium*. J. Cell Sci. 65: 223-231

Potter J.D., Strang-Brown P., Walker P.L. and Iida S. (1983) Ca^{2+} binding to calmodulin. Methods Enzymol. 102: 135-143

Schultz J.E., Grünemund R., von Hirschhausen R. and Schönefeld U. (1984) Ionic regulation of cyclic AMP levels in *Paramecium tetraurelia*. FEBS Lett., 167: 113-116

Conclusions

A synopsis of geometry, mechanisms and controls of ciliary motion, as they have been addressed in Section I/2, suggests that the axonemal molecular architecture incorporates crucial elements of function (such as active sliding, longitudinal and transverse translocation of active sliding) in a unique structural framework. An increasing preciseness in the assessment of three-dimensionality of active sliding may profoundly modify current models of initiation, expression and propagation of bending waves, and of a presumed "switch-type" alternation between two phases of the two-dimensional cycle (Satir, 1985). Membrane-controlled modulation of the ciliary cycle is mediated via Ca^{2+} (i.e., a "one-dimensional" intraciliary messenger). In view of (1) this central result of ciliary physiology and (2) the existence of spatially and temporally diverse motion in cilia and flagella, axonemal chemo-mechanical transduction promises to reveal an unforeseen topological diversity.

Acknowledgement. This work was supported by the Deutsche Forschungsgemeinschaft, Forschergruppe Konzell 3.

REFERENCES

Brokaw C.J. (1974) Movement of the flagellum of some marine invertebrate spermatozoa: 93-109. In: M.A. Sleigh (ed.) Cilia and flagella. Academic Press, London, New York

Gibbons I.R. (1975) The molecular basis of flagellar motility in sea urchin spermatozoa: 207-231. In: S. Inoué and R.E. Stephens (eds.) Molecules and cell movement. Raven Press, New York

Hines M. and Blum J.J. (1985) On the contribution of dynein-like activity to twisting in a three-dimensional sliding filament model. Biophys. J. 47: 705-708

Ishijima S. and Mohri H. (1985) A quantitative description of flagellar movement in golden hamster spermatozoa. J. Exp. Biol. 114: 463-475

Machemer H. (1972a) Properties of polarized ciliary beat in *Paramecium*. Acta Protozool. 11: 295-300

Machemer H. (1972b) Ciliary activity and the origin of metachrony in *Paramecium*: Effects of increased viscosity. J. Exp. Biol. 57: 239-259

Machemer H. (1977) Motor activity and bioelectric control of cilia. Fortschr. Zool. 24: 195-210

Machemer H. and Eckert R. (1975) Ciliary frequency and orientational responses to clamped voltage steps in *Paramecium*. J. Comp. Physiol. 104: 247-260

Machemer H. and Eckert R. (1977) Electromechanical coupling of ciliary activity in *Paramecium*. Fortschr. Zool. 24: 211-215

Machemer H. and Sugino K. (1986) Parameters of the ciliary cycle under membrane voltage control. Cell Motility Cytoskel. 6: 89-95

Omoto C.K. and Brokaw C.J. (1983) Quantitative analysis of axonemal bends and twists in the quiescent state of *Ciona* sperm flagella. Cell Motility 3: 247-259

Pape H.C. and Machemer H. (1986) Electrical properties and membrane currents in the ciliate *Didinium*. J. Comp. Physiol. A 158: 111-124

Pitelka D. (1968) Fibrillar systems in protozoa: 280-388. In: T.T. Chen (ed.) Research in protozoology. Pergamon Press, Oxford

Sale W.S. and Satir P. (1977) Direction of active sliding of microtubules in *Tetrahymena* cilia. Proc. Natl. Acad. Sci. USA 74: 2045-2049

Satir P. (1985) Switching mechanisms in the control of ciliary motility. Modern Cell Biology 4: 1-46

Satir P., Wais-Steider J., Lebduska S., Nasr A. and Avolio J. (1981) The mechanochemical cycle of the dynein arm. Cell Motility 1: 303-327

Sugino K. and Machemer H. (1987) Axial-view recording: An approach to assess the third dimension of the ciliary cycle. J. Theor. Biol. 125: 67-82

Sugino K. and Machemer H. (1990) Depolarization-controlled parameters of the ciliary cycle and axonemal function. Cell Motil. Cytoskeleton 16: (in press)

Sugino K. and Naitoh Y. (1982) Simulated cross-bridge patterns corresponding to ciliary beating in *Paramecium*. Nature 295: 609-611

Tamm S.L. and Tamm S. (1981) Ciliary reversal without rotation of axonemal structures in ctenophore comb plates. J. Cell Biol. 89: 495-509

MOTION OF BODY PARTS:
ITS GENERATION AND CONTROL IN HIGHER ANIMALS

(Coordinators: Hans Scharstein and Wolfram Zarnack)

Introduction

The following four contributions deal with the question how the general behavior of higher animals (summarized here in contrast to single cell organisms) is based on the properties of the central nervous system (CNS), the sensory equipment, the muscular activity and the sceleton. The term 'general behavior' includes all sorts of locomotion, communication like courtship etc. (cf. Kandel 1979, Selverston 1985).

All behavior of higher animals is manifested by the movements of limbs and other body parts as wings, tails etc. So we use behavior and movement synonymously in this article. Higher animals have a great repertoire of movements. For instance, a locust can fly, walk, hop and swim. It walks, stridulates, swims and kicks with its legs. Thus, the single study has to be restricted to a certain type of movement which is analysed by means of specific experiments.

Several types of movement are periodical per se, e.g. running, stridulation and wing-beat of insects, and have become important objects of neurophysiology. But generally, the duration of a certain movement is short and non-repetitive, thus making it difficult to observe in detail all processes which are involved in the motion under study. However, many movements are based on reflexes by which an animal can be forced to perform repeatedly the same motion: the experimenter stimulates the animal in a specific manner and records the wanted movement of a certain body part, i.e. the animals response (output) to the stimulus (input).

The concept of reflexes includes that the animal reacts deterministic, i.e. the movement occurs whenever the animal is stimulated. But there is more and more evidence, that spontaneous, random activity is an integral part of the general behavior (see Zarnack et al.,this section) and has to be taken into account in modelling.

The final result of a physiological study would be a complete description of the causal connections between all elements of the system. As a first approach, stock-taking of the behavioral repertoire should be done. The analysis starts with a quantitative description of the movement. Then one tries to find out the inner structure in the measured data, to extract invariants and rules which govern the diversity of possible behavior (see Müller, this section). The result is sometimes a model of behavior which is a simplified but generally sufficient description of the normal behavioral repertoire (see Dean and Cruse, this section). This model may be an input-output description (concept of black box) in a cybernetic sense.

But such a black-box model is not the endpoint of our investigation. Physiologists want to know how the system is organized inside, on which paths the information flows, where the interaction takes place. As already mentioned, the aim is to describe and to understand the system performance on the basis of the existing organs, especially the CNS. The description is not necessarily a one-to-one connection diagram (circuitry) of all elements but a scheme of all necessary parts at an adequate level of abstraction.

Higher animals are able to react extremely flexible and adaptive to any disturbation. If one works on this exiting capability, two general problems arise: (i) With only few exeptions, the animal has to be fastened to any support for neurophysiological and physical measurements. Such suspension blocks many of the animal's degrees of freedom and, as a consequence, disturbs its normal behavior. This impairs especially studies on the plasticity of the behavior. (ii) Due to the flexibility, one has to collect a huge amount of data which has to be managed by computerized data aquisition and evaluation. For this reason, there are only few useful paradigms in this field of investigation which are carried out preferably with insects (the long legs of stick insects demand for analysis of coordination).

Because of the enumerated difficulties, only in a few cases a rather complete understanding of a certain behavior is achieved. We reference to the stereotype movements of the stomatogastric system of crayfish which is controlled by 14 neurons (Selverston et al. 1983) and to the behavior of Aplysia (Kandel 1979). However, in the great majority the central nervous systems are very complex so that it is impossible until now to formulate similar comprehensive descriptions. This is also valid for the concept of a central pattern generator of locust flight (Wilson 1961) being controversely discussed until now (cf. Pearson 1985, Stevenson and Kutsch 1987, Ronacher et al. 1988) although a number of about 600 identified interneurons of the locust's CNS have been already described (Burrows 1977, Robertson and Pearson 1983, Rowell 1988).

Consequently, this section does not given a representative overview of the state of investigation (for this purpose cf. Elsner and Singer 1989, Gewecke and Wendler 1985), but a rather pointillistic look at questions and methods concerned with the behavior of higher animals. The four contributions of this section deal with stereotype movements which are carefully described. The contributions may be understood as examples of the different stages of the outlined process of analysis.

Two articles deal with arthropod walking. (i) **Dean** and **Cruse** discuss the current state of control models of legged walking before they explain their model of the 6-legged walking of the stick-insect: a set of 6 relaxation oscillators, coupled by a number of neural and sensory mechanisms. (ii) **Müller** analyses the phase relationships of the perturbed leg-oscillators of walking crayfish.

The investigations of **Zarnack et al.** on locust flight enclose two sections. The first one deals with a rather comprehensive wing kinematics as a basis for the calculation of unsteady aerodynamic forces. In the second one it is shown, how the quality of flight is influenced by the properties of flight simulators. First hints about a non-machine-like behavior of flying insects and the consequences for neuroethological studies with locusts under experimental constraints are discussed.

The last article deals with vertebrate behavior. **Behrend** analyses the directed probing movement of an electric fish. Electrical stimulation of different nuclei- large groups of interconnected neurons- induces different motion patterns which are precisely measured and compared to the motion pattern the fish performs on its own. On the base of this data together with a calculation of the mechanical constraints of a model fish, a hypothesis is put forward how the nervous elements govern the patterning of motion.

MODELLING THE CONTROL OF WALKING IN INSECTS

Jeffrey Dean, Holk Cruse
Abteilung für Theoretische Biologie und Biokybernetik, Universität Bielefeld,
Postfach 8640, D-4800 Bielefeld 1, FRG

Abstract

The current state of control models for legged locomotion is discussed first in terms of general control requirements in order to illustrate the complexity of the problem and then in terms of a specific, kinematic model for leg coordination in the stick insect. The kinematic model is used to demonstrate that the coordinating mechanisms deduced from behavioral experiments are sufficient to qualitatively simulate normal step patterns. However, quantitative discrepancies indicate an important role for the dynamic factors not yet incorporated in the model.

1. Control Requirements for Legged Locomotion

Legs easily outperform wheels in climbing or travelling over uneven ground. This advantage is the reason why numerous laboratories devote much time and effort to designing walking machines and testing diverse control algorithms. However, progress has been slow despite numerous working prototypes--animals which walk using a variety of structures and mechanisms. Before examining the control of walking in the stick insect, it is useful to consider the general problem in order to appreciate why learning to walk has been so hard for machines and where a theoretical analysis runs into difficulties.

The first task to be solved by a walking system is easy to define: the legs must always be positioned to resist the effect of gravity and to maintain the distance between body and substrate within working limits. As a first goal in learning to walk upright, this task can be formulated as keeping the center of gravity above an area of support defined by the locations of the feet contacting the ground. Fulfilling this condition provides static stability even if the feet cannot actively grasp the substrate: the walker can halt its movement at any point and not fall over.

However, it is in the nature of legged locomotion that the feet vary their positions with respect to the center of gravity as they propel the body forward. Moreover, because the length of a leg is finite, each leg must be periodically lifted and returned to where it can begin a new stance. Thus, each leg performs rhythmic step movements in which it alternates between stance, when it provides support and propulsion, and swing, when it returns to the starting point for the next stance.

This rhythmic stepping causes the relation between the center of gravity and the support area to vary continuously. Therefore, the concept of static stability must be augmented to include the notion of safety margins and the realization that maximum stability is only possible for slow speeds. The faster the system needs to move, the smaller the number of legs which can be kept in contact with the substrate at any moment. To increase speed still more, as in the faster gaits of vertebrates, static stability is completely abandoned in favor of dynamic stability. A fast-moving vertebrate uses the inertia of its body and limbs to pendulate from one unstable position to another.

Besides ensuring adequate support, the walking system must provide forward propulsion and allow for changes in speed and direction. These functions involve adjusting the magnitude and direction of the forces the legs apply to the substrate.

Thus, the criteria for successful walking are few and only basic physics is required. Nevertheless, theoretical approaches to walking have advanced more slowly than one might have expected. The problem lies in the complexity of the total system. First, the six legs of an insect theoretically allow nearly 40 million gaits--different sequences (stepfall patterns) of swing and stance in the six legs (McGhee, 1976). Second, each leg has three major joints which must be coordinated for proper stepping. Finally, legs in stance are mechanically coupled through the substrate and the body itself possesses additional joints. As a result, the relationship between what any one leg does and the global performance of the system is indirect. The set of possible step movements for each leg is enormous and what movements are appropriate for maintaining stability or achieving a change in direction depend upon the actions of the other legs.

2. Approaches to Modelling Leg Coordination of Walking Insects

Before turning to a specific model for the stick insect, it is worthwhile to briefly review previous theoretical approaches to insect walking.

A deductive approach was followed by McGhee and his colleagues in a formal analysis of gaits for four and six-legged walkers (McGhee, 1976). The set of possible gaits, generated by combinatorial analysis of the corresponding finite state machine, was first reduced using the plausible assumption that all legs make similar movements. Then the remaining gaits were evaluated using a stability criterion defined as the minimum distance between the vertical projection from the center of gravity onto the substrate and the boundary of the area of support spanned by the feet. The gaits for slow walking identified by this procedure were not new: they were the same metachronal gaits typically used by many arthropods.

Relying upon the evaluation of a global parameter, such as stability, to evaluate gaits creates difficulties when step patterns must be generated because it requires a central processor which monitors the configuration and movement of all the
legs. The complexity of biological walking systems would place an enormous computational burden upon such a central processor. In fact, behavioral observations
discussed below indicate that control is decentralized, so that each leg, and possibly
each leg joint or set of joints, has its own semi-autonomous controller, referred to
here as the step pattern generator. These controllers interact with each other to
generate appropriate step patterns.

The inductive approach, which begins with observed gaits and attempts to
understand how these are generated, has a longer history. It has been applied to both
physiological and functional control mechanisms. The latter application uses models
on an abstract level to try to understand the control principles underlying walking
(e.g. Box 3). The ultimate goal is a comprehensive model which provides a quantitative formulation of the control algorithms for specifying leg movements to achieve a
particular behavioral performance. This is the approach followed here. The former
application attempts to synthesize physiological information on neural properties and
connectivity. The goal is to understand the neural mechanisms producing the observed pattern of muscle activation.

Only initial steps have been taken toward these goals. In the absence of a
comprehensive model, attention has focussed on the generation of step rhythms.
Models at the behavioral or functional level begin with coordination rules deduced
from observations of walking insects. The model serves as a kind of shorthand for the
behavioral description. For example, the observation that swings of adjacent legs
tend not to overlap is used to postulate an inhibitory influence between the step pattern generators of adjacent legs. Such models can be used to test the completeness of
hypotheses derived from behavioral data or, where relevant parameters cannot be
precisely measured, evaluate the consequences of assuming different values for such
parameters. This is the method used here and in numerous earlier models (review
Cruse and Graham, 1985; Graham, 1985).

Models of the physiological mechanisms explicitly consider the neural origin
of rhythmic stepping. They require information on neuronal properties and interconnections. Sufficient physiological information is available to construct realistic
neural models for some rhythmic behaviors but not for walking. Efforts in this direction have been limited to the demonstration that simple arrangements of two or
more neurons or neuron pools can produce alternating activity, which is taken to represent swing and stance (e.g. Brown, 1911, Wilson and Waldron, 1968) or a simple
step cycle (Szkeley, 1968).

3. The Control of Leg Movement in the Stick Insect
3.1. The Nature of the Step Pattern Generator for Individual Legs

Numerous investigators have shown that different legs may step with different frequencies (e.g. von Holst, 1935; Wendler, 1964; Foth and Bässler, 1985). Therefore, it appears that each leg possesses its own step pattern generator. Coordination of the legs, the focus of this paper, represents the coordination of these step pattern generators. Lower levels of control, which organize the movements of the different joints within a leg, and higher levels of control, which govern walking speed and direction, will not be considered.

An important issue for neurobiologists was the question of what anatomical elements participate in the step pattern generator. Simple neuronal models have often been advanced with the implication that central neural circuits can generate motor activity patterns sufficient for walking. Autonomous central pattern generators (CPGs)--neural circuits in the central nervous system (CNS) which can produce much or all of the normal motor output in the absence of patterned sensory input--have been characterized for many rhythmic behaviors (review Delcomyn, 1980). A central pattern generator for walking with this degree of autonomy has not yet been demonstrated conclusively.

Abstract descriptive models derived from behavioral observations do not explicitly consider the identity of the step pattern generator (e.g. Box 3). When applied to normal step patterns, models of this kind often make no formal distinction between an autonomous CPG and one dependent on feedback from the periphery (peripheral oscillator models). However, several control tasks, such as responding to external disturbances or maintaining gait continuity in starting and stopping, are more easily handled within the framework of a peripheral oscillator model. Most recent behavioral results have emphasized the role of sensory information in modulating the step pattern (review Bässler, 1987; Cruse, 1990).

These experiments show that the step pattern generator includes the peripheral sensory and motor elements. Changing the afferent input, by either altering the external conditions or manipulating leg proprioceptors, leads to changes in the spatial and temporal pattern of stepping. The step pattern generator is affected by the position of the leg and by the load it experiences. The force developed by the leg also reflects the load. Thus, leg movement is not merely the expression of endogenous activity in the CNS; the neuromuscular system of the leg itself is an important element in the step pattern generator. Further experiments suggest that while information on position and load is used to determine the transitions between swing and stance, the control system assumes the characteristics of a velocity controller in carrying out each movement (Dean and Cruse, 1986; Weiland and Koch, 1987).

These results, plus the absence of a robust CPG, have led most investigators to include the peripheral elements, both sensory and motor, as an integral part of the pattern generator and therefore an important factor in the interactions between leg controllers. One consequence is that state variables such as leg position have been explicitly included in many recent models. A second consequence relates to possible analytical treatments. At the current level of understanding, a leg can be in one of two discrete states, stance or swing, in which different coordinating mechanisms are active. This formulation does not lend itself to simulating the step rhythm with smooth functions. In contrast, the neural correlates of these states, the activity in populations of neurons, make state transitions which are rapid but not instantaneous. As discussed below, the introduction of force as a state variable in models may create a continuous variable at the behavioral level and provide a bridge between these two viewpoints.

3.2. The Nature of the Mechanisms Mediating Coordination of the Legs

The ability of the step pattern generator to vary its activity in accord with the local state of the leg establishes a non-neural link between the legs. Through the mechanical coupling, the action of each leg affects the load on the other legs and this, in turn, can activate intraleg mechanisms which influence force development and step pattern.

However, the mechanical influences experienced by one leg are not tightly linked to specific actions by other legs, so they do not provide good information for global coordination. In fact, insects continue to produce coordinated stepping when mechanical coupling is reduced, as in supported walking on a treadwheel, or eliminated, as in supported walking on a slippery surface. This finding demonstrates the existence of coordinating mechanisms mediated by the central nervous system. These mechanisms are the focus of the model studies described here. However, intraleg mechanisms responding to the current mechanical state continue to play an important role in leg coordination, as shown below. Some intersegmental mechanisms mediated indirectly by mechanical coupling are equivalent to influences transmitted through the CNS, particularly during slow walking.

Neurally mediated coordinating mechanisms appear to act only between adjacent ipsilateral legs and between the contralateral legs of each segment. Experimental results for the stick insect have identified six different mechanisms contributing to coordination among the legs either by influencing the step end-points (the anterior extreme position or AEP and the posterior extreme position or PEP) or by modulating the force exerted during stepping (reviews Graham, 1985; Cruse, 1990). The position and timing of the transition from stance to swing is crucial for maintaining adequate stability because this transition ends the support phase of the leg. The most important timing influences affect this transition (Fig. 1).

The three ipsilateral influences are asymmetric: within each pair of legs, one is the controlling leg or sender and the other is the controlled leg or receiver. One

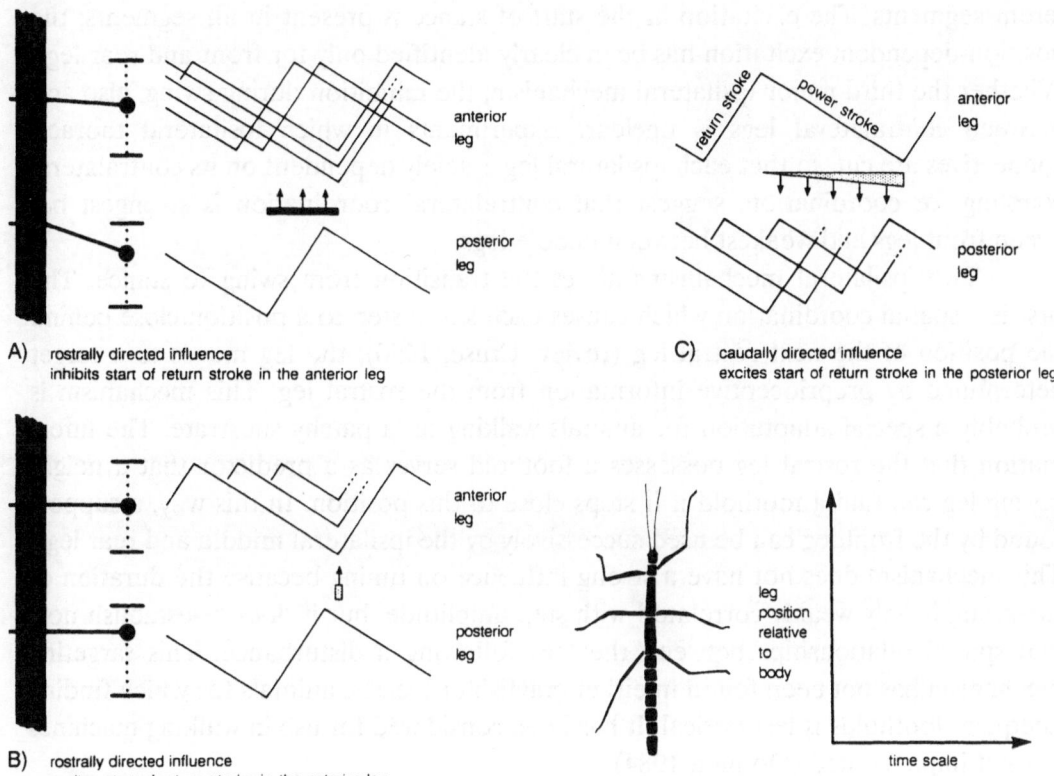

A) rostrally directed influence
inhibits start of return stroke in the anterior leg

C) caudally directed influence
excites start of return stroke in the posterior leg

B) rostrally directed influence
excites start of return stroke in the anterior leg

Figure 1. Coupling mechanisms between adjacent ipsilateral legs of the stick insect measured in behavioral experiments. The range of movement of the two legs is illustrated at the left. The single trace in each part plots the movement of the controlling leg against time; upward change corresponds to forward movement of the leg. The multiple traces illustrate the way the influence modifies the movement of the controlled leg for different initial configurations. The duration and intensity of the influences are indicated qualitatively by the length and thickness of the bar or wedge: solid and open figures represent inhibition and excitation, respectively.

influence passes from front to rear. The farther the controlling leg retracts, the more it excites the adjacent caudal leg to begin a swing (Fig. 1c). Two others are rostrally directed. When the controlling leg is in swing, it inhibits the adjacent rostral leg from beginning a swing (Fig. 1a). When the controlling leg has completed its swing and begins active retraction, it excites the rostral leg to begin a swing (Fig. 1b). In the free-walking animal, both these rostrally directed influences are probably augmented by local mechanisms in the controlled leg: intrinsic responses to the load changes presumed to result from the actions of the controlling leg act in the same direction as these influences mediated through the CNS.

Contralateral coordination is weaker and less easily measured. The contralateral mechanisms have the same form as the ipsilateral mechanisms, but the interactions are symmetric. Both excitatory mechanisms identified for ipsilateral leg pairs also act within contralateral leg pairs. The strength of the coupling varies in different segments. The excitation at the start of stance is present in all segments; the position-dependent excitation has been clearly identified only for front and rear legs. Whether the third major ipsilateral mechanism, the inhibition during swing, also acts between contralateral legs is unclear. Experiments in which ipsilateral thoracic connectives are cut, so that each ipsilateral leg is solely dependent on its contralateral coupling for coordination, suggest that contralateral coordination is strongest between front legs and weakest between middle legs.

Two ipsilateral mechanisms affect the transition from swing to stance. The first is a spatial coordination which causes each leg to step to a position close behind the position of the next rostral leg (review Cruse, 1990); the leg moves to a target determined by proprioceptive information from the rostral leg. This mechanism is probably a special adaptation for animals walking on a patchy substrate. The information that the rostral leg possesses a foothold serves as a predictor that a neighboring leg can find a foothold if it steps close to this position. In this way, a support found by the front leg can be used successively by the ipsilateral middle and rear legs. This mechanism does not have a strong influence on timing because the duration of the swing is only weakly correlated with step amplitude, but it does re-establish normal spatial relationships between the legs following a disturbance. This targeting mechanism has not been found in either crayfish or locusts, animals for which finding adequate footholds is less critical. It has been considered for use in walking machines but not implemented (Donner, 1984).

The second influence on the transition from stance to swing serves to correct errors in leg placement. If a leg steps onto its rostral neighbor, then the tactile input from the rostral leg triggers a reflex causing the caudal leg to lift and step slightly to the rear (Graham, 1979).

A final influence affects the force generated during the stance. When the controlling leg experiences a greater resistance and increases the force of its power stroke, then it also excites adjacent legs to exert more force. This interaction is active in both directions in all pairs of adjacent legs except the two rear legs.

4. The Structure of the Model
4.1. The Nature of the Step Pattern Generator for Individual Legs

To test the sufficiency of the mechanisms described above, the four timing influences were combined in a computer model which will be described in detail elsewhere. Here the model will be described briefly and then used to illustrate several features of the coordinating mechanisms.

In constructing a model (Box 3), it is first necessary to select decision rules for switching between swing and stance. Like other kinematic models with an explicit peripheral referent, the present model uses leg position for this purpose. The model

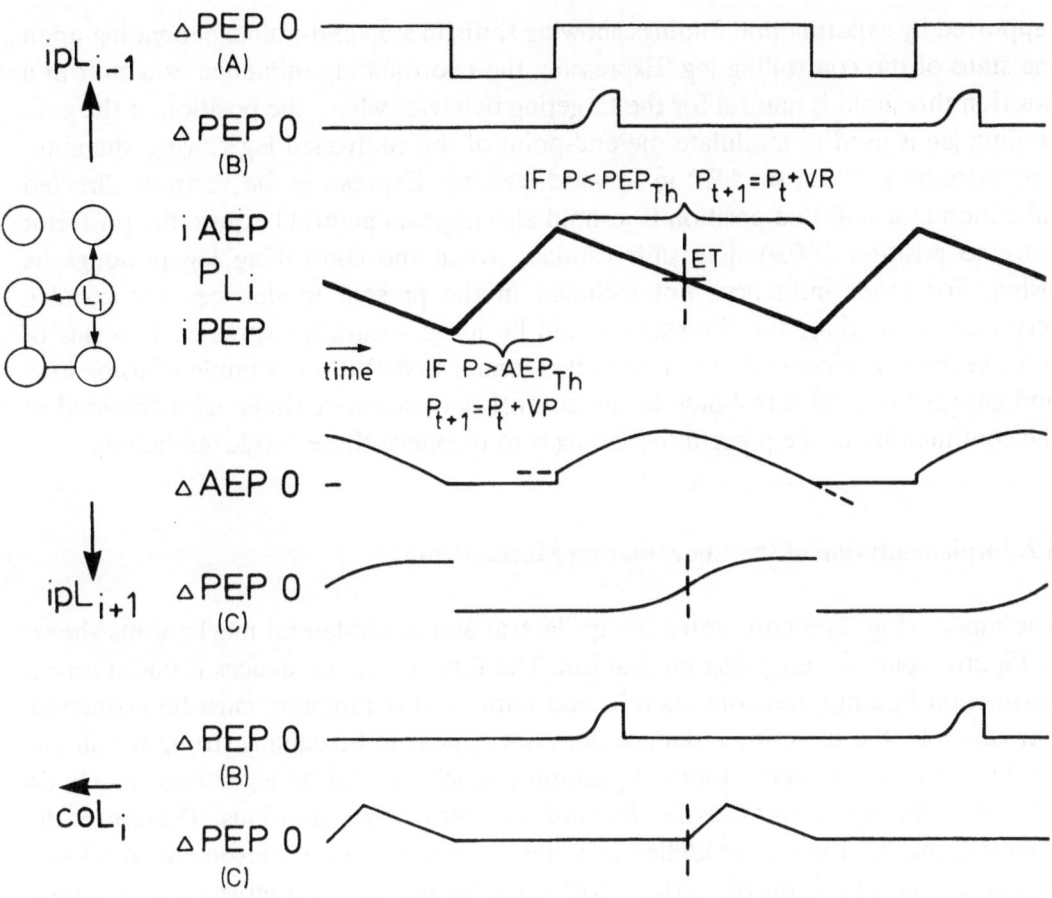

Figure 2. The step pattern generator for each leg, illustrated here for leg Li, is simulated as a relaxation oscillator in which the variable corresponds to leg position (P) and the two states correspond to stance and swing. In successive time steps, the leg position is compared with the current threshold to determine whether the leg should continue moving in the same direction or reverse direction; then the appropriate velocity input (VR, VP) is added. All six step pattern generators have the same intrinsic thresholds for ending stance (iPEP) and swing (iAEP). Each leg sends signals of different types to adjacent ipsilateral (ipL) and contralateral legs (coL). These signals depend upon the state, the leg position and the velocity inputs of the sending leg; they additively change the indicated thresholds of the receiving leg. The letters in parentheses denote the corresponding behavioral effects shown in Figure 1. The position ET, which determines when the position-dependent influences are neutral, moves rostrally with increasing retraction speed. The dashed lines for the influence on AEP indicate the course of the effect for stance positions rostral to the iAEP and caudal to the iPEP.

leg functions as a relaxation oscillator with threshold positions defining the endpoints of the two movements (Fig. 2). One simplification is the use of constant velocities during swing and stance to replace the forces actually developed by the muscles. This makes the transitions between phases instantaneous and not smooth. Coordinating influences are expressed as changes in the thresholds. This choice of state variable has the advantage that leg position is more easily measured than load. It is supported by experimental findings showing shifts in step end-points depending upon the state of the controlling leg. Expressing the coordinating influence as a shift in a position threshold is natural for the targeting behavior where the position of the controlling leg is used to modulate the end-point of the controlled leg's swing, the anterior extreme position or AEP in forward walking. Expressing the rostrally directed inhibition as a shift in a position threshold also appears natural because the posterior extreme position (PEP) may shift caudally when the controlling leg prolongs its swing. For other influences not included in the present model, e.g. the intraleg responses to loading, the threshold would be more naturally expressed in terms of force vectors. The control system used by Donner (1984) is an example of using load and changes in load thresholds as the control parameter. A challenge addressed in the continuation of the present modelling is to reconcile these two formulations

4.2. Implementation of the Coordinating Mechanisms

The model (Fig. 2) incorporates the ipsilateral and contralateral mechanisms shown in Figure 1 plus the targeting mechanism. The form of the influences is based on experimental findings, but some details and numerical parameters must be estimated. For example, the targeting influence does not appear to be equally strong for all target leg positions: extreme forward positions, which the caudal leg cannot reach, do not cause the leg to move as far forward as intermediate positions. Therefore, the targeting mechanism was modelled as a linear combination of movement to a fixed, slightly forward position when the target leg is far forward and movement to a position a constant distance behind the target leg when this leg is at or caudal to its mean PEP.

    The rostrally directed inhibition (Fig. 1a) is represented by a step change in the PEP threshold. The size of this step cannot be determined from the experimental data because the disturbance involved in physically blocking the swing of one leg probably influences the behavior of the other legs. The size of the step determines the efficacy of the inhibition. Intuitively, it needs to be large enough that if two ipsilateral legs simultaneously approach their intrinsic PEP thresholds and the caudal leg begins its swing first, then the inhibition should keep the rostral leg in stance until this swing is completed. During this interval the rostral leg is compelled by the constant stance velocity in the model to continue moving to the rear. It appears that the inhibition active in fast walking must be nearly equal to the full step amplitude, but in fact less is required. The PEP change measured in the animal will be reduced because a real leg will begin to resist the forward movement and force a decrease in the retraction velocity. This effect is ignored in the present model. Furthermore, the threshold change required for adequate inhibition can be reduced in both model and insect because the excitation from the rostral leg will normally cause the caudal leg to advance its swing.

The strength of the other influences is also incompletely determined by the experimental results. The rostrally directed excitation (Fig. 1b) and the corresponding contralateral excitation are thought to be coupled to the beginning of active retraction in the controlling leg. In the stick insect, but not in the current model, the active retraction may begin with a delay after the leg steps onto the ground. To simulate this interval the influence used in the model includes a delay inversely proportional to retraction velocity.

The caudally directed influence (Fig. 1c) and the corresponding contralateral excitation present greater uncertainties. This mechanism should delay or advance the start of a swing by the caudal leg so that the swing is completed about the time that the rostral leg reaches its intrinsic PEP threshold. Therefore, the position of the controlling leg for which the effect is neutral depends upon the retraction velocity: when the insect walks faster, the swing must be triggered at more rostral positions of the controlling leg. Under the simplifying assumption of constant retraction and protraction velocities, the PEP of the controlled leg required for the leg to arrive at its AEP just as the controlling leg reaches its PEP can be determined as a function of the retraction velocity and the position of the controlling leg. The relation is a line between the following two configurations of controlled leg (the receiver, r) and controlling leg (the sender, s): (AEPr, PEPs) and (PEPr, POs), where PO is given by the distance the controlling leg retracts during a full-length swing of the controlled leg. (This line follows the margin of the shaded triangles in Figure 4a and continues to the AEP of the controlled leg.) For slow retraction speeds, POs is near the PEPs; for equal protraction and retraction speeds, it is at the AEPs. An excitatory coupling following this minimal requirement, similar to the contralateral influence in Figure 2 or 4a, does not enforce strict alternation; it allows quite asymmetric phase relations because some leg configurations are not modified. The amount of the allowable asymmetry increases with decreasing walking speed.

For the symmetric coupling existing between contralateral legs, strict alternation can be achieved by letting PO approach the midpoint between AEP and PEP, rather than the PEP, as walking speed decreases. This change means that the mechanism begins to excite a swing in the controlled leg at more rostral positions of the controlling leg. The contralateral phase values observed in the stick insect lie between symmetric alteration and the extreme asymmetry allowed by the minimal form required to prevent overlapping swings.

The effects of using different forms for the threshold shift have also been investigated. The occurrence of in-phase steps by contralateral legs and the symmetry of the coupling suggest that the threshold shift for the controlled leg is never greater than the distance of the controlling leg from its own PEP. Thus, the threshold shift must have a maximum somewhere between the AEP and the PEP of the controlling leg.

In contrast, the coupling between ipsilateral legs is asymmetric and in-phase steps by adjacent legs are rare. Therefore, the relation between the threshold shift and the position of the controlling leg presumably differs from that of the contra-lateral mechanism. In particular, the excitation from the controlling leg appears to increase monotonically as the leg moves farther to the rear. To prevent this influence totally dominating other influences, the ipsilateral relation is given a sigmoid form (Fig. 2).

Ipsilateral phase relations typically show a single peak corresponding to asymmetric stepping. Therefore, as a first approximation, the relation between neu-tral position and retraction speed of the controlling leg can be represented by the minimum necessary to prevent overlapping swings. This is not completely realistic because there is a delay, during which both legs retract together, between the end of the caudal leg's swing and the start of the rostral leg's swing. Moreover, the quantita-tive relation between ipsilateral lag intervals and step period presents further compli-cations which can be approximated but not completely resolved within the framework of the present model (see below).

5. Step Patterns Produced by the Model
5.1. General Features

The present model combining several coordinating influences qualitatively repro-duces the temporal and spatial organization of step coordination in stick insects (Fig. 3). Unlike models based on a single mechanism (e.g. Graham, 1977; Cruse, 1979), the present model does not require a hierarchy of intrinsic rhythms in the segmental step pattern generators. It rapidly assumes a stable coordination from any starting configuration in a natural way. The coordination of ipsilateral legs is charac-terized by metachronal sequences of steps progressing from rear to front. When the contralateral position-dependent excitation uses a speed dependence like that used ipsilaterally, it does not enforce symmetric alternation. Histograms of contralateral phase values reveal a broad band of permitted phases. The width of this band can be reduced by letting the excitation begin at more forward positions of the controlling leg. Phase relationships observed in the animal suggest that the actual mechanism is somewhat more restrictive than the minimum requirement but not sufficiently restrictive to enforce symmetric alternation. Nevertheless, the model uses a similar speed-dependence for both ipsilateral and contralateral coupling on the grounds that the physiological mechanisms would be simpler if both influences share a common source. One consequence of this arrangement is a tendency for diagonal leg pairs to step nearly simultaneously, a coordination which has often been observed but is not a fixed rule (Graham, 1972; Hughes, 1952). Here, this pattern results from diagonal leg pairs being subject to a common influence from a third leg, as in the model of Gra-ham (1977); it does not represent a coordinating mechanism connecting legs in diag-onal pairs, as postulated by Cruse (1980).

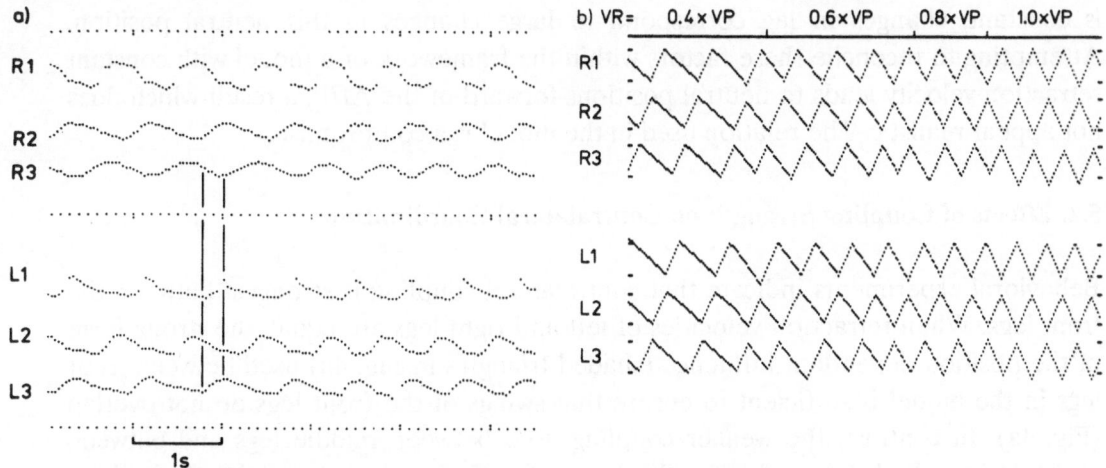

Figure 3. Comparison of the step patterns produced by the model (b) with those of the stick insect (a). Each trace represents the position of a leg; the legs are designated as left or right and numbered from front to back. Upward change in a trace corresponds to forward movement of the leg. The results for the simulation show step patterns for several different retraction velocities (VR), expressed as ratios of the protraction velocity (VP).

Two differences between model and insect are worth mentioning. First, the simulated leg movements have the sharp, saw-tooth appearance characteristic of a relaxation oscillator, whereas the steps of the insect are more rounded. The latter correspond to rhythmic changes in forward velocity such that the velocity is slower when one or more legs are near a transition between swing and stance and faster when legs are in the middle of their retraction (Fig. 3b). The reduction in speed at the transitions may simply reflect low-pass characteristics of the skeletomuscular system, but it could also reflect neural delays necessary for processing local and inter-segmental coordination signals.

The second, related difference is that the natural speed dependence of the lag between steps by adjacent ipsilateral legs is only approximately duplicated by the model: the lag does not increase sufficiently for slow speeds. In the adult stick insect, the speed dependence lies between that of constant lag and that of constant phase. Swing duration is virtually independent of step period (Wendler, 1964; Graham, 1972), so the speed dependence primarily affects the duration of the stance overlap after the caudal leg completes its swing. Because retraction velocity is reduced during this interval, its duration has a relatively small effect on the spatial configuration of the legs: the mean AEP and PEP do not vary measurably with velocity. In the model, the neutral position for the ipsilateral position-dependent mechanism must change in order to allow for variable amounts of stance overlap. Because the retraction velocity is constant, changes in lag correspond to large changes in this neutral position. Attempting to reconcile these factors within the framework of a model with constant retraction velocity leads to neutral positions forward of the AEP, a result which does not appear realistic. The relation used in the model is a compromise.

5.2. Effects of Coupling Strength on Contralateral Coordination

Behavioral experiments indicate that contralateral coupling is strongest between the front legs. When retraction velocities of left and right legs are equal, the strong form of the position-dependent influence (shaded triangles in Fig. 4a) used between front legs in the model is sufficient to ensure that swings of the front legs do not overlap (Fig. 4a). In contrast, the weaker coupling used between middle legs and between rear legs (e.g. shaded areas in Fig. 4b) does not suffice to prevent overlapping swings. However, the ipsilateral position-dependent mechanism, because it modulates the transition from stance to swing over a wide range of leg configurations, provides a tight ipsilateral coupling and imposes the alternation of the front legs on the other leg pairs. In order to reveal the effect of contralateral coupling in the other segments, the normal coordination must be perturbed by either placing the legs in unusual configurations or varying the retraction velocity of one or more legs.

One unnatural configuration is with symmetric positions of left and right legs in each segment. When a stick insect starts from such a configuration, segmental leg pairs often make several in-phase steps before switching to alternation. The model cannot produce sustained in-phase stepping unless the contralateral position-dependent influence is weaker than that of Figure 4a.

How the stick insect recovers from in-phase stepping is not well-studied. Here the model can generate predictive hypotheses. For example, when the hierarchy of coupling strengths is such that the contralateral coupling is strongest at the front, then the recovery begins at the front and propagates to the rear (Fig. 5a). Recovery is gradual in the example shown because the metachronal sequence facilitated by the ipsilateral configurations prevents the contralateral influence between the front legs from effecting a correction within one step. Nevertheless, alternation is achieved within a few steps. To allow still longer sequences of in-phase stepping, the strength of the contralateral coupling must be reduced. If the form of the influence is slightly modified or the amplitude is allowed to vary, then the switch from in-phase to alternate stepping in the front legs can occur abruptly in a manner often seen in the animal.

When the contralateral coupling is strongest at the rear, then alternation is established more slowly but the pattern is the same (Fig. 5b). The ipsilateral mechanisms used here again cause the recovery to depend upon the direction of change in the front legs. Recovery is slower because the coupling between the front legs is weaker. The strong coupling between the rear legs may mean that this pair achieves alternation about the same time as the front legs and before the middle legs.

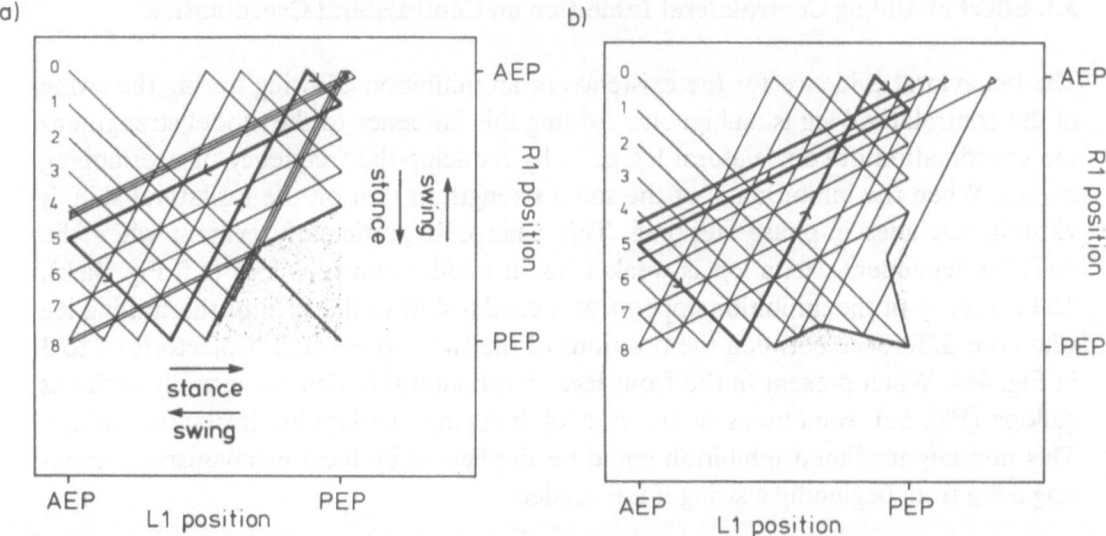

Figure 4. Phase-plane representation of the coordination of a single contralateral pair of legs. The figure illustrates the effect of combining the position-dependent excitation with inhibition during the swing. In this simulation, one leg begins at its intrinsic AEP and the other leg at one of 9 positions from AEP to PEP. The ensuing step cycle is followed until the first leg returns to its AEP in order to measure the change in the spatial configuration produced by the coordinating mechanisms. Stable coordinations correspond to closed trajectories--those which return to the starting point; these are not stable attractors because the mechanism is neutral for some configurations. For leg configurations in the shaded areas, the position-dependent excitation from the controlling leg (the trailing leg--that farther from its intrinsic PEP), shifts the PEP threshold for the controlled (leading) leg rostrally to the margin of the shaded area. This influence advances the step of the controlled leg in order to reduce (b) or avoid (a) swing overlap. The strong form alone (a) is sufficient to prevent overlapping swings and adding inhibition during the swing of the leading leg has no effect. Weaker coupling (b) does not excite a swing soon enough to prevent overlapping swings. Adding inhibition prevents overlapping swings by forcing caudal shifts in the PEP of the trailing leg for some starting configurations (e.g. trajectories 1-3).

The strength of the contralateral coupling between the middle legs appears less important. Keeping this coupling weak simplifies the use of an additive threshold computation because the middle leg PEP threshold is also subjected to three ipsilateral influences whereas the front leg PEP is subjected to only two and the rear leg PEP to just one.

The same qualitative result occurs when the contralateral coupling is measured by letting the legs of the two sides step with different frequencies. When the contralateral coupling is strongest at the front, the front legs make fewer overlapping swings than the middle and rear legs and vice versa.

5.3. Effect of Adding Contralateral Inhibition on Contralateral Coordination

The behavioral evidence for the existence of an inhibition of swing during the swing of the contralateral leg is ambiguous. Adding this influence to the model strengthens the coordination of contralateral leg pairs by reducing the frequency of overlapping swings. When the inhibition is of the same strength as that on the ipsilateral side, it virtually excludes in-phase stepping. This change is particularly evident when the position-dependent excitation is weaker, as in middle and rear legs (c.f. Fig. 4a,b). The presence of the inhibition appears as a caudal shift in the PEP of the trailing leg when the difference between the positions of the two legs is small (trajectories 1 to 3 in Fig. 4b). When present in the front legs, contralateral inhibition virtually excludes gallops (Fig. 5c), sometimes at the cost of inducing overlapping ipsilateral swings. This neurally mediated inhibition could be duplicated by local mechanisms preventing a leg from beginning a swing if it is loaded.

6. Conclusion

In the introduction, several tasks were described which a successful walking machine must solve. The most basic is the provision of adequate support. For a large, heavy machine, falling is a failure and intensive effort must be invested to achieve fail-safe operation. For small animals, falling may have less serious consequences; therefore, the control system may accept a greater risk of falling in return for advantages in other respects. Both insects and smaller mammals often do stumble and fall, particularly when moving fast. The stick insect, the subject of the current study, probably is an animal for which falling is a constant risk and carries a high penalty. Crayfish use a different control strategy (see Müller, this section; Cruse, 1990) which appears appropriate where the consequences of inadequate support are less serious. The tolerance of such failures must be born in mind when considering the function of biological control systems and their possible use as models for technical systems.

Nevertheless, the control principles used by insects offer several advantages. One principle is the generation of an appropriate coordination through a combination of several mechanisms. Stability is obtained by combining local mechanisms intrinsic to each leg controller with coordination mechanisms mediated by the central nervous system. The local mechanisms appear necessary because the algorithms for the distributed control reduce but do not totally exclude the possibility that instabilities will occur, for example, through overlapping swings by adjacent legs. A second principle is that the central coordinating mechanisms themselves are redundant: for example, several different mechanisms encourage alternate stepping in adjacent legs. Some mechanisms, like the forward directed inhibition, serve as checks to prevent

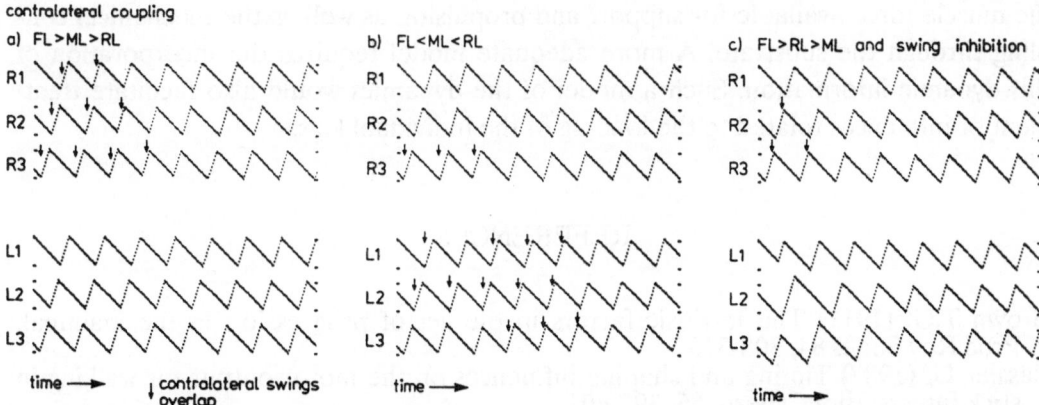

contralateral coupling

a) FL>ML>RL   b) FL<ML<RL   c) FL>RL>ML and swing inhibition

R1   R2   R3   L1   L2   L3

time ⟶   ↓ contralateral swings overlap   time ⟶   time ⟶

Figure 5. Simulations of walks beginning with bilaterally symmetrical leg positions. This configuration leads to in-phase stepping (gallops). For the ipsilateral coupling used in the present model, the switch by contralateral legs from overlapping swings, marked by arrows, to alternation is more rapid when the contralateral coupling is strongest at the front (a). The strength of the contralateral coupling is expressed as a fraction of the minimum form required to prevent overlapping swings (e.g. Fig. 4a); the values for front, middle and rear legs were as follows: a) 1.0, 0.5, and 0.25; b) 0.25, 0.5, and 1.0, and c) 1.0, 0.25, and 0.5 with inhibition during the swing of the contralateral leg.

impending instabilities. Others, like the two excitatory mechanisms, are promotive in the sense that they establish leg configurations which are unlikely to lead to instabilities.

Some of these mechanisms have been incorporated into more recent, decentralized controllers for walking machines. Acknowledging a debt to the biological literature, Donner (1984) implemented a control system with rostrally directed inhibitory and excitatory influences for ipsilateral coordination. The resulting step patterns possess several features characteristic of insect walking but do show some irregularities. The present kinematic model demonstrates that the coordinating mechanisms identified in the stick insect suffice to produce robust and stable walking patterns. In particular, the caudally directed position-dependent influence, which modulates stepping over a wide range of leg configurations, provides a good, predictive mechanism for ipsilateral coordination. The elegance of walking machines can presumably be improved by incorporating these additional mechanisms.

However, the results also show the limitations of a kinematic model. Not all of the behavioral findings can be explained and several mechanisms are not easily incorporated. The mechanisms inhibiting a swing, both local and intersegmental, require a certain plasticity in the motor pattern so that modulations in step timing can be accommodated within the physical constraints on leg position and force. In the animal, this plasticity is provided by changes in velocity which reflect changes in the muscle force available for support and propulsion as well as the mechanical coupling through the substrate. A more adequate model requires the incorporation of this dynamic information. Such a model of the dynamics would also facilitate treatment of influences related to the loading of the individual legs.

REFERENCES

Brown T.G. (1911) The intrinsic factors in the act of progression in the mammal. Proc.Roy.Soc.B 84, 308-319

Bässler U. (1987) Timing and shaping influences on the motor output for walking in stick insects. Biol.Cybern. 55, 397-401

Cruse H. (1979) A new model describing the coordination pattern of the legs of a walking stick insect. Biol.Cybern. 32, 1-7

Cruse H. (1980) A quantitative model of walking incorporating central and peripheral influences. II. The connections between the different legs. Biol.Cybern. 37, 137-144

Cruse H. (1990) What mechanisms coordinate leg movement in walking arthropods? TINS 13, 15-21

Cruse H., Graham D. (1985) Models for the analysis of walking in arthropods. In: B.M.H.Bush, F.Clarac (eds) Co-ordination of motor behaviour: 283-301 SEB Seminar 24. University Press, Cambridge

Dean J., Cruse H. (1986) Evidence for the control of velocity as well as position in leg protraction and retraction by the stick insect. In: H.Heuer, C.Fromm (eds) Generation and modulation of action patterns: 263-274. Exp.Brain Res.Series 15. Springer, Heidelberg

Delcomyn F. (1980) Neural basis of rhythmic behavior in animals. Science 210, 492-498

Donner M.D. (1984) Control of walking: local control and real time systems. PhD Thesis. Carnegie-Mellon University, Pittsburg

Foth E., Bässler U. (1985) Leg movements of stick insects walking with five legs on a treadwheel and with one leg on a motor-driven belt. II. Leg coordination when step-frequencies differ from leg to leg. Biol.Cybern. 51, 319-324

Graham D. (1972) A behavioral analysis of the temporal organisation of walking movements in the 1st instar and adult stick insect (Carausius morosus). J.Comp.Physiol. 81, 23-52

Graham D. (1977) Simulation of a model for the coordination of leg movement in free walking insects. Biol.Cybern. 26, 187-198

Graham D. (1979) Effects of circum-oesophageal lesion on the behaviour of the stick insect Carausius morosus. II. Changes in walking co-ordination. Biol.Cybern. 32, 147-152

Graham D. (1985) Pattern and control of walking in insects. Adv.Insect Physiol. 18, 31-140

Holst E. von (1935) Die Koordination der Bewegung bei den Arthropoden in Abhängigkeit von ~ontralen und peripheren Bedingungen. Biol.Rev. 10, 234-261

Hughes G.M. (195   The coordination of insect movements. I. The walking movements of insects. J.exp.Biol. 29, 267-284

McGhee R.B. (1976) Robot locomotion. In: R.M.Herman, S.Grillner, P.S.G.Stein, D.G.Stuart (eds) Neural control of locomotion: 237-264. Plenum, NY

Szekely G. (1965) Logical network for controlling limb movements in urodela. Acta Physiol.Acad.Scien.Hungaricase 27, 285-289

Weiland G., Koch U.T. (1987) Sensory feedback during active movements of stick insects. J.exp.Biol. 133, 137-156

Wendler, G. (1964) Laufen und Stehen der Stabheuschrecke <u>Carausius</u> <u>morosus</u>: Sinnesborstenfelder in den Beingelenken als Glieder von Regelkreisen. Z.vergl.Physiol. 48, 198-250

Wilson D.M., Waldron I. (1968) Models for the generation of the motor output pattern in flying locusts. IEEE Proc. 56, 1058-1064

BOX 2

COMPUTER SIMULATION OF STEP MOVEMENTS AND INTERLEG COORDINATION USING RELAXATION OSCILLATORS

Jeffrey Dean, Uwe Müller, Holk Cruse

The step pattern generator for leg L_i is simulated as a relaxation oscillator with two states, corresponding to stance or retraction and swing or protraction, and two intrinsic thresholds (TH) determining the transitions between states (Fig. 1). The state variable and variable of integration can be interpreted as the movement phase and the position of the leg or as the type and magnitude of activity in circuits within the central nervous system which control leg muscles. In most models of arthropod stepping, interleg coordination mechanisms modify one or both thresholds of the controlled leg (the receiver), as in the example outlined below and the models presented by Dean and Cruse and by Müller (this section), but they may also affect the rate of change during one or both states. These signals depend upon the state, the variable of integration and the velocity inputs of the controlling leg (the sender).

Digital computer models decompose the simulation into a series of discrete time intervals. In successive time intervals, the effective thresholds (TH')

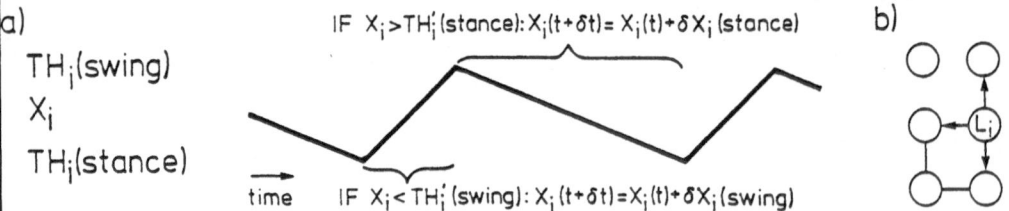

a)
TH_i(swing)
X_i
TH_i(stance)

IF $X_i > TH_i'$ (stance): $X_i(t+\delta t) = X_i(t) + \delta X_i$ (stance)

time   IF $X_i < TH_i'$ (swing): $X_i(t+\delta t) = X_i(t) + \delta X_i$ (swing)

b)

Figure 1. a) The oscillator consists of a state variable (S), a variable of integration (X), two inputs defining the rate of change of X during each state, and two intrinsic thresholds (TH(S)) used in determining the transition from one state to the next. The two inputs have opposite sign but can differ in absolute value. b) Each leg exchanges coordinating signals with other legs: only influences sent to adjacent legs are indicated.

BOX 2

are computed. Then the variable of integration is compared with the effective threshold to determine whether to continue the same state or change states. Finally, the appropriate velocity input is added.

Sample program structure

```
/* Set up and initialize variables. Here, TH(stance) < TH(swing),
so δX(stance) < 0. Define function(s), f_ijk(state, position,
velocity), for coordinating influence(s) k of leg j on leg i */


/* Loop for calculating changes in leg position and state */
REPEAT UNTIL t => t end
     FOR i = 1 to NLEGS
          TH'(S_i) = TH(S_i) + Σ     Σ f_ijk(S_j, X_j, δX_j)
                              j<>i  k

          IF S_i = "stance" THEN
               IF X_i < TH'(S_i) THEN S_i = "swing"
          ELSE
               IF X_i > TH'(S_i) THEN S_i = "stance"
          X_i(t + δt) = X_i(t) + δX_i(S_i)
     END
     t = t + δt
END
```

COORDINATION OF CONTRALATERAL LEGS IN WALKING CRAYFISH

Uwe Müller

Fakultät für Biologie, Universität Bielefeld

Postfach 8640, D-4800 Bielefeld 1, FRG

Abstract

In the experiments a crayfish walked with the legs of the left and right side on sepa-rate motor-driven belts. Even when the belts were driven with different speeds, the animal tried to synchronize the leg movements of both sides. The resulting effect can be described as *relative coordination*. A model calculation is presented which is suffi-cient to describe the interactions between the legs.

1. Introduction

Walking of an animal depends on the coordinated interaction of all involved legs. An analogy exists between the biological phenomenon and a more technical approach, the theory of coupled nonlinear oscillators. In this analogy each leg can be regarded as an autonomous *relaxation oscillator* (von Holst, 1943; Bässler, 1986). Such a relaxa-tion oscillator switches between two well-defined states. In the biological example these two states were the *power* - and the *return stroke*. In forward walking, during the power stroke the leg has ground contact, and moves posteriorly. During the *return stroke* the leg is lifted off the ground and moves anteriorly. After a disturbance of its oscillation such a nonlinear oscillator returns back to a stable *limit cycle* (Pavlidis, 1973).

The movements of different legs have to be synchronized in order to produce a coordinated locomotion. This is the task of neuronal *coordinating mechanisms*. According to the analogy between biology and engineering, the legs, including their interactions, can be regarded as a coupled system of relaxation oscillators. By means of these coupling mechanisms, some sort of information about the *states* of the oscil-lators were interchanged among each other. When all oscillators are on their limit cycles, actions of the coordinating mechanisms were of course not observable, but they can be revealed as soon as disturbances were applied experimentally.

An analysis of the reactions of coupled oscillators to external disturbances can be carried out by evaluation of their *phase response curves* (PRC; Pavlidis, 1973; Stein, 1976). In a PRC the changes in the period of a measured oscillator (y, see Fig.1,

Trace 2) were plotted against the phase, with respect to a reference oscillator (x, see Fig.1, Trace 1). The method used here relies on the possibility to induce walks, in which the legs on the two sides of the animal step with slightly different frequencies. With this method all possible phase relations can be generated and the leg's reactions studied.

2. Results

A crayfish walked with the legs of each side on separate, motor-driven belts. The movements of all eight walking legs were recorded using *position electrodes* as described in Cruse and Müller (1986). Representatively only the interactions of the 4th and 5th leg pairs will be discussed here. The anterior leg pairs generally showed similar but more variable behaviour. Fig.1 shows the recordings of the movements of these legs in parallel to the longitudinal body axis. Slightly different belt speeds induced a shift in the phase relations between contralateral legs from one step to the next. Nevertheless, the animal tried to synchronize the leg movements of both sides in order to produce coordinated stepping. This is done by modulating the amplitude and period of the legs.

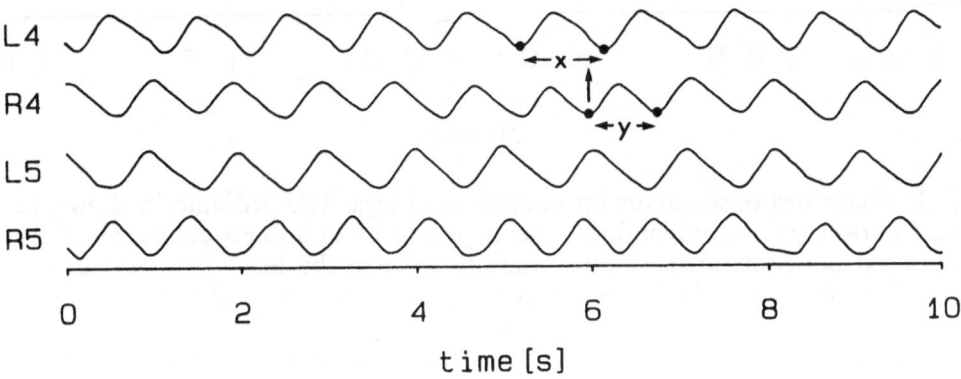

Figure 1. Stepping pattern of legs 4 and 5 of the left (L) and right (R) side of a crayfish. The abscissa is time. Upward deflection indicates a return stroke, downward deflection a power stroke. The period is defined as the time interval from the beginning of one return stroke to that of the next. The phase is defined as the beginning of the return stroke of the measured leg (y) within the period of the reference leg (x). The right belt speed is adjusted 5% faster than the left one.

Fig.2 shows the measured PRCs for L4 measured in R4 and vice versa. In both measurements all phase values are represented, but a preferred mean phase is appa-

rent, as is visible in the density of the measured values. The period of the faster
(right) leg, which is shown in Fig.2b, was decreased, when the phase was above or
below 0.4. When the phase was near 0.4 the period was not influenced. The apparent
modulation of the period led to a proper coordination, though the average stepping
frequency was different for both sides. The leg of the slower (left) side, which is
measured in Fig.2a, showed no deviation from the mean period for any phase. In this
example an extreme unidirectional influence acted from the leg of the slower walking
side onto the faster walking leg of the other side. Normally reciprocal influences,
though with different strengths, acted in both directions. Relative coordination could
be observed only under the experimental conditions described above. When the
more dominant side was walking with the higher frequency, absolute coordination
appeared in a certain range of speed ratio.

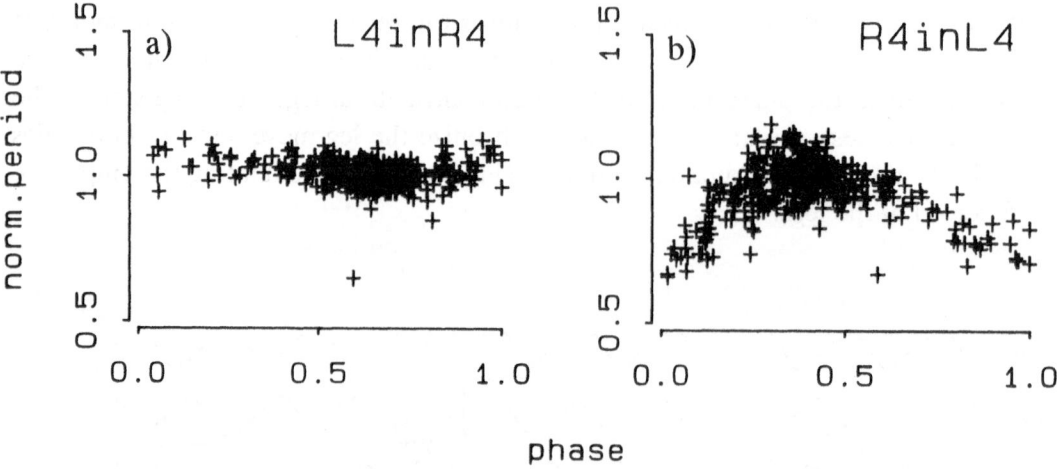

Figure 2a,b. Phase response curves for contralateral legs. The ordinate in shows the
normalized period (measured period : mean period of uninfluenced steps) of the
measured leg dependent on the phase relation between the legs. The ratio of belt
speeds (R:L) is 1.05. In (a) the left leg (L4) is measured and the right leg (R4) is the
reference leg. In (b) the opposite case is drawn. Mean phase in (a): 0.64; in (b): 0.36.
Concentration parameter in (a): 0.60; in (b): 0.56. Mean phases and concentration
parameters were determined using circular statistics (Batschelet, 1982).

Separate measurements of the *durations* of *return* - (RSD) and *power stroke* (PSD)
showed that the described compensation was produced by a spatial modulation of the
anterior extreme position (AEP). This occurs at that time when the leg switches from
return - to power stroke. Correspondingly the position at that time when the leg switches from power - to return stroke is called the *posterior extreme position* (PEP). Fig.3
shows the influence of the reference leg on the measured leg. This can be calculated
directly from the measured PRCs. Thus, depending on the phase, both, RSD and

PSD of the influenced leg will be more or less shortened. Although steps of the measured leg started at very different phase intervals within the period of the reference leg (as shown in the first step), they are concentrated at a fixed phase interval at 0.4 within the next two steps. This leads to *relative* or *gliding coordination*, as described by von Holst (1943).

A model calculation was developed to investigate whether influences as described above are sufficient to stabilize a proper coordination between the contralateral legs. The basic part of the model consists of 4 relaxation oscillators (see Box 3) which oscillate between the limits of 0 and an intrinsic upper threshold $TH(AEP)_i$. Preliminary experiments showed that the RSD is constant, whereas the PSD is a function of the belt speed. This was implemented in the model, by choosing the intrinsic frequencies, based on the slope of the power stroke. The output values of the oscillators, x_j, correspond to leg position. The PRCs were generated as a function of x_j, weighted with a coupling factor, c_{ij}, which determines the strength of coupling. The coupling influences act on each $TH(AEP)_i$ so that it can be described as follows:

(1)
$$TH(AEP)_i = 1 - \sum_{j=1}^{4} PRC(x_i) * c_{ij}$$

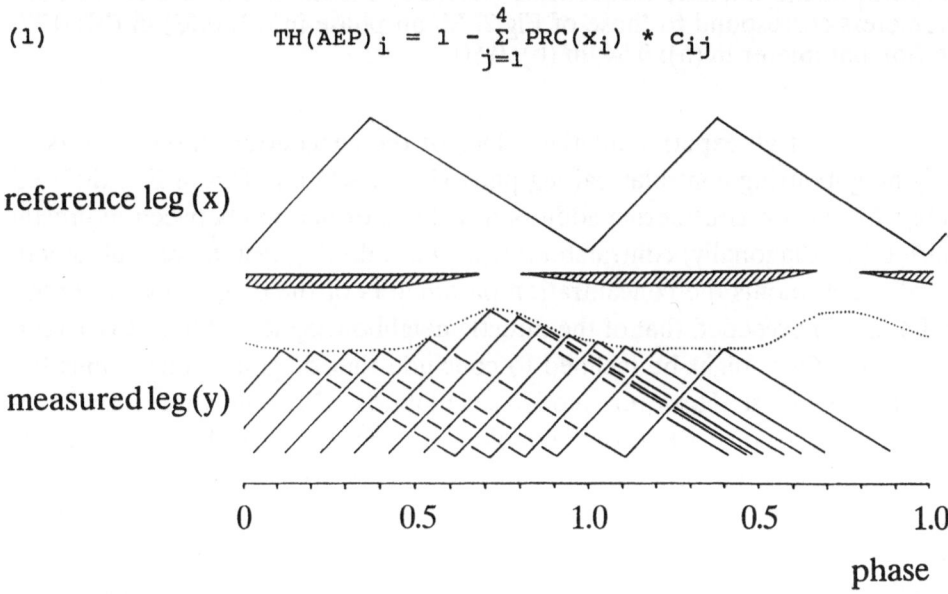

Figure 3. Influence of a dominant oscillator (x) on a subdominant one (y). The hatched area indicates the time interval and strength of the influence.

A simulation of a single experiment is presented in Fig.4. The values of c_{ij} can be adjusted to the experimental results. In the case shown in Fig.4 mean phases and concentration parameters were of the same order of magnitude as those of the experiment shown in Fig.2.

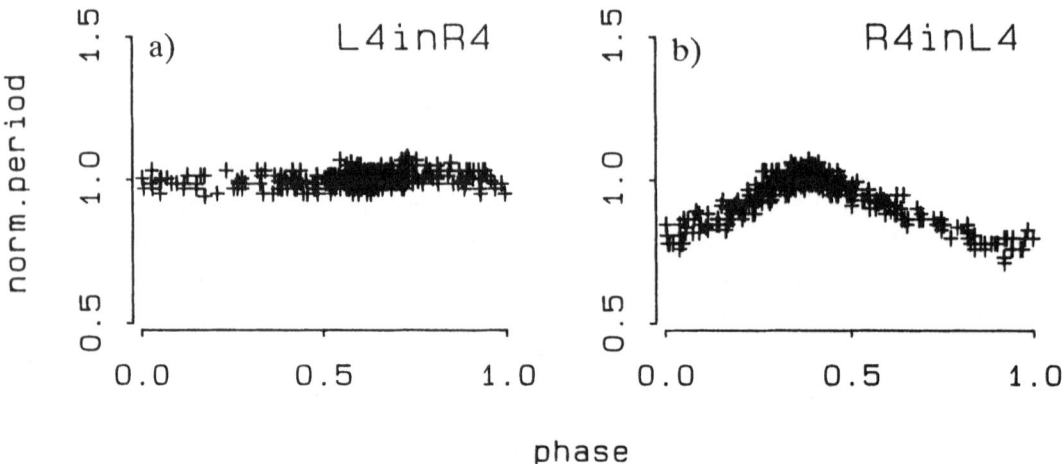

Figure 4a,b. Phase response curves for the simulation of the experiment shown in Fig.2. The ratio of the intrinsic frequencies (OR:OL) for the oscillators is 1.05. All other parameters correspond to those of Fig.2. Mean phase in (a): 0.63; in (b): 0.37. Concentration parameter in (a): 0.54; in (b): 0.50.

In more than 50% of all experiments the values of the concentration parameters of the directly neighbouring contralateral leg pairs (L4 in R4) and (L5 in R5) differed considerably. Moreover, considering additionally the coordination between indirectly neigh-bouring (i.e. diagonally) contralateral legs, the following results were obtained: in 42% of all experiments the concentration parameters of the diagonally neighbouring legs (L4 in R5) exceeded that of the directly neighbouring legs (L5 in R5), as can be seen in Fig.6a. One might be inclined to conclude that diagonal connections between walking legs exist. However, model calculations with only three oscillators showed unexpected results: when only couplings in one direction between directly neighbouring ipsilateral (L4 on L5) and directly neighbouring contralateral (L4 on R4) oscillators were implemented (as is shown in Fig.5a), the concentration parameter of the diagonally neighbouring contralateral oscillators (L5 in R4) was always slightly smaller than that of the directly neighbouring contralateral ones (L4 in R4). When a similar, but weaker coupling in the opposite direction was added (Fig5b, inset), the concentration parameter of the diagonally neighbouring oscillators (L5 in R4) became stronger than the directly neighbouring ones (L4 in R4), as is shown in Fig.5b.

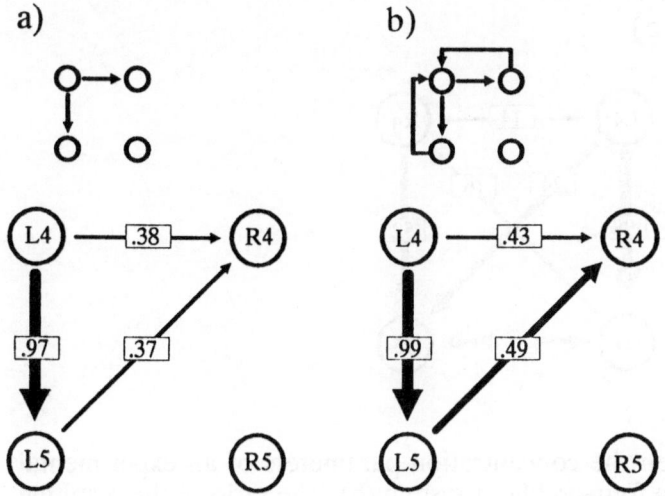

Figure 5a,b. Concentration parameters of a 3 oscillator system (a) with unidirectional and (b) with bidirectional coupling. The ratio of the intrinsic frequencies (R4:L4 and R5:L5) is 1.05. The insets above show the implemented couplings. The additional couplings acting from L5 and R4 to L4 in (b) amount to 20% of the coupling acting from L4 in the opposite direction.

Introducing mutual coupling in a system of four coupled oscillators led to the following results:

I. When the couplings between both front and both rear oscillators were equal, all four contralateral concentration parameters were nearly identical.

II. When, in contrast, the front oscillators were somewhat weaker coupled than the rear ones, then over a wide range of speed ratios the rear contralateral concentration parameter (L5 in R5) had the highest value and was higher than its corresponding diagonal contralateral one (L4 in R5).

III. When however the coupling of the front oscillators was decreased still further, the diagonal contralateral concentration parameter (L4 in R5) had the highest values over a wide range of speeds.

This is shown in a simulation of the experiment of Fig.6a which is presented in Fig.6b. The contralateral coupling influence of L4 on R4 amounted to 20% of the coupling influence of L5 on R5. Strong ipsilateral couplings acted from the front - to the rear oscillators. Additional weak couplings in the opposite direction existed between ipsilateral - and weak mutual coupling between the direct contralateral oscillators. No couplings existed between the diagonal contralateral oscillators. The strength of the measured diagonal concentration parameters in Fig.6a, however, suggests a diagonal connection between L4 and R5.

a)   b)

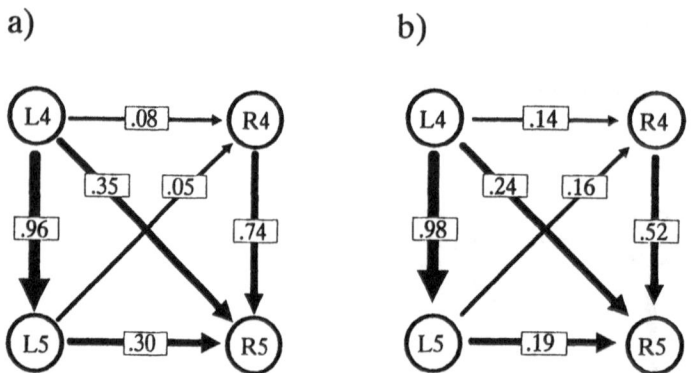

Figure 6a,b. Comparison between the concentration parameters of an experimental walk (a) and a simulation using a four-oscillator system (b). The ratio of the resulting frequencies (R:L) in (a) is 1.25; in (b) 1.18 for (R4:L4) and 1.41 for (R5:L5).

3. Discussion

The results of this paper demonstrate the possibility of *"virtual couplings"*. They can occur in a system, consisting of at least three coupled oscillators, when the two following conditions were fulfilled: there has to be I. a gradient of strength of coupling and II. a bidirectionally coupling between direct neighbouring oscillators. Under this particularly conditions strong coordination between two indirect neighbouring oscillators, with values above the corresponding direct neighbouring oscillators can result, although no direct coupling is apparent. This means that a high concentration parameter is not necessarily based on a corresponding direct influence. As a consequence it is difficult to draw conclusions about the existing coordinating influences between the regarded oscillatory systems on the basis of a purely behavioural analysis. To obtain the most parsimonious hypothesis, the performance of a model calculation is an appropriate method. The hypotheses derived from such a model calculations then can be the subject of tests by means of neurobiological methods.

Acknowledgements: This work was supported by DFG (CR 58/6-2)

REFERENCES

Batschelet E. (1981) Circular Statistics in Biology. London, New York: Academic Press.

Bässler U. (1986) On the definition of central pattern generator and its sensory control. Biol.Cybern. 54, 65-69.

Cruse H. and Müller U. (1986) Two coupling mechanisms which determine the coordination of ipsilateral legs in the walking crayfish. J.exp.Biol. 121, 349-369.

Holst E. von (1943) Über relative Koordination bei Arthropoden. Pflügers Archiv 246, 874-865.

Pavlidis T. (1973) Biological Oscillators: Their Mathematical Analysis. New York: Academic Press.

Stein P.S.G. (1976) Mechanisms of interlimb phase control, pp. 465-487. In R.M. Herman, S. Grillner, P.S.G. Stein and D.G. Stuart (Edit.) Neural Control of Locomotion. New York: Plenum Press.

FLIGHT MANOEUVRES OF LOCUSTS

Wolfram Zarnack, Gabriele Reuse and Thomas Schwenne
1. Zoologisches Institut der Universität, Berliner Str. 28, D-3400 Göttingen

Abstract

Solutions of non-linear partial differential equations are necessary for the calculation of unsteady aerodynamic forces produced by flying animals. There is a new aerodynamic concept (Send 1989) which could solve the unsteady problems of insect flight. According to this concept, nine kinematic parameters of a single wing are aerodynamically important. We will present our way to analyse the kinematic parameters of locust flight.

The flight system of locusts (*Schistocerca gregaria*) integrates several feedbacks inside and outside the animal. When a locust is fastened to any support, outside feedbacks are normally opened. Visual (outside) feedbacks can be closed by means of flight simulators. In roll manoeuvres of locusts performed in slow and fast flight simulators, long flight sequences occured with tonic responses to visual stimuli. But the locusts were able to change between tonic and phasic responses if they had no success with their current intention. Locusts with very stereotype wing-beat performed random flight behaviour too. With slow simulators, the reaction time was always of several seconds. With a quick simulator, the reaction time was less than 0.1 s.

1. Wing movement and unsteady aerodynamics of insect flight

The basic property of actively flying (i.e. not gliding) animals is a more or less regular wing-beat. Therefore, such a flight system can be described as an oscillator. This definition is useful for aerodynamic studies (see below) but not sufficient for physiological research, because flight control and behaviour are effected by changes of the kinematic parameters. An animal's flight system, consisting of wings, articulations, arrangements of the flight muscles etc., is always very complex. To understand the physiological function of its parts, the activation pattern of the muscles and the movement pattern of the wings, it is necessary to know the time-dependent aerodynamic forces, torques and power produced by the wing-beat.

The aerodynamic quantities must be calculated on the base of the actual wing movements. The aerodynamic processes are described by the equation of STOKES-NAVIER, a well known non-linear partial differential equation, a solution of which depends on the boundary conditions. In the present case, these are the velocities at the surfaces of the moving wings. In insects, the boundary conditions are not as complicated as for instance in birds: Not to mention that the wing-beat is almost ideally periodic, the wing surfaces are generally coherent and get less deformation during a wing-beat cycle than those of birds. Therefore, apart from the biological relevance, kinematic studies of insect flight could contribute to solve unsteady aerodynamic problems as zoo-physiological observations have already done (cf. Lighthill 1973).

It depends on the reduced frequency $\Omega^* = 2\pi fc/2u_0$ (c chord length of the wing, f beat frequency, u_0 flight speed) whether unsteady aerodynamic theories have to be applied or not. In insects, the reduced frequency is generally greater than 0.1. Thus, the flight is effected by unsteady aerodynamics. In this case, it is most difficult to solve the equation of STOKES-NAVIER.

In the past, there has been little progress to overcome these difficulties. But there is a new theory for calculating the unsteady aerodynamic forces and torques (Send 1989). In contrast to the results of Küssner (1936), no thrust is genera-ted by a pure wing flap. Basing on a kinematical analysis of Zanker (1987), lift and power of *Drosophila* flight could be calculated with an error of less than 10 %. According to this calculation, about 35 % of the total lift depends on unsteady aerodynamics (Send, in press).

We hope to apply this theory successfully on the flight of locusts. Locusts have two pairs of different wings the shape of which (especially of the hindwings) is periodically more or less changing. The reduced frequency is $\Omega^* = 0.17 \pm 10$ % for the forewings and $\Omega^* = 0.38 \pm 10$ % for the hindwings at f = 20 Hz, $u_0 = 3$ ms^{-1} and medium span.

The wing-beat of each wing is a result of three independent rotative generally non-harmonic oscillations:

(i)   pitching, rotation around the wing's longitudinal axis, i.e. pronation and supination
(ii)   flapping, i.e. up and down movement
(iii)   lagging, i.e. forward and backward movement.

We record the wing-beat of locusts by two methods: (i) A high-frequency stereo-photogrammetry (series of 50 pairs of stereo-pictures at time intervals of 2ms) is used to analyse precisely the absolute positions and shapes of the wings (Zarnack 1972, 1983). This method is restricted to short flight sequences. (ii) For a continuous electrical recording without temporal restrictions, we use a threedimensional inductive recording technique (Koch 1977, Koch and Elliott 1983, Zarnack 1978, 1988, Schwenne and Zarnack 1987, Waldmann and Zarnack 1988).

At the present state of development, the aerodynamic theory of Send can be applied to a plane airfoil oscillating harmonically in three dimensions. It does not consider high frequency terms of the real wing movement of locusts and interferences occuring between fore- and hindwing of one side (Zarnack 1983, Schwenne and Zarnack 1987). In accordance to the theory, we describe the oscillations by the following equations:

$$(1) \qquad \alpha(t) = \alpha_s + \alpha_0 \cos(\Omega t) \qquad \text{(pitching)}$$
$$(2) \qquad h(t) = h_s + h_0 \cos(\Omega t + \tau) \qquad \text{(flapping)}$$
$$(3) \qquad s(t) = s_s + s_0 \cos(\Omega t + \sigma) \qquad \text{(lagging)}$$

(α_s stationary angle of attack; s_s, h_s co-ordinates of the center of oscillation; α_0, h_0, s_0 amplitudes; τ and σ phase angles relative to pitching, Ω frequency)

The nine parameters, α_s, s_s, h_s etc., have different aerodynamic effect. For instance, lift increases with increasing h_0 but decreases with increasing α_0. Thus, even under this rough approximation, the aerodynamic forces of a single wing are influenced by nine independent kinematic parameters. However, great lift and thrust are achieved only in restricted ranges of parameter values. For instance, lift sufficient for level flight is achieved only then if the phase angles τ and σ are in the range of 90°. Therefore, not all arbitrary combinations of values of the kinematic parameters are realized in animal flight. In locusts, we have to consider 18 independent kinematic parameters even in case of symmetric movements on either side during straight flight due to the different shapes and movements of fore- and hindwings.

In flight manoeuvres, the kinematic parameters are generally time-dependent. In *Drosophila*, only few parameters change during visual induced flight manoeuvres (Götz and Wandel 1984) whereas in other diptera, many parameters are varied (Nachtigall 1983). In locusts, certain alterations of some kinematic parameters occur very often, e.g. changes in α_s and α_0 (Zarnack 1988, Waldmann and Zarnack 1988, Reuse and Zarnack 1988). Nevertheless, each of the four wings generally performs its own movement pattern, resulting in 36 independent kinematic parameters which have to be considered for a sufficient description of a locust's flight mode (in preparation).

2. Physiological studies on flight control

One general aim of zoo-neurophysiological research is to understand how an animal's locomotion is controlled by its central nervous system. As mentioned above, the wing-beat of several flying insects, e.g. flies and migratory locusts, is almost ideally periodic and can be elicited under laboratory conditions. For this and other reasons, the flight of these insects has become an important object of zoo-neurophysiological studies.

Some general difficulties in analysing and modeling locomotion and behaviour of arthropods have been already described (Scharstein and Zarnack, in this volume). Further difficulties arise due to the great number of parameters which control neuronal activation and movement patterns. For instance, each of the four wings of a migratory locust is driven by ten muscles, many of which are provided with several independent motor units. In the present contribution, only those problems will be concerned which are due to the experimental conditions used in studies of locust flight behaviour.

2.1 Locust flight system

Fig. 1 gives a general view of the flight system. It is assumed (cf. Wilson 1961, Burrows 1977, Rowell 1988), that a neuronal central pattern generator produces spontaneously or triggered by external stimuli a basic neuronal rhythmic activation pattern. This is modulated in an integrating neuromotoric network by various sensory inputs of proprioceptors, optical sensors and windreceptors (cf. Möhl 1989a). The flight muscles are activated by the excitation pattern of the motoneurons. Due to the muscle contractions, movements and shape deformations of the wings occur. By these, aerodynamic forces and torques are generated which move the whole animal, effecting its attitude and speed in space.

In addition to serial couplings, feedback couplings occur at the following parts of the flight system: (i) The wing movements are sensed by proprioceptors like wing stretch receptors, the output of which modulates the activation pattern of the integrating neuromotoric network. (ii) The movements are effected by muscle contractions and also by the accelerated air. (iii) The aerodynamic quantities depend on wing movement and further on attitude and speed of the whole body. (iv) The inputs of aerodynamic and optical sensors are influenced by attitude and speed of the flying animal.

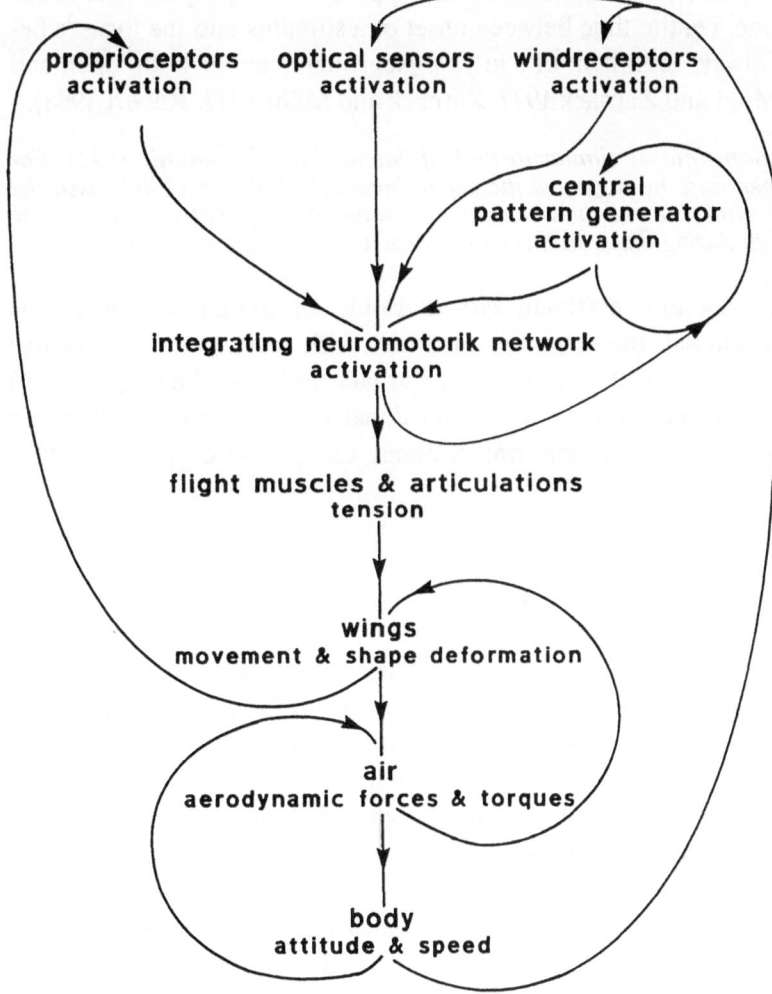

Fig. 1 Diagram of the flight system.

2.2 Flight manoeuvres under open and closed loop conditions

The flight system is not a simple chain of elements of different functions but integrates several feedbacks. According to a basic cybernetic concept, the characteristics of a feedback system can be studied by opening the feedbacks. This is very difficult if not impossible for those inside the animal. The external feedbacks are opened automatically when the animal is fastened to a support like a rotative actuator, force transducer etc. Then, the animal cannot control its attitude and speed in space and the information flow from the effectors to the extero-sensors like eyes, wind hairs, antennae etc. is interrupted.

2.2.1 Earlier results on the reaction time of flight manoeuvres

Many neurophysiological and kinematical studies in locust flight were made under open loop conditions. In yaw experiments, the response was approximately proportional to the stimulus. The reaction time, i.e. the time between onset of a stimulus and the locust's behavioural response, was always less than 0.1 s in yaw experiments under both, open and closed loop conditions (Möhl and Zarnack 1977, Zarnack and Möhl 1977, Robert 1988).

The concept of 'reaction time' is similar to that of 'latency' (cf. Burkhardt 1971). For neurophysiological phenoma, he suggested the use of 'latency'. Waldron (1967) used the term 'latency' for the time interval between the activation of different muscles in the cyclic activation pattern during flight. In this case, cause and effect are unknown.

In roll experiments, the responses to stimuli with sinusoidal or triangular time course were approximately proportional, the responses to stimuli with rectangular time course were phasic with a reaction time < 0.1 s (Möhl and Zarnack 1977). In Thüring's (1986) roll-experiments, locusts showed more or less proportional responses to stimuli with a trapeziform time course. The reaction time was of about 1 s and more (Thüring, 1986, Figs. 2-4).

2.2.2. Roll manoeuvres under closed loop conditions

Migratory locusts are able to fly over long distances. Their great flight stability depends crucially on the maintenance of the horizontal equilibrium. Therefore, a migratory locust should be able to react very quickly on disturbances of its equilibrium. To study flight stability, we caused roll manoeuvres around the longitudinal axis of the locust's body. The flight manoeuvres were induced optically by means of flight simulators. We used different types of them briefly described in the following.

Artificial horizon. A simple artificial horizon (Fig. 2a) which was simulated by a cylinder (diameter 25cm, length 25cm) colored black over half its inside circumference and white over the other. The animal was inside on the cylinder axis and flew in a laminar airstream. The cylinder rested on two rollers, one of which could be rotated by a motor.

By its operation both the tilt angle of the horizon and the tilting speed could be adjusted. The roll angle of the animal was measured by a special meter which allowed the locusts to follow the tilted horizon. Some locusts flew with the dorsal side downwards when the white part of the horizon was below. Thus, locusts were able to perform roll manoeuvres, i.e. they could compensate for the visual disturbance. Therefore, the animals were under closed loop conditions though quick reflexes might have been damped by the relative great moment of inertia of the angular transducer.

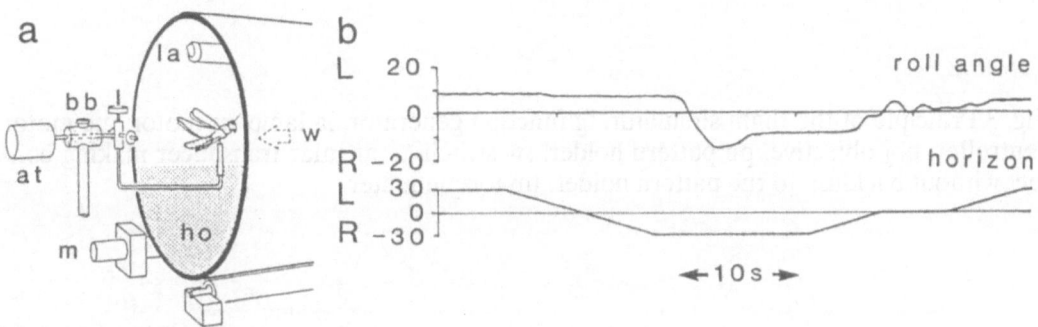

Fig. 2 **a** Arrangement of the artificial horizon. at angular transducer, bb ball bearing, ho horizon, la lamp, m motor,w wind. **b** Roll angle of a locust and tilt angle of the horizon during roll. Scaling of ordinates in degree. L left turn, R right turn (after Schmidt and Zarnack 1987).

In these experiments, the reaction time of the roll manoeuvres was of about 10 s and more (Fig. 2b, cf. Schmidt and Zarnack 1987) confirming earlier observations (Waldron 1967, Wilson 1968). However, the reaction time was much greater than in yaw experiments. This was a very surprising result, because short latencies should be necessary for a stable migratory flight as was mentioned above. Therefore, we supposed that the great latencies during roll manoeuvres were only due to the experimental conditions.

Quick flight simulator. To prove this assumption, we developed a flight simulator similar to that used by Götz (1987) and Robert (1988) supplemented with a quick servomechanism: From behind the locust into the inside of a homogeneous colored cylinder (Fig. 3), an optical pattern was projected which simulated the horizon. The animals were fastened to a rigid torque meter so that they could not change their attitude in space. The horizon could be tilted by a quick servomechanism, one input of which was the roll torque produced by the animal. During an experiment, the horizon was tilted alternately by the bias of a function generator. When the animal intended to follow the tilted horizon it produced a roll torque by means of which the tilt angle was quickly reduced. Thus, roll manoeuvres could be simulated similar to those occuring during free flight. Concerning the visual information flow, in this case the animal flew under closed loop conditions. By opening the switch, it could be forced to fly under open loop condition.

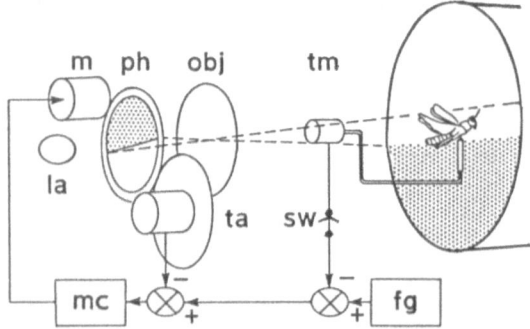

Fig. 3 Principle of the flight simulator. fg function generator, la lamp, m motor, mc motor controller, obj objective, ph pattern holder, sw switch, ta angular transducer making contact without backlash to the pattern holder, tm torque meter.

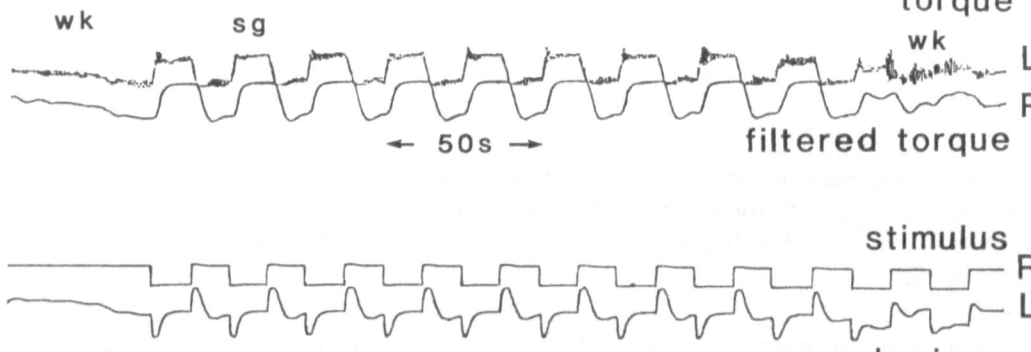

Fig. 4 Roll torque, filtered roll torque, stimulus and tilt angle of the horizon during roll manoeuvres in the quick flight simulator under closed loop conditions. The amplitude of the tilt angle is 90°_{p-p} (p-p peak to peak) depending on stimulus and roll torque. sg strong response to stimulus, wk weak response. Filtering of the torque signal is necessary for stable operation of the simulator. L left turn, R right turn.

Fig. 4 shows roll manoeuvres which are rather symmetric to either side. But even under closed loop conditions, locusts which performed stable flight of long duration (>1h) did not always follow the tilt of the horizon. Flight sequences with strong response to the stimulus (sg, Fig. 4) alternated with others with weak or no response (wk, Fig. 4). Similar results were achieved by Preiss and Gewecke (1988). In case of a response to a stimulus, in the quick flight simulator the reaction time of the roll manoeuvres (Fig. 4) was only about 0.5 s which was tested separately. In the very quick simulator (a further improvement of the quick simulator by optimizing its delay), the reaction time was reduced (Fig. 5a) to about 0.1 s.

In all roll experiments under closed loop conditions (Figs 3-5a), the locusts produced an almost tonic (i.e. time-independ) torque with an only small phasic (i.e. time-dependent) portion. Thus, the locusts reacted like a proportional controller. Therefore, inspite of the several feedbacks (Fig. 1), linear concepts for modeling of the flight system seemed to be applicable (but see below).

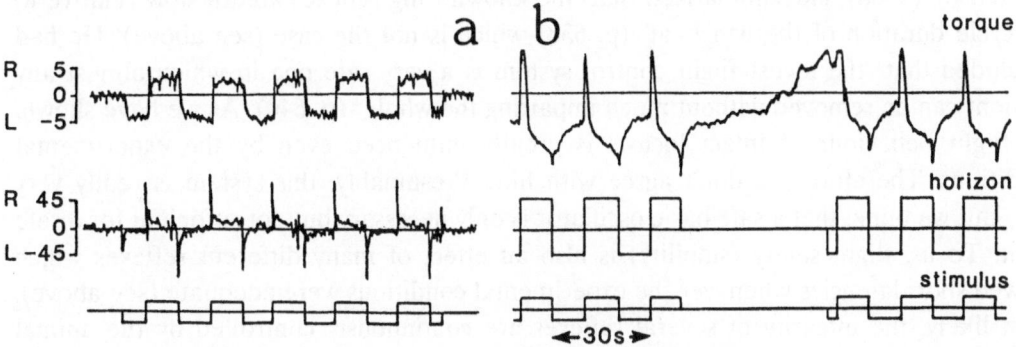

Fig. 5 Roll torque, tilt angle of the horizon and stimulus during roll manoeuvres of one individual in the very quick simulator with optimized delays. The stimulus corresponds to maximum tilt angles of 90°_{p-p} under open loop conditions (see b). Scaling of roll torque in 10^{-5} Nm. Scaling of tilt angle in degree. L left turn, R right turn. a Closed loop conditions. b Open loop conditions.

2.2.3 Roll manoeuvres under open loop conditions

In a further experiment, the same individual as before (Fig. 5a) was forced to fly under open loop conditions, after it had flown some time under closed loop conditions. Now an asymmetric phasic response occured without any tonic portion (Fig. 5b). After tilting the horizon to the right, the animal reduced quickly the initially great right turning torque and generated then even a left turning one (<0). This means that the animal intended to roll at first to the right side but then to the left side though the horizon remained tilted to the right.

2.2.4 Conclusions

This paper is a qualitative comparative study but not a cybernetic analysis of roll manoeuvres. Our intension was to demonstrate some qualitative results on the flight behaviour of locusts which may be of general interest for the analysis of its flight system.

The locusts had flown very regularly. However in these roll manoeuvres, the reaction time varied between several seconds (Fig. 2b) and 0.1 s (Figs 4 and 5), depending on the type of simulator. The reaction time may be further influenced by the different rise times of the stimuli. Nevertheless, locusts are able to react very quickly to disturbances around the roll axis as they do to those around the yaw axis under open loop conditions (Möhl and Zarnack 1977) and closed loop conditions (Robert 1988). This result is further in accordance with the observation of quick (<0.1s) changes in lift (Zarnack 1969, Fig. 4), of fast (within few cycles of wing-beat) influence of proprioceptive feedback Wendler (1974, p. 173) but contrary to the results of Waldron (1967).

Wilson (1968) had summarized that "the known wing reflexes are all slow relative to the cycle duration of the wingbeat" (p. 638) which is not the case (see above). He had concluded that "the locust flight control system is a very safe one in which almost any element can be removed without much impairing the whole" (p. 640). As we have shown, the flight behaviour of intact locusts is greatly influenced even by the experimental conditions. Therefore, we don't agree with him. Presumably, this system is really very safe. But we think, that a safe basic oscillator is only necessary but not sufficient for a safe flight. To us, flight safety (stability) is also an effect of many different reflexes which showed short latencies whenever the experimental conditions were adequate (see above). Most likely, the integrity of several reflexes are continuously controlled by the animal (Möhl 1989b).

We don't state that all sensory input must be intact for a natural flight behaviour. However, it seems to be impossible to get reliable results concerning "real" flight with crude experimental conditions. To us, rhythmic activity of some flight motoneurons or muscles does not imply neither "flight" nor current activity of a control reaction, because in our experiments, the very regular flying individual did not always respond to the very important visual stimulus (Fig. 4, wk and sg). Further, it switched from tonic to phasic response (Figs 5 a and b). The different behaviour under closed loop and the following open loop conditions was perhaps the result of learning which occured during the previous flight under closed loop conditions. Most recently, Möhl (1988) had found short-term learning during flight control in locusts.

Obviously, even very stereotype and stable flying locusts act like real animals and not like machines. Similar to other animals (Heisenberg 1983), they act often more or less spontaneously or randomly as easily can be observed in their biotop too. Heisenberg and Wolf (1988) and Möhl (1989b) thought, that biological noise is important for flight steering and adaptation to new situations. In flies, different behaviour between open- and closed-loop conditions was also found (Heisenberg and Wolf 1988). Möhl (1989b) has found similar results in other experiments with locusts.

The non-deterministic behaviour should be taken into account when interpretating neurophysiological experiments concerning "flight behaviour" with insects unable to perform "normal" behaviour and with preparations of insects often reduced to only one or two thoracic ganglia. In many of such experiments, no observation of any behavioural component is possible and the investigator is not informed about the current state of the nervous system. It seems to be very questionable if any standardization of such preparations is possible.

Finally, some remarks on the concept of 'reaction time' may be necessary. Burkhardt (1971) explained this term as follows. "Bei einer Verhaltensreaktion des gesamten Organismus setzt sich diese Reaktionszeit aus der Latenzzeit im Rezeptor, den afferenten Leitungszeiten, der intrazentralen Leitungszeit, den intrazentralen Latenzen, der efferenten Leitungszeit und der Latenzzeit des Erfolgsorgans zusammen" (p. 872). According to this, the reaction time of the roll manoeuvres should be more or less constant but was not. Therefore, new conceptional considerations about the analysis of the locusts flight system may be necessary.

Acknowledgements

Thanks are due to Professors Dr. N. Elsner and Dr. K.G. Götz for continuous encouragement to this project, and Dr. B. Hedwig for critical reading of the manuscript. Supported by the Deutsche Forschungsgemeinschaft Za 86/5-1 and 6-1.

REFERENCES

Burkhardt D. (1971) Wörterbuch der Neurophysiologie. Jena: VEB Gustav Fischer Verlag, 2. Auflage

Burrows M (1977) Flight mechanisms of the locust. In: Hoyle E (ed) Identified neurons and behaviour of arthropods. Plenum Press, New York London

Götz K.G. (1987) Relapse to 'preprogrammed' visual flight-control in a muscular subsystem of the drosophila mutant 'small optic lobes'. J. Neurogenetics 4: 133-135

Götz K.G. and Wandel U. (1984) Optomotor control of the force of flight in *Drosophila* and *Musca*. II. Covariance of lift and thrust in still air. Biol. Cybern. 51: 135-139

Heisenberg M. (1983) Initiale Aktivität und Willkürverhalten bei Tieren. Naturwissenschaften 70: 70-78

Heisenberg M. and Wolf R. (1988) Reafferent control of optomotor yaw torque in *Drosophila melanogaster*. J. Comp. Physiol. A 163: 373-388

Koch U. (1977) A miniature movement detector applied to recording of wingbeat in locust. Fortschr. Zool. 24 H2/3: 327-332

Koch U.T. and Elliott C.J.H. (1983) Miniature angle detectors - Principle and improved evaluation methods. In: Nachtigall W (ed) BIONA-report vol.1: 41-50, Akad Wiss Mainz, Gustav Fischer, Stuttgart New York

Küssner H.G. (1936) Zusammenfassender Bericht über den instationären Auftrieb von Flügeln. In: Luftfahrtforschung, Bd. 13 (1936), p. 410-424

Lighthill M.J. (1973) On the Weis-Fogh mechanism of lift generation. J. Fluid Mech 60: 1-17

Möhl B. (1988) Short-term learning during flight control in *Locusta migratoria*. J. Comp. Physiol. A 163: 803-812

Möhl B. (1989a) Sense organs and the control of flight. In: Goldsworthy G.J. and Wheeler C.H. (eds.) Insect flight. pp.75-97. Boca Raton (Florida): CRC Press Inc.

Möhl B. (1989b) Function-oriented plasticity of a sensorimotor pathway in the locust flight system. Naturwissenschaften 76: 130-132

Möhl B. and Zarnack W. (1977) Activity of the direct downstroke flight muscles of *Locusta migratoria* L. during steering behavior in flight. II. Dynamics of the time shift and changes in burst length. J. Comp. Physiol. A 118: 235-247

Nachtigall W. (1983) Untersuchungen zum Flug der Dipteren. Stationäre Luftkraftmessungen an Flügeln, Flügelbewegungen beim Steigflug und stationäre Meßgrößen beim Flug vor dem Windkanal. In: Nachtigall W. (ed.) BIONA-report vol.1: 51-60. Akad Wiss Mainz. Stuttgart, New York: Gustav Fischer

Preiss R. and Gewecke M. (1988) Visually-induced wind compensation in the migratory flight of the desert locust, *Schistocerca gregaria*. In: Elsner N. and Barth F.G. (eds.) Sense Organs - Interfaces between environment and behaviour. Poster 38. Stuttgart New York: Georg Thieme Verlag

Reuse G. and Zarnack W. (1988) Wing beat velocities and muscle activity correlated with hindwing movements in locusts. In: Elsner N. and Barth F.G. (eds.) Sense Organs - Interfaces between environment and behaviour. Poster 124. Stuttgart New York: Georg Thieme Verlag

Robert D. (1988) Visual steering under closed-loop conditions by flying locusts: flexibility of optomotor response and mechanisms of correlational steering. J. Comp. Physiol. A 164:15-24

Rowell C.H.F. (1988) Mechanisms of steering in flight by locusts. In: Selverston A.I. (ed.) Model Neuronal Networks and Behavior. pp. 21-35. Plenum Press. New York

Scharstein H. and Zarnack W. (1989) Physiology of motion and its control in higher animals. In: Alt W. and Hoffmann G. (eds.) Biological Motion. Berlin Heidelberg New York: Springer

Schmidt J. and Zarnack W. (1987) The motor pattern of locusts during visually induced rolling in long-term flight. Biol. Cybern. 56: 397-410

Schwenne T. and Zarnack W. (1987) Movements of the hindwings of Locusta migratoria, measured with miniature coils. J. Comp. Physiol. A 160: 657-666

Send W. (1989) Unsteady lift and moment coefficients of an engine nacelle, pp. 159-168. In: Proc of the European Forum on aeroelasticity and structural dynamics. Bonn: DGLR

Thüring D.A. (1986) Variability of motor output during flight steering in locusts. J. Comp. Physiol. A 158: 653-664

Waldmann B. and Zarnack W. (1988) Forewing movements and motor activity during rollmanoeuvers in flying desert locust. Biol. Cybern. 59: 325-335

Waldron I. (1967) Neural mechanismen by which controlling inputs influence motor output in the flying locust. J. Exp. Biol. 47: 213-228

Wendler G. (1974) The influence of proprioceptive feedback on locust flight co-ordination. J. Comp. Physiol. 88: 173-200

Wilson D.M. (1961) The central nervous control of flight in a locust. J. exp. Biol. 38: 471-490

Wilson D. (1968) Inherent asymmetry and reflex modulation of the locust flight motor pattern. J. Exp. Biol. 48: 631-641

Zanker J.M. (1987) Über die Flugkrafterzeugung und Flugkraftsteuerung der Frucht-fliege *Drosophila melanogaster*. Diss FB Biologie Universität Tübingen

Zarnack W. (1972) Flugbiophysik der Wanderheuschrecke (*Locusta migratoria* L.) I. Die Bewegungen der Vorderflügel. J. Comp. Physiol. 78: 356-395

Zarnack W. (1978) A transducer recording continuously 3-dimensional rotations of biological objects. J. Comp. Physiol. A 126: 161-168

Zarnack W. (1983) Untersuchungen zum Flug von Wanderheuschrecken. Die Bewegun-gen, räumlichen Lagebeziehungen sowie Formen und Profile von Vorder- und Hinterflügeln. In: Nachtigall W. (ed.) BIONA-report vol.1: 79-102, Akad Wiss Mainz. Stuttgart New York: Gustav Fischer

Zarnack W. (1988) The effect of forewing depressor activity on wing movement during locust flight. Biol. Cybern. 59: 55-70

Zarnack W. and Möhl B. (1977) Activity of the direct downstroke flight muscles of Locusta migratoria L. during steering behaviour in flight. I. Pattern of time shift. J. Comp. Physiol. A 118: 215-233

HOW A FISH'S BRAIN MAY MOVE A FISH'S BODY

K. Behrend
Zoologisches Institut Sekt. Biophysik
Universität Mainz
Saarstr. 21 6500 Mainz, FRG

Abstract

The brainstem of the electric fish *Eigenmannia virescens* contains a small number of neural networks projecting directly onto the motoneuron pools of the spinal cord such that each segment is reached by at least one neuron of all the networks. Quantitative data of movement patterns, recorded from freely moving animals as well as elicited by electrical stimulation of the single networks, suggest that each network contains a full program for a particular movement, e. g. bending of the body axis to one side. By virtue of the structural relationship the programs are executable with a small number of segments. Mixing of small pieces of programs along the body axis allows for a far larger variety of movement patterns seen in the animals than may be expected from the small number of networks.

Introduction

Coordinated movement of higher animals, e.g. insects or vertebrates, may be viewed as the transformation of the intended motor operation into the actual contractions of the right muscles at the right time with the appropriate amplitude. If the task is the optimizing of some sensory input, e. g. bringing an object into better view, it is a transformation of a geometrical distribution of stimulus parameters into time dependent motor activity. The necessary interaction of a brain and the physical machinery comprising the motion apparatus spoil the attempt to describe the motion in the manner appropriate for one-celled organisms. It necessitates a description at several different levels, some of which are not susceptible to rigorous mathematical formulation. Of this type is the structural organization of the interaction between brain and motion apparatus, representing the "hardware" constraints of the neural network. Again, for reasons of manageability, this task may be divided into a description of the brain part that forms the motor act involving all the information processing units, and a description of another brain part that actually moves the body parts. The latter is addressed in the following and the example chosen is a fish with some unique features, the electric fish *Eigenmannia virescens* . Starting with a brief description of the motor machinery and a rather unrefined but reasonable working model of it, I will go on and show how the neuronal organization relates to this machinery and last based on quantitative measurements of movements discuss some of its functional properties.

I. The motor apparatus of Eigenmannia

Eigenmannia (Fig. 1) is a peculiar fish insofar as it does not locomote in the manner usual for fish, i.e. using a tail fin. Rather, it swims with the undulating elongated anal fin the muscles of which are separate from the trunk musculature. The latter is predominantly used for positioning the body, especially the tail end. This is because the fish continuously produces electric discharges that generate a dipole field around its body with the tip of the tail as the sink of the dipole. Moving the tip of the tail redistributes the field over its front body in a way that increases the local field disturbances caused by objects of conductivity differing from the surrounding water. The fish measures these local disturbances with specialized receptors and so acquires a kind of "electrical image" of its surroundings. The positioning motion, called "probing", consists of a bending of the caudal two thirds of the body towards an object of interest (Fig. 2). Like all fish, *Eigenmannia* has

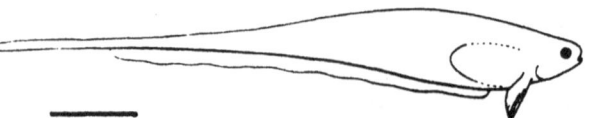

FIG. 1 Lateral aspect of *Eigenmannia virescens*. Note the long anal fin which is constantly undulating and allows for swift forwards as well as backwards motion. Seen from above, the body is very small (see Fig. 2 and 4). Scaled to appr. ½ nat. size

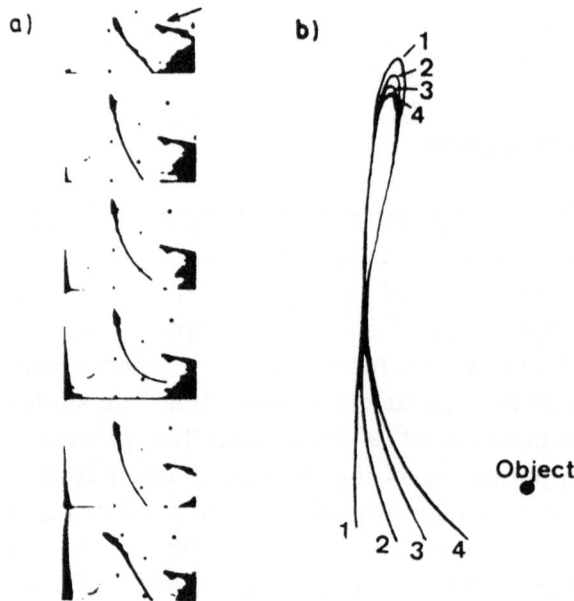

FIG. 2 Probing motion of *Eigenmannia* a) from top to bottom: the fish approaches an object (identifiable by the alligator clip), stops and starts bending its body towards the rod. Time between frames: 120 msec b) drawings of of such a sequence of bending with higher time resolution (40 msec between positions)

a segmented body and the probing motion can be viewed as the result of a concerted action of the segmental muscle masses. To get an idea of the mechanical properties of such a body, a simple model has been developed (see BOX). To simulate the bending motion the three coefficients e_i, f_i and m_i had to be chosen. They were adjusted so that the model mimicked the motion of a fish that had been curarized to block active muscle contraction and then moved sinusoidally with a clamp attached to its head. It turned out that, for friction (f_i) and inertia (m_i) coefficients, a smooth decline of the values according to the lateral aspect of the fish (see Fig.1) gave the best results.

BOX 3

MODEL FISH

Konstantin Behrend

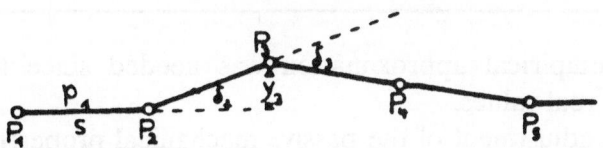

A number of segments of length s is connected by joints (P_i). Each joint is thought to contain an elastic force EF_i that keeps the angle δ_i between segments zero degrees (see Fig.). Three moments act on each segment: the moment of elasticity (EM_i), the friction torque (FT_i) and the moment of inertia (IM_i). They can be calculated as

(1) $EM_i = e_i \delta_i - e_{i+1} \delta_{i+1}$

(2) $FT_i = s f_i (\omega_i + \omega_{i-1})$

(3) $IM_i = 1/3 s^2 m_i (\dot{\omega}_i - \dot{\omega}_{i-1})$

where e_i is the coefficient of elasticity at P_i, f_i is the coefficient of friction, m_i the mass, ω_i the angular velocity and $\dot{\omega}_i$ acceleration of segment p_i. For $i = n$ (the tail end), equation (1) is reduced to

(1) $EM_n = e_n \delta_n$

Also, at the head ($i = 1$) the following constraints hold: P_1 is kept fixed for all simulations and to adapt the model to the passive mechanical properties of a curarized animal p_1 is moved sinusoidally by an external force around P_1. Segment p_2 is then the first freely moving segment and equations (2) and (3) for p_2 are reduced to

(2) $FT_2 = s f_2 \omega_2 + \omega_1$

(3) $IM_2 = 1/3 s^2 m_2 \dot{\omega}_2 - \dot{\omega}_1$

For each segment

---

(4) EM + FT + IM = 0   **BOX 3**

holds. Solved for IM one writes

(3) $-EM_i - FT_i = 1/3s^2 m_i(\dot{\omega}_i - \dot{\omega}_{i-1})$

and solved for acceleration $_i$, the result is

(5) $\dot{\omega}_i = -3/(s^2 m_i)(EM_i + FT_i) + \dot{\omega}_{i-1}$

Acceleration of p_i results from δ_i, velocity and acceleration of p_{i-1}. Starting with a known position y_i of all P_i and $\omega_i = 0$ for all i, the δ_i are calculated, then the moments (EM_i, FT_i, IM_i) and finally the accelerations $(\dot{\omega}_i)$.

From this the new velocity

$\omega_i = \omega_{i\,old} + \dot{\omega}_i\, dt$ (dt is the time step)

and the new position

$y_i = y_{i\,old} + \omega_i$ is derived.

This procedure of empirical approximation was needed since there are no measurements of the real values.

After this process of adjustment of the passive mechanical properties the model was endowed with "muscles", i.e. the effect of muscle contraction is mimicked in terms of a time dependent force working against the elasticity at each P_i, and the EM term is expanded to yield

$EM_i(t) = e_i \delta_i - g(P_i)\sin(2\pi ft - \varphi_i)$

   see BOX < equ.(1) >

The time course of force developement was chosen to be sinusoidal for reasons of "first approximation" - simplicity only (e.g. phase relations of the segmental activity are easily obtained). Playing with $g(P_i)$ and φ_i leads to an idea of the connection between segmental activity patterns and resulting overall motion and I shall come back to this later.

The onset of activity and the time course of force in the single segments will result in the desired position over time of the body and those parameters are controlled by the nervous system.

II. The nervous system.

One part of control of movement execution is done locally at the segmental level in the spinal cord (Grillner et al., 1987). The coordination of these spinal circuits is a result of distributed activity of several brain centers comprising primary sensory as well as associative parts of the brain. This activity is eventually funneled to the spinal cord through what is known as the "final common pathway"

(Sherrington, 1906). In *Eigen-mannia* this comprises predominantly neurons in the brainstem reticular formation. These neurons are organized in seven separate nuclei. The neurons of these nuclei send their axons to the spinal segmental circuits. How many segments are innervated seems to depend upon the particular nucleus and varies from one to maybe four. In addition to the reticular nuclei, one nucleus with a specialized sensory input projects to the spinal cord, namely a vestibular center representing gravitational and accelerational parameters. The projection scheme is the same as in the reticular centers. From a spinal segmental view, it means that each segment is reached by at least one representative of eight different sources in the brainstem (Behrend & Donicht, 1990).

III. Quantitative description of the movement

For a quantitative description of the movement, the curvature at equidistant points on the fish body was used as the basic measure. This is based on the observation that when the segmental muscle mass contracts the angle between two adjacent segments changes and the curvature at that particular point increases. Wether the velocity of this angular change (the first time derivative of curvature) or the angular acceleration (the second time deri-

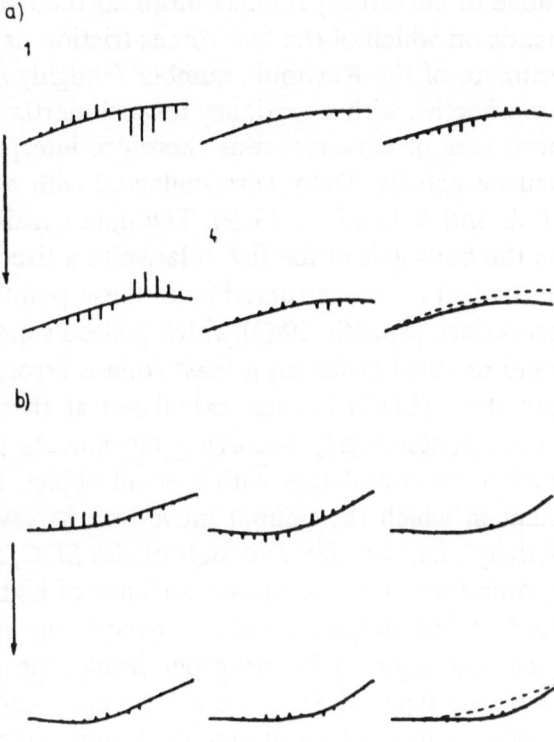

FIG. 3 Movement pattern of a freely moving animal. The fish is represented by the image of the spline function used to approximate the real data (head to left). A stimulus was applied at the position of the small square. The vertical bars represent (relative) size and direction of the second time derivative of curvature at the particular body co-ordinates. A bar to one side means that between the segments neighboring this point the curvature will start to increase in the direction of the bar, i.e. the muscles will contract. a) and b) taken from the same animal during one stimulation epoch about 400 msec apart. Time runs from top to bottom (arrow) in columns from left to right. The broken line in the sixth frame shows the starting position. Time between frames: 40 msec

vative of curvature) reflects more correctly the onset of motoneuron activity depends on which of the two forces friction or inertia dominate the movement. An estimate of the Reynolds number (roughly several hundred) shows that the two are effective with a tendency toward inertia (Nachtigall, 1981). The second time derivative of curvature was therefore interpreted to indicate the onset of motoneuron activity. Data were collected with a specially developed optical method (Behrend & Donicht, 1989). The data consist of a series of evenly spaced points on the body axis of the fish relative to a fixed coordinate system. The middle axis of the fish is reconstructed from these points using a cubic spline approximation procedure (Spaeth, 1983) which proved superior to a simple polynomial fit of second or third order on a least square error base. The second time derivative of curvature (SDC) is then calculated at these coordinates. Data were collected from spontaneously occurring movements as well as from probing movements elicited by stimulation with a small object. Fig. 3 shows the result of an experiment in which the animal moved on its own. The pattern of "motoneuronpool activity", i.e. the size and sign of the SDC, turned out to be surprisingly simple and uniform: there is always a center of highest activity, i.e. greatest SDC, with a kind of bell-shaped decrease towards the neighboring elements and within the time resolution of 20 msec per frame, the activity appears simultaneously. The two main types of SDC are a C- shaped and a S-shaped one and they appear in varying combinations of amplitude and center of highest SDC.

To describe the resulting motion a fish-centered co-ordinate system is used, i.e. the long axis of the straight fish represents the X - axis with the Y - axis being orthogonal to it, and the origin at the caudal end of the head (Fig. 4). The overall motion of the body may be described as the deviation y at a particular x over time t: $y_i := f(x_i, t)$ $i = 1,...k$
it turns out that, except for a scaling factor, $f(x_i, t)$ is the same for all x, which is merely a reflection of the mechanical properties (see I) and muscle contraction dynamics. What is different is the phase. Using the tip of the tail as reference, one finds that it leads the motion, and the wave of

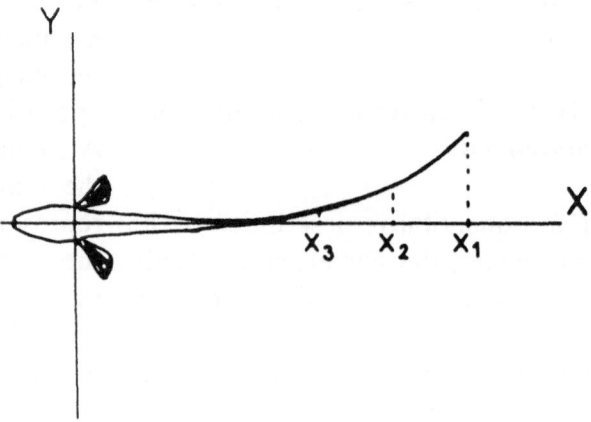

FIG. 4 Co-ordinate system to measure the overall motion. Tip of snout and pectoral fin insertion define the origin. The tip of the tail is taken as reference and index counting starts there.

lateral bending travels towards the
head, i.e. y_{i-1} leads y_i (Behrend,
1984). Although there is a 20 msec
uncertainty, the data suggest that, as
stated above, no phase difference
exists in the activation at the segmental level. Using the model fish a
good approximation of the described overall motion indeed emerges,
when the phase φ_i of the sine function used to simulate the time
course of segmental muscle
contraction is zero at all nodes. This
corroborates the observation and,
taking for granted that all segments
involved in the bending act simultaneously, one can turn to the question of how such a pattern emerges
from the brain structures acting on
the segmental motoneuron pools.

One way of studying this question is
to stimulate the brain centers directly with a train of electric pulses
(5 μA; pulse width 70 μsec) via a
small microelectrode and monitor
the resulting movement pattern with
the same optical setup as in the
free-swimming animal. The result is
surprising: in the three nuclei of the
reticular formation so far investigated, three different movement patterns were elicited (Behrend & Donicht, 1989), all seen in the normal
probing sequences. Increase of stimulus strength (i.e. increase of stimulus pulse frequency) led to an increase in movement amplitude and
again the distribution of SDC mimicked the one found when the
animal performed on its own, i.e.

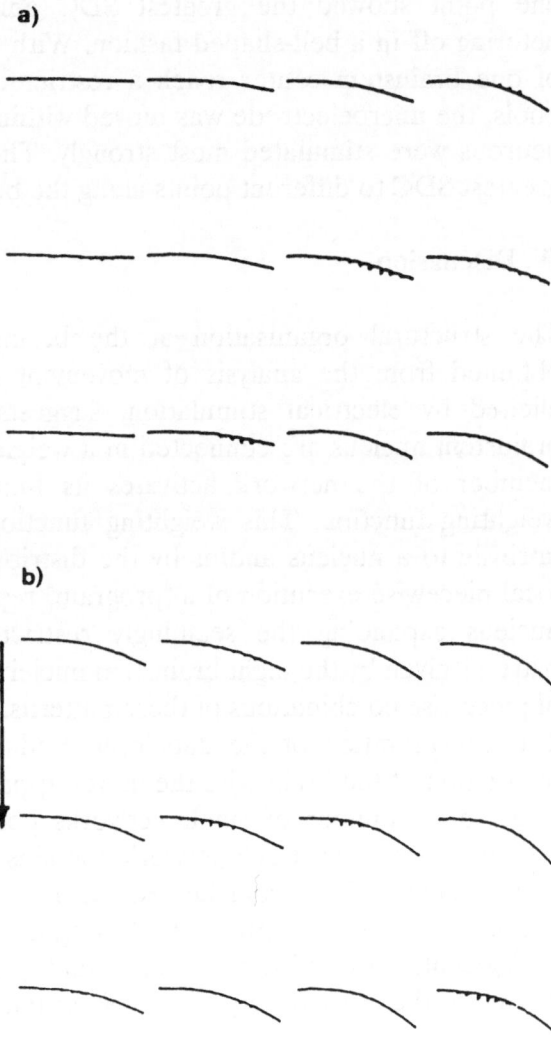

FIG. 5 Movement patterns seen with stimulation at different sites within one
brainstem center (*nfr medius*). The center
of activity is clearly shifted along the body
axis. The two recordings show both the
same time period, starting 420 msec after
stimulation. The two sites are ca. 110 μm
apart within the nucleus. All other details
as in Fig. 3

one point showed the greatest SDC amplitude with the neighboring ones petering off in a bell-shaped fashion. With the finding in mind that the neurons of one brainstem center reach a restricted number of segmental motoneuron pools, the microelectrode was moved within one nucleus, meaning that different neurons were stimulated most strongly. The result was a shift of the center of greatest SDC to different points along the body axis (Fig. 5) as was expected.

IV. Discussion

The structural organisation at the brainstem level together with the data obtained from the analysis of movement performance, be it spontaneous or elicited by electrical stimulation, suggests that the neurons of a particular brainstem nucleus are connected in a weighted manner such that activity of one member of the network activates its functional neighbors according to the weighting function. This weighting function may be achieved by connections intrinsic to a nucleus and/or by the distribution of input fibers. It allows for a local piecewise execution of a "program" represented in the particular brainstem nucleus expanding the seemingly restricted number of possible movement patterns given by the eight brainstem nuclei to an almost indefinite variety by way of piecewise combinations of these patterns.

The interpretation of the data presented can be summed up as follows. The interaction of the brain with the motor apparatus in *Eigenmannia* is organized in terms of a number of small networks comprised by nuclei in the brainstem reticular formation. Each network contains a complete program for moving the body segments in a coordinated manner and when activated as a whole all segments perform the program. A program may be executed on a small number of segments only and the mixing of such program segments along the body axis allows for the great variety of movement patterns seen in the animals.

Literature

Grillner S, Wallén P, Dale N, Brodin L, Buchanan J, Hill R: Transmitters, membrane properties and network circuitry in the control of locomotion in lamprey. TINS 1987; 10:34-41

Behrend K: Cerebellar influence on the time structure of movement in the electric fish Eigenmannia. Neurosci 1984; 13,1:171-178

Behrend K, Donicht M: Neuronal organisation of motor activity in the electric fish Eigenmannia. In: Personnaz L, Dreyfus G (ed) Neural Networks frm Models to Application I.D.S.E.T. Paris, 1989

Behrend K, Donicht M: Descending connections from the brainstem to the spinal cord in the electric fish *Eigenmannia* . Quantitative description based on retrograde horseradish peroxidase and fluorescent dye transport. Brain Behav Evol 1990; 35/4:227-239

Nachtigall W: Hydromechanics and Biology. In: Biophysics of Structure and Mechanism. Springer Verlag Heidelberg Berlin New York 1981; 8:1-22

Sherrington CS: The integrative action of the nervous system. Yale Univ. Press, New Haven 1976 (2nd ed. 1947)

Spaeth H: Spline Algorithmen zur Konstruktion glatter Kurven und Flächen. R. Oldenbourg Verlag München-Wien 1983

Conclusions

Comparing methods and results of this chapter with those of the other ones, one will find a very different structure of the contributions. Two aspects are, in our opinion, the reason.

First, the ability of an animal to compensate for almost any disturbation and to adapt to new situations lies in a more or less deterministic performance of the movement system the specificity of which depends on the specific behavior. With other words, the extreme flexibility of higher animals can be explained physiologically only by the complex pattern of movements and forces. For a sufficiently precise description, many time-variing kinematic parameter are generally necessary, in case of locust flight at least 18 which are controlled by the activity of 20 muscles on each side. This holds for the most simple case of unaccelerated straight level flight.

Therefore, it is generally not sufficient to approximate e.g. a wing-beat by a simple sinusoidal movement or to construct a model network generating such an oscillation. We need very specific models of motion which describe and explain with sufficient precision the different behavioral modes of a species. Only then the physiological connections can be understood. This is in contrast to the approach of explaining for example the cytoplasma-flow inside an amoeba by a rather general solution of biophysical or biochemical equations.

The second principal difference seems to lie in the apparatus higher animals use to solve their complex problems: they have invented a universal machine for problem-solving, the CNS. Our question is how generation and control of motion is solved in a special case (leg coordination or wing movement). To answer the question we have to find out which sort of program evolution has written for that problem but not to solve a physical equation under special boudary conditions. This situation leads naturally to a high diversity of different solutions. Our hope is, however, to understand mainly the general principles after having understood some special solutions nature has found.

REFERENCES

Burrows M. (1977) Flight mechanisms of the locust. In: Hoyle E. (ed.) Identified neurons and behavior of arthropods: 339-356. Plenum Press, New York, London

Elsner N. and Singer W. (1989) Dynamics and plasticity in neuronal systems. Proceedings of the 17th Göttingen Neurobiology Conference. Georg Thieme Verlag, Stuttgart, New York

Gewecke M. and Wendler G. (1985) Insect locomotion. Proceedings of the symposium 4.5 from the XVII. International Congress of Entomology held at the University of Hamburg 1984. Paul Parey Verlag, Berlin, Hamburg

Kandel E.R. (1979) Behavioral biology of Aplysia. A contribution to the comparative study of opisthobranch molluscs. Freeman, San Francisco

Pearson K.G. (1985) Are there central pattern generators for walking and flight in insects? In: W.J.P. Barnes and M.H. Gladden (eds.) Feedback and motor control in invertebrates and vertebrates: 307-315. Croom Held, London

Robertson R.M. and Pearson K.G. (1983) Interneurons in the flight system of the locust: distribution, connections and resetting properties. J.Comp. Neurology 215: 33-50

Ronacher B., Wolf H. and Reichert H. (1988) Locust flight behavior after hemisection of individual thorathic ganglia: evidence for hemiganglionic premotor centers. J.Comp.Physiol.A 163:749-759

Rowell C.H.F. (1988) Mechanisms of steering in flight by locusts. In:J. Camhi (ed.) Neuroethology: A multi-author review. Experientia 44:389-395

Selverston A.I. (1985) Model neural networks and behavior. Plenum Press, New York

Selverston A.I., Miller J.P. and Wadepuhl M. (1983) Cooperative mechanisms for the production of rhythmic movements. Symp.Soc.Exp.Biol. 37:55-87

Stevenson P.A. and Kutsch W. (1987) A reconsideration of the central pattern generator concept for locust flight. J.Comp.Physiol.A 161:115-129

Wilson D.M. (1961) The central nervous control of flight in a locust. J.Exp.Biol. 38:471-490

PART II

LOCOMOTION OF SINGLE ORGANISMS

CHARACTERISTICS OF SEARCH PATHS

(Coordinator: Gerhard Hoffmann)

Introduction

As we step up from the level of cell or body parts to a new organisational level of biological systems, the single organism, the kind of motions we observe changes. Relative movements of its parts are the presupposition of every active movement of an organism and their efficiency determines what movements it can achieve, and where and how fast it can move (Section I/3). Yet, as the construction of a car does not determine where the driver will steer, there is a wide range of possible motion paths available to an animal within these limits.

The wide spectrum of motions that single organisms use for different functions is illustrated in this part. We present also several principally different theoretical approaches and techniques available for the analysis of both structure and function of the observed phenomena.

In the first section general methods are given, which can be used for the analysis of motion paths with a complicated structure. Such paths are observed for instance when an animal is searching around for an important resource. The contributions in the second section deal with the orientation responses of animals in a physical gradient, that helps them to orient themselves within their environment. In the last section the classical question of identifying taxis and kinesis, two basic orientation mechanisms, is taken up.

In search for food, a hiding place, or a sexual partner an animal moves often very large distances on complicated paths. The contributions in the present section address the question, how these search paths can be quantitatively described. This question is fascinating both from a biological and a mathematical point of view. The reason is that the survival of an animal in many situations depends on its ability to find an important resource as fast as possible, but it cannot reach it on a straight way because it has got not enough information about its position. Therefore it moves around on a path whose structure is a complicated mixture of random and systematic elements which are difficult to disentangle and describe.

The section starts with a fundamental paper of **Wolfgang Alt**, who presents an elegant method for the analysis of complex movement paths based on the quantification of correlations. Several papers in this book show the wide applicability of this approach. One example is the paper of **Scharstein** following it. By a very careful experimental and mathematical analysis he tries to describe the movements paths of insects walking around without the help of orientation cues.

Klafter, White and Levandowsky use a different model. They compare the movement of microzooplankton in search for food to a <u>Levy</u> walk, a special type of a random walk with discrete random turns and straight path segments the length of which is given by the Levy stable law.

Bovet and Benhamou give first examples of the complexities of orientation reactions that higher animals show and discuss how an animal could change the structure of its path by changing its "sinuosity". The problem of an animal which has to leave its burrow in search for food and needs to return there afterward is taken up by the paper of **Hoffmann**. The mathematical description of these mechanisms is complicated by the fact that the animal controls its current movement path on the basis of information about its previous movements which it stores for extended periods of time.

Focardi et.al. analyze the same behaviour from a different point of view. They compare characteristic parameters of the foraging excursion of intertidal chitons to the result of a simulation model.

The problems of a tiny beetle larvae searching for its prey on big shrubs are analysed by **Hoffmann** in the last contribution of this section. Using methods developed for the theory of optimal search, he shows that the beetle larvae are able to give their search a structure which increases its success.

CORRELATION ANALYSIS OF TWO-DIMENSIONAL LOCOMOTION PATHS

Wolfgang Alt
Abteilung Theoretische Biologie
Botanisches Institut der Universität
Kirschallee 1, D-5300 Bonn 1, Germany

Abstract

For general stochastic locomotion processes in the plane we try to describe the spatial and temporal characteristics of search paths by analyzing the (complex) autocorrelation function of the velocity vector. Under quite general hypotheses we derive expressions for the mean squared displacement, area search intensity, directional persistence length and time, and discuss the appearence of a rotational or a directional drift. Using simple stochastic rules for endogenous locomotory control mechanisms and possible influences by external directional cues we model the locomotion of fibroblasts or leukocytes, the circling paths of gametes and the meander search pattern of isopods, for example. The appropriate correlation analysis can be performed for observed and simulated path data and compared with the theoretical results.

1. Introduction: Search Characteristics

The broad spectrum of various locomotion patterns as they appear in grazing, dispersal, mating or homing can be subsumed under the general concept of searching: namely for food, space, hosts, sexual partners or just for a favorable environment. Depending on its sensory capacities, a moving organism is perceiving more or less information about location (and some other properties) of its search 'goal'. However, in cases where external stimuli become effective, they often do not direct the organism's motion in a determined way (as in topotaxis). They rather perturb an endogenous locomotory activity in such a way that the resulting locomotion is 'indirectly' led towards the 'goal' with a certain probability (as in klinotaxis or in certain cases of kinesis, see the glossary in Box 14). Such external influences might operate on totally different time scales:

- on the 'short' time scale of the locomotion itself, when e.g. deviation from a favorable orientation is detected and the moving direction is more or less re-adjusted;

- on 'longer' time scales, during which adaptation to a new environment, e.g. a lower food density, is achieved;

- on a 'slow' evolutionary scale having brought up the nowadays search pattern as optimal with repect to the 'usual' environment and some physiological limitations of the searching organism, e.g. when showing a 'meander' locomotion pattern.

In all cases the effectivity of a search depends on an interplay between spatial and temporal characteristics of the locomotion path itself (e.g. its directional persistence or its turtuosity) and those of the environment (e.g. spatial distribution of food). In order to distinguish such search characteristics in a quantitative manner, this article uses autocorrelation analysis of observed or simulated 2-dimensional tracks in connection with stochastic models for the directional control of locomotion. The following questions are addressed: (a) Which informations can be gained about endogenous processes regulating locomotory activity; (b) to what extend are these endogenous activities influenced by some (known or unknown) external factors; (c) how can the concepts of area search intensity and directional response be quantified?

2. Velocity Autocorrelation Function

Any continuous and piecewise smooth 2-dimensional path $Z_t = (X_t, Y_t)$ is completely determined by its velocity vector V_t defined via the differential displacement $dZ_t = V_t dt$ or, equivalently, by its speed $S_t = |V_t|$ and the angle α_t between x-axis and locomotion direction

$$(1) \qquad\qquad \Theta_t = (\cos\alpha_t, \sin\alpha_t) =: e^{i\cdot\alpha_t}.$$

In this complex vector notation the velocity is

$$(2) \qquad\qquad V_t = S_t \cdot e^{i\cdot\alpha_t},$$

and denoting by $V_t^* = S_t \cdot e^{-i\cdot\alpha_t}$ its complex conjugate vector we can consider the complex product

$$(3) \qquad\qquad V_{t+\tau} \cdot V_t^* = S_{t+\tau} \cdot S_t \cdot e^{i\cdot(\alpha_{t+\tau}-\alpha_t)}.$$

Its real part, proportional to $\cos(\alpha_{t+\tau} - \alpha_t)$, is the usual scalar product between the instantaneous velocity vector V_t and the velocity vector $V_{t+\tau}$ at a time τ later. It describes the degree of parallelity of these two vectors. Analogously the imaginary part, proportional to $\sin(\alpha_{t+\tau} - \alpha_t)$, is the scalar product between the orthogonal vector $V_t^\perp$ and $V_{t+\tau}$ describing the tendency to turn to the left or right for increasing time intervals τ.

Furtheron we consider only such paths which are realizations of a stationary stochastic process V_t in the plane. This implies that the random vector V_t has a fixed probability density being invariant under time shifts, with a constant mean velocity $\underline{V}$.

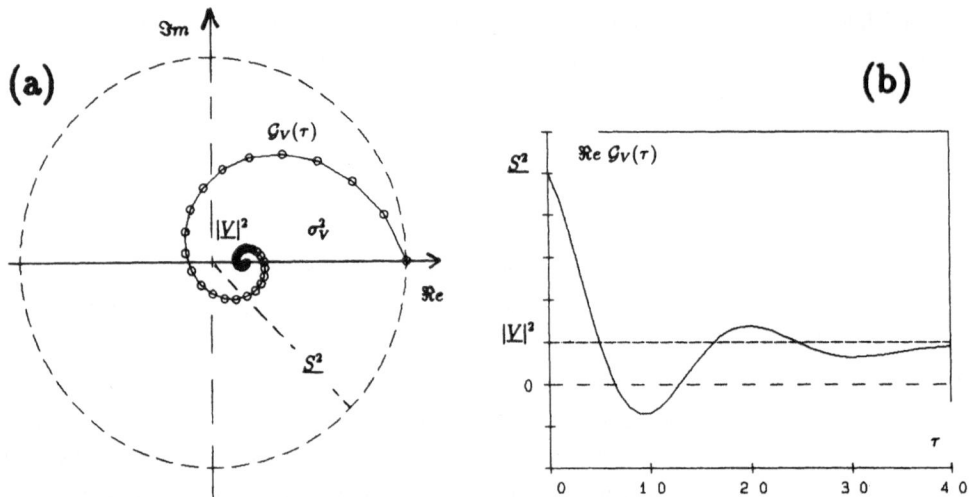

Figure 1: (a) Trace of the autocorrelation function $\mathcal{G}_V(\tau)$ in the complex plane for $0 \leq \tau < \infty$, with marks at regular steps $\Delta \tau = 1$; (b) the graph of its real part $\Re\, \mathcal{G}_V(\tau) = |\underline{V}|^2 + \sigma_V^2 \cdot \cos \underline{\omega} \tau \cdot e^{-\tau/T_V}$ calculated for the generalized fibroblast model according to formula (6), with positive mean angular velocity $\underline{\omega}$. The weighted area between the graph and its asymptote in (b) gives twice the motility coefficient μ, see eqs. (9) and (10). Compare also the analogous Fig.3 in (Scharstein, this section).

Then the expected mean value of the complex product defined in (3), the so-called complex velocity autocorrelation function,

$$(4) \qquad \mathcal{G}_V(\tau) \; := \; \langle V_{t+\tau} \cdot V_t^\star \rangle$$
$$= \; (\langle S_{t+\tau} \cdot S_t \cdot \cos(\alpha_{t+\tau} - \alpha_t) \rangle \, , \; \langle S_{t+\tau} \cdot S_t \cdot \sin(\alpha_{t+\tau} - \alpha_t) \rangle)$$

depends only on the correlation time τ . In general, the value of this function at both the starting ($\tau = 0$) and the asymptotic end point ($\tau = \infty$) is real:

$$\mathcal{G}_V(0) \; = \; \langle |V_t|^2 \rangle \; = \; \underline{S}^2 \, , \; \text{the mean squared speed,}$$
$$\mathcal{G}_V(\infty) \; = \; |\langle V_t \rangle|^2 \; = \; |\underline{V}|^2 \, , \; \text{the squared modulus of the mean speed velocity vector.}$$

However, in-between the function can be non-real with modulus $|\mathcal{G}_V(\tau)| \leq \underline{S}^2$, see Fig. 1.

Example 1 (Graham Dunn's fibroblasts):
The very slow translocation of the 'center of mass' of fibroblasts is due to relatively rapid protrusions of surrounding lamellipods in 'random' directions. Let b denote their mean 'amplitude' and $T_V = 1/a$ their mean retraction time. Also, allow a possible (externally induced) mean drift velocity $\underline{V}$. Then the resulting velocity process can be modelled in a simplified way by a stochastic differential equation (Dunn and Brown, 1987):

$$(5) \qquad\qquad\qquad dV_t \; = \; a \cdot (\underline{V} - V_t) dt + b \cdot dW_t,$$

where W_t describes the white noise Wiener process. See Boxes 11 and 12 for general properties of such 'SDE's. Even for complex parameters a and b (with $\Re\, a > 0$) this model makes sense, and the autocorrelation function can be computed as

$$
(6) \qquad \mathcal{G}_V(\tau) \;=\; |\underline{V}|^2 + \frac{|b|^2}{2\Re\, a} \cdot e^{-a\cdot\tau},
$$

and plotted as shown in Fig.1. In the generalized model above the velocity V_t performs a stationary Gauss process fluctuating around a mean velocity $\underline{V}$ with $\underline{\text{variance}}$ given by $\sigma_V^2 = |b|^2/2 \cdot \Re\, a$, mean $\underline{\text{relaxation time}}$ $T_V = 1/\Re\, a$ and constant $\underline{\text{rotational drift}}$, i.e. $\underline{\text{mean angular velocity}}$ $\underline{\omega} = -\Im m\, a$. In the fibroblast case, where a is real, we have $\underline{\omega} = 0$, and the autocorrelation function is real.

3. Mean Squared Displacement and Area Search Intensity

In order to characterize the motility of an organism one wants to measure its $\underline{\text{dispersal}}$ after releasing it from some point Z_0 at time $t = 0$. This is achieved by the $\underline{\text{mean squared}}$ $\underline{\text{displacement}}$ function

$$
\begin{aligned}
(7) \qquad \mathcal{D}^2(t) \;&:=\; \langle |Z_t - Z_0|^2 \rangle \\
&=\; 2 \cdot \int_0^t \int_0^{\tau_1} \Re\, \mathcal{G}_V(\tau)\, d\tau\, d\tau_1.
\end{aligned}
$$

Thus, this well-known function equals the twofold integral of the real part of the velocity autocorrelation function $\mathcal{G}_V$ defined in eq.(4) and its asymptotic properties can be computed as

$$
(8) \qquad \mathcal{D}^2(t) \;=\; \begin{cases} \underline{S^2 \cdot t^2} & \text{for small } t \geq 0, \\ |\underline{V}|^2 \cdot t^2 + 4\mu \cdot t & \text{for large } t, \end{cases}
$$

where the so-called $\underline{\text{random motility}}$ coefficient

$$
(9) \qquad \mu \;:=\; \frac{1}{2} \cdot \int_0^\infty (\Re\, \mathcal{G}_V(\tau) - |\underline{V}|^2) d\tau
$$

is given as half the weighted area between the real part of the autocorrelation function and its asymptote (see Fig.1b). It is always positive.

In the general case of Example 1 we obtain from eq.(6) :

$$
(10) \qquad \mu \;=\; \frac{\sigma_V \cdot T_V}{2} \cdot \frac{1}{1 + \underline{\omega}^2 T_V^2}.
$$

The first factor group corresponds to the well-known formula for diffusive Brownian motion. However, the random motility μ is progressively reduced by the denominator of the second term, which grows with the mean rotation $\underline{\omega} \cdot T_V$ performed during a 'characteristic' time. In case of a centered locomotion with no drifts (translocational: $\underline{V} = 0$; rotational: $\underline{\omega} = 0$) equation (7) yields an explicit expression for the mean squared displacement

$$(11) \qquad \mathcal{D}^2(t) \;=\; 2 \cdot \sigma_V^2 \cdot T_V \cdot \left(t - T_V \cdot (1 - e^{-t/T_V}) \right).$$

This well-known formula, which similarly holds for other correlated velocity processes (see Othmer et al., 1988) has been widely used to fit experimental data, for example in (de Boisfleury-Chevance et al. 1988), see Fig.2. However, the experimental curves often have differing shapes, in particular, the lymphocyte curve shows reduced mean squared displacement ('dispersal') over a considerably long time, suggesting that lymphocytes perform a more intense random search within the visited area, when compared to curves of granulocytes or monocytes.

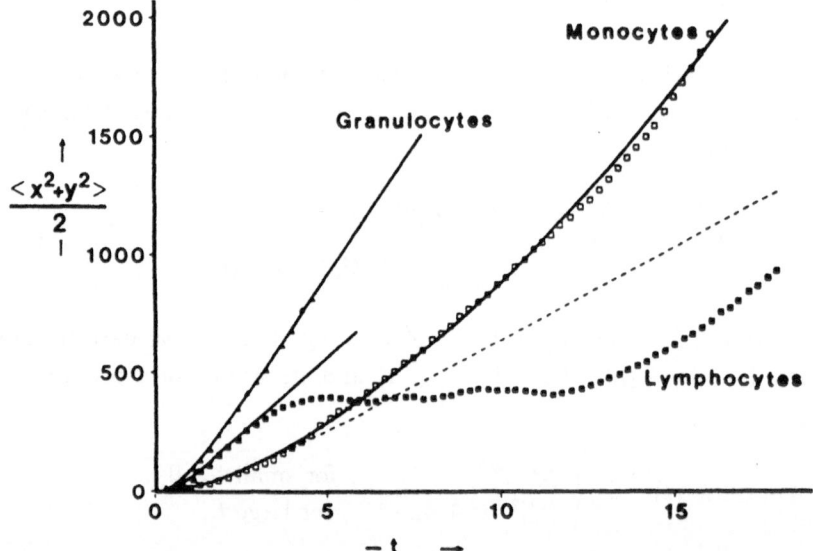

Figure 2: Experimental data of half the mean squared displacement function $\mathcal{D}^2(t)$ in μm^2 over time t in min, for different white blood cells, fitted with more or less success by theoretical curves of type (11). From (de Boisfleury-Chevance et al. 1988: Figure 6).

The general antagonistic relation between dispersal activity and area search intensity might be quantified in the following way. Assume that the organism has a 'detection width' W along its locomotion track, so that the mean encountered area after a time t is $A(t) = W \cdot \underline{S} \cdot t$, with mean speed $\underline{S} = \langle S_t \rangle$. Notice that this function continuously counts multiply visited regions. Deviding it by the so-called mean encircled area, which may be approximately defined as $F(t) = \pi \cdot max_{\tau \le t} \, \mathcal{D}^2(\tau)$, yields a rough measure of the area search intensity up to time t:

$$(12) \qquad\qquad\qquad I(t) \; := \; \frac{A(t)}{F(t)}.$$

Supposing a vanishing mean drift ($\underline{V} = 0$) we conclude from eqs.(8) and (9) that, in 2-D only, I(t) approaches a finite value for $t \to \infty$, the asymptotic search intensity

$$(13) \qquad\qquad I_\infty \;=\; \frac{W \cdot \underline{S}}{4\pi \cdot \mu} \;=\; \frac{W}{2\pi \cdot R_\infty},$$

if we define the asymptotic turtuosity radius by

$$(14) \qquad\qquad R_\infty \;:=\; \frac{2 \cdot \mu}{\underline{S}} \;=\; \frac{1}{\underline{S}} \cdot \int_0^\infty \Re e\, \mathcal{G}_V(\tau)d\tau.$$

This general definition means that the asymptotic search intensity I_∞ is just the ratio between the searcher's detection scope W and the circumference $2\pi R_\infty$ of a circle with curvature (or 'sinuosity')

$$(15) \qquad\qquad 1/R_\infty \;=\; \frac{\underline{S}}{2 \cdot \mu}.$$

Therefore we refer to this quantity as the asymptotic sinuosity or turtuosity of the random path. For further explanation see eqs.(27)-(29) and, for comparison, also Bovet and Benhamou, later in this section). Summerizing we can state that the (asymptotic) search intensity is directly proportional to path turtuosity, thus inversely proportional to random motility (or 'dispersal activity').

4. Persistence of Direction

In many observations of swimming or migrating organisms the endogenous regulation of velocity $V_t = S_t \cdot \Theta_t = S_t \cdot e^{i \cdot \alpha_t}$ seems to be performed by at least two (more or less) independent processes:

- by controling the locomotion speed S_t which is either fluctuating around some positive mean speed $\underline{S}$, or vanishing $S_t = 0$, characterizing a resting state;

- by controling the change of direction $\Theta_t = e^{i \cdot \alpha_t}$, i.e. $d\alpha_t = \omega_t \cdot dt$ with angular turning rate (or angular velocity) ω_t .

Example 2 (Annete Geller's and my favored male *Ectocarpus siliculosus* gametes): While the swimming speed of a male gamete moving along a surface is determined by the beating frequency of its front flagellum, its angular turning is modified by an asymmetry of flagellar beating with respect to its body axis (Geller and Müller, 1981). In the simplest model for the resulting locomotion behavior (see Fig.3a) both speed S_t and angular turning rate ω_t can be considered as mutually independent, stationary Gauss processes having means $\underline{S}$ and $\underline{\omega}$, variances σ_S^2 and σ_ω^2 as well as relaxation times T_S and T_ω , respectively, see eq. (17) below.

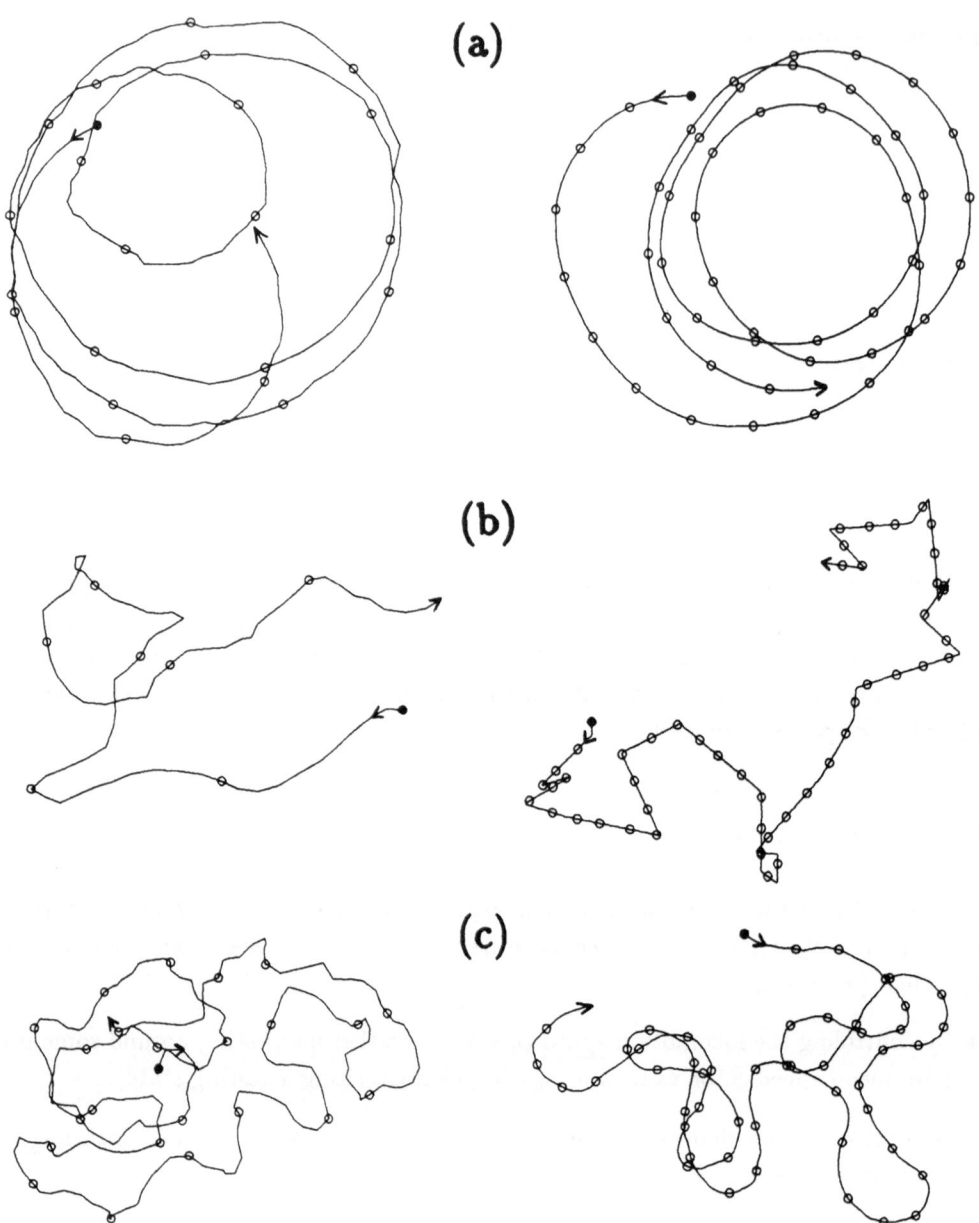

Figure 3: Typical 2-dimensional locomotion paths of (a) a 'circling' male *Ectocarpus siliculosus* gamete of cell length $8\mu m$, data recorded by Annette Geller (cp. Müller and IWF, 1982), (b) a 'piecewise smoothly migrating' neutrophilic leukocyte of cell diameter $10\mu m$ (Gruler 1989), and (c) a 'meandering' desert isopod, $5mm$, same animal as shown by (Hoffmann, later in this section: Fig.2). On the right hand sides, corresponding simulated paths according to the stochastic models in Example 2, eqs.(16),(17): (a) fluctuating turning process with $\omega > 0, \Omega_t \equiv 0$; (b) angular jump process with $\omega_t \equiv 0$, and (c) fluctuating turning process as in (a), but with ω replaced by a periodic turning rate with zero mean.

Other examples as 3-dimensional flagellar swimming (e.g. of *E.coli*) or insect walking (e.g. of ants, isopods or beatles, see the other contributions in this section) or faster cell migration (e.g. of *dictyostelium* and leukocytes) often show longer pieces of more or less smooth paths, occasionally interrupted by sudden sharp turns, sometimes connected with shorter resting states. Again using the simplest modeling assumptions for this general situation, the angular turning might be written as a sum

$$(16) \qquad\qquad d\alpha_t \;=\; \omega_t \cdot dt + \Omega_t$$

of a stationary fluctuation process ω_t as in Example 2 (simulations shown in Fig.3a and c):

$$(17) \qquad\qquad d\omega_t \;=\; \frac{1}{T_\omega}(\underline{\omega} - \omega_t) \cdot t + \sqrt{\frac{2\sigma_\omega^2}{T_\omega}} \cdot dW_t$$

and an independent jump process Ω_t characterized by probability distributions $f(\tau)$ for waiting times τ between sudden turns, and $k(\varphi)$ for corresponding turn angles φ or direction jumps $e^{i \cdot \varphi}$ (a simulation is shown in Fig.3b).

Under these (or even more general) hypotheses the complex velocity autocorrelation function can be written as a product of three terms

$$(18) \qquad \mathcal{G}_V(\tau) \;=\; \langle S_{t+\tau} \cdot S_t \rangle \cdot \langle e^{i \cdot \int_0^\tau \omega_t dt} \rangle \cdot \langle e^{i \cdot \int_0^\tau \Omega_t} \rangle$$
$$\qquad\qquad\quad =\; G_S(\tau) \cdot \mathcal{G}_{\Theta,\omega}(\tau) \cdot \mathcal{G}_{\Theta,\Omega}(\tau)$$

where

$$(19) \qquad G_S(\tau) \;=\; \underline{S}^2 + \sigma_S^2 \cdot \exp(-\tau/T_S),$$
$$(20) \qquad \mathcal{G}_{\Theta,\omega}(\tau) \;=\; \exp\left(-\sigma_\omega^2 \cdot T_\omega \left(t - T_\omega \cdot (1 - e^{-t/T_\omega})\right) + i \cdot \underline{\omega}\tau\right)$$

and where the Laplace transform of $\mathcal{G}_{\Theta,\Omega}(\tau)$ can, in general, be computed (Alt, 1988) by using the Laplace Transform of $f(\tau)$ and the so-called directional persistence index (or mean cosine) of the angular jump process Ω_t

$$(21) \qquad\qquad \psi \;:=\; \int_{-\pi}^{\pi} k(\varphi) \, \cos\varphi \, d\varphi.$$

In particular, for an exponential 'smooth run' time distribution $f(\tau) = 1/T_\Omega \cdot e^{-\tau/T_\Omega}$ and a symmetric turn angle distribution $k(\varphi)$ we derive, with the so-called persistence time

$$(22) \qquad\qquad T \;:=\; T_\Omega/(1 - \psi),$$

the simple 'exponential' autocorrelation formula

$$(23) \qquad\qquad \mathcal{G}_{\Theta,\Omega}(\tau) \;=\; e^{-\tau/T}.$$

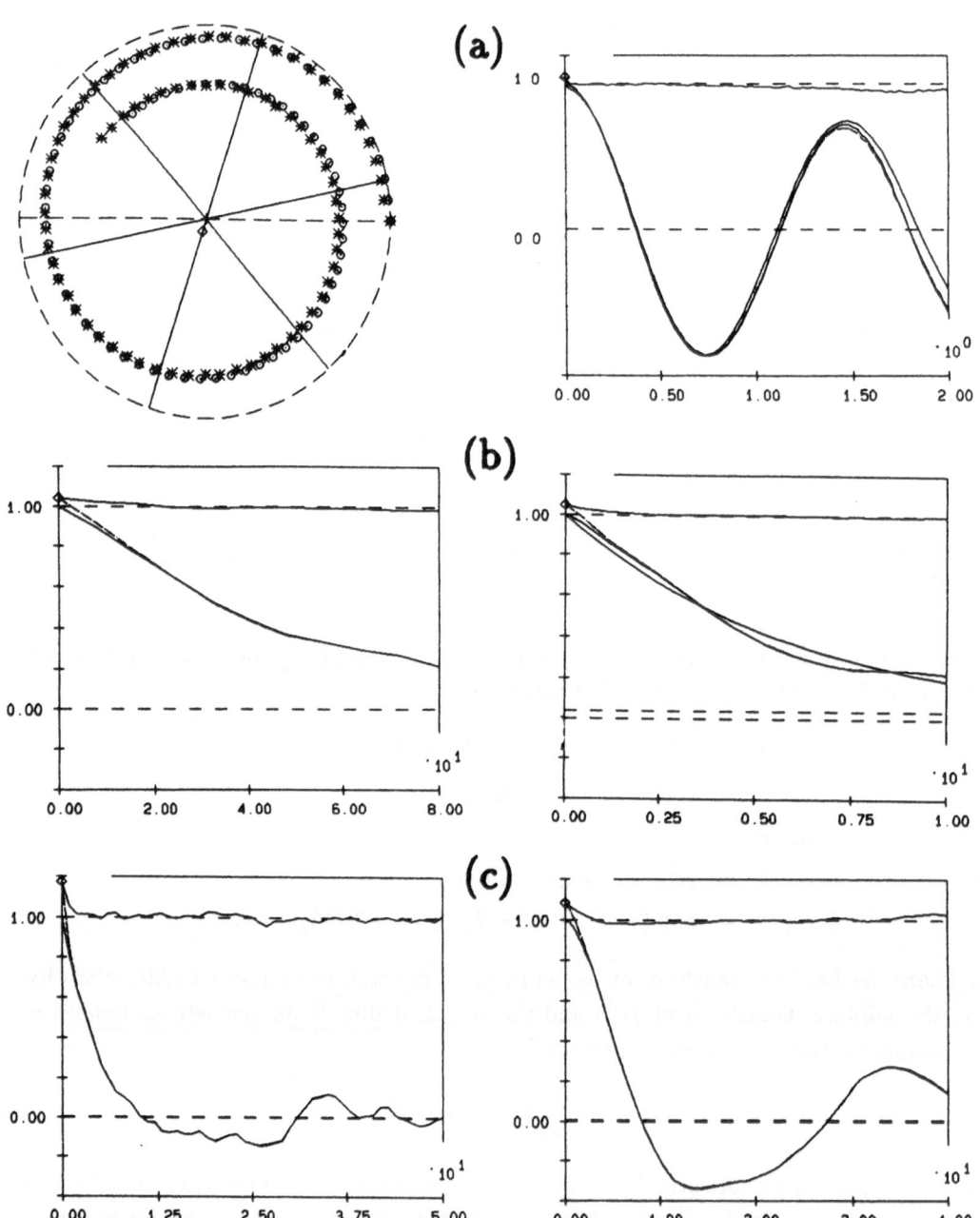

Figure 4: Autocorrelation analysis of the paths plotted in Fig.3 comparing observed (left) with corresponding simulated paths (right), except in plot (a: left): upper curves show the speed autocorrelation G_S (19), lower curves the real parts of the velocity (broken line) and directional (full line) autocorrelation functions according to (18). In addition, (a: left) plots the (discrete) time course of the full (complex) function $\mathcal{G}_\Theta$ for the observed gamete's ($*$) and the fitted simulation path ($\odot$), the deviations of the corresponding real parts are also visualized in plot (a: right). Moreover, in (b: right) the theoretically expected exponential (23) is plotted. Notice the deviations from the simulated result, obviously due to the short lenght of the path in Fig.3b .

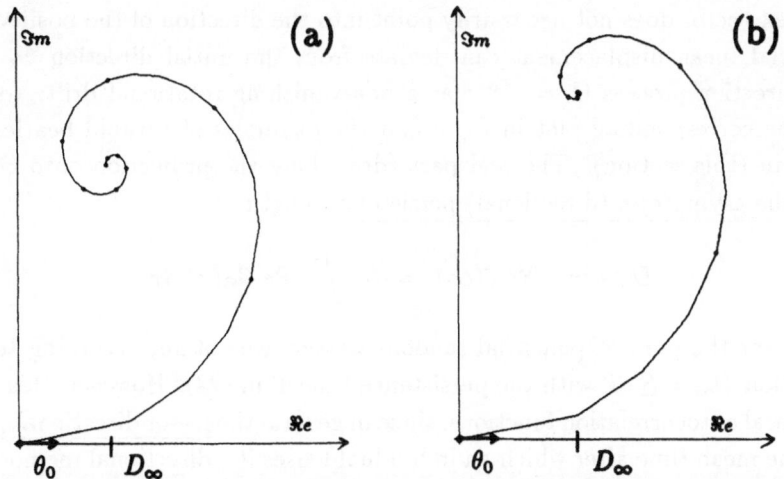

Figure 5: Mean displacement $\zeta(\tau)$ in eq.(26) of a 'theoretical' random walker with starting direction θ_0, plotted for increasing times τ (marks at regular time steps) for the case of a 'circling' movement, as shown by the gamete in Fig.3a. The directional autocorrelation function $\mathcal{G}_{\Theta,\omega}(\tau)$ in eq.(20) is numerically integrated for parameters (a) corresponding to the fitted data in Fig.4a yielding almost uncorrelated turning fluctuations, see also (30), whereas in (b) with significant first order correlations in ω_t.

With the general assumptions above, the autocorrelation of the direction process $\Theta_t = e^{i \cdot \alpha_t}$ can (theoretically) be obtained by using (18) and dividing through by the speed autocorrelation function, yielding the (complex) <u>directional autocorrelation</u> function:

$$(24) \qquad \mathcal{G}_\Theta(\tau) \quad := \quad \mathcal{G}_V(\tau)/\mathcal{G}_S(\tau) \quad = \quad \langle \Theta_{t+\tau} \cdot \Theta_t^* \rangle.$$

Its real part

$$(25) \qquad \Re\, \mathcal{G}_\Theta(\tau) \quad = \quad \langle \Re\, (\Theta_{t+\tau} \cdot \Theta_t^*) \rangle \quad = \quad \langle \cos(\alpha_{t+\tau} - \alpha_t) \rangle$$

exactly measures the expected <u>mean directional persistence</u> after a locomotion time τ, i.e. the <u>mean cosine of the angular deviation</u>, as it is plotted and fitted e.g. for leukocyte path data in (Gruler and Bültmann 1984: Fig.8). We refer to the contribution by Scharstein (following in this section) where definition (24) is used for evaluation of observed locomotion tracks.

The directional autocorrelation function (24) can also be used to evaluate the expected <u>mean displacement relative to an initial direction</u> Θ_t after a time τ, namely

$$(26) \qquad \zeta(\tau) \quad := \quad \langle\, (Z_{t+\tau} - Z_t) \cdot \Theta_t \,\rangle$$
$$= \quad \underline{S} \cdot \int_0^\tau \mathcal{G}_\Theta(\tau_1) d\tau_1.$$

In the homogeneous case without any preferred direction ($\underline{V} = 0$) the function $\zeta(\tau)$ approaches a finite asymptote for large τ:

$$\zeta(\infty) \quad = \quad \underline{S} \cdot \int_0^\infty \mathcal{G}_\Theta(\tau_1) d\tau_1.$$

This complex vector does not necessarily point into the direction of the positive real axis (i.e. the final mean displacement can deviate from the initial direction Θ_t), namely when the direction process $\Theta_t = e^{i \cdot \alpha_t}$ has a non-vanishing rotational drift: $\langle \alpha_t \rangle \simeq \underline{\omega} \cdot t$. Compare the corresponding plot in Fig.5 and the examples of carabid beatles analyzed by Scharstein (this section). The real part (describing the projection onto Θ_t) can be defined as the asymptotic (directional) persistence length

$$(27) \qquad\qquad D_\infty \;\; := \;\; \mathfrak{Re}\; \zeta(\infty) \;\; = \;\; \underline{S} \cdot \int_0^\infty \mathfrak{Re}\; \mathcal{G}_\Theta(\tau)\; d\tau.$$

Notice that for the pure 'exponential random walker' we obtain, according to (23), the simple relation $D_\infty = \underline{S} \cdot T$ with the persistence time T in (22). However, this only holds for exponential autocorrelation functions, since in general the mean directional persistence time, i.e. the mean time after which an individual looses its 'directional memory' , should be derived from the persistence distribution (25) as

$$(28) \qquad\qquad P_{mean} \;\; = \;\; \int_0^\infty \tau \cdot \mathfrak{Re}\; \mathcal{G}_\Theta(\tau) d\tau \Big/ \int_0^\infty \mathfrak{Re}\; \mathcal{G}_\Theta(\tau) d\tau.$$

If speed fluctuations are negligable (σ_S and/or T_S small compared to $\underline{S}$ resp. to P_{mean}), as it is the case for leukocyte migration (Gruler, section II/2: Fig.2b), then we can assume $\mathcal{G}_V = \underline{S}^2 \cdot \mathcal{G}_\Theta$ for positive correlation times. From this we derive that $D_\infty = R_\infty$, i.e. the (asymptotic) persistence length equals the (asymptotic) turtuosity radius defined earlier in (14). Consequently, an asymptotic formula for the mean squared displacement results, which improves formula (8) with $\underline{V} = 0$: For large t we have

$$(29) \qquad\qquad \mathcal{D}^2(t) \;\; = \;\; 2 \cdot R_\infty \cdot \underline{S} \cdot (t - P_{mean} + \epsilon_{(t \to \infty)}).$$

This formula should be used to fit data as those of white blood cells in Fig.2: From the slope and intercept of the empirical $\mathcal{D}^2(t)$ - asymptote one can simultaneously estimate, without reference to any particular model, both parameters motility $\mu = 1/2 \cdot R_\infty \cdot \underline{S}$ and persistence time P_{mean}, thus, with already known mean speed $\underline{S}$, one also obtains an estimate of the path turtuosity $1/R_\infty$.

As mentioned above, only in cases of exponential autocorrelation functions $\mathcal{G}_\Theta$ we have $R_\infty = \underline{S} \cdot P_{mean}$. For example, this is fulfilled for the fluctuating angular turning process in (17) with arbitrary mean angular velocity $\underline{\omega}$, but only in the 'diffusion limit' where correlations are negligible. This means that $T_\omega \to 0$ and $\sigma_\omega \to \infty$ with finite angular diffusion coefficient

$$(30) \qquad\qquad \mu_\omega \;\; = \;\; \frac{1}{2} \cdot \sigma_\omega^2 \cdot T_\omega.$$

Then the mean persistence time is just

$$(31) \qquad\qquad P_{mean} = \frac{1}{2\mu_\omega + \underline{\omega}^2/2\mu_\omega},$$

and with eq.(14) the random motility coefficient is $\mu = \underline{S}^2 \cdot P_{mean}/2$. However, in realistic situations of correlated angular turning the corresponding formulas are more complicated (cp. Tranquillo, section II/2).

5. Tests for Directional Orientation, Alignment or Rotation

Given a locomotion path of finite length, recorded by a finite number of points (thus approximating the real track by, say, n discrete translocation steps), the question is how we can decide whether the path does (not) show

- a significant 'external' influence in form of a <u>directional drift</u>, i.e. <u>orientation</u> ($\underline{V} \neq 0$), or in form of an <u>alignment</u> or a <u>guidance</u> ($\underline{V} = 0$), see (b) below and the glossary in Box 14 for further explanation ;

- a significant 'endogenous' bias in form of a positive or negative <u>rotational drift</u>, i.e. mean angular velocity, leading e.g. to 'loops';

- some other 'endogenous' systematic or stochastic activity pattern, e.g. periodic turning or '<u>meanders</u>'.

(a) <u>Rayleigh test</u>:
The theoretically expected mean velocity is $\underline{V} = \underline{S} \cdot \underline{\Theta}$ with the so-called Rayleigh vector or 'mean direction' $\underline{\Theta} := \langle \Theta_t \rangle = (\langle \cos \alpha_t \rangle, \langle \sin \alpha_t \rangle)$. Suppose we already know that $\underline{S} = \langle S_t \rangle$ is significantly positive, then a statistical <u>test for directional orientation</u> consists in estimating the squared modulus of the Rayleigh vector from below:

$$(32) \qquad\qquad |\underline{\Theta}|^2 \geq \epsilon_{(n\to\infty)} \cdot \underline{S^2}/(\underline{S})^2,$$

where the small number ϵ depends on the confidence level and the number n of recorded steps. Notice, however, that both $\underline{S}$ and $\underline{S}^2 = \mathcal{G}_V(0)$ carry additional discretization errors (see Box 4).

(b) <u>Circular distributions</u>:
Histograms of the measured angles $\alpha_t \bmod 2\pi$ can give a more distinctive picture: a unimodal circular distribution indicates directional orientation with respect to $\underline{\Theta} \neq 0$, cp. Figs. 3 and 8 in (Häder, 1985), whereas a symmetric bimodal and antipodal distribution with $\underline{\Theta} = 0$ suggests the possibility of guidance. However, also for large observation number n, these 'stationary' histograms cannot say anything about the dynamics of a possible directional or tensorial response. This is achieved by evaluating the (complex) <u>circular distribution of the 'mean direction'</u> after a correlation time τ

$$(33) \qquad\qquad \mathcal{H}(\tau, \theta)d\theta = \langle \Theta_{t+\tau} \,|\, \Theta_t \in \theta + d\theta \rangle, \quad \theta \epsilon S^1, \tau \geq 0.$$

For any give direction $\Theta_t = \theta$ on the unit circle the (complex) function $\mathcal{H}(\tau, \theta)$ describes the expected tendency how the mean of locomotion directions $\Theta_{t+\tau}$ evolves with time τ :

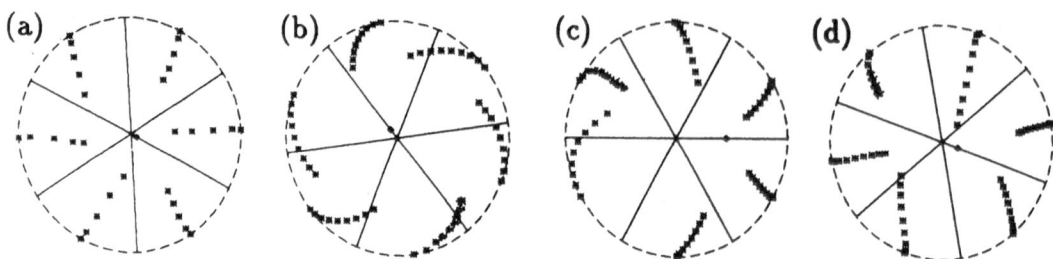

Figure 6: Typical possibilities for the circular distribution of mean directions (33) plotted as vector histograms for several discrete correlation times $\tau_j \geq 0$. Outputs of a 'classified' autocorrelation analysis applied to simulations of the four different prototypes of (a) isotropic motion, (b) rotational drift, (c) directional orientation, and (d) alignment, according to model equation (16) with appropriate variations of $\underline{\omega}$. For further details see paragraph (b).

It starts with $\mathcal{H}(\tau, \theta) = \theta$ and extends into the unit circle for increasing correlation times. In Fig.6 we show corresponding 'empirical' vector histograms, classified in 6 segments of the unit circle. Generally we can formulate the following criterion:

- ($\imath$) if $\mathcal{H}(\tau, \theta)$ for 'all' θ tends centripetally in direction $-\theta$, then we can conclude an isotropic locomotory behavior (Fig.6a), whereas if $\mathcal{H}(\tau, \theta)$ always tends to one side of θ , we will observe a corresponding rotational drift (Fig.6b);

- ($\imath\imath$) if $\mathcal{H}(\tau, \theta)$ for 'all' θ tends towards the significantly non-zero Rayleigh vector $\underline{\Theta}$ (depicted by $\diamond$), then we can quantify this as directional orientation (Fig.6c);

- ($\imath\imath\imath$) in the alternative case $\underline{\Theta} \approx 0$ we might nevertheless detect a tendency of $\mathcal{H}(\tau, \theta)$ towards a certain direction θ_0, but only for initial θ already pointing into the θ_0-direction, and a tendency towards $-\theta_0$ for opposite θ's; then we can conclude alignment of the path with, or guidance by the line $\pm\theta_0$ (Fig.6d).

Notice that theoretically the mean of $\mathcal{H}(\tau, \theta)$ rotated by θ just gives the directional autocorrelation function:

$$(34) \qquad \int \mathcal{H}(\tau, \theta) \cdot \theta^* \, \mu(d\theta) \;\; = \;\; \mathcal{G}_\Theta(\tau),$$

where $\mu(d\theta)$ denotes the stationary measure of the directional process on the unit circle.

(c) 'Real' directional autocorrelation function:
Suppose that, by negation of (b $\bullet$ ($\imath\imath$) and ($\imath\imath\imath$)) directional and tensorial (external or, may be, internal) influences onto the locomotory control have been excluded, then according to (b $\bullet$ ($\imath$)) a possible endogenous rotational drift might be quantified by estimating the mean angular turning rate $\underline{\omega} = \langle \omega_t \rangle$. One procedure is to compute the discretized values of $\omega_t = d\alpha_t/dt$ and take into account the related error (in analogy to Box 4). Another check for $\underline{\omega}$ is to plot the compensated autocorrelation function

$$(35) \qquad G_\Theta(\tau) \;\; := \;\; \mathcal{G}_\Theta(\tau) \cdot e^{-i \cdot \underline{\omega} \cdot \tau}.$$

and see if it follows the real axis, at least approximately for moderate τ, since the theory claims that according to equations (18)-(20) this function is 'real'. An equivalent procedure is to estimate $\underline{\omega}$ as the regression slope of the angular value $\arg \mathcal{G}_\Theta(\tau)$ over τ. Thus, if these estimates are satisfactory, we might use the 'real' (part of the) 'empirical' function $G_\Theta(\tau)$ for characterizing special features of the endogenous locomotory activities. An alternative, more robust procedure is the direct investigation of the modulus $|\mathcal{G}_\Theta(\tau)|$ as used by Scharstein (following contribution).

Example 3 (Gerhard Hoffmann's desert isopods):
When searching for their burrow after going astray (see the article by Hoffmann later in this section) the isopods perform their random walk in such a way that they frequently return near to their starting point. The latter phenomenon could be achieved by an angular drift $|\underline{\omega}|$ which, in analogy to the loop search of gametes, would effectively increase the area search intensity (compare Fig.4a). However, a meander structure seems to represent the major characteristics of a long search of isopods: Typical path data (as the example in Fig.3c: left) show low values of $|\underline{\omega}|$, but a real autocorrelation function $G_\Theta(\tau)$ taking negative values for correlation times around 25 sec (see Fig.4c: left). This possibly means that, on the average, isopods have a strong tendency to bend their path backwards within that time. A similar directional autocorrelation picture appears for the simulated path with periodically changing $\underline{\omega}(t)$ (see Fig.4c: right). However, the path itself (Fig.3c: right) looks too smooth (with short time directional correlations) compared to the isopod's path, which has less correlations on a short time scale (steep exponential decay in Fig.4c: left), but nevertheless strong negative correlations in the long run. For a further characterization of such meander paths detailed models have to be developed and simulated, which might distinguish between systematic oscillations of the angular turning rate and possible stochastic fluctuations with autocorrelations of higher order.

6. Preview to Further Analyses: Characterization of Locomotion Paths

From the modeling, analysis and simulation of the locomotion of gametes, leukocytes and isopods we have to admit, that different hypotheses on the locomotory control can lead to quite similar directional autocorrelation functions. In particular, the common exponential-like decay for small correlation times can originate from a fluctuating velocity V_t as in eqs.(5),(6), from jumping angles α_t with exponential waiting times, eqs.(21)-(23), and from uncorrelated fluctuations of the turning rate ω_t, see eqs. (30),(31). Therefore, in order to characterize different types of observed search paths, it seems necessary to extract some more informations out of the data. One technique currently being developed tries to detect and mark such points (and times) where a moving individual stops ($S_t = 0$) or performs a sudden turn (jump in α_t). Thereafter, the velocity process can be seperately analyzed for each 'smooth' path segment between the markers, and the resulting 'segmented' autocorrelation function can be compared with the one for the whole path. By considering also histograms of moving or resting times and of sudden turn angles, we hopefully are able to distinguish between different locomotory activity patterns and to characterize stochastic and systematic features of an underlying endogenous regulation process.

Acknowledgements

For providing these theoretical investigations with stimulating ideas during long discussions before and at the Workshop, but also afterwards, as well as for delivering 'real' data of locomotion tracks, on disks or paper, I want to thank many friends and colleagues, in particular V. Calenbuhr, Graham Dunn, Annete Geller, Gerhard Hoffmann, Kolja Wawrowsky, Hans Gruler, Hansuli Keller and Hans Scharstein.

REFERENCES

Alt W. (1988) Modelling of motility in biological systems, pp.15-30. In: ICIAM'87 Proceedings. SIAM, Philadelphia pp. 15-30

De Boisfleury-Chevance A., Rapp P. and Gruler H. (1989) Locomotion of white blood cells. A biophysical analysis. Blood Cells 15: 315-333

Dunn G.A. and Brown A.F. (1987) A unified approach to analysing cell motility. J.Cell Sci. Suppl. 8: 81-102

Geller A. and Müller D.G. (1981) Analysis of the flagellar beat pattern of male *Ectocarpus siliculosus* gametes in relation to chemotactic stimulation by female cells. J.Exp. Biol. 92: 53-66

Gruler H. (1989) Biophysics of leukocytes: Neutrophil chemotaxis, characteristics and mechanisms. In: (M.B.Hallet ed.) The Cellular Biochemistry and Physiology of Neutrophils. CRC Press, Boca Raton pp. 63-95

Gruler H. and Bültmann B.D. (1984) Analysis of cell movement. Blood Cells 10: 61-77

Häder D.-P. (1985) Computer aided studies of photoinduced behaviour . In: (G.Colombetti et al. eds.) Sensory Perception and Transduction in Aneural Organisms.

Müller D.G. and Inst.Wiss. Film (1982) Pheromonwirkungen bei der Befruchtung von Braunalgen. Film C 1242 by IWF, Göttingen

Othmer H.G., Dunbar S.R. and Alt W. (1988) Models of dispersal in biological systems. J.Math.Biol. 26: 263-298

PATHS OF CARABID BEETLES WALKING IN THE ABSENCE OF ORIENTING STIMULI AND THE TIME STRUCTURE OF THEIR MOTOR OUTPUT

Hans Scharstein

Lehrstuhl Tierphysiologie

Zoologisches Institut der Universität

Weyertal 119, D-5000 Köln 41 FRG

Abstract

The locomotor activity of carabid beetles walking in the absence of orienting stimuli is analysed. Under such conditions the animals walk with no preferred direction and typically show a circular walking pattern which results from a motor asymmetry. Three parameters (speed, direction and angular velocity) characterising the paths are evaluated. Changes of speed and of walking direction are mutually independent. The performed autocorrelation analysis reveals a complex directional autocorrelation function with a monotone decay and a discontinuity at the origin, the curve starts with a non-zero slope. Two endogeneous processes of directional changes could be responsible for this shape of the autocorrelation function: either a broad band 'noisy' course of angular velocity or an uncorrelated jump process of directional changes.

1. Introduction

The path of a walking animal can be considered as an interplay between the activity of the motor pattern generator, its endogeneous modulation (start-stop pattern, circling, meandering) and the influence of exogeneous spatial information used for orientation. By ruling out all orientational cues in a compensated walking paradigm, the following investigations provide some insight in the endogeneous mechanisms generating the walking behaviour of beetles.

2. Experimental situation

The animals (5 species of night-active carabid beetles: *Agonum assimile*, *Abax ovalis*, *Abax parallelus*, *Molops piceus* and *Nebria brevicollis*) walk on a locomotion compensator, the "Kramer sphere", which is a two-dimensional actively driven treadmill (Kramer, 1975; Wendler and Scharstein, 1986). An animal, marked with a reflecting foil walks on the top of a ball of 30 cm diameter. An infrared illuminating and sensing system measures the position of the reflecting spot. If the animal walks in one direction, the measured deviation from the central position drives one or both of the servomotors at the equator of the sphere in a way that the animal is brought back to the top of the apparatus. The two-dimensional (x,y-) displacements of the ball are sampled with a resolution of 1.3 mm. Depending on the speed of the animal the positional data occur in time intervals in the range of about 25 ms to 50 ms. The resulting time series are recorded on magtape and digitized off line. The paths of the animals are reconstructed from the x,y-data at equidistant time intervals Δt of in general 0.5 s, in cases where a higher resolution is needed of 0.1 s.

To study the unoriented walk of beetles, the animals have to walk in darkness on the horizontal top of the apparatus. A typical example for the reconstructed path of an individual is shown in Fig.1A. As often in the absence of directional cues, the animal displays a motoric asymmetry which results in a circular walking pattern.

From the x- and y-coordinates of the path, the following measures of the animals activity

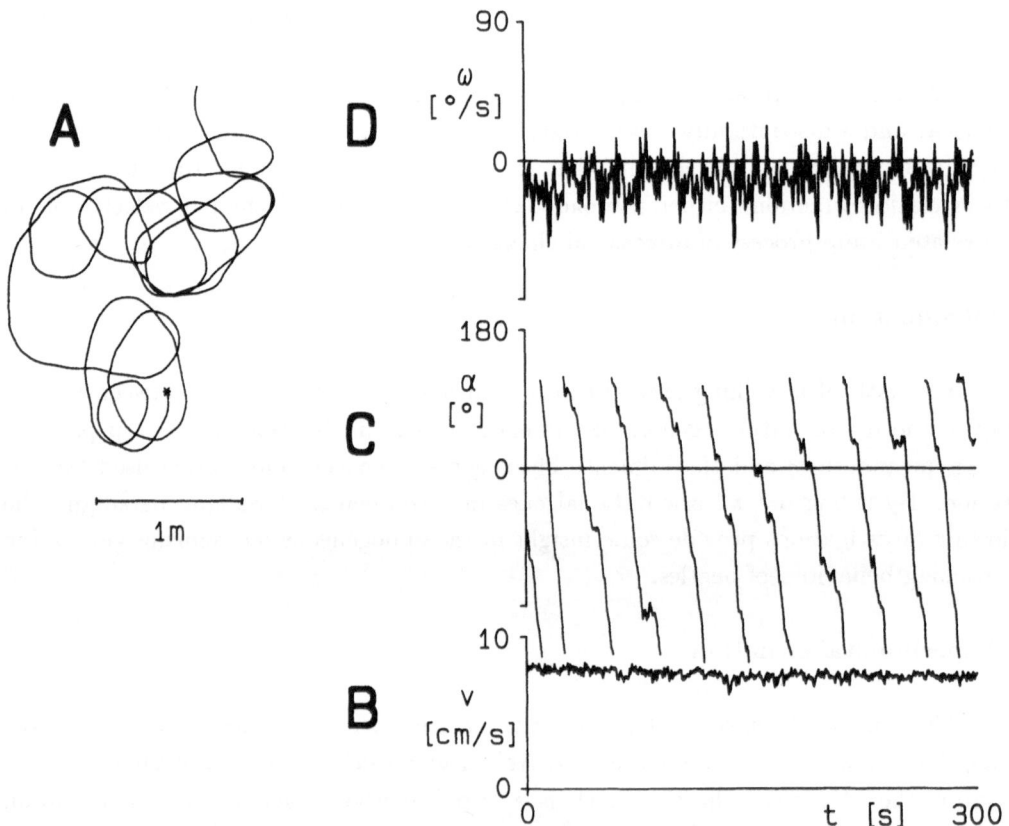

Figure 1: A 5-min walk of an unoriented *Nebria brevicollis*. (A): Reconstructed path of the animal. Right: Diagrams of (B) speed $v(t)$, (C) walking direction $\alpha(t)$ and (D) angular velocity $\omega(t)$ evaluated in equidistant time intervals $\Delta t = 0.5s$.

are calculated: speed $v(t)$, walking direction $\alpha(t)$ and angular velocity $\omega(t)$. The sampling interval Δt is 0.5 s. During the 5 min interval chosen for analysis the beetle walked without interruption with a constant speed of approx. 7 cm/s (Fig.1B). The walking direction (Fig.1C) changed continuously (clockwise). This is also illustrated by the constant mean angular velocity (Fig.1D) on which stochastic changes are superimposed.

3. Characteristics of unoriented walking

As a basic measure for the behaviour three frequency distributions (walking direction, speed and angular velocity) and their mutual dependencies are analysed (Fig.2). Fig.2A shows the distribution of walking directions. The graph demonstrates a homo-

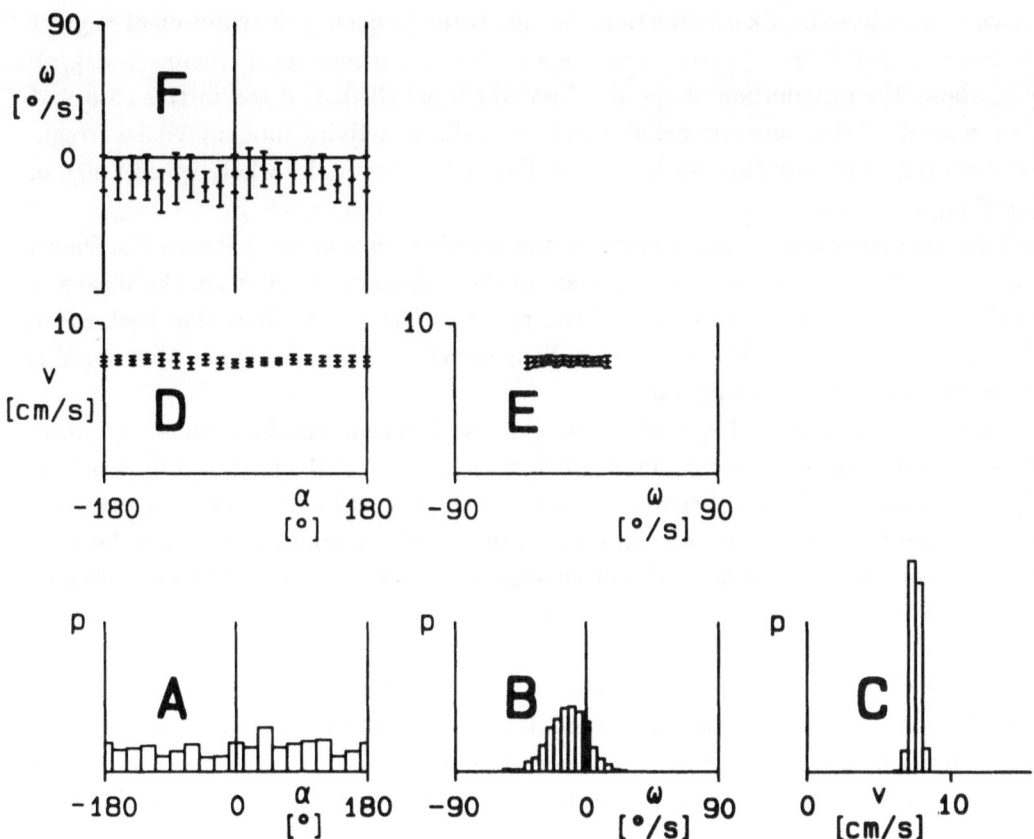

Figure 2: Statistical analysis of walking direction, speed and angular velocity of the walk shown in Fig.1. Frequency distribution of (A) walking directions $p(\alpha)$, (B) angular velocities $p(\omega)$ and (C) speed values $p(v)$, measured in 600 intervals of 0.5s duration. (D) dependence of the mean speed values $(+/-SD)$ and of (F) mean angular velocities $(+/-SD)$ on walking direction. (E) Dependence of mean speed values $(+/-SD)$ on angular velocity.

geneous angular distribution, indicating that the animal has no preference to walk in a certain direction. Because the graph represents the frequency of the directions of unit vectors, the concentration and the mean direction of the distribution are calculated by circular statistics.

The length $|R|$ of the mean vector of the distribution is given by

$$(1) \qquad |R| = 1/N\sqrt{(\sum sin\alpha_i)^2 + (\sum cos\alpha_i)^2}$$

and the mean direction $\bar{\alpha}$ by

$$(2) \qquad \bar{\alpha} = arctan \sum sin\alpha_i / \sum cos\alpha_i$$

$|R|$ lies within the interval $0 \leq |R| \leq 1$, giving the mean cosine of the deviation of the single vectors from the mean direction. In Fig.2A $|R|$ has a value of 0.1, which indicates the lack of a preferred walking direction. In Fig.2B the frequency distribution of angular velocities is shown being typically bell-shaped with a non-zero mean (here -13 deg/s). Fig.2C shows the distribution of speed values which is bell-shaped too in this case, but not in general. Often animals exibit bursts of walking activity interrupted by irregu- larily occuring stops, resulting in frequency distributions over the whole speed range of $0 \leq v \leq v_{max}$.

The three diagrams D-F in Fig 2 examine the possible correlations between the shown parameters. For this question, in each class of the frequency distribution the means of the other two parameters are calculated and plotted. The results show that both speed and angular velocity are independent on walking direction (2D and 2F) and that speed is independent on angular velocity (2E).

One parameter in Fig.2 is of special interest for the following considerations: the mean angular velocity which is not zero in the distribution in Fig. 2B. As shown in detail for grain weevils by Wendler and Scharstein (1986), each population of an insect species dis- plays a spectrum of motor asymmetries which, in absence of orientational cues, leads to different mean angular velocities. The mean angular velocity of an individual may change slowly over a long time and is superimposed by independent, quick changes of the angular velocity.

This example has illustrated the basic features of the walking pattern of unoriented in- sects: The animals walk with varying speed interrupted by stops of varying duration. Due to internal asymmetries, they exibit a slowly varying or constant mean angular velocity which is superimposed by momentary stochastic changes which are independent from the mean. Walking speed and angular velocity vary independently.

4. Autocorrelation analysis of walking direction

Knowing now the frequency distributions of speed, walking direction and angular velocity, we turn to the time structure of the walking pattern. Autocorrelation analysis provides us with an intuitive understanding of the process generating the animals move- ment and is also very useful for the analysis of the spatial distribution of animals in their

habitat resulting from the individual motor pattern. A detailed description of the mathematical background is found in Alt (1988) and Alt, this section.

The path of an animal can be considered as the time course of a two- dimensional vector.
For that reason the autocorrelation function (ACF) is also two-dimensional and may be
derived in complex notation.

The path of the animal is described as the time course of the (complex) velocity vector

$$(3) \qquad\qquad V(t) = S(t) \cdot e^{i \cdot \alpha(t)}$$

where $S(t)$ is the speed and $\alpha(t)$ the walking direction of the animal which are, as shown,
mutually independent.

From the complex product

$$(4) \qquad\qquad V(t+\tau) \cdot V^*(t) = S(t+\tau) \cdot S(t) \cdot e^{i \cdot (\alpha(t+\tau) - \alpha(t))}$$

the complex ACF can be derived which is the expected value of the complex product

$$(5) \qquad\qquad \mathcal{G}_V(\tau) = \langle V(t+\tau) \cdot V^*(t) \rangle$$

The asterisk $*$ denotes the complex conjugate value. The direction process $\Theta(t) = e^{i \cdot \alpha(t)}$
and its ACF $\mathcal{G}_\Theta(\tau)$, which is of special interest, can be obtained by dividing equ.(5) by
the speed autocorrelation $G_S(\tau)$ (Alt,this section)

$$(6) \qquad\qquad \mathcal{G}_\Theta(\tau) = \mathcal{G}_V(\tau)/G_S(\tau)$$

This holds under the constraint, that speed and angular change process are mutually
independent. The complex ACF $\mathcal{G}_\Theta(\tau)$ may be written as

$$(7) \qquad\qquad \begin{aligned} Re(\mathcal{G}_\Theta) &= \langle cos(\alpha(t+\tau) - \alpha(t)) \rangle \\ Im(\mathcal{G}_\Theta) &= \langle sin(\alpha(t+\tau) - \alpha(t)) \rangle \end{aligned}$$

That means, that the real part is the expected value of the cosines of the angular changes
in the correlation time τ, whereas the imaginary part is the expected value of the sines
of the angular changes. As in chapter 3, in the case of the calculation of the mean vector
of an angular distribution, the angular ACF is the mean vector of the angular *changes*
relative to the starting direction which occured in the correlation time τ.

Instead of its real and imaginary part we can take its modulus $|\mathcal{G}_\Theta(\tau)|$ and its phase $\varphi(\tau)$
(corresponding to length and mean direction of the mean vector). The modulus $|\mathcal{G}_\Theta(\tau)|$
describes the angular spread relative to the starting direction after the correlation time
τ, the phase $\varphi(\tau)$ the increasing total turn which sums up over time τ. Fig.3 shows as
an example the directional ACF of the experiment in Figs. 1 and 2 and is explained in
detail in the following chapter.

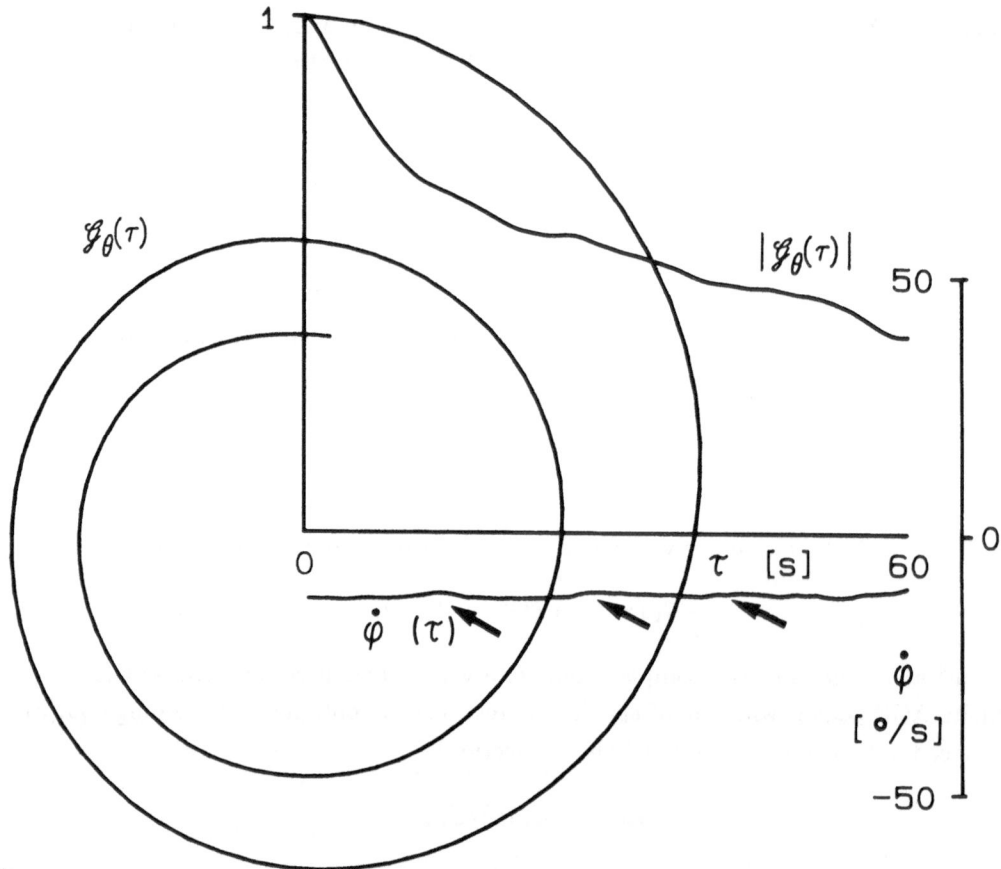

Figure 3: Angular auto-correlation function of the data in Figs.1 and 2. $\mathcal{G}_\Theta(\tau)$: ACF of the direction process $\Theta(t)$ in the complex plane. To achieve correspondence with the starting point of the modulus curve, the complex plane is rotated counterclockwise by 90 deg. Note that the real axis is now the ordinate! $|\mathcal{G}_\Theta(\tau)|$: modulus of the complex ACF. $\dot{\varphi}(\tau)$: time derivative of the phase of the complex ACF .The corresponding scale is given on the right side in deg/s. Further explanations see text.

5. Interplay of stochastic and systematic parts of the direction process

In Fig.3 the first 60 s of the complex ACF $\mathcal{G}_\Theta(\tau)$ and two graphs derived from it are shown. As the angular differences at $\tau = 0$ are zero, the cosines of them are equal to 1, the sines equal to 0 (see eq.7). Therefore $\mathcal{G}_\Theta(\tau)$ starts with a value 1. For a better correspondence with the modulus $|\mathcal{G}_\Theta(\tau)|$, the curve has been rotated about 90 deg to the left, the real axis being now the ordinate. $\mathcal{G}_\Theta(\tau)$ exhibits a clockwise turning spiral with a slowly decreasing distance from the origin. The turning corresponds to the (clockwise) circular movement of the animal (see Fig.1A), which is summed to an increasing mean directional change relative to the starting direction. The decreasing radius, the time course of which is plotted as modulus curve $|\mathcal{G}_\Theta(\tau)|$, shows the decreasing mean vector length which is equivalent to the spreading out of the distribution of angular differences. The modulus can be used as a measure of the remaining information about the starting direction which in our example obviously decreases more and more due to the stochastic process of angular changes.

Since the phase $\varphi(\tau)$ of the complex ACF is equivalent to the increasing mean angular change which sums up with correlation time, the time derivative $\dot{\varphi}(\tau)$ gives the mean angular velocity as a function of correlation time. It is not necessarily constant: regular changes (see arrows) may signal systematic influences of weak orienting stimuli which were hidden in the broad range of angular velocities of the original data (see Fig.1D and Fig.2F).

6. The correlation function at $\tau = 0$: which process generates the angular changes?

The considerations have so far concentrated on the process generating the walking direction $\alpha(t)$ as a function of time. While walking, however, the motor system generates angular *changes* instead of the walking direction. How can the underlying angular velocity process be characterized?

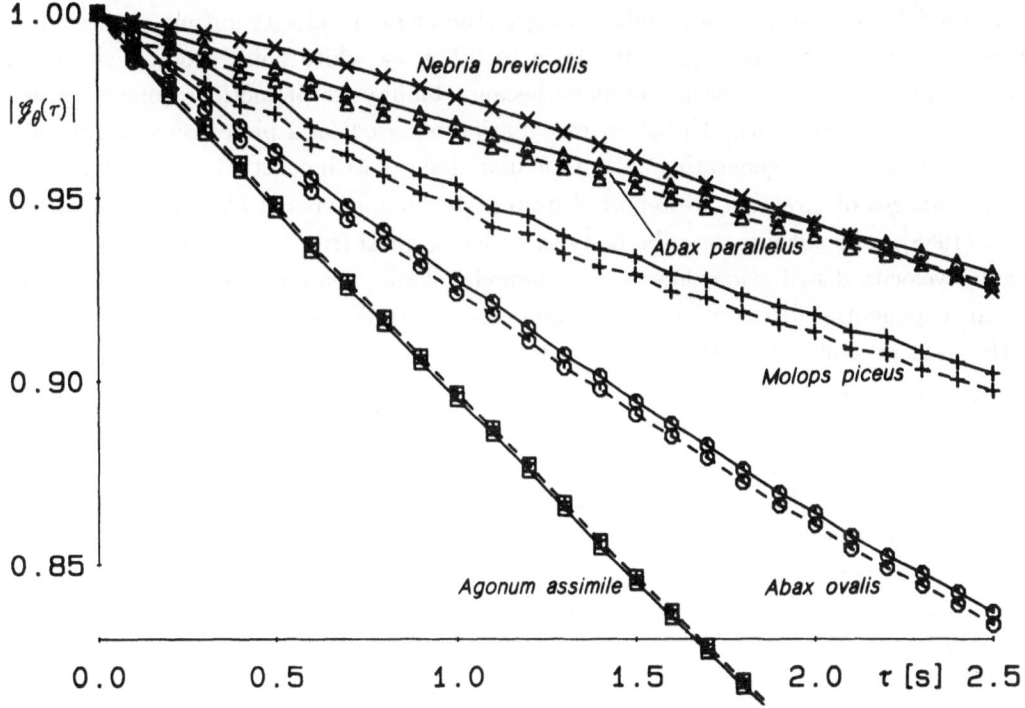

Figure 4: Upper part of the modulus ACF of experiments with 5 different carabid species. Dashed lines: uncorrected values. Solid lines: corrected for positional and discretisation errors. Note the expanded scale of the axes.

As Alt(1988) has shown, the way the ACF leaves its maximum at $\tau = 0$ provides a hint: if the process generating the directional changes is an uncorrelated jump process, the directional autocorrelation function $\mathcal{G}_\Theta(\tau)$ will decay exponentially with a discontinuous first derivative at the origin. If the angular changes have a low pass characteristic (e. g. an Ornstein-Uhlenbeck process), the angular ACF has zero slope at $\tau = 0$. To find a hint as to what is probable, this part of the ACF has been analyzed with a higher resolution (Fig.4).

In this case the data are evaluated with a sampling interval $\Delta t = 0.1s$ and the positional and the discretisation errors are corrected following the description given in Box 4. The ACFs of nearly all of the tested species show a rather straight descending course with no hint to a quadratic maximum. Only the function obtained with *Nebria brevicollis* has a tendency to positive curvature but does not show a horizontal slope at $\tau = 0$. This is the same experiment as shown in the figures before, this animal walks with an extremely constant speed, making rarely any stop. It is not clear, however, if that is the reason for the differing shape of this curve.

Conclusion

The straight decay of the autocorrelation function may be explained by two different models. The process of angular changes (the angular velocity output of the motor system) could be - in the time scale down to 0.1 s - a white noise process without a low-pass limitation. This seems unlikely because biophysics of motion alone does not allow unlimited fast turns. The other explanation is based on a piecewise straight path with a Poisson process generating abrupt angular changes with constant probability. Such abrupt changes of direction are reported from flight manoeuvres of *Drosophila* by Mayer et. al.(1988). The authors describe body saccades in quasi free flying flies. The reported angular velocity distributions have a clear bimodal shape, being composed of a Gaussian and an exponential distribution. The exponential part is responsible for the saccades. In the data presented here the angular velocity distribution has a pure Gaussian shape (Fig.2B). Furthermore the paths of the animals (Fig.1A) as well as the time course of walking direction (Fig.1C) seem to be too smooth for such a model. However, the high rotatory bias in this experiments may have masked some of the properties of the data. If one tries to analyse directly the angular velocity process (using the angular differences of consecutive intervals), the measuring errors become unbearable high. For that reason a decision between the two models of angular turning cannot be made at the moment. Improvements of resolution in time and in position or direction will be necessary to solve this problem.

REFERENCES

Alt W. (1988) Modelling of motility in biological systems, 15-30. In: ICIAM'87 Proceedings. Philadelphia: SIAM

Kramer E. (1975) Orientation of the male silkmoth to the sex attractant bombykol. In: D.A. Denton and J.P. Coghlan (eds.) Olfaction and taste 5:329-335. Academic press, New York

Mayer M., Vogtmann K., Bausenwein B., Wolf R., Heisenberg M. (1988) Flight control during 'free yaw turns' in *Drosophila melanogaster*. J. Comp. Physiol. A 163: 389-399

Wendler G., Scharstein H. (1986) The orientation of grain weevils (*Sitophilus granarius*): Influence of spontaneous turning tendencies and gravitational stimuli. J. Comp. Physiol. A 159: 337-389

BOX 4

THE INFLUENCE OF DISCRETE POSITION MEASUREMENTS
ON THE CORRELATION ANALYSIS OF 2–DIMENSIONAL TRACKS

Hans Scharstein, Wolfgang Alt

When measuring the position $Z(t) = (x(t), y(t))$ along a continuous, piecewise smooth (rectifiable) track at times $t_i = t_o + i \cdot \Delta t$ with regular <u>time steps of size Δt</u>, the resulting <u>discrete time series</u>

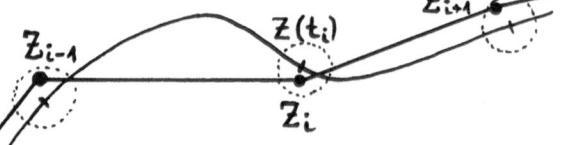

(1) $\qquad Z_i = Z(t_i) + \epsilon_i$

contains errors ϵ_i due to incorrect positioning which can be assumed as independently and normally distributed with variance σ^2.

A simple approximation of the track velocity $V(t) = dZ(t)/dt$ is

(2) $\qquad V_i = \dfrac{1}{\Delta t}(Z_i - Z_{i-1}) = S_i \cdot (\cos\alpha_i, \sin\alpha_i)$

with approximating directional angle α_i and displacement speed $S_i = |V_i|$. Since

(3) $\qquad V_i = \dfrac{1}{\Delta t} \int\limits_{t_i - \Delta t}^{t_i} V(t)\, dt + \dfrac{1}{\Delta t}(\epsilon_i - \epsilon_{i-1})$

the piecewise averaged speed will in general be underestimated by S_i, even for constant track speed $S = |V(t)|$, see eqs. (6) and (7) below. Also, the discrete angles α_i are only averaged measures of the movement direction, which should be taken carefully, too. However, under these general conditions the <u>theoretical velocity autocorrelation</u> function $\mathcal{G}_V(\tau)$ defined in (Alt, this section, eq.(4)) is explicitly connected to the analogously defined "discrete" <u>empirical velocity autocorrelation</u> function

$\qquad \widetilde{\mathcal{G}}_V(k) := \langle V_{i+k} \cdot V_i^* \rangle$ $\quad$ by the following formula (for $k \geq 0$):

(4)

$\qquad = (\mathcal{G}_V * h_{\Delta t})(k \cdot \Delta t) + \dfrac{\sigma^2}{\Delta t^2}(2 \cdot \delta_{k=0} - \delta_{k=1})$

where V^* denotes the complex conjugate of the vector V. The first term in (4), furtheron denoted by $\widehat{\mathcal{G}}_V(k)$, is the convolution with the unit triangle function $h_{\Delta t}$ of width Δt and "smoothens" $\mathcal{G}_V$ over this band width (see ○ in Fig.1), whereas the other terms provide a positive correction at k=0 and a negative one at k=1, proportional to the positioning variance σ^2 divided by Δt^2 (see * in Fig.1).

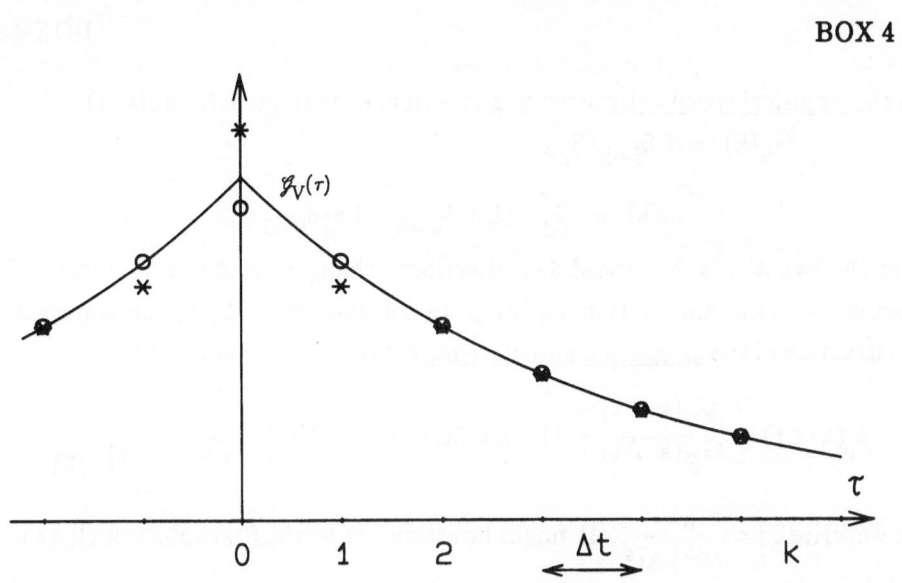

BOX 4

Figure 1. Example of a velocity autocorrelation function of exponential type with marked errors due to temporal discretization (○) and to position measurement (*). See formula (4).

Deminuishing the step size Δt clearly reduces the temporal discretization error, but on the other hand it leads to an error increase due to unprecize spatial positioning. For a fixed small variance σ^2 the <u>optimal choice</u> Δt_{opt} <u>of step size</u> which minimizes the "total" squared discretization error (for values at k=0 and k=1 only) can be determined as $\Delta t_{opt} \simeq (\kappa/|g_1|)^{1/3} \cdot \sigma^{2/3}$ with $g_1 := \mathscr{G}_V'(0)$ and a fixed constant $\kappa = 4 \cdot (\sqrt{17/8} - 1)$. Thus, knowledge of the <u>positioning error</u> σ in path tracking suggests a certain preferred interval between sampling times.

In any case, for measured data points with given σ and Δt we want to reconstruct the theoretical autocorrelation functions as far as possible. A reasonable first approximation of the modulus of the autocorrelation function for small correlation times τ is $G_V(\tau) = |\mathscr{G}_V(\tau)| = \langle S^2 \rangle \cdot \exp(-\tau/\Pi)$ with an "initial" <u>persistence time</u> Π , see the examples in (Scharstein, this section, Fig.4) . For small $\Delta t/\Pi$ and $\hat{\sigma}^2/\Delta t^2 \ll \langle S^2 \rangle$ the following approximations hold

(5) $\hat{G}_V(k) \simeq \begin{cases} (1 - \Delta t/3\Pi) \cdot G_V(0) & \text{for } k = 0 \\ G_V(k \cdot \Delta t) & \text{for } k \geq 1 \end{cases}$

and

(6) $\hat{G}_S(k) \simeq (1 - \Delta t/3\Pi) \cdot G_S(k \cdot \Delta t)$ for $k \geq 0$

BOX 4

with the <u>empirical speed autocorrelation</u> function, in analogy to formula (4)

(7) $\widetilde{G}_S(k) := \langle S_{i+k} \cdot S_i \rangle$

$\simeq \widehat{G}_S(k) + \frac{\sigma^2}{\Delta t} \cdot (1 + \delta_{k=0} - 0.5 \cdot \delta_{k=1})$.

Under the hypothesis that speed and directional changes are mutually independent processes , see (Alt, this section: eq.(24)), we conclude from (5)–(7) an approximate reconstruction of the <u>directional autocorrelation</u> function (for small k≥1):

(8) $\mathscr{G}_\theta(k \cdot \Delta t) = \frac{\mathscr{G}_V(k \cdot \Delta t)}{G_S(k \cdot \Delta t)} \simeq (1 - \Delta t/3\Pi) \cdot (1 + \tilde{\sigma}^2 + \frac{1}{2}\tilde{\sigma}^2 \cdot \delta_{k=1}) \cdot \frac{\widetilde{\mathscr{G}}_V(k)}{\widetilde{G}_S(k)}$

Here we write $\tilde{\sigma}^2 = \dfrac{\sigma^2}{\langle S^2 \rangle \Delta t^2}$. Π might be estimated by the first values of $\widehat{G}_V(k)$.

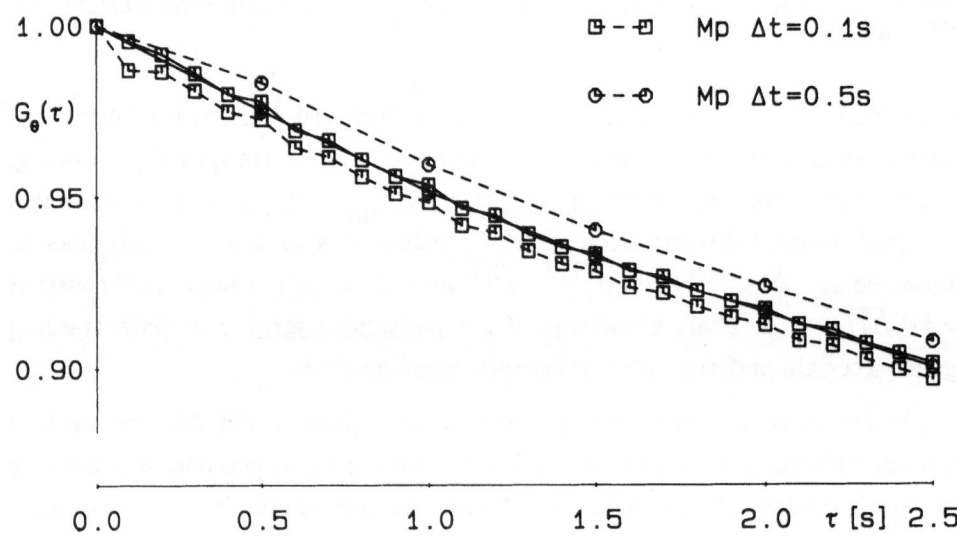

Figure 2. Initial part of the empirical directional autocorrelation function from experiment Mp (Scharstein, this section) for sampling intervals $\Delta t = 0.1$ s and $\Delta t = 0.5$ s. Dotted lines: raw data; solid lines: corrected values according to formula (8). Notice the interrupted ordinate. A typical <u>positional error</u> is the low value of the $\Delta t=0.1$s curve at k=1. For the curve with larger $\Delta t=0.5$s the <u>smoothing error</u> predominates, showing increased values at all points k≥1, compare also Fig.1 . After correction both curves coincide almost exactly, suggesting that the above assumptions are justified.

MICROZOOPLANKTON FEEDING BEHAVIOR AND THE LEVY WALK

J. Klafter
School of Chemistry
Tel Aviv University
Tel Aviv, 69978 Israel

B. S. White
Exxon Research and Engineering Co.
Annandale, NJ 08801

M. Levandowsky
Haskins Laboratories, Pace University
New York, NY 10038

Abstract

We propose a Levy random walk model for the grazing behavior of swimming microzooplankton. In this model the path of the organism consists of straight line segments which are traversed at constant speed. The segments are at random angles, which correspond to sudden changes in direction, and the lengths of the line segments are random, with probability density given by the Levy stable law. The Levy law is chosen because of its role in the generalized Central Limit Theorem: it is a possible limit law for sums of independent identically-distributed random variables. We show that for values of the Levy parameter β below 1, the small amount of overlap in successive excursions due to the relatively sparse distribution of turning events would lead to a more efficient feeding pattern than for a Brownian motion or a Levy walk with larger value of β. Possible physiological mechanisms are discussed, as well as an approach to testing for Levy walk behavior patterns.

1. Introduction

Many single-celled organisms exhibit a generalized random walk type of motility, in which episodes of approximately straight locomotion are inter- rupted randomly by reorientation events.   Such behavior has often been ob- served in swimming bacteria (*e.g.* Berg, 1983) and protists (Jennings, 1906; Fraenkel and Gunn, 1940) and also in the movements of some ameboid cells. It has given rise to models in which the basic concept has been a Gaussian diffusion, as in the familiar diffusion equation (Patlak, 1953; Keller and Segel, 1971; Alt, 1980; Okubo, 1980; Alt and Lauffenburger, 1985). This model arises naturally from a Gaussian approximation if the straight line segments are of random length, with finite variance.

In recent years the ecological importance of the motile behavior of swimming protists in relation to their role in the planktonic food webs of natural waters has been the subject of much discussion (e.g. Sherr et al, 1986; Porter et al, 1985; Sieburth, 1984).   In this connection, Fenchel (1987) has suggested that bacterivorous microzooplankters would not survive in the more oligotrophic open waters, based on assumptions about motility and feeding.   In this, a maximum grazing rate was calculated, based on esti- mates of maximum pressure gradients sustainable across a row of filtering cilia.   Fenchel's model assumes, implicitly, that grazing is non-selective, whereas recent studies suggest that behavioral responses to chemical signals from food organisms may be very important in the feeding of micro- zooplankters (Buskey et al, 1988; Verity, 1988).

In this paper, we consider another aspect of swimming behavior, which may also be significant in relation to feeding efficiency, and thus to sur- vival of phagotrophic microzooplankters in waters where food is the limiting factor.   As it swims, regardless of whether the organism is grazing non- selectively or searching for chemical signals, it is effectively sweeping out a 3-dimensional volume.   (The crossectional area of the sampled cylin- drical volume will depend on whether it can detect chemical or other signals from nearby prey, or simply blunders into food as it progresses, as assumed in Fenchel's model).   The 3-dimensional volume may be thought of as sausage shaped. Thus, for a random walk where the lengths of successive excursions have a normal distribution, modeled by a Wiener process, it has become customary among mathematicians to refer to this volume as the "Wiener sausage" (e.g., Donsker and Varadhan, 1975).

From the viewpoint of a grazing microzooplankter, the problem with the Wiener sausage is that the sampled volume has a great deal of overlap, due to the frequency of turning, so that the organism following such a random walk would continually be resampling water it had sampled before. It is therefore interesting to investigate other types of sausage, based on non-normally distributed random walks. In this paper we shall look at the properties of a family of such random walks in which the amount of overlap is less than in the Wiener sausage.

2. Search Strategies.

We consider the strategy of a microorganism searching for food in three dimensional space. Let ρ be the mean number of food particles per unit volume of space. Neglecting the effects of clumping, we assume that the food is distributed randomly according to the Poisson distribution

$$(1) \qquad P\left\{ N \text{ particles in volume V} \right\} = \frac{(\rho V)^N}{N!} e^{-\rho V}$$

We assume that food is sparse and that the chemoreceptive range of the organism is small. That is, food will be detected only if it lies within a small sphere of radius ϵ about the organism's central position. We neglect the effect of depletion of resources by grazing, and concentrate instead on the inter-feeding time, T, when searching randomly about in space among fixed food particles distributed according to equation (1).

Let the stochastic process $X(t) \epsilon R^3$ be the position of the organism at time t, and let T be the random variable denoting the time to next feeding. It is assumed that $X(0)$ is at the origin of space, which is chosen independently of the distribution of food. Of vital importance is the probability

$P(T>t)$, that the organism goes hungry for time t. If $P(T>t)$) does not van-
ish sufficiently rapidly as $t\to\infty$ the organism is likely to starve. Thus a
species must adapt to keep $P(T>t)$ small for large t. Equivalently, the random volume V_t that is explored in time t must, with high probability, be
sufficiently large.

For species below a certain size, the vagaries of molecular collisions
will determine the search strategy, i.e., its members will be subjected to
Brownian motion. Thus the random trajectories $X(t)$ may be described by the
Wiener process. The search volume V_t is the volume of an ϵ-neighborhood
about the organism's trajectory. It is a tortuously self-intersecting tube
about the Wiener process (dubbed the "Wiener sausage") and its asymptotic
behavior for large t has been determined in the rigorous work of Donsker and
Varadhan (1975). We consider here species large enough to influence the
search pattern to some extent, through their adaptive swimming and grazing
behavior. In particular, we consider still waters, so that the effects of
fluid motion, e.g. turbulence, may be neglected.

Our organisms may, of course, choose the Wiener strategy, among other
options of mindless simplicity. This behavior would emerge, typically, from
execution of many small steps, each with finite variance. The result is a
Gaussian distribution of $X(t)$ dictated by the Central Limit Theorem. There
are, however, limit laws other than the Gaussian which can result from sums
of independent identically-distributed random variables (Breiman 1968). If
the additional requirement of non-directionality is imposed, the totality of
such limit distributions in one dimension is described by the two-parameter
family (β,δ) of symmetric Levy stable laws, with $\delta>0$ and $0<\beta\leq 2$. Here β is
the key parameter, henceforth called "The" Levy parameter, while δ merely
gives the spatial scale. For $\beta=2$, the Levy reduces to a Gaussian, with variance 2δ. For $0<\beta\leq 2$ the characteristic function (Fourier transform of the
probability density $f(r)$) is

(2) $\hat{f}(\omega) = \exp(-\delta |\omega|^{\beta})$

A stochastic process related to (2) for $\beta\neq 2$ is that of the Levy flight.
An organism moving according to this process would sit idly for a random

time (with exponential probability distribution) after which it would jump instantaneously to another point at a random distance determined by the Levy distribution. The volume V_t of the related sausage has likewise been studied by Donsker and Varadhan (1975), who give a rigorous theory for its asymptotic behavior. We consider here a heuristic theory for the Levy walk, a process similar to the Levy flight, but with finite (constant) speed along the trajectories between points of the Levy flight.

Unlike the Wiener case, an organism performing a Levy walk retains directional information for some time. That is, it decides to travel resolutely with constant velocity in a random direction and for a random time predetermined by a Levy stable law. Since it moves some distance along a straight line segment, its path is less likely to be self-overlapping, and hence we expect the volume V_t might be somewhat larger.

Since for large r the Levy stable law has, for $0<\beta<2$, the asymptotic form $f(r) \sim const/r^{1+\beta}$, Levy steps with $0<\beta<2$ are likely to be large (they have infinite variance), with smaller values of β yielding more of the very long straight line segments. The choice of β therefore determines how likely the organism will be to go to great lengths to avoid resampling territory already explored. To assess the possible strategies, we need to calculate $P\{T>t\}$.

Let $R(t) = \sup_{0<s<t} |X(s)|$ be the largest value of $|X|$ up to time t, and let $g_{R(t)}(r)$ be its probability density. Then from the law of conditional probability

$$(3) \qquad P\{T>t\} = \int_0^\infty P\{ T>t \mid R(t)=r\} \, g_{R(t)}(r) \, dr$$

To calculate $P\{T>t \mid R(t)=r\}$ we assume that for large t,r a significant fraction of the sphere of radius r is explored. This is true both for the Brownian and the Levy flight cases, and we expect it to hold here also, since the walker explores more of space than the flyer. Since the distribution of food is Poisson, the probability of finding no food in a volume proportional to r^3 is

(4) $P\{\ T{>}t\ |\ R(t){=}r\} \sim \exp\{-\mu\ r^3\}$

where μ is a constant involving both the prevalence of food, ρ, and the likely fraction of volume explored.

Let τ_r be the time a Levy walker first exits from a sphere of radius r about the origin. The cumulative probability distribution of R(t) is then

(5) $P\{\ R(t){<}r\} = P\{\tau_r{>}t\}$

and so

(6) $g_{R(t)}(r) = \dfrac{\partial}{\partial r}\ P\{\tau_r{>}t\}$

We assume that the exit time τ_r has, for large t, asymptotically an exponential distribution

(7) $P\{\tau_r{>}t\} \sim C(r)\ \exp\{-t/t_0(r)\}$

where $t_0(r)$ is a characteristic time for distance of order r. Equation (7) is again by analogy to the Brownian and Levy flight cases, where it can be proved ($1/t_0(r)$ can, in those case, be related to an eigenvalue problem for a deterministic operator related to the relevant stochastic process). Here we verify equation (7) by Monte Carlo simulations of a Levy walk, as described in section 3.

We next use a heuristic scaling to find the r-dependence of $t_0(r)$ for large r. It is known that (see accompanying Box) the mean square displacement satisfies, for large t

(8) $< X^2(t) > \sim \begin{cases} \text{const x } t^2 & \text{for } 0{<}\beta{\le}1 \\ \text{const x } t^{3-\beta} & \text{for } 1{\le}\beta{\le}2 \end{cases}$

We therefore assume that the scaling for $t_0(r)$ is governed by the inverse of equation (8), i.e.

$$t_0(r) \sim \text{const} \times r^\gamma \quad, \text{ with}$$

(9)

$$\gamma = \begin{cases} 1 & \text{for } 0<\beta\leq 1 \\ 2/(3-\beta) & \text{for } 1\leq\beta\leq 2 \end{cases}$$

We now have the terms for substitution into (3).

(10)
$$P\{T>t\} \sim \int_0^\infty C_0(r) \exp\{-\mu \, r^3 - Ct/r^\gamma\} \, dr$$

Where $C_0(r) \sim -C(r) \, d(t/t_0(r))/dr$. The integral (10) may be computed asymptotically for large t by neglecting the pre-exponential factor and using Laplace's method. the result is that

(11)
$$\log (P\{T>t\}) \sim -\text{const} \times t^\alpha$$

where the exponent $\alpha = 3/(3+\gamma)$ is given by

(12)
$$\alpha = \begin{cases} 3/4 & \text{for } 0<\beta\leq 1 \\ (9-3\beta)/(11-3\beta) & \text{for } 1\leq\beta\leq 2 \end{cases}$$

The largest value, $\alpha=3/4$, and hence the best strategy, is attained for any β with $0<\beta\leq 1$. Hence we predict that only β's in this range will be observed in nature. This is also the range of β's for which the Levy step has infinite mean.

3. Monte Carlo Simulations.

Equations (7) and (9) were tested by Monte Carlo simulations of Levy random walk in three dimensions with $\beta=1.5$. We estimated the probability distributions of the exit times from spheres of radius $r=8,16,32,64$, and 128 using histograms of the exit times obtained from the results of 50,000 simulations for each value of r. For each simulation, independent Levy random variables were generated by the method described below. Each walk was constructed by joining straight line segments with random lengths given by this Levy distribution and with random directions given by unit vectors uniformly distributed on the unit sphere. Each walk was continued until a straight line segment was determined to intersect the sphere of radius r. Since the walker was assumed to travel with unit speed, the exit time was recorded as the total length of the path up to the time of intersection with the sphere of radius r.

To generate Levy random variables, we first computed the Levy probability density function by fast Fourier transform of the characteristic function given by equation (2). This function was integrated numerically to obtain the cumulative distribution function $F(t)$. The inverse function F^{-1} of F was computed on a uniform mesh using linear interpolation. Since $F(t) \to 1$ as $t \to \infty$, $F^{-1}(U)$ goes rapidly to infinity as U goes to one. Therefore to obtain acceptable accuracy near $U=1$ we used the inverse of the asymptotic expression

(13) $1 - F(t) \sim 1/(8\sqrt{2\pi}\, t^{1.5})$, as $t \to \infty$

Now using a well-known method, a Levy random variable can be obtained by the transformation $F^{-1}(U)$, where U is uniformly distributed on $[0,1]$. We thereby obtained Levy random variables using a uniform random number generator.

We first tested equation (7) by plotting, for each value of r, $\text{Log}[P(\tau_r > t)]$ versus t. A typical plot, that for $r=128$, is shown in Figure 1. As predicted by equation (7) it is well fitted by a straight line for large values of t. The slope of the regression line is approximately $-1/t_0$.

We next tested equation (9) using the estimates of $t_0(r)$ which were obtained for the five chosen values of r. As shown in Figure 2, the graph of $t_0(r)$ versus r is well approximated by a straight line on a log-log plot, thus verifying a power law dependence. The slope obtained by linear regression was 1.55, in error by about 16% with the predicted value of 1.33.

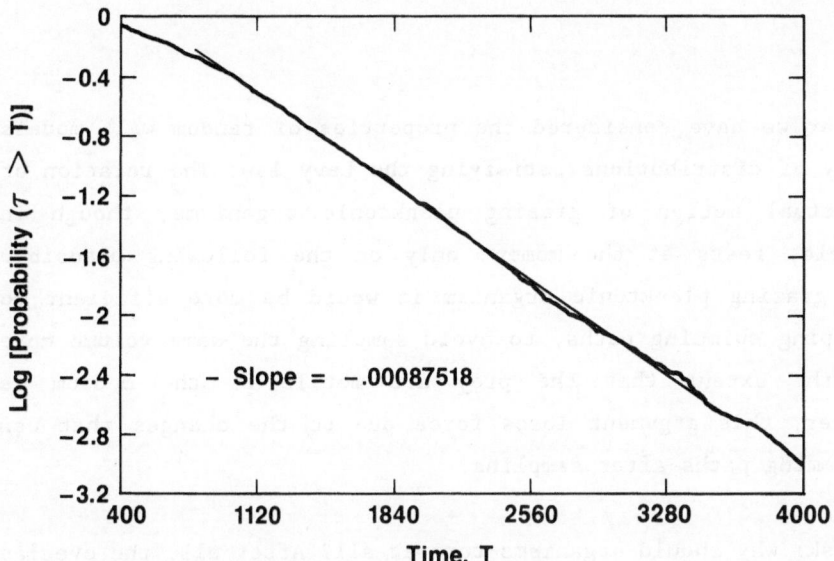

Figure 1. Probability distribution of exit time, τ_r, from a sphere of radius r=128.

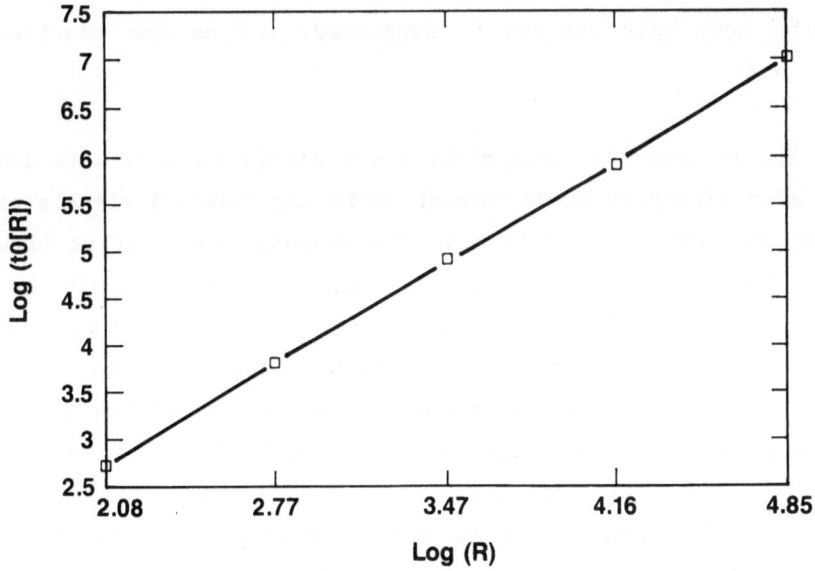

Figure 2. Verification of power law behavior for time scale $t_0(r)$.

4. Discussion

In this paper we have considered the properties of random walk models based on a family of distributions satisfying the Levy law. The relation of these to the actual motion of grazing planktonic organisms, though in principle testable, rests at the moment only on the following plausible argument: for a grazing planktonic organism it would be more efficient to minimize overlapping swimming paths, to avoid sampling the same volume more than once. To the extent that the prey are motile or the medium is turbulent, however, this argument loses force due to the changes that can occur in the swimming paths after sampling.

One might ask: why should organisms turn at all? After all, the overlap in swimming paths would be minimal if there were no changes in direction. To address this point, we need to consider other aspects of the biology of the grazer. It seems reasonable to think of its behavior as having two parts: first the search for food, and then, when the presence of food is detected, the behavior involved in capturing and feeding on the prey. It is for the first part that the Levy walk pattern is suggested. Let us now consider briefly the second part.

Detection of the presence of food might occur simply through a random collision, but is also likely to occur through diffusing chemical signals or (for swimming prey) mechanical vibrations in the medium. From what we know of the chemosensory behavior of certain ciliate protozoans, for example, response to chemical signals by these swimming unicellular grazers might take the form of a change in both swimming speed and turning frequency, in such a way that they tend to accumulate in the vicinity of the source of the signal (Levandowsky et. al. 1984; Hellung-Larsen et. al. 1986). This type of response is sometimes called a "kinesis" (Fraenkel and Gunn 1940), and the underlying physiology and biochemistry of the mechanism of directional change has been the subject of much study (e.g. Naitoh and Eckert 1974).

With all this in mind then, we suggest the following picture of the grazing plankter:

(1) In the absence of any signal, it tends to suppress the turning mechanism and swim in long straight paths, as described by the Levy walk model.

(2) When a signal is detected indicating the presence of food, a different swimming mode comes into play. Both turning frequency and swimming speed may now be modified as the grazer starts a detailed hunt for the prey.

A plausible variation on this strategy, based on the observation that many planktonic algal prey organisms, for example, have a "patchy" distribution, would be for the grazer to follow a Levy walk swimming pattern until it comes to a patch of food organisms, whereupon it changes to a different pattern. In this case, the Levy walk pattern becomes a strategy for finding regions of high food density, and once in such regions, a different behavior would be used.

In this model, then, the Levy walk is generated by suppressing most of the turning behavior of the organism. The rare turns that do occur are in a sense "vestiges" of another behavioral pattern that is largely suppressed. In more physiological detail, one might plausibly hypothesize that turns are related to random changes of membrane permeability to calcium ions, and that the threshold for such events is raised during the Levy walk swimming pattern.

We have assumed throughout that the food organisms themselves do not swim. This would be appropriate in the case of diatoms or cyanobacteria, both of which can be dominant in phytoplankton. In the case of flagellates, however, the swimming activity of the prey would tend to blur the "memory" of the predator's search strategy, as the prey swims into the sampled path. Thus the adaptive value of the Levy walk strategy would decrease in the presence of motile prey organisms.

To test the Levy walk model, one might utilize the rate of population spreading or diffusion predicted by the model, when the parameter β is less than one. From equation (8), the root mean square displacement is proportional to time for β in this range. Hence, if a population of such organisms is injected into a small volume, the rms displacement of the dispersing cloud of organisms should advance with constant velocity. This should be observably different from, say, a population dispersing according to Brownian motion, where the rms displacement would be proportional to the square root of time. Such a population would advance with dwindling velocity proportional to $t^{-1/2}$.

REFERENCES

Alt, W. (1980) Biased random walk models for chemotaxis and related diffusion approximations. J. Math. Biol. 9: 147-177.

Alt, W. & Lauffenburger, D.A. (1985) Transient behavior of a chemotaxis system modelling certain types of tissue inflammation. J. Math. Biol. 24: 691-722.

Berg, H.C. (1983) Random Walks in Biology Princeton University Press, Princeton.

Breiman, L. (1968) Probability, Addison-Wesley, Reading

Buskey, E.J. & D. Stoecker (1988) Locomotory patterns of the planktonic ciliate Favella sp.: adaptations for remaining in food patches. Bull. Mar. Sci. 43: 788-796.

Donsker, M. & Varadhan, S. (1975) Asymptotics for the Wiener sausage. Comm. Pure Appl. Math. 28: 525-565.

Fenchel, T. (1987) Ecology of protozoa, Springer-Verlag, Berlin.

Fraenkel, G.S. & D.L. Gunn (1940, reprinted 1961) The orientation of animals, Dover, N.Y.

Hellung-Larsen, P., Leick, V., and Tommerup, N. (1986) Chemoattraction in Tetrahymena: on the role of chemokinesis. Biol. Bull. 170: 357-367.

Jennings, (1906) Behavior of the lower organisms, Columbia, New York.

Keller, E.F. & Segel, L. (1971) Model of chemotaxis. J. Theor. Biol. 30: 225-234.

Levandowsky, M., Cheng, T., Kehr, A., Kim, J., Gardner, L., Silvern, L., Tsang, L., Lai, G., Chung, C., and Prakash, E. (1984) Chemosensory responses to amino acids and certain amines by the ciliate Tetrahymena Biol. Bull. 167: 322-330.

Levandowsky, M., Klafter, J. and White, B. S. (1988) Swimming behavior and chemosensory responses in the protistan microzooplankton as a function of the hydrodynamic regime. bull. Mar. Sci. 43: 758-763.

Levandowsky, M., Klafter, J. and White, B. S. (1988) Feeding and swimming behavior in grazing microzooplankton, J. Protozool. 35: 243-245.

Naitoh, Y., and Eckert, R. (1974) The control of ciliary activity in protozoa, In: M. Sleigh (ed.) Cilia and Flagella: 305-352, Academic Press, Orlando.

Okubo, A. (1980) Diffusion and ecological problems: mathematical models. Springer-Verlag, Berlin.

Patlak, C.S. (1953) Random walk with persistence and bias, Bull. Math. Biophys. 15:311-338.

Porter, K.G., Sherr, E.B., Sherr, B.F., Pace, M. & Sanders, E. (1985)

Protozoa in planktonic food-webs. J. Protozool. 32: 409-414.

Schlesinger, M.F. & Klafter, J. (1985) Levy walks versus Levy flights. In: Stanley, H.E. & Ostrowski, N. (eds.) On Growth and Form: 279-283. Martinus Nijhof Rub. Amsterdam.

Sherr, E.B., Sherr, B.F. & Paffenhofer, G-A. (1986) Phagotrophic protozoa as food for metazoa: a "missing" trophic link in marine food webs? Mar. Microb. Food Webs 1:61-80.

Sieburth, J. McN. (1984) Protozoan bacterivory in pelgic marine waters. In: Hobbie, J.E. & Williams, P.J. (eds.) Heterotroophic Activity in the Sea: 405-444, Plenum.

Verity, Peter (1988) Chemosensory behavior in planktonic ciliates. Bull Mar. Sci. 43: 772-82.

BOX 5

MEAN SQUARE DISPLACEMENT FOR UNCORRELATED RANDOM WALKS

J. Klafter, B. S. White, and M. Levandowsky

We here show how to calculate the mean square displacement for an organism which executes an uncorrelated random walk, in particular a Levy walk, for large times t. The path, $X(t)$, of the organism is composed of straight line segments of random length which the organism traverses at constant speed. After each segment it turns off randomly in another direction and proceeds at constant speed along another straight line path. Let the mean square displacement of $X(t)$ be defined by

$$(1) \qquad D(t) = E[|X(t)|^2 | X(t_0)=0]$$

with $t_0=0$ denoting a turning event, and let $t_1,t_2,\ldots$ be the random turning times, independent identically distributed random variables with Levy probability density function, $f(t)$. We will derive an integral equation of "renewal" type by considering expectations conditional on t_1.

First, we consider $t<t_1$. Since in this event the path is along a single straight line

$$(2) \qquad E[|X(t)|^2 | t_1=\xi] = t^2, \text{ for } t<\xi$$

Next consider $t>t_1$. First note that since $X(t)-X(t_1)$ is at a random angle to $X(t)$, we have

$$(3) \qquad E[(X(t)-X(\xi))\bullet X(\xi) | t_1=\xi] = 0 \text{ , for } t>\xi$$

Also, if $t_1=\xi<t$ then $X(t)-X(\xi)$ has the same statistics as $X(t-\xi)$, so that

$$(4) \qquad E[|X(t)-X(\xi)|^2 | t_1=\xi] = D(t-\xi) \text{ for } t>\xi$$

BOX 5

Now combining (3) and (4) we have that

$$(5) \qquad E[|\ X(t)\ |^2\ |\ t_1{=}\xi] = \xi^2 + D(t{-}\xi), \text{ for } t{>}\xi$$

By conditioning on t_1 we obtain that

$$(6) \qquad D(t) = \int_0^\infty E[|\ X(t)\ |^2\ |\ t_1{=}\xi]\ f(\xi)\ d\xi$$

By putting (2), (5) into (6) we obtain the integral equation

$$(7) \qquad D(t) = q(t) + \int_0^t D(t{-}\xi)\ f(\xi)\ d\xi$$

where

$$(8) \qquad q(t) = t^2 \int_t^\infty f(\xi)\ d\xi + \int_0^t \xi^2\ f(\xi)\ d\xi$$

We denote Laplace transforms by tildes, with s the variable dual to t. then denoting by prime derivative with respect to s we have from (8)

$$(9) \qquad \tilde{q}(s) = 2\frac{\tilde{f}'(s)}{s^2} + 2\frac{(1{-}\tilde{f}(s))}{s^3}$$

From Laplace transform of equation (7)

$$(10) \qquad \tilde{D}(s) = \frac{\tilde{q}(s)}{(1{-}\tilde{f}(s))}$$

and therefore

$$(11) \qquad \tilde{D}(s) = \frac{2}{s^3} \left[1 + \frac{s\ \tilde{f}'}{(1{-}\tilde{f})} \right]$$

From equation (11) we can now deduce the large t behavior of D(t), by

BOX 5
considering the behavior of $\tilde{D}(s)$ for small s.

If $0<\beta<1$, then $f(t)\sim const/t^{1+\beta}$ implies that $\tilde{f}(s)\sim 1-Cs^\beta$ for some constant C. Putting this into (11) then yields

$$(12) \qquad\qquad \tilde{D}(s) \sim \frac{2(1-\beta)}{s^3} \qquad \text{as } s\to 0 \text{ for } 0<\beta<1$$

Therefore $D(t)\sim const\ t^2$ as $t\to\infty$, $0<\beta<1$, as asserted.

If $1<\beta<2$ the mean of t_1 is finite, i.e.

$$(13) \qquad\qquad C_1 = E[t_1] = \int_0^\infty t\, f(t).dt < \infty \ , \quad \text{for } 1<\beta<2$$

Therefore for $1<\beta<2$, $\tilde{f}(s) \sim 1-C_1 s+Cs^\beta$ as $s\to 0$. Putting this into (11) then yields

$$(14) \qquad\qquad \tilde{D}(s) \sim \frac{2C(\beta-1)}{C_1} \frac{1}{s^{4-\beta}} \qquad \text{as } s\to 0,\ 1<\beta<2$$

Thus $D(t)\sim const\ t^{3-\beta}$ as $t\to\infty$, $1<\beta<2$ as asserted.

ADAPTATION AND ORIENTATION IN ANIMALS' MOVEMENTS: RANDOM WALK, KINESIS, TAXIS AND PATH-INTEGRATION

P. Bovet   S. Benhamou

Equipe Modélisation des Déplacements Orientés
Laboratoire de Neurosciences Fonctionnelles
C.N.R.S., BP71, F-13402 Marseille Cedex 9

Abstract

A first order correlated random walk model was developed to represent animals' movements. This model makes it possible to formalize the stochastic concept of sinuosity of an animal's search path and to determine some basic properties such as its diffusion. This model can be used to analyse actual animals' movements, or to develop more complex movement models integrating cybernetic controls of the sinuosity and the velocity as a function of environmental stimulations. Models of this kind show how animals can orient themselves in a stimulation gradient field or exploit patchy environments using simple klino- and ortho-kinetic mechanisms which have been reformulated. When movement regulation depends on the local value of the field potential (absolute mode), klinokinesis and orthokinesis can both be seen to be elementary space-use mechanisms, allowing the animal to adapt its movements to the environmental conditions. When movement regulation depends on variations in the field potential (differential mode), klinokinesis can be seen to be an elementary spatial orientation mechanism, whereas orthokinesis seems to have no biological relevance. Contrary to differential klinokinesis, taxis is an orientation mechanism based on the determination of the gradient direction. The properties of differential kinesis and taxis are compared and statistical tools for distinguishing between them by means of path analysis are proposed. Finally, the first order correlated random walk model was used to quantify navigational errors involved in path-integration as a function of the type and the magnitude of estimation errors about route-based information.

1. Introduction

Animals (as well as bacteria and free cells) often exhibit random search paths (cf Bovet & Benhamou 1988). This intrinsic randomness does not however prevent them from efficiently orienting towards specific goals and/or aggregating in the the most suitable areas of their environment. To understand space-use mechanisms (allowing the animal to regulate the time it spends in the various areas of the environment) and orientation mechanisms (allowing the animal to move towards a specific goal), it is necessary to first model the search paths. Afterwards it is important to determine which environmental cues are relevant to animals and which kinetic parameters they have to regulate to be efficient. In the various models presented here we have attempted to formalize some of the mechanisms involved in movement control.

As in introduction to what follows, let us call "stimulation field", an environment containing an intensity factor, the field potential in mathematical terms. A heterogeneous environment is a stimulation field where the potential is much higher in some randomly localized parts than in others. When the potential varies nearly monotonously either along an axis or as a function of the distance to the centre of the field, the environment will be referred to as an axial or a radial stimulation gradient field, respectively, the gradient direction corresponding to the direction in which the largest local increase in the field potential occurs. A heterogeneous stimulation gradient field can be defined by combining the two elementary types of field defined above, but this is out of our present scope.

2. The First Order Correlated Random Walk Model: Definition of the Sinuosity of Search Paths

The simple random walk model used in physics to represent the movements of brownian particles is not suitable for representing the search paths of most animals because their cephalo-caudal polarization and their bilateral symmetry cause a tendency to go forward that this model does not account for. A first order correlated random walk model is able to account for these basic properties. Any search path can then be represented as a sequence of steps with constant length P, and changes of direction α randomly drawn from a normal distribution (wrapped around the circle) with a null mean and a standard deviation σ (0.1 rad$\leq\sigma\leq$1.2 rad): the correlation between the direction of successive steps is expressed by $r = E(\cos \alpha) = \exp(-\sigma^2/2)$. Dividing a path into discrete steps is a useful but very often arbitrary way of representing it. Consequently, the sinuosity of an actual path (which constitutes a measure of the "amount of turning" associated with a given path length) should not depend on the discretization step used to analyse it. We have shown that the expected standard deviation of the distribution of changes of direction σ_R obtained after rediscretization with a step length R (with R>P/2 and R=P) of a path simulated with parameters σ and P is given by $\sigma_R = 0.85\,\sigma/\sqrt{(R/P)}$ while $\sigma_R \leq 1.2$ rad. Separating what belongs to the model (σ and P) from what relates to the rediscretization (σ_R and R), we define sinuosity as (Bovet & Benhamou 1988):

(1) $S = \sigma/\sqrt{P} = 1.18\,\sigma_R/\sqrt{R}$

The first part of this formula is a theoretical expression for the sinuosity, whereas the second part is a practical expression for quantifying the sinuosity of actual paths.

The expected square diffusion distance $E(D^2)$ of an N-step first order correlated random walk with variable step length δ and correlation parameter r, with any 0-centred distribution of changes of direction, is given by (Kareiva & Shigesada 1983; Marsh & Jones 1988):

(2) $E(D^2) = N.E(\delta^2) + 2r/(1-r) \, E^2(\delta) \, (N - (1-r^N)/(1-r))$

When the step length is constant $(E(\delta^2)=E^2(\delta)=P^2)$ and N is large, this formula can be reduced to $E(D^2)=N.P^2(1+r)/(1-r)$. One can then demonstrate (see Bovet & Benhamou 1988) that the expected value and the standard deviation of the diffusion distance are equal to $E(D)=(\pi.E(D^2)/4)^{1/2}$ and $sd(D)=(4/\pi-1)^{1/2}E(D)$. Assuming that changes of direction are drawn from a normal distribution with a null mean and a standard deviation σ $(r=\exp(-\sigma^2/2)$, the diffusion distance of a search path with the sinuosity $S=\sigma/\sqrt{P}$ and the length $L=N.P$ can be statistically characterized by (Bovet & Benhamou 1988):

(3) $E(D) = 1.77 \, \sqrt{L}/S$
(4) $sd(D) = 0.92 \, \sqrt{L}/S$

A dynamic expression for the expected diffusion is obtained by replacing the path length by the product of the travel time and the mean velocity.

3. Redefinition of Klinokinesis

The original definitions of klinokinesis and orthokinesis (Gunn et al 1937; Fraenkel & Gunn 1940) have led to great confusion about the properties of these two mechanisms, because they were both based on spatio-temporal concepts, the rate of change of direction and the velocity intensity (i.e. the speed), respectively, when first coined to account for animal aggregation in the most suitable areas of the environment. As the rate of change of direction of a given path varies with the animal's velocity it becomes impossible to distinguish between the respective roles of the spatial and temporal components of the movements. We have therefore proposed to redefine klinokinesis by substituting the purely spatial concept of sinuosity for the previous spatio-temporal concept of rate of change of direction, and to keep the original definition of orthokinesis (Benhamou & Bovet 1989): klinokinesis and orthokinesis are now defined as mechanisms regulating the sinuosity and the velocity intensity, respectively, of a random movement as a function of either the local value of the field potential (in the absolute mode, i.e. "without sensory adaptation"), or its variation (in the differential mode, i.e. "with sensory adaptation").

4. Movement Adaptation to the Spatial Heterogeneity of the Environment: Absolute Klinokinesis and Orthokinesis

Absolute klinokinesis and orthokinesis are elementary space-use mechanisms. They allow animals (as well as bacteria and free cells) to adapt their movements to the spatial heterogeneity of the environment. An animal regulating its sinuosity (klinokinesis) and/or its velocity (orthokinesis) as a function of the suitability of the area where it is moving can thus increase the time spent in the most suitable areas of the environment, and reduce the time spent travelling between them (Benhamou & Bovet 1989). For this purpose, the animal only needs to move with a sinuosity as high and/or a velocity as low as the suitability of the area crossed is high in comparison with the mean suitability of the environment. This is simply a consequence of the diffusion law: the time spent in a given area is proportional to the square of the sinuosity and inversely proportional to the velocity.

One important application of absolute klinokinesis and orthokinesis relates to the adaptation of foraging behaviour to the contagious distribution of the local density of prey items occurring in patchy environments: increasing the sinuosity after finding a prey item leads an animal to concentrate its search effort in higher prey density areas, and decreasing the velocity leads it to increase both the probability of detecting encountered prey items (Gendron & Staddon 1983; Knoppien & Reddingius 1985) and the time devoted to feeding (Arditi & Dacorogna 1988). Absolute klinokinesis and orthokinesis can thus be used to formalize the "area-restricted searching" behaviour mentioned in the framework of optimal foraging theory (references in Benhamou & Bovet 1989).

5. Indirect Orientation in a Stimulation Gradient Field: Differential Klinokinesis

Differential klinokinesis constitutes a sinuosity regulating mechanism which can be used by animals (and by bacteria and free cells) to efficiently orient themselves in a stimulation gradient field without determining the gradient direction: they need only to measure the variations in the field potential occurring during their movements. In an axial gradient field, this mechanism can be formulated as (Benhamou & Bovet 1989):

$$(5) \qquad S = S_b \, (1-k.\cos \Theta)$$

where S is the local sinuosity of the path, S_b the basic sinuosity, k (ranging between -1 and 1) the klinokinetic factor and Θ the orientation of the ongoing step in relation to the gradient direction (cos Θ expresses the normalization between -1 and 1 of the intensity variation perceived during a step: no perception of the gradient direction is therefore required). Using computer simulations, we have shown that this elementary mechanism efficiently leads to a drift by the animals in the gradient direction. The expected drift is practically equal to the product of the path length L and the klinokinetic factor k:

(6) $E(D_k) = k.L$

whether the sinuosity regulation is carried out by varying the standard deviation of the distribution of changes of direction or the step length (Benhamou & Bovet 1989).

A theoretical application of differential klinokinesis was made by Bovet (1984) to account for the light-orientation behaviour of organisms endowed with primitive eyes which are able to determine local illumination, but not the light direction. Other applications of differential klinokinesis, involving a radial stimulation gradient field, have been made by Benhamou (1989) to account for the olfatory orientation of a mammal in its home range based on its own scent marks, and by Jamon (1987) to account for the search-loop behaviour exhibited by ants (Wehner & Srinivasan 1981) and woodlice (Hoffmann 1984). From the original definitions of kineses, Rohlf & Davenport (1969) have shown that both klinokinesis and orthokinesis are able, in the differential mode, to lead to an aggregation of animals in the most suitable part of a stimulation gradient field. In actual fact, these two mechanisms both involve variations in the sinuosity, and correspond to equivalent forms of differential klinokinesis. Differential orthokinesis along a constant sinuosity path seems to have no biological applications.

6. Direct Orientation in a Stimulation Gradient Field: Taxis

Unlike diferential klinokinesis, taxis involves reliance by the organism on the gradient direction to orient itself in a stimulation gradient field. Perfect determination of the gradient direction obviously leads to a straight, well oriented movement, as in the particular case of visually guided movements towards a conspicuous goal by animals with highly evolved eyes (from a mathematical point of view, a conspicuous goal can be seen as the centre of a radial stimulation gradient field, the potential being the distance from the goal). In many cases, however, the steering of movement may be subject to random variations both because of fluctuations in the field potential due to the turbulence of the environment and because of gradient direction perception errors due to the inaccuracy of receptors. When the perceived gradient direction is subject to large fluctuations, a purely tactic mechanism systematically forcing the organism to move in the perceived gradient direction is not very efficient because it leads to a very chaotic path. One plausible solution for preventing the organism from taking excessively large changes of direction consists of incorporating a strong forward persistence. In any case, the performances (as expressed by the ratio between the straight line distance measured along the gradient axis and the actual path length) achieved on the basis of taxis and differential klinokinesis can often be very similar.

It is possible to differentiate between these two mechanisms by performing a statistical analysis of the path. For the sake of clarity we restricted ourselves to the case where only a purely tactic mechanism may be involved. Here, the analysis consists of compararing the distribution of changes of direction between successive steps and the distribution of step directions in relation to the gradient direction. These two distributions are centred on 0, but the dispersion of the former is likely to be higher than that of the latter in the case of taxis, and lower in the case of differential klinokinesis. Furthermore, a large negative auto-correlation between changes of directions is to be expected in the case of taxis, but not in that of differential klinokinesis, because of the successive steering corrections. A statistical response is not always very clear, however, and a physiological response, when possible, is more reliable. For example, oriented movements by magnetic bacteria in relation to the earth's magnetic field clearly correspond to taxis, whereas oriented movements of bacteria in a chemical gradient field can involve either taxis or differential klinokinesis, since bacteria are probably able (depending on what specific receptors they are provided with) to perceive variations in the field potential (i.e. the chemical concentration) and/or the gradient direction.

7. Orientation relying on an Egocentric Spatial Memory: Route-based Navigation (Path-integration)

Another type of orientation is exhibited by an animal which is able to return to a particular place in the absence of any gradient or landmark indicating the location of this place. This animal has to memorize the location of this place by means of an egocentric coding process, i.e by relating the distance and the direction of the place to the animal's own location and orientation by processing the information collected "en route" about its ongoing path. This type of orientation mechanism has been formulated as inertial navigation (Barlow 1964), path-integration (Mittelstaedt & Mittelstaedt 1980, 1982) and route-based navigation (Baker 1981). It requires the animal to continuously update the egocentric coding of the location of its goal (distance and direction) during its movements (Potegal 1982). After the ith change of direction (rotation) α_i, estimated by $\hat{\alpha}_i$, and the $i+1^{th}$ step with length (translation) P_{i+1}, estimated by $\hat{P}_{i+1}$ (after $i+1$ steps, the animal has performed only i changes of direction), the new egocentric coding of the place where the animal intends to return is recurrently expressed by the estimations of the distance to this place, D_{i+1}, and its direction in relation to the animal's antero-posterior axis, Ω_{i+1} (Benhamou et al 1990):

(7) $\hat{D}_{i+1} = (\hat{D}_i^2 + \hat{P}_{i+1}^2 - 2\,\hat{D}_i.\hat{P}_{i+1}.\cos(\hat{\Omega}_i - \hat{\alpha}_i))^{1/2}$

(8) $\hat{\Omega}_{i+1} = \arctan(\sin(\hat{\Omega}_i - \hat{\alpha}_i)/(\cos(\hat{\Omega}_i - \hat{\alpha}_i) - \hat{P}_{i+1}/\hat{D}_i)) + k\pi$

with $k=0$ when the denominator is positive and $k=1$ otherwise.

Because of the inaccuracy of animals' measuring systems, the memorized location of the place is subject to random errors which tend to accumulate during the process because of its iterative nature. We performed computer simulations in order to quantify the accuracy of the egocentric coding of the place where the animal intends to return as a function of the sinuosity and length of the outward path and the type and magnitude of route-based information estimation errors: the coding accuracy was found to be very sensitive to the errors occurring in idiothetic (without external reference) estimations of rotatory movements but not in estimations of translatory movements or in allothetic (compass-based) estimations of rotatory movements. It is therefore not surprising that woodlice (Hoffmann 1984) and ants (Wehner & Wehner 1986), which have been shown to use a sun-compass, are able to achieve better orientational performances than mammals (Etienne et al 1986, 1988), which have to idiothetically estimate their changes of direction during their movements.

8. Conclusion: Aggregation should not be confused with Orientation !

Two different types of mechanism have been described here: space-use mechanisms (absolute klinokinesis and orthokinesis) and orientation mechanisms (differential klinokinesis, taxis, and route-based navigation). Unfortunately, these two types of mechanism have not always been clearly differentiated in the past, because orientation has frequently been confused with aggregation. Aggregation can indeed be a population effect caused by individual orientations, provided the individuals move independently (do not attempt to avoid each other) and orient themselves to the same areas of the environment. Aggregation has therefore often been used as an index to orientational performance. Aggregation can occur however in the absence of any orientation mechanism, as in the case of absolute klinokinesis and orthokinesis. For example, consider two-compartment experimental tank containing micro-organisms which are free to move from one compartment to the other. When using absolute klinokinesis and/or orthokinesis, the micro-organisms will spend more time in the compartment with the more suitable environmental conditions, and therefore aggregate there. This aggregation is a simple statistical effect, and not the expression of an orientation mechanism: after a transition phase following the introduction of organims into the tank, the barycentre of the population stabilizes somewhere in the more favorable compartment, but it is not subject to a drift which increases with time, as in the case of differential klinokinesis in a stimulation gradient field. The confusion between aggregation and orientation is at the origin of the old misleading question as to the need for sensory adaptation to explain animal aggregation in terms of kineses: differential klinokinesis as well as absolute klinokinesis and orthokinesis lead to animal aggregation, but for very different reasons, namely oriented drift and unoriented diffusion control.

REFERENCES

Arditi R. & Dacorogna B. (1988) Optimal foraging on arbitrary food distributions and the definition of habitat patches. Am. Nat. 131: 837-846.

Baker R.R. (1981) Human navigation and the sixth sense. London: Hodden and Stoughton.

Barlow J.S. (1964) Inertial navigation as a basis for animal navigation. J. Theor. Biol. 6: 76-117.

Benhamou S. (1989) An olfactory orientation model for mammals' movements in their home ranges. J. Theor. Biol. 139: 379-388.

Benhamou S. & Bovet P. (1989) How animals use their environment: a new look at kinesis. Anim. Behav. 38: 375-383.

Benhamou S., Sauvé J.P. & Bovet P. (1990) Spatial memory in large scale movements: efficiency and limitation of the egocentric coding process. J. Theor. Biol. 145, 1-12.

Bovet P. (1984) Modèles clinocinétiques, pp. 232-236. In: P. Clement & R. Ramousse (Edit.), La vision chez les invertébrés. Paris: C.N.R.S.

Bovet P. & Benhamou S. (1988) Spatial analysis of animals' movements using a correlated random walk model. J. Theor. Biol. 131: 419-433.

Etienne A., Maurer R. & Saucy F. (1988) Limitations in the assessment of path dependent information. Behaviour 106: 81-111.

Etienne A., Maurer M., Saucy F. & Teroni E. (1986) Short distance homing in the golden hamster after a passive outward journey. Anim. Behav. 34: 696-715.

Fraenkel G.S. & Gunn D.L. (1940, revised 1961) The orientation of animals; kineses, taxes, and compass reactions. New-York: Dover Publications.

Gendron R.P. & Staddon J.E.R (1983) Searching for cryptic prey: the effect of search rate. Am. Nat. 121: 172-186.

Gunn D.L., Kennedy J.S. & Pielou D.P. (1937) Classification of taxes and kineses. Nature 140: 1064.

Hoffmann G. (1984) Orientation behaviour of the desert woodlouse Hemilepistus reaumuri: adaptations to ecological and physiological problems. Symp. Zool. Soc. Lond. 53: 405-422.

Jamon M. (1987) Effectiveness and limitation of random search in homing behavior, pp. 284-294. In: P. Ellen & C. Thinus-Blanc (Edit.), Cognitive processes and spatial orientation in animal and man, vol 1: Experimental psychology and ethology. Dordrecht: Martinus Nijhoff Publishers.

Kareiva P.M. & Shigesada N. (1983) Analyzing insect movement as a correlated random walk. Oecologia 56: 234-238.

Knoppien P. & Reddingius J. (1985) Predators with two modes of searching: a mathematical model. J. Theor. Biol. 114: 273-301.

Marsh L.M. & Jones R.E. (1988) The form and consequences of random walk movement models. J. Theor. Biol. 133: 113-131.

Mittelstaedt H. & Mittelstaedt M.L. (1982) Homing by path integration, pp. 290-297. In: F. Papi & H.G. Wallraff (Edit.), Avian navigation. Berlin Heidelberg: Springer-Verlag.

Mittelstaedt M.L. & Mittelstaedt H. (1980) Homing by path integration in a mammal. Naturwissenschaften 67: 566.

Potegal M. (1982) Vestibular and neostriatal contributions to spatial orientation, pp. 361-387. In: M. Potegal (Edit.), Spatial abilities, development and physiological foundations. London: Academic Press.

Rohlf F.J. & Davenport D. (1969) Simulation of simple models of animal behavior with a digital computer. J. Theor. Biol. 23: 400-424.

Wehner R. & Srinivasan M.V. (1981) Searching behaviour of desert ants, genus Cataglyphis. J. Comp. Physiol. A 142: 315-338.

Wehner R. & Wehner S. (1986). Path integration in desert ants. Approaching a long-standing puzzle in insect navigation. Monitore Zool. Ital. 20: 309-331.

THE SITE INDEPENDENT INFORMATION THAT AN ISOPOD USES FOR HOMING

G. Hoffmann, Zoologisches Institut II,
Röntgenring 10, D-8700 Würzburg, FRG

Abstract

In order to return to a given point animals use two different sources of information. One source are landmarks or trails. Specifically for long range homing animals prefer to use site independent information which they get by measuring and concatenating their active displacements, and store until they need it.

A characteristic example is a desert isopod (*Hemilepistus reaumuri*) which after searching for food has to return to the burrow in which it lives together with its family. Its orientation behaviour has been analyzed experimentally and theoretically in order to clarify structure and extent of the information made available by path concatenation, and the method how an isopod uses this information in order to return as fast as possible.

The information that an isopod uses is not restricted to a single estimated value of the burrow position. It takes into account that this relative orientation method necessarily is deficient in precision, and uses information about the probability of the respective orientation errors. The information about the burrow position available to an isopod can be best described as a probability density with which the burrow is situated in different positions relative to its own position.

A homing isopod always moves to the nearest point where the probability density is maximal. This hypothesis explains the behaviour of an isopod both on the straight part of the return path and further on when it has to search for the burrow entrance.

1 Introduction

Many vertebrates and invertebrates are able to return to a point where they have been before, even if they are kept from using landmarks or trails that they produced for orientation. A typical example is the desert isopod *Hemilepistus reaumuri* (Crustacea, Isopoda). In search for food an isopod leaves its burrow deep in the soil. After having found food it has to return to its burrow, since it is the only shelter in which it can survive (Linsenmair and Linsenmair 1971). The orientation problems that an isopod faces in this situation depend on the season and the place where it decided to build its burrow. In the following we will analyse the orientation behaviour of isopods living in a typical habitat which carry food home for their offspring, which remains in the burrow for about two weeks after birth before it forages for itself.

By displacing such an isopod which is about to home, it is very easy to show that it does not need information deduced from its immediate surroundings for orientation. Without hesitation it runs straight to the spot where the burrow would be had it been displaced with it, and then starts to search for it (Hoffmann 1984).

Beginning with Jander (1957) there were several attempts to explain such a behaviour (Mittelstaedt 1978, 1985, Wehner and Wehner 1986, Müller and Wehner 1988). All existing theories start from the assumption that an animal could register direction and length of its active displacements and concatenate them. They describe this concatenation process, often called <u>path integration</u>, by giving formulas for an estimated value of the direction to the burrow. This value is thought to describe the whole information available to the animal for orientation. Only Mittelstaedt (1978, 1985) implicitly gives a rule for the estimated value of the homing distance.

An animal using such limited information to control its return path would be in a very difficult situation when it moved straight through the estimated distance into the estimated direction without detecting its target. At this point, it would have no further information available where it should go next. However, not finding its target at the end of the straight part of the return path is a typical problem for the desert isopods and probably also of other animals using this type of orientation. Up to now their exists no theory which could describe how an animal is able to home by displacement concatenation in such a situation.

This is the reason why in this contribution it is first tried to determine structure and extent of the information that an isopod gets by displacement concatenation. Then it shall be shown how an isopod uses this information in order to return as fast as it is possible with the restricted information available.

2 Results

2.1 How an isopod forages and returns home afterward

If an isopod would move straight into one direction during its search for food it could return home rather easily. It only had to turn around and move into the opposite direction for the same distance it walked away from the burrow. In fact an isopod restricts its orientation problems by keeping the main direction once chosen when it leaves the burrow. Nevertheless, it continuously and rapidly turns left and right while moving forward and probing the ground with the antennae in search for food. Even if one disregards these countless small turns, the path of an isopod is rather meandrous (Fig. 1). The path length of a foraging excursion (101 $\pm$ 120.7 cm, n=89) is 1.6 ($\pm$ 0.57) times as long as the beeline distance between the burrow and the place where the isopod found food.

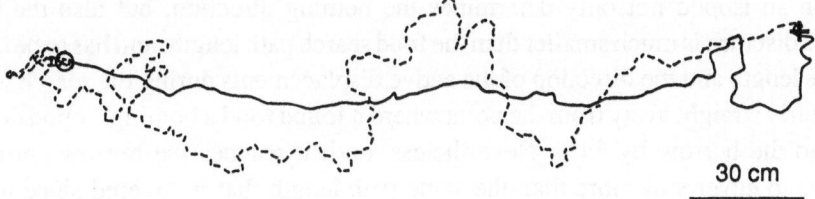

Figure 1: The search for food of a desert isopod (- - -) and its return path (——) to the burrow (+). The place where it found food to carry it home is marked by o.

To us it seems to be a very simple task to return to a point at a distance of 56 cm, the average beeline distance that a homing isopod has to overcome. To an isopod, however, this is a most difficult task. It detects the burrow entrance only by contact with the antennae from a distance of 20 mm (Hoffmann 1983). An isopod that has to return from a distance of 56 cm and deviates by more than 2.3 degrees from the true homing direction will pass by the burrow entrance without detecting it.

The orientation mechanisms of the desert isopods meet these high demands. The direction of the first, straight part of the return path (Fig. 1) deviates on average by only 0.9 degrees from the true homing direction. With the same precision an isopod displaced when it is about to carry food home heads for the point where the burrow would be had it been displaced with the isopod (Fig. 2a). This observation shows that a desert isopod is able to determine its homing direction very precisely by displacement concatenation only.

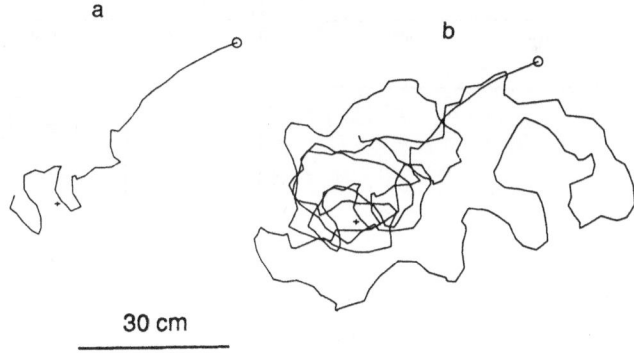

Figure 2a: Return path of a desert isopod after displacement by c. 7 m without rotation. The cross (+) marks the point at which the burrow entrance would be if it had been displaced along with the animal. **b:** The subsequent search for the burrow entrance.

If an isopod chooses the wrong homing direction it does not move straight forever. When it has walked for a distance which approximately equals the true homing distance, it turns abruptly and starts searching for the burrow entrance. The very same behaviour can be observed in isopods displaced during foraging (Fig. 2a). It shows that by displacement concatenation an isopod not only determines the homing direction, but also the homing distance. This distance is much smaller than the food search path length, and has to be deduced both from the length and the direction of the active displacements during the search for food.

Moving nearly straight away from the point where it found food a homing isopod decreases its distance to the burrow by 84%. Nevertheless, until it reaches the burrow entrance on average it has to cover still more than the same path length that it covered since it started homeward. This observation could give the wrong impression that after the end of the straight part of the return path the isopod has got no further information about the burrow position to direct its course and therefore simply wanders around similar to molecule in brownian motion until it runs into its burrow. However, this first impression is misleading as can be shown by artificially introducing orientation errors which the isopod has to compensate for. If one displaces an isopod by distance between 20- 200 cm it returns faster to the burrow than is to be expected if it searched in a brownian way (Fig. 3).

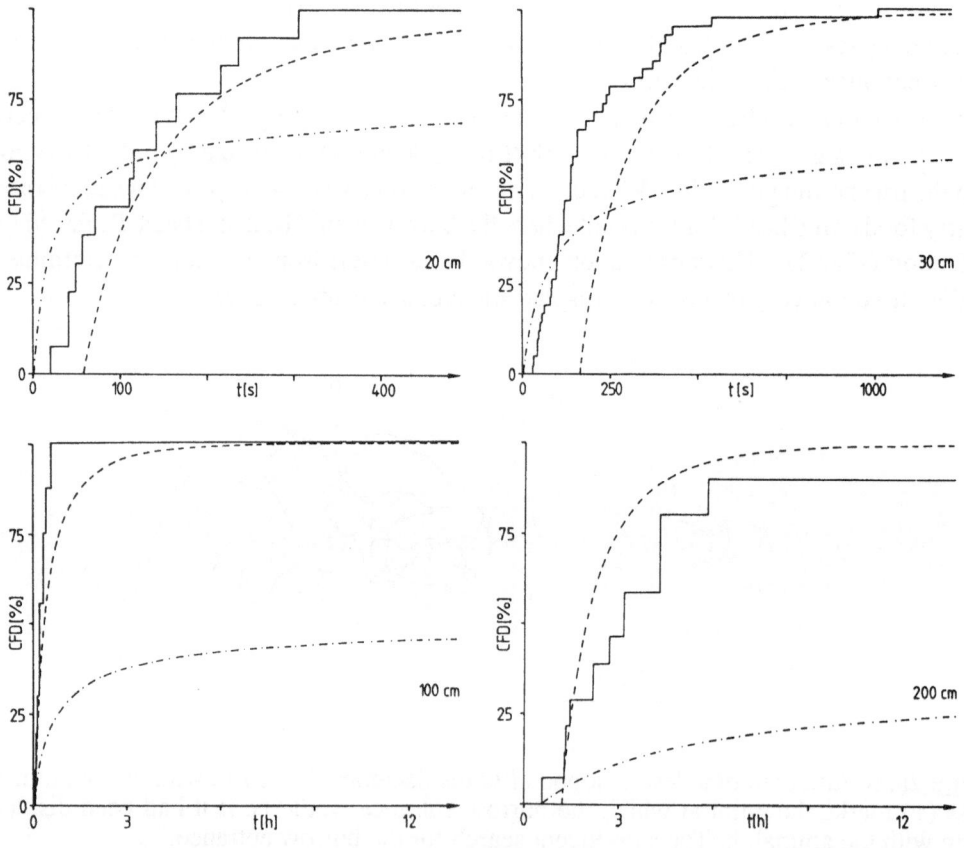

Figure 3: Success of the search of isopods for their burrow (——) which have been displaced from a point where they found food in comparison to the success of a brownian random search (-----) and a systematic search (- - -). Shown is the cumulative frequency with which an isopod detects its burrow by time t (note the different time scales). The displacement distance is given in each figure.

2.2 Structure and extent of the stored information

After having described the <u>behaviour</u> of an isopod during its search for food and the return to the burrow we now proceed to the main questions of this paper: How can we describe the information processing that is connected with this orientation behaviour, what is the structure of the stored information, what amount of information is stored, and how does an isopod use it to steer the movement direction in each moment?

By moving straight from the point where it found food to the point where it starts searching for the burrow entrance an isopod displaced on its way home shows us that it has got rather precise information both about the direction and the distance to the burrow. However this information is not always correct. The point where an isopod starts its search scatters around the true position of the burrow entrance (Fig. 4). Large errors are rare, small ones are common.

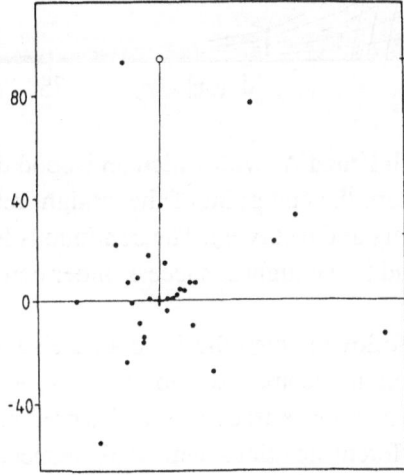

<u>Figure 4</u>: Position (●) of an isopod at the end of the straight part of the return path relative to its burrow (+). The return path has been turned in such a way around the burrow that its begin lays exactly above the burrow. The average distance between this begin and the burrow is shown by the symbol o, the respective standard deviation by the marks on the right. The isopods have been displaced far from the burrow. Therefore + marks the position of the burrow if it had been displaced with the isopod.

The astonishing fact is that in the following period an isopod does not wander around like a molecule in brownian motion, or search everywhere in its surrounding with equal intensity for the burrow entrance. During its search it seems to take into account the frequency that the burrow is in a given distance. It searches most intensively in the neighbourhood of the end point of the straight return path and with decreasing intensity further out (Fig. 5).

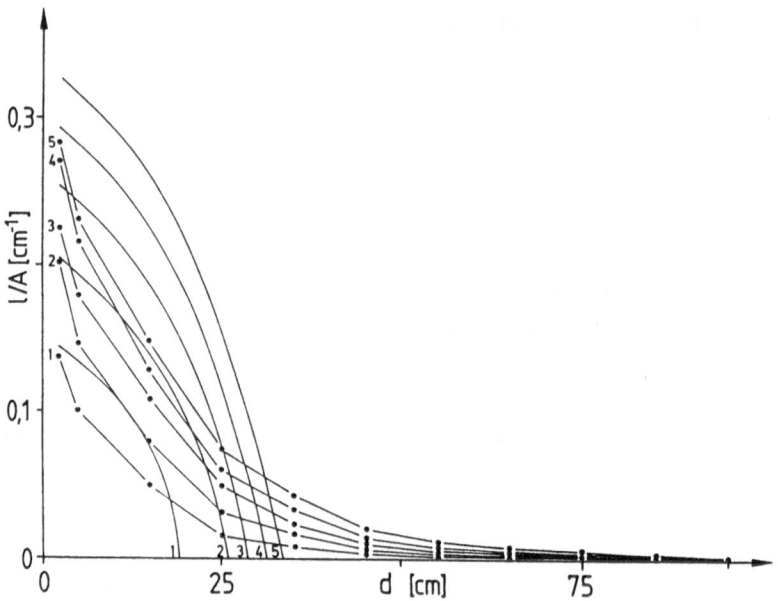

<u>Figure 5</u>: Intensity (●, path length l/ area A) with which an isopod displaced during homing searches in a given distance d from the end point of the straight return path within each of the first 5 minutes (cf. the numbers at the curves). The continuous line shows the respective search intensity which would lead to the highest success under conditions given in the text.

This observation leads to the following hypothesis about the structure of the information that an isopod uses for orientation: it encompasses not only a single direction and distance, but a whole set of positions. Each position is associated with a preference value. This structure of the information, a set of different positions with their respective preferences, is best described by an <u>orientation map</u>, on which the preference values are shown for a given area (Fig. 6a).

At any moment the observation of a homing isopod shows us only a minute part of the information available to it: the direction to the nearest point with the highest preference value. After having found food it first runs straight to one point, the approximate position of the burrow, and then searches around this point in a systematic way, changing its movement direction continuously.

To check our hypothesis about the structure of the information that an isopod uses for orientation we have to specify the orientation map quantitatively. For this purpose we assume that the preference value of an isopod for a given area corresponds to the probability (observed by us) that the burrow is in a given position relative to the isopod. At the end of the straight part of the return path this probability can be easily measured (Fig 4, 6b). It corresponds quite well to the twodimensional gaussian distribution

(1)
$$p(x) = \frac{1}{2\pi\sigma^2}\exp(-x^2/2\sigma^2)$$

with s = 13.2 cm (cf. Box 6, Fig.1a).

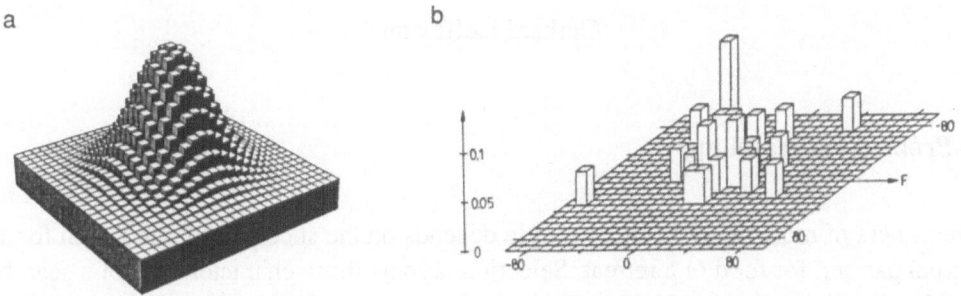

a   b

Figure 6a: Orientation map representing the information of an isopod about the position of its burrow at the end of the straight part of the return path. It shows the preference of the isopod for a given square. <u>b</u>: frequency density with which the burrow entrance actually is in a given square with an area of 64 cm^2 relative to the end of the straight part of the return path. F marks the direction to the point from where the isopod returned.

However, if an isopod wants to return as fast as possible it has to take into account that this probability changes continuously. By searching in a given area in vain the isopod decreases the probability that the burrow entrance is there. Since it could just oversee it, this probability becomes never zero. According to our observations the detection function

(2) $$b(z) = 1 - \exp(-cz)$$

with c = 22.2 (cm^2/s) describes quite well the conditional probability that the isopod will have found the burrow entrance if it is in a given area, and the isopod searches there with intensity z (s/cm^2).

Now our hypothesis has been stated in such a mathematical form that it can be tested experimentally. By a method described in Box 6 the best solution of the orientation problem which an isopod faces at the end of the straight part of the return path can be calculated. As figure 3 shows an isopod returns as fast as it is possible if its information about the burrow position corresponds to equation (1), and it takes into account its detection errors as described by the detection function (2).

BOX 6

HOW TO DESCRIBE AN EFFICIENT SEARCH ?

Gerhard Hoffmann

1 Problems of searching

The fitness of most living beings strongly depends on the success of their search for a sexual partner, for food or a retreat. Selection favours those characteristics of a search that increase its success and leaves unaffected characteristics which do not alter it. The relation between a given search problem, the search strategy of the animal and its success shall be clarified.

In most cases the information that an animal has got about its target's position relative to its own, and that it can use for the organization of its search is not exact. Rather there are many positions, where the target could be, although with a varying probability. Therefore the information available to the animal is described by a probability density, the target density $p(x)$, defined on a search space X. It gives the probability density that the target is in a point x relative to the initial position of the animal $(x = 0)$ (Fig. 1a).

It is evident that the success of a search depends on the fact how intensely the animal searches at the places where its target is with the highest probability. In all cases where there are areas in which the target can be anywhere with equal probability, details of the search path in these areas do (on average) not influence the success of the search, as long as the animal searches these areas evenly. Therefore we describe the spatio-temporal structure of the search path rather coarsely by the level of the search intensity $z(x,T)$, disregarding details of the search path which do not influence the success. The search intensity is best defined by stating that the integral

$$(1) \qquad\qquad T_A = \int_A z(x,T)dx$$

shall give that part T_A of the total search time T during which the animal searched within an area A.

An animal may search repeatedly in one place without detecting its target, even when it is within reach of the sense organs, and will find it only after repeated attempts at the very same place. A predator searching for camouflaged prey is a typical example. To describe this observation, the detection function $b(z)$ is used. It gives the conditional probability that the animal will detect the target, when it searches at the position of the

BOX 6

target with the intensity z . We will assume, that this probability increases monotonously with the search intensity, but with a monotonously decreasing rate. This condition is fulfilled by the often used (homogeneous) detection function $b(z) = 1 - e^{-cz}$ (Fig. 1b).

Now we can calculate the success of given search strategy $z(x, T)$ with a total duration

(2) $$T = \int_X z(x, T)dx.$$

The probability f that the animal will have found its target at point x by this time is given by

(3) $$f = p(x)b(z(x, T)),$$

and the probability that it will have found its target at all is given by

(4) $$P[z] = \int_X p(x)b(z(x, T))dx.$$

2 Calculating the most successful search strategy

To find the best solution of the search problem of an animal, we have to calculate the search intensity $z^*(x, T)$ for which the probability of success $P[z^*]$ is maximal under the given conditions and the restriction that the total search time T is fixed. If searching in one place does not influence the success of searching in another place, the usual Lagrange Multiplier technique for solving optimization problems with constraints can be used for this purpose. It leads to the local equation for $z^*(x, T)$

(5) $$\frac{\partial}{\partial z}(p(x)b(z) - \lambda z)\,|_{(z=z^*)} = 0$$

$$p(x)\frac{\partial}{\partial z}b(z)\,|_{(z=z^*)} = \lambda.$$

Let $\rho(x, z) := p(x)\frac{\partial}{\partial z}b(z)$. It is called marginal rate of return, since it is the rate with which the probability increases that the animal will find its target by searching in point x, if it has searched there already with intensity z .$\rho(x, z)$ is a strictly monotonously decreasing function of z for $z > 0$ (Fig. 1b).

For a given value of λ the corresponding optimal value of the search intensity $z^*(x)$ is given by

(6) $$z^*(x, \lambda) = \rho^{-1}(x, \lambda) := \begin{cases} Inverse\ function\ of\ \rho(x, z)\ for\ 0 < \lambda \le \rho(x, 0) \\ 0\ for\ \lambda > \rho(x, 0) \end{cases}.$$

BOX 6

For $b(z) = 1 - e^{-cz}$, for example, $z^* = \frac{1}{c}\ln\left(\frac{cp(x)}{\lambda}\right)$. The total search duration $T(\lambda)$ resulting from $z^*(x,\lambda)$ can be calculated by means of equation 2.

(7) $$T(\lambda) = \int_X z^*(x,\lambda)dx.$$

It is a strictly decreasing function of λ.

Now the necessary equations for calculating the optimal search intensity are available. First equation 7 can be used to calculate $\lambda(T)$ for a given search problem and duration T. Then the optimal search intensity $z^*(x,T)$ can be calculated by means of equation 6.

It is important to note that under the given conditions the best search strategy calculated for a given search time T_1 is also the best first part of a search with a longer duration T_2. Therefore an animal searching with intensity $z^*(x,T)$ will both maximize the probability of success up to an arbitrary time T and minimize the average search time until it finds its target.

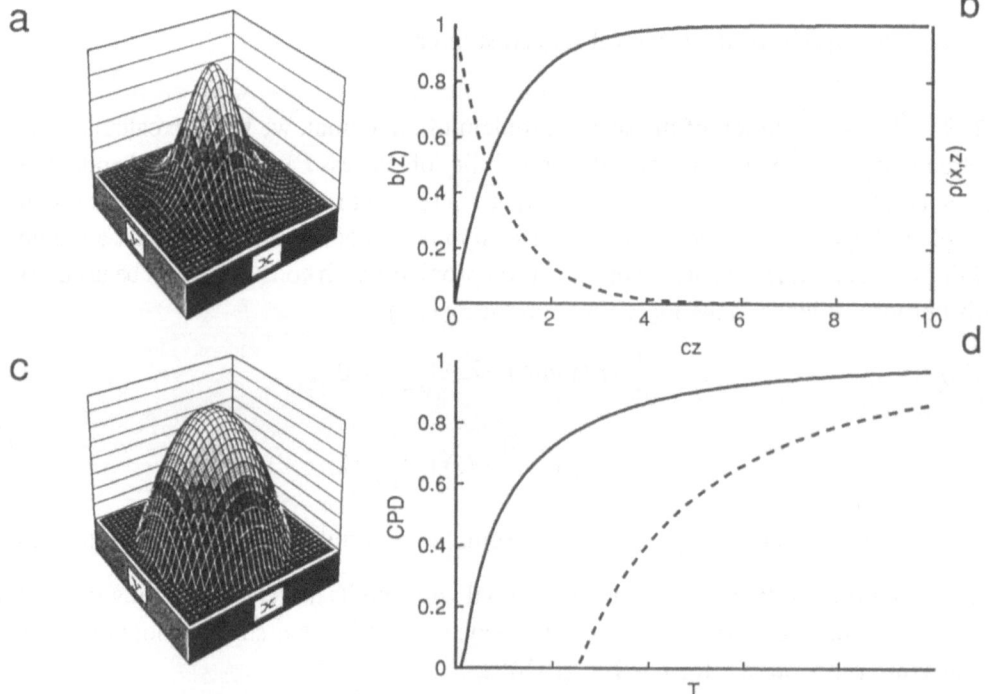

Figure 1: Examples of functions characteristic of an efficient search. a: target density $p(x) = 1/2\pi\sigma^2 \exp(-x^2/2\sigma^2)$. b: detection function $b(z) = 1 - \exp(-cz)$ and marginal rate of return $\rho(x,z)$. c: optimal solution $z^*(x,T)$ of the given search problem for a search duration of $T = 50$. d: probability of success of this search strategy in case the target is at a distance of 1 σ(————),2 σ(- - - -) (for details see text).

BOX 6

3 Comparison between the observed and the most successful search

Which observable characteristics of an animals search are suitable for a quantitative test of the hypothesis, that it searches as effectively as possible under the given conditions? Measuring the time course of the probability of success of a search in different experimental search situations is the best test. It can be easily calculated for every experimental situation by means of the equation $P[z^*] = b(z^*(x_0, T))$ if the target the animal is searching for has been placed to the point x_0 by the experimenter (Fig. 1d).

Our hypothesis makes several predictions also about the spatiotemporal structure of the search which can be tested:

(1) Up to a given time T the animal must not search in certain areas, namely everywhere, where the probability to detect the target is too low $(\rho(x,0) < \lambda(T))$.

(2) The area in which the animal should search $(\rho(x,0) \geq \lambda(T))$ increases steadily, but ever more slowly, since $\lambda(T)$ decreases with a decreasing rate. Inside this area the animal should search with a (measurable) intensity, which increases with time as $\lambda(T)$ decreases (Fig. 1c). The consequence is, that the animal should return repeatedly to areas where it searched already before, because it could have "overseen" its target. It should search there with a decreasing additional intensity per time increment.

(3) The last characteristic to be mentioned is due to the fact, that the calculated best form of the search can only be approximated by an animal, since it cannot search at the same time in every place where $\rho(x,0) \geq \lambda(T)$. Therefore the animal should move from its actual place to those areas where $\rho(x,z)$ is maximal. From this rule rather precise predictions can be deduced where the animal should move next.

---

3 Discussion

3.1 Advantages and drawbacks of the displacement concatenation mechanism

If, in order to return to a given point, an animal has to overcome a distance which is much larger than the range of its sense organs, learning the relative positions and characteristics of a series of landmarks becomes a difficult problem. To use trails for this task pays only when a large group of individuals of social animals moves between two fixed points. Probably this is the reason why so many animals orient by displacement concatenation while homing over large distances.

The advantages of displacement concatenation are opposed by difficulties necessarily connected with this type of relative orientation. An animal deduces the relative position of its target from the measurement and concatenation of displacements. Therefore errors tend to accumulate, and cannot be compensated for by the displacement concatenation mechanism itself. Fig. 4 shows that this can lead to dangerous errors.

An isopod that would not take these errors into account, and would search in a brownian way for the burrow entrance would not survive more than a few foraging excursions (Hoffmann 1984). Only by searching effectively it is able to return fast, even if it went totally astray on the straight part of the return path.

To coordinate a long time search is an even more arduous orientation task for an isopod than to return near to the burrow entrance after foraging. The minimal requirement is that it has to keep track of its position relative to starting point of the search. To improve the efficiency of its search it has, to some degree at least, to remember where it searched already before in vain.

As many other animals the desert isopods diminish their orientation problems in part by using a landmark that they produce themselves. Each family living in a single burrow (Linsenmair 1987) carries its faeces out of the burrow entrance, and puts it in a ring around the burrow entrance. This ring has an outer diameter of ca. 20 cm. An isopod detects it by means of the same contact chemoreceptors that it uses for the detection of the burrow entrance. Since it is much larger than the minute entrance, it is an important orientation cue to a homing isopod which went coarsely astray. Soon after an isopod has reached the ring of faeces it locates the burrow entrance within (Hoffmann 1985a,b).

3.2 Comparison between different theories of homing by path concatenation

The first attempt to describe mathematically the ability of animals to return to a given point by means of path concatenation is due to Jander (1957). He gave a rule, how a foraging ant could get information about its homing direction by averaging its movement directions arithmetically over the time while its moves away from the nest. Wehner and Wehner (1986) much later took up Janders idea again and applied it to the analysis of the homing behaviour of the desert ant Cataglyphis. Recently Müller and Wehner (1988) modified this theory by using a better approximations of the trigonometric functions describing the true homing direction.

Mittelstaedt (1978,1985) uses a different approach. His theory is based on the assumption that an animal is able to add up its displacements in a vectorial form. In this way an animal could get precise information not only about its direction to the point where it wants to return, but also about its distance. This is a major difference to the theories of Jander (1957), Wehner and Wehner (1986). It is most important from a biological point of view, since an animal which moves straight forever into the direction once chosen and misses its target, would return much safer by searching around in a brownian way from the very beginning.

All theories described so far fail totally to describe the return path of an animal in those cases where it does not reach the right point on a straight way. Since explicitly or implicitly they assume that the information made available by path concatenation is restricted to at most a single angle and distance, this information is used up at the end of the straight return path. No information would be available to an isopod to direct its further course.

It has been assumed until recently that an animal could solve the remaining "minor" orientation problems, until it really is back home, by landmark orientation and an unspecified kind of random search (Wehner and Flatt 1972). However, already the first quantitative check proved that this assumption is wrong, at least in the case of the desert isopods (Hoffmann 1978, 1983).

The hypothesis presented here about the information made available to an animal by the path concatenation mechanisms, and the way how the animal could use this information to direct its course on the straight part of the return path and during its search overcomes this difficulties. It assumes that not only one, but several possible positions of the target are stored, and each position is associated with a preference. This information does not require a large amount of storage capacity in the brain of an animal. Further experiments will show how the preferences for a given point are connected with length, average direction, and number of bends of the foraging path, and how they change during the return.

Acknowledgements: Supported by the Deutsche Forschungsgemeinschaft (Ho 831/3). I thank Prof. K.E. Linsenmair for suggestions, Christine Mahsberg and Annedore Buhler for technical assistance.

REFERENCES

Hoffmann, G. (1978) Experimentelle und theoretische Analyse eines adaptiven Orientier-ungsverhaltens: Die 'optimale' Suche der Wüstenassel Hemilepistus reaumuri, Audouin und Savigny (Crustacea, Isopoda, Oniscoidea) nach ihrer Höhle, Dissertation, Universität Regensburg.

Hoffmann, G. (1983) The random elements in the systematic search behaviour of the desert isopod Hemilepistus reaumuri. Behav. Ecol. Sociobiol. 13, 81-92.

Hoffmann, G. (1984) Homing by systematic search. In: D. Varju, H.-U. Schnitzler (eds.): Localization and orientation in biology and engineering, Springer Verlag, 192-199 .

Hoffmann, G. (1985a) The influence of landmarks on the systematic search behaviour of the desert isopod Hemilepistus reaumuri I. Role of the landmark made by the animal. Behav. Ecol. Sociobiol. 17, 325-334.

Hoffmann, G. (1985b) The influence of landmarks on the systematic search behaviour of the desert isopod Hemilepistus reaumuri II. Problems with similar landmarks and their solution. Behav. Ecol. Sociobiol. 17, 335-348.

Jander, R. (1957) Die optische Richtungsorientierung der Roten Waldameise (Formica rufa L.). Z. vergl. Physiol. 40, 162-238.

Linsenmair, K.E. (1987) Kin recognition in subsocial arthropods, in particular in the desert isopod Hemilepistus reaumuri. In: D.J.C. Fletcher, C.D. Michener (eds) Kin recognition in animals: 121-208. John Wiley & Sons Ltd.

Linsenmair, K.E. and C. Linsenmair (1971) Paarbildung und Paarzusammenhalt bei der monogamen Wüstenassel Hemilepistus reaumuri (Crustacea, Isopoda, Oniscoidea). Z. Tierpsychol. 29, 134-155.

Mittelstaedt, H. (1978) Kybernetische Analyse von Orientierungsleistungen. In: G. Hauske, E. Butenand (eds.): Kybernetik 1977, R. Oldenburg Verlag, München Wien, 144-195 .

Mittelstaedt, H. (1985) Analytical cybernetics of spider navigation. In: F.G. Barth (eds.): Neurobiology of Arachnids, Springer, Berlin, 298-316.

Müller, M. and R. Wehner (1988) Path integration in desert ants, Cataglyphis fortis. Proc. Natl. Acad. Sci. USA 85, 5287-5290.

Wehner, R. And I. Flatt (1972) The visual orientation of desert ants, Cataglyphis bicolor, by means of terrestrial cues. In: R. Wehner (eds.): Information processing in the visual system of arthropods, Springer Verlag, 295-302 .

Wehner, R. And S. Wehner (1986) Path integration in desert ants. Approaching a long-standing puzzle in insect navigation. Monit. Zool. Ital. 20, 309-331.

A SIMULATION MODEL OF FORAGING EXCURSIONS
IN INTERTIDAL CHITONS

Stefano Focardi
Istituto Nazionale di Biologia
della Selvaggina
via Ca' Fornacetta 9
40064 Ozzano dell' Emilia
Italy

Giacomo Santini
Guido Chelazzi
Dipartimento di Biologia Animale
e Genetica
via Romana 17, 50125 Firenze
Italy

Abstract

A simulation model of the foraging path of intertidal chitons
(Acanthopleura brevispinosa, A. gemmata and A. granulata) (Mollusca:
Polyplacophora) is developed which takes into account different
orientation mechanisms and the spatial distribution of food. The outputs
of the model are in agreement with the empirical data on individual
movements obtained from field records of animals' paths and suggest
that the model is an useful tool to study the orientation mechanisms of
algal-grazer molluscs and to understand the causal relationship between the
habitat structure, the individual movements and the very variable
bioeconomical strategies which have been described in the different
species.

1. Introduction

Intertidal chitons and gastropods are good subjects for the study of
the dynamic interaction between primary consumers and resources. Their
activity is organized into discrete behavioural modules which include a
resting and a feeding phase. The selection of a suitable resting position
protects them from the abiotic and biotic stress factors, while during
the activity phase the animals exploit the algal grounds. Locomotion

functionally links these two stages and in this respect its temporal and
spatial organization is a major factor in the adaptation of these animals
to the intertidal environment (Underwood, 1979; Chelazzi, Focardi and
Deneubourg, 1989).

A comprehensive field study of the temporal and spatial organization
of activity has been conducted on three tropical chitons (Acanthopleura
spp.) - quite similar in morpho-physiology and ecology - including detailed
analyses of the individual foraging paths (Chelazzi, Focardi and
Deneubourg, 1983; Chelazzi, Della Santina and Parpagnoli, 1987; Focardi and
Chelazzi, 1990).

In the present paper we formulate a Monte Carlo model for the study
of Acanthopleura locomotion, in order to complement the empirical analysis
of their spatial behaviour, and as a method of theoretical evaluation of
the etho-physiological and eco-ethological aspects of the
grazer-algae interactions. In fact, the mathematical modelization and
simulation of animal movement are able to predict the spatial patterns of a
population, at least when simple locomotion and orienting mechanisms are
involved (e.g. Okubo 1980; Van Houten and Van Houten 1982; Doucet and
Drost 1985; Doucet and Wilschut 1987). Despite its intrinsic behavioural
simplicity, our study case is more complex than most of those subjected to a
similar analysis because it includes different orientation mechanisms and a
realistic structure of the habitat.

2. The model

The analysis of the foraging excursions of the three Acanthopleura
chitons has shown that, under natural conditions, the foraging path is
composed by four different functional phases: (i) search for the first
food-patch, (ii) feeding, (iii) search for the second and successive
patches, (iv) recover of the new resting position.

The four phases are characterized by different values of some motion parameters (Table 1). The numerical values of the parameters used in the simulations were estimated from the empirical data sets. The random variables - hereafter denoted by the symbol $\hat{}$ - are extracted from pertinent look-up tables.

The movement of chitons is modelized as a discrete random process (Patlack 1953; Rohlf and Davenport 1969; Bovet 1983; Doucet andDrost 1985): each step of costant lenght s, is characterized by an angular turn, $\hat{\delta}_i$ with respect to the orientation of the previous step. The step lenght has been set equal to 1 cm, approximately corresponding to the smaller movement which can be detected by our recording system. The rectangular coordinates of the animal's position (head) at the i-th step are:

$$X_i = X_{i-1} + s \cos(\hat{\delta}_i),$$

(1)

$$Y_i = Y_{i-1} + s \sin(\hat{\delta}_i).$$

$$i = 1, \ldots, \hat{N}$$

The total number of steps in the path, $\hat{N}$, and the $\hat{N}-k$ steps of the fourth phase are independently assigned.

The animal moves over an unbounded surface starting from the home of coordinates $X_0 = 0$, $Y_0 = 0$.

The animal is assumed to stop when either

(2a)
$$\sqrt{(X_i - X_0)^2 + (Y_i - Y_0)^2} \leq 3 \text{ cm};$$

with probability of stopping γ, or

(2b)
$$i = \hat{N}.$$

The motion mechanism described above is modified by more complex orientation systems which become active in different part of the path when the motivational state of the chiton changes: trail-following, directional component (or taxis) and trail search behaviour.

1. <u>Trail-following</u>. Experiments have shown that at least <u>A. gemmata</u> has a short range detection of the trail previosly laid (Chelazzi, Della Santina and Parpagnoli, 1987) and in the model the detection distance is taken about equal to the width of animals' body (2 cm). The minimal distance d_i between the i-th step and the trail previously laid is:

$$d_i = \min \left\{ \sqrt{(X_i - X_j)^2 + (Y_i - Y_j)^2} \right\}$$

$$j=1,\ldots\ldots,i-8$$

Note that the last part of the trail (here set equal to 8 steps) is not accessible to the chiton because it is hidden under its own body.

If $d_i \leq 2$ cm the simulated animal has a probability ε of following it and a probability ϕ $(1 - \phi)$ to take the homeward (outward) direction.

While retrailing, the animal has a mean probability per step, τ, to leave the trail and to resume the normal movement pattern (eq. 1).

2. <u>Directional component</u>. The presence of a preferential direction of movement, $\hat{\Theta}$, active in phase I only, is simulated using a modification of the algorithm developed by Rohlf and Davenport (1969) (Fig.1):

(4a) $P\left\{ \alpha_{i+1} = \alpha_i - \hat{\delta}_{i+1} \right\} = Q = 0.5 + \beta \sin(\alpha_i)$

(4b) $P\left\{ \alpha_{i+1} = \alpha_i + \hat{\delta}_{i+1} \right\} = 1 - Q.$

where β is the tendency of the chiton to follow an assigned direction of movement $\hat{\Theta}$. For the three species the numerical values of β has been estimated by the model: the selected values (Table 2) maximize the

Table 1. Values of species- and phase-specific parameters. The mean δ value of is $0°$ by definition and only the standard deviations are reported.

Parameters		Species	Phase			
			I	II	III	IV
ε	probability of beginning trail-following	gemmata	.423	.718	.750	.888
		brevispinosa	.592	.647	.727	.875
		granulata	.346	.625	.667	.655
τ	probability of ending trail-following	gemmata	.313	.067	.051	.011
		brevispinosa	.257	.049	.087	.028
		granulata	.244	.177	.118	.094
ϕ	probability of re-trailing homeward	gemmata	.273	.036	.0	.042
		brevispinosa	.207	.0	.0	.114
		granulata	.222	.1	.346	.289
$\hat{\delta}$	angle of turn	gemmata	23.06	35.13	23.9	20.5
		brevispinosa	36.06	41.32	36.69	32.51
		granulata	37.48	44.32	34.28	35.32

similarity of simulated and empirical paths in phase I.

3. <u>Trail</u> <u>search</u> <u>behaviour</u>. The analysis of empirical paths has evidenced an asimmetry in the frequency of clockwise and counterclockwise turns in most of chitons: in some animals there is a higher fraction of clockwise than counterclockwise turns (destrorse chitons), or viceversa (sinistrorse). The consequence is that animals perform an U- or spiral-shaped movement which leads to cross the previously-laid trail. In the model this search mechanism becomes active after that the animal has exploited a certain number of food patches ($\hat{f}$). To each animal is initially attributed of being sinistrorse or destrorse with a probability π of making a turn in the preferred direction.

In the environment surrounding the home a certain number M of food patches is initially distributed. Each patch, supposed

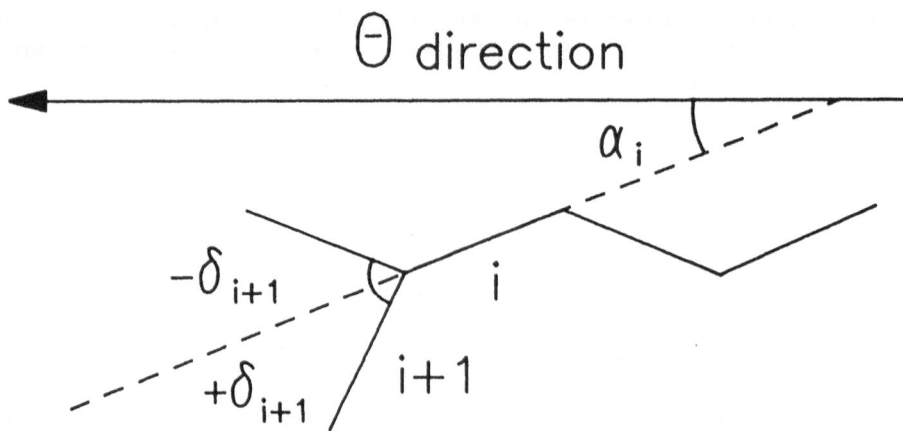

Figure 1. The algorithm given in eq. 4a,b yields an higher probability to have a turn towards the assigned p referential direction. α is the deviation of the i-th step from Θ

circular, is characterized by the radius $\hat{D}$, its distance from the starting point, $\hat{1}$, and the relative orientation, $\hat{\Theta}$. The species-specific value of M has been selected so that the simulated chitons have an encouter rate with the patches equal to the one observed in the field. Note that the model is scarcely sensible to the M value.

The numerical values used in the model are summarized in Table 2. With the exception discussed above all the parameter values have been estimated from the empirical data.

The simulation model, which is available from the Authors, has been written in Fortran 77 ESV and run on the 64 bit machine Honeywell-Bull Very Large System DPS 90 of the Centro di Calcolo Elettronico of the University of Florence.

Table 2. Mean values of species-specific parameters. For random variables the standard deviation is given in brackets. Lenghts in cm.

----

		gemmata	brevispinosa	granulata
β	tendency of following the preferencial direction i	.5	.3	.4
M	number of food patches used in the simulations	100	100	400
γ	probability of stopping when the home is found	1.	1.	.54
π	probability of turning	.839	.662	.704
$\hat{f}$	number of patches visited before beginning the return	1.7(.7)	1.8(.9)	2.7(1.2)
$\hat{\Theta}$	preferential direction in phase I (0 =landward)	173.2 (44.4)	142.6 (53.6)	211.9 (73.1)
$\hat{l}$	distance patch-starting point of the foraging excursion	60.9 (30.9)	27.2 (17.4)	22.2 (16.8)
$\hat{D}$	patch diameter	15.5 (14.4)	5.3 (6.5)	2.3 (3.4)
$\hat{N}$	excursion lenght	214.5 (100)	127.2 (57.3)	98.0 (66)

----

4. Results and Discussion

A qualitative comparison between empirical and simulated paths is shown in Fig. 2a,b,c. They look alike also in the details: for instance the two paths of A. gemmata show the U-shaped search of the trail, and a very good trail- following during the return. A. brevispinosa exibits a similar pattern but in this species the trail-following is less precise than in the former, while the homing performance remains good. In A. granulata the paths are completely different: the trail-following is scarce and the

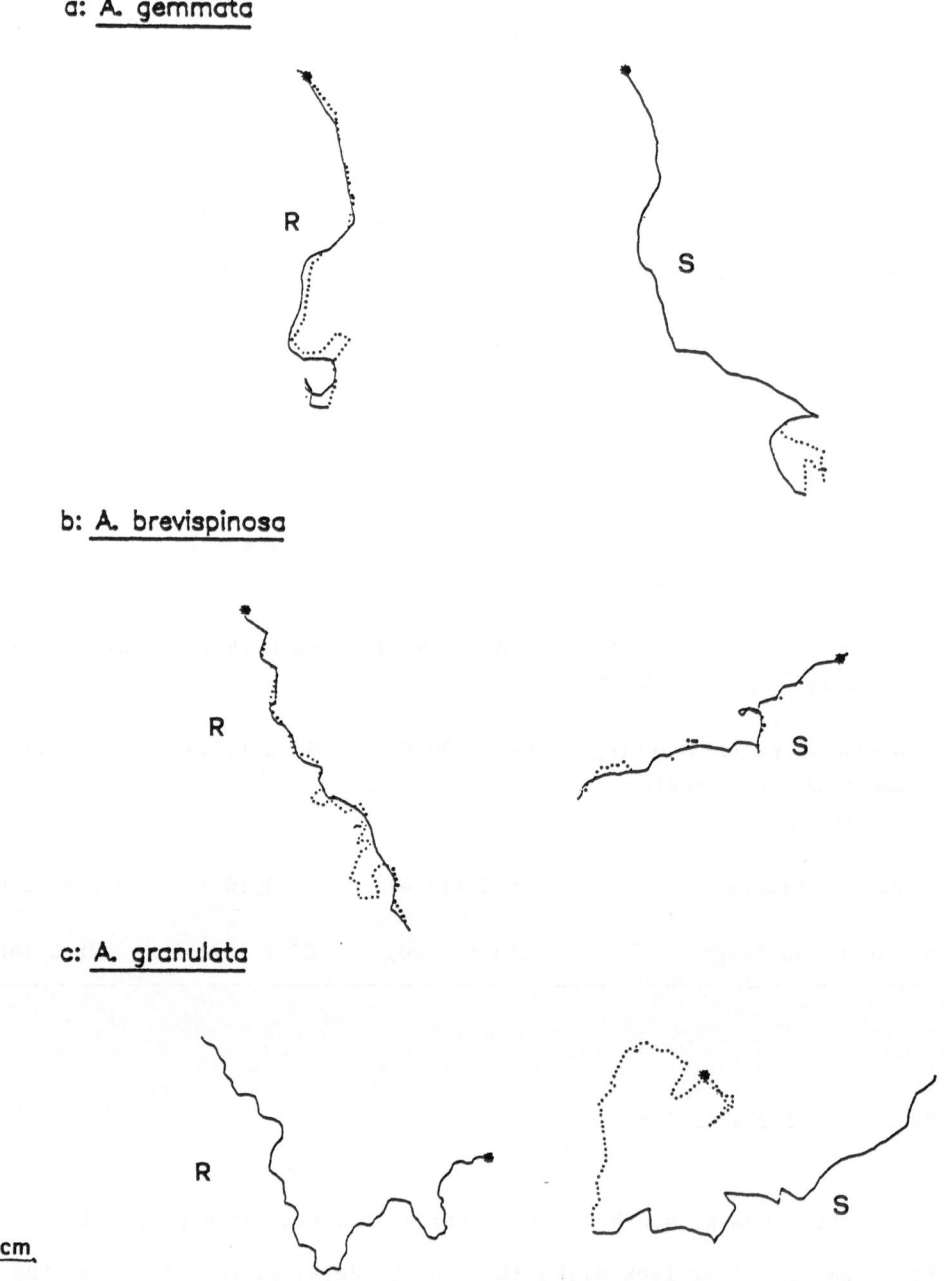

a: A. gemmata

R

S

b: A. brevispinosa

R

S

c: A. granulata

R

S

10 cm

Figure 2. Simulated (S) and actual (R) paths of the three species. H indicates the first and the star the final point of the excursions. Dots represent fractions of the return branch which are not overlapped to the outward trail

feeding loop becomes open.

A more quantitative fitting of model's outputs has been made comparing the empirical and simulated distributions of some global path parameters:

1. the maximal distance from the starting point (MD);

2. the homing error (RHE), given by the ratio between the distance between starting and final points of the path and MD;

3. the trail-following (TF), estimated by computing the fraction of the path which is closer than 2 cm to a previous path segment.

The plots (Figs. 3a,b,c) and the relative statistics (Table 3) show a very good agreement between empirical data and simulations for A. gemmata and A. brevispinosa while the fitting relative to A. granulata is less good but yet non-significant. The statistical agreement between simulated and actual global parameters is a good proof that the model is biologically sounded. The use of complex models with a large number of parameters, such as that used in the present study, is especially allowed by the good estimates of their values which was possible to compute on the basis of

Table 3. Results of the statistical comparison between the frequency distributions of movement parameters obtained from the empirical (97 A. gemmata, 76 A. brevispinosa and 136 A. granulata) and simulated excursions (1000 chitons per species).

		X	df	P
	gemmata	7.02	3	> 0.5
TRAIL-FOLLOWING	brevispinosa	5.27	5	> 0.1
	granulata	10.50	4	.1 > P > 0.05
	gemmata	7.46	5	> 0.1
HOMING ERROR	brevispinosa	3.28	3	> 0.1
	granulata	11.13	8	> 0.05
	gemmata	13.31	5	> 0.05
MAXIMUM DISTANCE	brevispinosa	8.30	4	> 0.05
	granulata	6.17	3	> 0.1

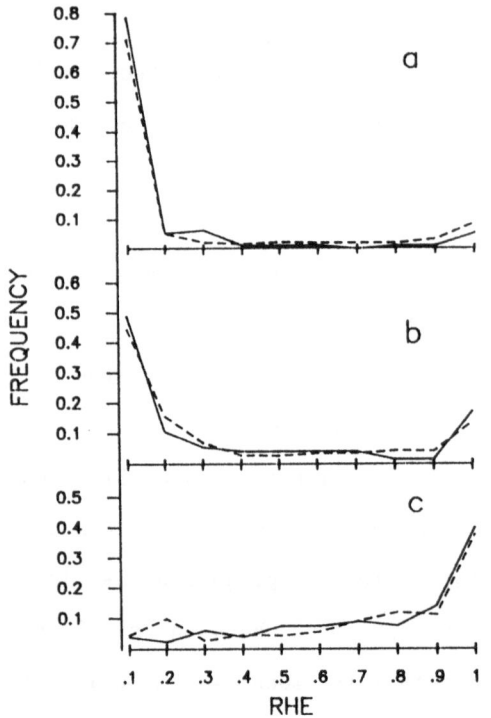

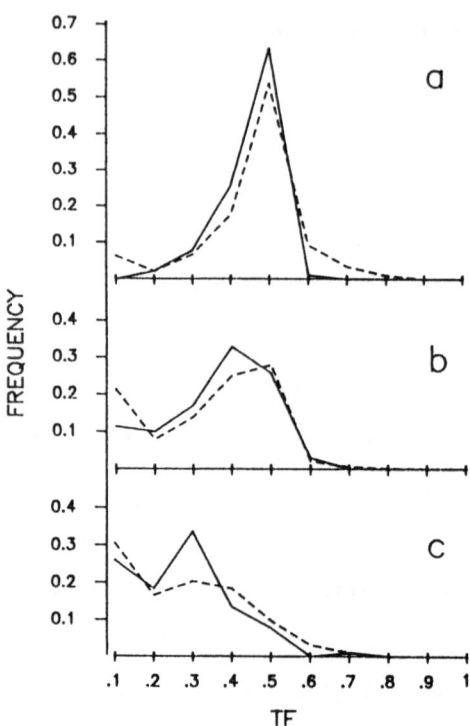

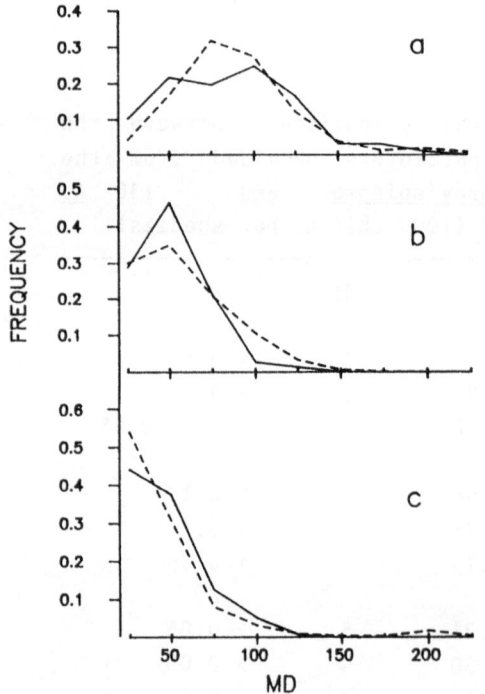

Figure 3. Comparison between the empirical (solid line) and simulated (broken line) distributions of the three global path parameters for _A. gemmata_ (a), _A. brevispinosa_ (b), _A. granulata_ (c). MD is in cm.

large samples of paths.

The model poses an upper limit to the proliferation of hypotheses about orientation mechanisms driving the feeding excursions of intertidal chitons and can be used as a guide for further experiments.

Moreover, the results showed that a basic behavioural algorithm is able to generate quite different and adaptive movement patterns for different parameter assignments.

In conclusion the model represents an useful link between individual behavior and spatial pattern of the population and open the possibility to understand the adaptive value of the behavioural program when it copes with different environmental situations.

REFERENCES

Bovet P.(1983) Optimal randomness in foraging movement: a central place model, pp.295-302. In: Cosnard M., Demongeot J. and Lebreton A.(Edits.) Rhythms in biology and other field of application. Berlin: Springer-Verlag.

Chelazzi G.,Focardi S. and Deneubourg J.L.(1983) A comparative study on the movement patterns of two sympatric tropical chitons (Mollusca: Polyplacofora). Mar.Biol. 74 : 115-125.

Chelazzi G.,Della Santina P. and Parpagnoli D.(1987) Trail-following in the chiton Acanthopleura gemmata: operational and ecological problems. Mar. Biol. 95 : 539-545.

Chelazzi G., Focardi S. and Deneubourg J.L.(1989) Analysis of movement patterns and orientation mechanisms in intertidal chitons and gastropods, pp. 173-184. In: Chelazzi G. and Vannini M. (Edts.). Behavioural adaptations to intertidal life. New York: Plenum Press, New York.

Doucet P.G. and Drost N.J.(1985) Theoretical studies on animal orientation.II. Directional displacement in kineses. J. Theor. Biol. 117 : 337-361.

Doucet P.G. and Wilschut A.N. (1987) Theoretical studies on animal orientation. III. A model for Kineses. J. Theor. Biol. 127 : 111-125.

Focardi S. and Chelazzi G. (1990) Ecological determinants of bioeconomics in three intertidal chitons (Acanthopleura spp.). J. Anim. Ecol. 59 : 347-362.

Okubo A. (1980) Diffusion and ecological problems: mathematical models. Berlin: Springer-Verlag.

Patlak C.S.(1953) Random walk with persistence and external bias. Bull.
    Math. Biophys. 15 : 431-476.
Rohlf F.J. and Davemport D.(1969) Simulation of simple models of animal
    behavior with a digital computer. J. Theor. Biol. 23 : 400-424.
Underwood A. J. (1979) The ecology of intertidal gastropods. Adv. Mar.
    Biol. 16 : 111-210.
Van Houten J. and Van Houten J.(1982) Computer simulation of Paramecium
    chemokinesis behaviour. J. Theor. Biol. 98 : 453-468.

HOW TO DESCRIBE THE SEARCH OF A PREDATOR FOR PATCHILY DISTRIBUTED PREY ?

G. Hoffmann, Zoologisches Institut II,
Röntgenring 10, D-8700 Würzburg, FRG

Abstract

Lady beetle larvae preying upon aphids have to search for their patchily distributed prey on plants, a complicated three-dimensional world. In an attempt to understand the structure of their search path on a single plant it is compared to a path which solves their search problems in the best possible way. The part of the plant where the larva starts searching, the sequence in which the larva searches from top to bottom on additional leaves, and the intensity with which the larva searches on different leaves before it leaves the plant corresponds well to the rules for an optimal search.

Animals searching so effectively must remember the places where they have searched already before. Lady beetle larvae solve this problem probably by recognizing a marker which they deposit during their search.

1 Introduction

The search problem of a coccinellid larva whose main prey are aphids is tremendous. It has to find a prey whose density is strongly fluctuating with time and place in a three-dimensional world with a complicated structure, and it detects its prey only by contact (Banks, 1957). If its search remains unsuccessful for too long a time, the larva will die. Therefore lady beetle larvae preying upon aphids are classical paradigms for predators whose search behaviour is shaped by strong selection against individuals which search less efficiently than conspecifics (summaries Curio, 1976 , Hodek, 1986).

The mother of a lady beetle larva, being able to overcome large distances between aphid aggregations in flight, somewhat facilitates the search problem for its offspring. If it has found a plant on which aphids abound, it lays egg batches with a higher rate, mostly near aphid aggregations (Evans and Dixon, 1986). This behaviour of its mother ensures that a larva hatching from its egg will probably find prey in its neighbourhood. Nevertheless, a larva often has to search for aphids, since it needs hundreds of them to finish its development. The faster it finds food, the faster the larva grows, and the earlier it becomes a beetle (Dixon, 1970).

Since the fitness of a lady beetle larva depends so strongly on the effectiveness of its search behaviour, it suggests itself to compare this behaviour to a strategy solving the same search problem in the best possible way. By this method we will try to answer the specific question, how well a larva does take into account that aphids are not evenly distributed over a plant, but prefer certain parts of it (Ibbotson and Kennedy, 1951). Does a larva change its search behaviour accordingly on different parts of a plant?

2 Results

2.1 The optimal solution for the search problem of a larva

The success of a search procedure of a larva depends on how well its search intensity in a given area corresponds to the probability that aphids are there. Before we can calculate the best procedure, we therefore need to describe the distribution of aphids quantitatively. For this purpose we use a Poisson point process with a spatially inhomogeneous intensity function $\delta(x)$, defined in the following way (Diggle, 1983):

1) In an arbitrary area A the number of aphids $N(A)$ sitting there is given by a Poisson distribution with the expectation $\int_A \delta(x)dx$ (the integral has to be calculated for the <u>surface</u> of A).

2) Under the condition $N(A) = n$, the n aphids are independently distributed over the area A. The probability density that a given aphid is at point x is proportional to $\delta(x)$.

The success of a given search procedure depends not only on the intensity with which a larva searches in places where aphids are. The problem is that the larva might not detect an aphid, even if it searches on the leaf where its prey sits. Imagine an area A within which an aphid may sit everywhere with equal probability. In this area the larva searches evenly for t_A seconds. The search intensity z is then constant within area A and given by $z = t_A/A$. The (conditional) probability $b(z)$ that the larva will find the aphid, called the <u>detection function</u>, is given by

(1) $$b(z) = 1 - e^{-ct_A/A}.$$

It is assumed in the following that the detection function does neither depend on the exact position of the aphid, nor on the fact that the larva has detected other aphids, and that it is the same for every aphid.

Now it is easy to determine the number of aphids that a larva will find with a given search strategy. Assume that the area X includes all points which the larva could reach within time T. If it searches there with intensity $z(x, T)$, on average it will find $E[N] = \int_X \delta(x)b(z(x, T))dx$

aphids.

To answer the question, how a lady beetle larva should search in order to find as many aphids as possible within a given time period T, we have to calculate the search procedure $z^*(x,T)$ for which $E[N]$ is maximal under the restriction that the total search time T is limited. The methods to solve this optimization problem are well known (Ahlswede and Wegener 1979, Hoffmann 1983b, Koopmann 1980, Stone 1975), and are summarized in Box 6.

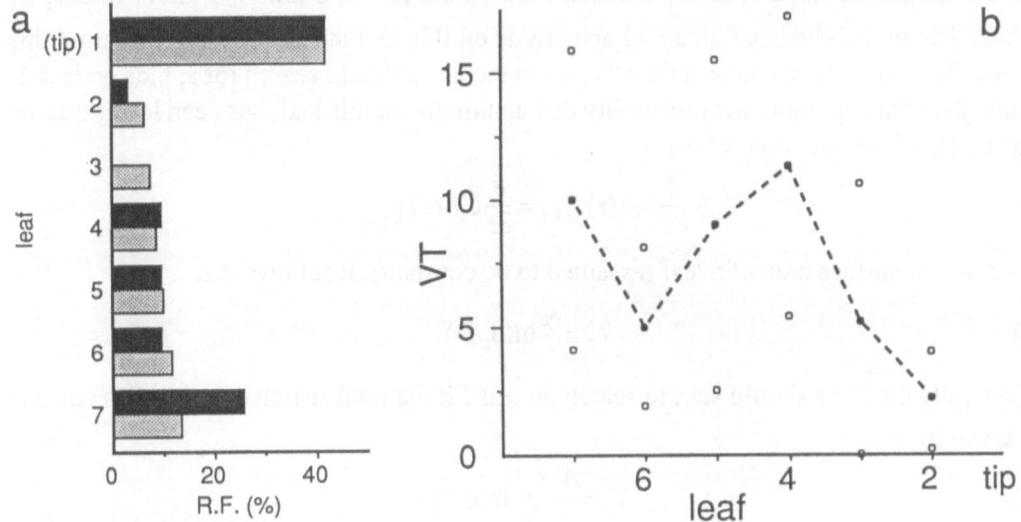

Figure 1a: ■ Frequency with which a larva first searches on a given leaf, or the tip (designated leaf 1) of a plant. 7 larvae of the species Propylaea quatuordecimpunctata searched on 5 plants with 6 leaves each. ▨ Respective expected frequency if the larva tries to move first to the tip, but mistakes with a probability of 13.5 % the petiole of a leaf for the stem. b: Total number VT of visits of other leaves before the first visit of a given leaf, beginning with and including the first visit of the tip. o mean, o standard deviation.

2.2 The behaviour of a larva on a plant

Imagine a larva which left one plant because it did find no further aphids on it. After walking on the soil surface it reaches a new plant with L leaves which emerge in different heights from a single stem. How should the larva proceed in its search to find as many aphids as possible on this plant?

Aphids tend to sit mostly on the youngest, i.e., the upper and outer parts of a plant (Foott, 1977). If we number the leaves from top to bottom (including the tip of the plant), then the average probability density δ_i $(i = 1,...,L+1)$ of aphids on leaf i is a strictly decreasing sequence. $E[N_i] = \int\limits_{leaf\ i} \delta_i dx$ is the expected number of aphids on leaf i .

A larva should move first to the place on the plant where aphids sit with the highest probability. Indeed in most cases the larva visits first the tip of the plant (Fig. 1a). However, it contends with difficulties discriminating between the stem and the petiole of a leaf. With a probability of about 13.5 % it mistakes a petiole for the shortest way to the uppermost part of the plant (Fig. 1a).

How should the larva proceed, if it searched on the tip for a time t_1 without finding an aphid? The probability that an aphid actually is on this leaf steadily decreases during this period. Therefore there comes a time T_2, when the larva should search for aphids on leaf 2, although in the beginning the probability that aphids are on this leaf, has been lower than on leaf 1. This time has come when

(2)
$$\frac{\partial}{\partial z}\delta_1 b(z)\,|_{z=z_1}=\frac{\partial}{\partial z}\delta_2 b(z)\,|_{z=0}$$

$z_1 = t_1/A_L$, A_L surface area of a leaf (assumed to be constant). It follows that

(3)
$$T_2 = \frac{A_L}{c}\ln(\delta_1/\delta_2)\,.$$

Generally the larva should start to search on leaf i if the total search duration T is equal to T_i given by

(4)
$$T_i = \frac{A_L}{c}\sum_{j=1}^{i-1}\ln(\delta_j/\delta_i)$$

It is important to note from a biological point of view that the time T_i depends only on the relation of the probability density that an aphid sits on leaf i to the probability density that an aphid sits on the leaves above. T_i depends neither on the absolute density of aphids on the plant, nor on the number of leaves below leaf i. This fact makes it much easier for a larva to fulfil the rules for an effective search, since otherwise it would have to take into account characteristics of the environment which are difficult to assess for it.

Seen from this perspective, it is easier to understand, how a lady beetle larva is able to approximately fulfil the rules for an effective search. It starts searching on a given leaf only after having searched on all higher leaves rather carefully (Fig. 1b). However, deviations from the rule that a larva should start searching on leaf i only after a total search time T_i which increases with the leaf number i can be observed. They are possibly due to the fact that during walking down the stem a larva sometimes misses the ramification of a leaf.

Up to a given total search time T the larva should search only on the $n(T)$ leaves above the first leaf i for which $T_i \geq T$. On these leaves the optimal overall search duration t_i is given by

(5)
$$t_i = \frac{T}{n(T)} - \frac{A}{cn(T)}\sum_{j=1}^{n(T)}\ln(\delta_j/\delta_i).$$

Consequently the larva should search for the longest overall time on the uppermost leaf, and for decreasing times on the lower leaves. Fig. 2a shows that, until a larva leaves a plant, it searches repeatedly on a single leaf, and visits upper leaves more often than lower ones. Since the duration t_V of a single visit is more or less constant, equation 5 is approximately fulfilled.

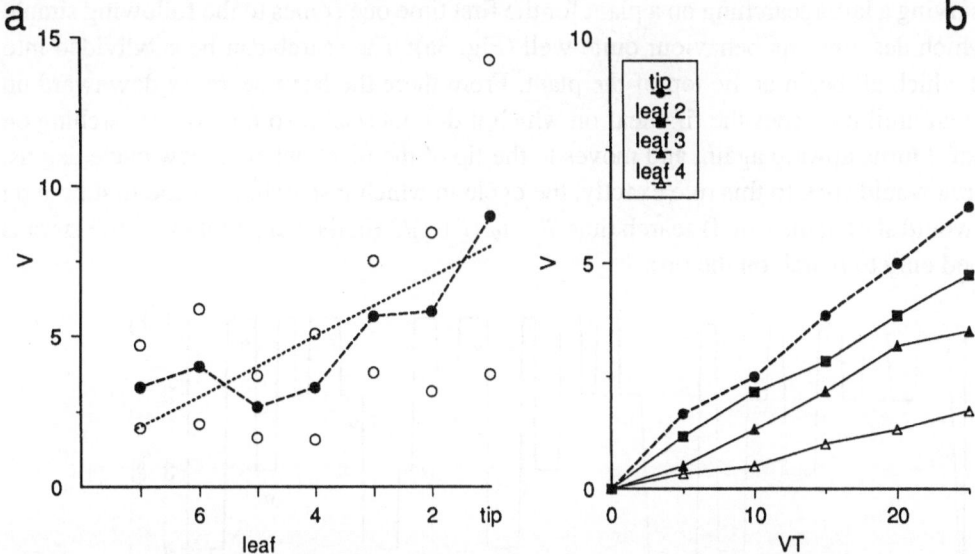

Figure 2a: Number V of visits on a given leaf before the larva leaves the plant. o mean, o standard deviation, - - - - expected value according to equation 7. b: Number V of visits on a given leaf as a function of VT, the total number of visits on all leaves, i.e., of the search duration.

The observation that a larva searches repeatedly on a given leaf seems to contradict the notion of an effective search. However, specifically this observation is one of the best proofs that a larva really searches effectively. The rules for an effective search require repeated visits of a given leaf, interrupted by visits of other leafs, because this results in a higher probability of an early success than a single long visit with the same total search duration. As an example consider the search on leaf 1. From time T_2 on, it is better to search not only on leaf 1, but also on leaf 2, with the same <u>additional</u> search intensity. A larva could fulfil this rule approximately by searching alternately for a given short time period on leaf one and two. Generally after some time $T_i \leq T < T_{i+1}$ it should search alternately on each of the i leaves for a short period. Fig. 2b shows that the larvae really follow this rule.

2.3 Search control

So far we only have shown that a larva searching for aphids on a plant in vain controls its search in such a way that it fulfils several rules for a search procedure which maximizes success. But still we don't know <u>how</u> it controls its search. It is long known that a lady beetle larva controls by means of light and gravity if it moves up or down on a plant (Fleschner, 1950). Yet this does not explain the astounding behaviour of a larva which searches on different leaves in a certain order which corresponds well to the a posteriori probability that aphids are on them.

Observing a larva searching on a plant for the first time one comes to the following simple rule which describes its behaviour quite well (Fig. 3a): The search can be subdivided into cycles which all begin at the top of the plant. From there the larva searches downward on every leaf until it reaches the first leaf on which it did not search so far. After searching on this leaf it turns upward again, and moves to the tip of the plant, where a new cycle begins. If a larva would stick to this rule exactly, the cycle in which it searched for the first time on leaf i would start at the (total) search time $T = t_V i(i + 1)/2$ (in the very first cycle the larva is assumed only to search on the tip).

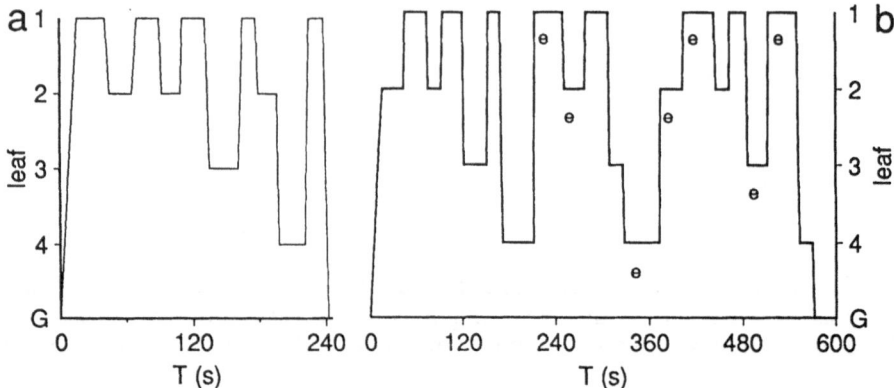

Figure 3: Vertical position of a larva (Coccinella septempunctata) on an artificial plant with 4 leaves made out of filter paper and no tip as a function of time T (G ground). a: Control experiment. b: Leaves marked with "e" have been exchanged before the larva returned to them.

The total search duration t_j on leaf j at this time is given by

(6) $t_j = t_V(i - j)$ $(j < i)$.

Under the assumption that the larva leaves the plant if it first should search on the (nonexistent) leaf $L + 2$ one gets

(7) $t_j = t_V(L + 3 - j)$

for the total search time on leaf j when the larva leaves the plant. This theoretical value corresponds quite well to the observed values ($L = 6$, Fig. 2a).

Giving a rule for the sequence with which a larva searches on a plant makes clearer, how it approximately fulfils the rules for an optimal search. Yet, it leads to the new question, how the larva remembers on which leaves it has searched already before. The answer to this question is given by the following experiment: larvae search on artificial plants made out of filter paper. At the time when they normally would leave the plant ($T \approx 180$ s, Fig. 3a), some of its leaves are exchanged. This exchange increases the time the larvae will search on the plant because they tend to move upward after a visit of an exchanged leaf, whereas they tend to move downward after a visit of a leaf on which they searched already before (Fig. 3b).

This observation directly shows that a larva recognizes such leaves, and uses this information to increase the efficiency of its search. Most probably, it chemically marks areas where it searches (Hoffmann, in prep.).

3 Discussion

3.1 What can we learn from the comparison between the search behaviour of an animal and the optimal solution of its search problems?

The search of a lady beetle larva for its prey has many characteristics in common with a search strategy with maximal success: The part of the plant where the larva starts searching (Fig. 1a), the sequence in which the larva searches from top to bottom on additional leaves (Fig. 1b), and the intensity with which the larva searches on different leaves before it leaves the plant (Fig. 2a). What can we conclude from this correspondence? Before we can answer this question, it is necessary to make quite plain which assumptions we have to make in order to calculate the optimal search strategy, and how these assumptions and the resulting predictions can be checked by observation.

Fundamental is the assumption that the process by which a larva detects its prey can be described by the detection function $b(z)$ (Eq. 1). This assumption is well justified by some general reasonings (Koopmann, 1980), and can be checked experimentally. Moreover optimal search strategies can be calculated also for a much wider class of detection functions (Stone 1975).

The second fundamental assumption necessary for the calculation of the optimal search plan concerns the information about the distribution of the prey available to a larva. It has to be carefully discriminated from the corresponding information available to us by observation of aphids. It suggests itself to assume that a coccinellid larva "knows" that aphids sit with the highest probability on the upper and outer parts of a plant. During evolution, or during its life the larvae could learn to associate certain characteristics of its environment with the presence of aphids. Nevertheless, this assumption, specified by the target density δ_i, cannot be checked directly by observation. The information about the prey distribution that a larva uses to control its search becomes observable only in the structure of its search path, mostly in its long term structure. To draw conclusions about this information from the path structure we need a further assumption, namely that the larva uses the information available to it as well as possible.

The advantage of the comparison between an observed search behaviour and an optimal search is that it elucidates the relation between functionally important aspects of the search path of a larva and the underlying search problem. It fills an important gap between approaches which try to describe the path structure on the basis of local path parameters, more or less disregarding its long range structure, and the available theories of optimal foraging, which only in very special cases give rules for the spatial structure of a food search (summaries: Krebs and Davies, 1984).

3.2 What kind of memory is necessary for an efficient search?

It was astounding to note that a lady beetle larva recognizes each single leaf of a plant on which it searched already before. Probably it uses a chemical marker for this purpose (Hoffmann, in prep.). Although this explains, how a larva memorizes where it searched before, it remains unclear, why it does so. Is this kind of memory really necessary for an efficient search?

Not every predator needs such a complicated "memory" for the places where it has searched already. Take for instance a plankton feeder (see, e.g., Klafter et al., section II/1). Its food is homogeneously distributed over large areas in two dimensions, and food density depends only on depth. Such a predator needs simply to swim straight to avoid places where it searched immediately before. The only thing it has to remember is one direction.

More complex is the memory problem for a predator searching for prey which in the beginning is distributed evenly within a limited area A , e.g., a prey patch. In this case, moving straight is a good solution only for a limited time. Cody (1971) showed by simulation that a search in the form of a discrete random walk is most effective, if the predator keeps approximately one direction for a couple of steps. Since he simulated the solution only for a very special case of the given search problem, it is not clear, how far his results can be generalized. The approach of Alt (section II/1) promises more general answers. He gives a method to calculate search intensity for a wider class of stochastic processes.

However, search procedures based on a random walk can have the same success as a systematic search, or even a better one only under special circumstances, e.g., if the search duration is very short (Hoffmann 1983a). The reason is that parts of the search path far apart in time have to be independent from their relative positions. With such a search strategy an animal would search in the same way when it returns to a place where it already searched before as when it searches in a new place. Papentin (1973) could show by simulations, that the success of a food search procedure increases strongly, if the animal is able to "remember" where it already searched before. If food is limited, individuals having this ability soon dominate in the population, and develop a search procedure including spirals and meanders.

The approach to the description of a search chosen in this paper makes clear to what extent long term memory influences the success of a search, in a way which is independent of a specific local structure of the search path. During an effective search a larva should search on leaf i with the intensity $z^*(i, T_1)$ up to time T_1, and with the intensity $z^*(i, T_2)$ up to time T_2 . Consequently in the time span between T_1 and T_2 the larva should search on leaf i with the additional intensity $z^*(i, T_2) - z^*(i, T_1)$. In simple cases, on a flat surface, an animal could give its search the corresponding structure without the ability to recognize a place where it searched already before by walking on a path with a built-in memory, like, e.g., a spiral, or a meander (see Hoffmann, section II/1). However, on such a complicated spatial structure as a plant, this seems to be very difficult a task. This could be the reason, why the coccinellid larvae use chemical marking of leaves as an orientation aid.

The increase of search effectiveness brought about by a memory for places in which an animal searched already before suggests that mechanisms as found in the coccinellid larvae could be observed also in other species. In fact there are hints that both protozoa (*Paramecium*, Markasin and Milschtein, 1975), and sediment feeders in the deep sea have similar abilities (Kitchell, 1979).

REFERENCES

Ahlswede, R., Wegener, I. (1979): Suchprobleme, Teubner Verlag, Stuttgart.

Banks, C. J. (1957): The behaviour of individual coccinellid larvae on plants. Brit. J. Anim. Behav. 5: 12-24.

Cody, M. L. (1971): Finch flocks in the Mohave desert. Theor. Popul. Biol. 2: 142-158.

Curio, E. (1976): The Ethology of Predation. In: D.S. Farner (Hrs.) Zoophysiology and Ecology. Springer-Verlag, Berlin Heidelberg New York.

Diggle, P. J. (1983): Statistical analysis of spatial point patterns. Academic Press, London.

Dixon, A.F.G. (1970): Factors limiting the effectiveness of the coccinellid beetle, Adalia bipunktata (L.), as a predator of the sycamore aphid, Drepanosiphum platanoides (Schr.). Anim. Ecol. 39: 739-751.

Evans, E.W. and A.F.G. Dixon (1986): Cues for oviposition by ladybird beetles (Coccinellidae): Response to aphids. J. Anim. Ecol. 55: 1027-1034.

Fleschner, C.A. (1950): Studies on searching capacity of the larvae of three predators of the Citrus red mite. Hilgardia 20: 233-265.

Foott, W.H (1977): Biology of corn field aphids Rhopalosiphum maidis (Homoptera: Aphididae) in south western Ontario.. Can. Entomol. 109: 1129-1135.

Hodek, I. (ed.)(1986): Ecology of Aphidophaga. Dr. W. Junk, Dordrecht, Netherlands.

Hoffmann, G. (1983a): Optimization of brownian search strategies. Biol.Cybern. 49: 21-31.

Hoffmann, G. (1983b): The search behaviour of the desert isopod Hemilepistus reaumuri as compared with a systematic search. Behav. Ecol. Sociobiol. 13: 93-106.

Ibbotson, A. and Kennedy, J.S. (1951) Aggregation in Aphis fabae Scop. I. Aggregation on plants. Annals of applied Biology 38, 65-78.

Kitchell, J. A. (1979): Deep-sea foraging pathways. An analysis of randomness and resource exploitation. Paleobiology 5107-125.

Koopman, B.O. (1980): Search and screening. Blackman , Washington, DC.

Krebs, J. R. and N. B. Davies (1984): Behavioural ecology. Blackwell Scientific Publications.

Markasin, V.S. and G.N. Milschtein (1975): Statistical analysis of paramecian movement. Zh. Obshch. Biol. 36(1): 119-125.

Papentin, F. (1973): A Darwinian evolutionary system. III. Experiments on the evolution of feeding patterns. J. theor. Biol. 39: 431-445.

Stone, L. D. (1975): Theory of optimal search. Academic Press.

Conclusions

The spectrum of orientation behaviours of single organisms described in this section and the mathematical methods used for their analysis gives a first impression of the attractiveness of this field of research. Some of the points mentioned in this section are taken up also in other sections. Noble, e.g. describes in section I/1 certain aspects of the food search of amoebic cells.

In view of the fact, that many species depend heavily on their ability to find food or other resources as fast as possible, it is a pity, that both experimental and theoretical work in this field is still scanty. Surely further research could bring up astonishing new facts, in the same way as the analysis of the food search behaviour of coccinelid beetles showed that they "remember" where they searched for food before by recognizing a marker which they deposit during the search (see Hoffmann, this section).

The discussion during the workshop have shown that it would be most promising to fill the gap between the classical form of description of search paths by means of a random walk model and the description by means of the theory of optimal search. The gap can be described in the following way: a stochastic analysis of movement paths without long term memory, based on its local characteristics, can lead to a very good description of many of its short and medium term characteristics. Paths of animals guided by a trail or similar to a simple geometric figure, like for instance a spiral, can also be well described. The path structure of stochastic movements however, with a long term memory, as the search of an animal for a place where it has been before, can only be analysed down to the level of the search intensity, i.e., the average search time per area.

REFERENCES AND FURTHER READING

Baker, R.R., (1978): The evolutionary ecology of animal migration, Hodder and Stoughton, London, Sydney, Auckland, Toronto.

Beugnon, G. (ed), (1986): Orientation in space, Ed. Privat, Toulouse.

Gauthreaux, S.A. (ed), (1981): Animal migration, orientation, and navigation, Academic Press, New York.

McCleave, J.D., G.P. Arnold, J.J. Dodson and W.H. Neill (eds), (1984): Mechanisms of migration in fishes, NATO Conference series IV: Marine scienes.

Schöne, H. (1984): Spatial orientation: The spatial control of behavior in animals and man, Princeton Series in Neurobiology and Behavior.

Section II/2

Examples of orientation responses to gradients

(Coordinator: Hans Gruler)

Introduction

The ability of a cell, of a microorganisms, an organisms, etc., to detect and to respond to its environment is a widely spread phenomenon in biology. The orientational response as well as the directed movement of cells, etc., are of crucial importance. To name a few examples: (i) **Immune system:** For example, the first defense line of mammalian against invaded microorganisms are polymorphonuclear leukocytes (=granulocytes). These cells are attracted to sites of inflammation to destroy by phagocytosis microorganisms and invaded cells. A disturbance of the immune system is dangerous to life. (ii) **Search for nutrients, light, etc:** Microorganisms like slime mould, pennate diatoms, bacteria, etc. use their active movement to find places with optimal concentrations of nutritiens, e.g. sugar. Phytoplankton optimize their position in respect to the light intensity. **(iii) Embryogenesis, wound healing, and tumor invasion:** The directed movement of cells during embryogenesis, wound healing, and tumor invasion has been well documented. In particular, there are many instances when embryonic cells migrate as individual cells during embryogenesis and consistently follow a precise pathway to find their final destination (e.g. neural crest, precardiac mesenchyme, to name a few). **(iv) Growth:** Roots have a complex growth behaviour. They have to find e.g. optimal condition for water and salt.

The environment of the cell, of the microorganisms, etc., is the cause of the external stimuli. Consequently different environmental conditions lead to different types of directional stimuli: **(i) concentration gradient**. The particle of interest can be quite different in nature: molecules like sugar, or ions like Ca^{++}, H^+, or even microorganisms like bacteria. **(ii) thermal gradient (iii) gravitational, magnetic and electric field (iv) light.** Light plays an important role in biological system. Different types of signals are used: The flux direction, the flux intensity, and the direction of polarization.

The list of examples is not complete but it gives an impression of the importance. Here in this section only a few typical examples are presented.

The first contribution in this section (**D.-P. Häder**) deals with
the problems of photoresponse of phytoplankton flagellates. Single
flagellates exposed to gravitational field and to different light
conditions, were tracked in real time by using image analysis. The
response functions, obtained from the trajectories at different conditions, show where the ultraviolet radiation damage photoorientation
and motility. A detailed knowledge of this process is of crucial
importance since the kill of oceanic phytoplankton by UV-light
affects the biomass production of the earth and contribute thus to
global climate changes. The second contribution (**H.C. Crenshaw**)
deals with the helical movement of spermatazoa of sea urchin. It
shows by means of mathematics, how it is possible to get a mean
movement in a gradient direction when the general type of movement
is a spiral, whose width and pitch adapts to the substrate concentration. **H. Petermann, D.G. Weiss, L. Bachmann, and N. Petersen** show
that the orientational response of magnetic bacteria is identical with
that of a compass needle in a viscous media. The magnetic bacteria
have an active movement but the orientational response is passive
so that the torque balance equation is valid. In granulocytes which
H. Gruler investigated, there the speed as well as the direction of
migration are determined by active cellular processes. He shows by
means of the trajectories obtained under different conditions that
the speed and the direction of migration are determined by a cellular
steering device and by a cellular automatic controller, respectively.
R.T. Tranquillo presents a microscopic model for chemical sensing.
A spatial recognition system is discussed for cells like granulocytes,
but also some evidence is given that these cells have a temporal
recognition system. The fluctuations in receptor binding is regarded
as the main noise source in chemotaxis. The growth and development
of capillary blood vessels is discussed by **C.L. Stokes, D.A. Lauffenburger, and S.W. Williams**. Their likely working hypothesis is that
chemotaxis is the mechanism for the directed growth of the vessel
tip. The last contribution of **V. Calenbuhr and J.-L. Deneubourg**
deals with ants. These animals have a sense of smell and have thus
the ability to find the smelling trail between a food source and the
nest. Artificial circular trails are used in the experiment

TRACKING OF FLAGELLATES BY IMAGE ANALYSIS

Donat-P. Häder

Institut für Botanik und Pharmazeutische Biologie der

Friedrich-Alexander-Universität, D-8520 Erlangen, FRG

Abstract

Oriented movement of flagellates with respect to the stimulus direction such as light or gravity is tracked in real time using image analysis. The organisms are observed in dark field in order to enhance the contrast using a CCD camera. Algorithms have been developed to follow the track of randomly selected organisms for a predefined period of time. Velocity and angular deviation can be extracted from the raw data. Histograms constructed from these data show that the organisms orient with respect to light and gravity often using antagonistic responses to accumulate in distinct horizons of suitable conditions. This can be verified by using a vertical plexiglass column inserted into a natural habitat from which samples are taken at regular time intervals along the length of the tube, which indicates that the populations undergo daily vertical migrations which serve to keep the photosynthetic organisms in suitable light conditions for photosynthesis and to avoid too bright irradiation which can photobleach or even kill the population.

Solar ultraviolet radiation has been found to damage photoorientation and motility in a number of phytoplankton flagellates even at currently observed levels. The mechanism of inhibition does not seem to involve DNA damage or photodynamic responses but rather affect the pigment composition of the cells directly. Any increase in the solar UV-B radiation due to a partion destruction of the stratospheric ozon layer by e.g. CFC gases may adversly affect the biomass production and contribute to global climate changes because the oceanic phytoplankton communities are the major sink for atmospheric CO_2.

1. Introduction

Many motile microorganisms are not randomly distributed in their habitat but rather optimize their position by orienting with respect to a number of external stimuli (Nultsch and Häder, 1988). The major stimuli are gravity (Bean, 1984; Kessler, 1985, 1986), thermal (Mizuno et al., 1984; Poff, 1985) and chemical (MacNab, 1985; Berg, 1985) gradients and the magnetic field of the earth (Ofer et al., 1984; Frankel, 1984; Stolz et al., 1986). Light certainly plays an important role not only for photosynthetic but also for nonphotosynthetic microorganisms (Foster and Smyth, 1980; Colombetti et al., 1982; Haupt, 1983).

Flagellates, like many other motile organisms, have developed three basically different strategies to respond to light and to actively search for optimal conditions for growth and survival:

1. *Photokinesis* describes a dependence of the linear velocity of an organism upon the ambient light intensity. The speed of movement in darkness is defined as a reference value, which can even be zero. Any increase in the velocity compared to the dark velocity is defined as positive photokinesis, while a lower value is defined as negative photokinesis. The response reaches a steady state after a certain exposure time and is independent of the direction of movement (Häder and Tevini, 1988).

2. In contrast, *phototaxis* is an oriented movement dependent on the direction of the impinging actinic light. Organisms may move toward the light source (positive phototaxis) or away from it (negative phototaxis) and, in addition, several cases of movement perpendicular to the light direction have been described (Nultsch and Häder, 1988, Rhiel et al., 1988a,b).

3. A sudden change in the fluence rate of the actinic light causes a *photophobic response*, which in contrast to photokinesis is a transient response. Unlike phototaxis it is independent of the direction of the actinic light. The change in the fluence rate can be spatial (e.g. when an organism moves from an area of one defined level of light intensity to another with a different light intensity) or temporal (e.g. when a organism experiences a change in the fluence rate). Organism have been found to respond photophobically to an increase in the fluence rate (step-up photophobic response), a decrease in the fluence rate (step-down) or both (Diehn et al., 1977).

Each of theses responses can cause a vectorial movement of the population or a non random distribution in the habitat (Häder, 1988b); e.g. a positive phototaxis causes an oriented movement of the population toward the ligh source and a positive photokinesis results in the accumulation of the organisms in shaded areas since the higher speed of movement in light causes them to spend less time in the bright areas of the habitat as compared to dark areas.

The basic differences between the three photoresponses described above require totally different approaches in the algorithms necessary to analyze and quantify them. In addition, the large differences between diffent species in speed, size and form also demand a variety of strategies and even hardware solutions to realize an automated and effective screening based on image analysis techniques. However, usually faster organisms are generally bigger than slower ones, so that the use of different magnification objectives allows to use the same strategies for tracking. Difficulties arise with very fast and small organisms such as *Chlamydomonas* or *Ectocarpus* zoospores. For these cells specifically effective algorithms had to be designed. This paper concentrates of the automatic tracking of flagellates in real time performed by a dedicated harware and software system based on video images digitized in real time.

2. Hardware configuration

The organisms are transferred into glass cuvettes, the inner dimensions of which are dependent on the speed and size of the cells studied (e.g. 40 x 8 x 0.17 mm) for the green flagellate *Euglena*. The cuvette is placed on the stage of a conventional light microscope using an appropriate objective magnification (between 1.25 and 25 times, de

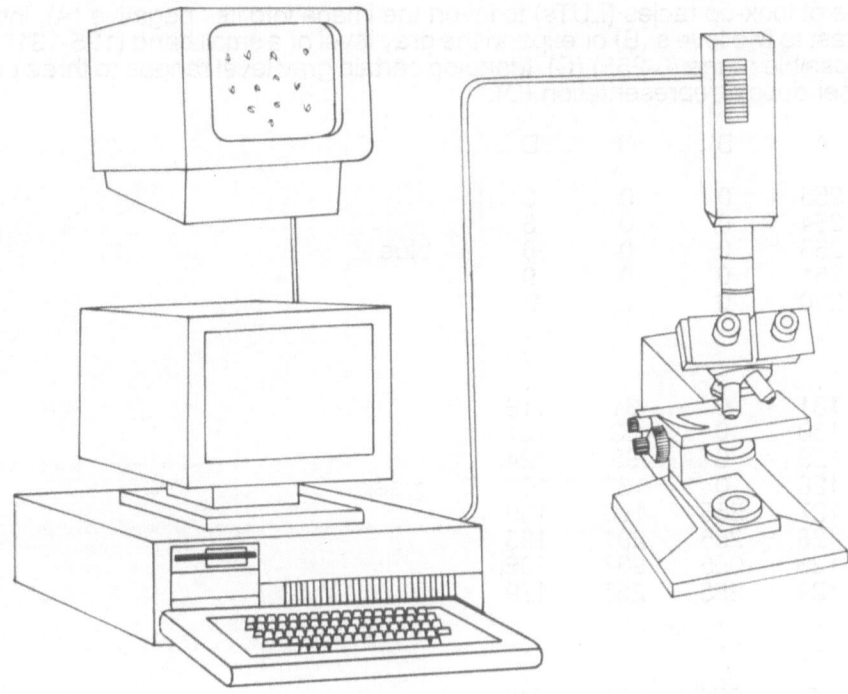

Fig. 1. Configuration of the image analysis system to track flagellates in real time. The cells move in a glass cuvette on the stage of a dark field microscope. The image is produced by an infrared irradiation and recorded by a CCD camera. The image is digitized and stored in a dedicated video RAM to which a microcomputer has access.

pending on the size and speed of the organisms). The contrast of the image is enhanced by using dark field irradiation, and an infrared monitoring radiation is used in order not to disturb the orientation with respect to the external stimulus, such as light or gravity, by the monitoring beam (Fig. 1). In addition, care has to be taken that the monitoring light does not activate the photosynthetic apparatus since the oxygen produced by organisms in the field of view could attract others from outsize this field disturbing the measurement.

The image of the moving cells is recorded by either a black and white video camera or a charged coupled device (CCD). The video signal is digitized (PIP-1024, Matrox, Quebec, Canada) in real time (20 ms per frame) using a plug-in card which occupies a long slot in an IBM AT compatible microcomputer (Tatung 7000, Teipei, Taiwan). The spatial resolution of the digitized image is 512 x 512 pixels at 256 possible gray levels. The image is stored in a dedicated RAM (read and write memory) localized on board of the digitizer card. The on-line video image is displayed on a black and white monitor and the digitized image is shown in pseudocolor representation on a color analog monitor (NEC Multisync II).

The microcomputer has access to the digitized image on a random pixel by pixel basis and can both read and write individual memory cells and thus analyze or manipulate the image using programable registers (Mayfield, 1984).

Table 1: Use of look-up tables (LUTs) to invert the image into its negative (A), increase the contrast to two levels (B) or expand the gray level of a small band (125-131) over the whole possible range (0-255) (C). Mapping certain gray level ranges to three colors results in pseudocolorrepresentation (D).

	A	B	C	D	
0	255	0	0	0	
1	254	0	0	3	
2	253	0	0	6	blue
3	251	0	0	9	
4	250	0	0	12	
.	.	.	.	.	
.	.	.	.	.	
.	.	.	.	.	
124	131	0	31	118	
125	130	0	63	121	
126	129	0	95	124	
127	128	0	137	127	green
128	127	255	169	130	
129	126	255	201	133	
130	125	255	233	136	
131	124	255	255	139	
.	.	.	.	.	
.	.	.	.	.	
.	.	.	.	.	
251	4	255	255	243	
252	3	255	255	246	
253	2	255	255	249	red
254	1	255	255	252	
255	0	255	255	255	

3. Image manpulation before analysis

Several mathematical techniques have been developed to manipulate the raw image after digitization and before analysis (Häder, 1988a). Look-up-tables (LUT) are often realized in hardware and allow a fast (real time) alteration of the image (Table 1). For instance, the image can be inverted into its negative: Let us assume we use 256 possible gray levels where 0 defines a dark black and 255 a bright white pixel; substituting a 255 for a 0, a 254 for a 1 etc. yields the exact negative of the image. Likewise the contrast can be enhanced by mapping certain gray levels to 0 and others to 255 or to expand the gray scale in a linear or logarithmic manner. Pseudocolor representation is achieved by mapping certain gray levels to predefined colors on the analog output device before D/A conversion. This is generally done by three parallel output LUTs (red, blue and green) (Bryan et al., 1985).

Simple mathematical operations such as adding a constant or multiplying all pixel values with a factor allows to make the image darker or brighter or enhance the contrast. Procedures such as smoothing based on matrix operations (Julez and Harmon, 1984) or Fast Fourier Transformation (FFT) allow to remove high frequency noise in the image (Häder and Lipson, 1986). Edge detection can be obtained by using Sobel or Laplace algorithms by which each pixel is replaced by a new one calculated from a matrix of the surrounding neighbors (Häder, 1988a).

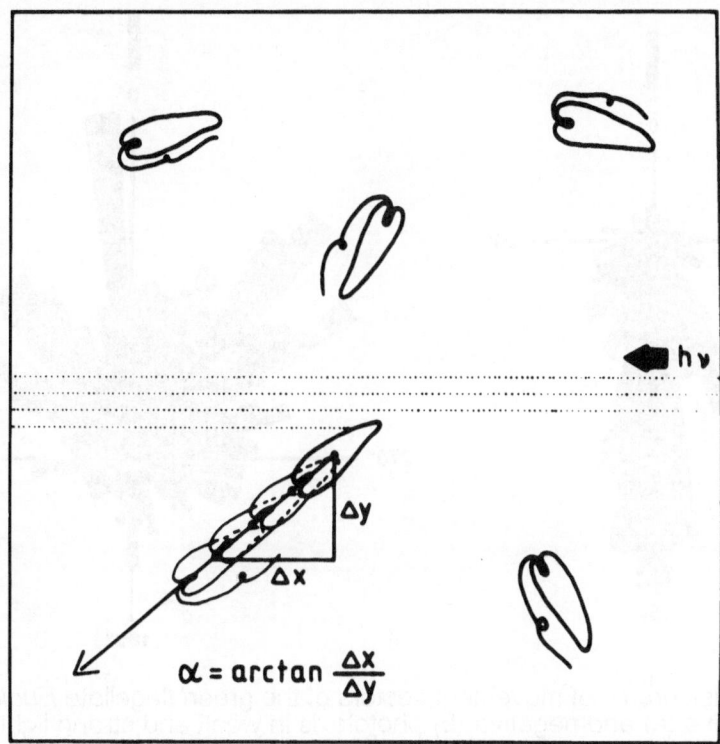

Fig. 2. Demonstration of the algorithm to detect and follow single organisms. The image is scanned until an object is found the brightness of which differs from the background by a predefined threshold. The outline is analyzed and the centroid calculated. This process is repeated for a predefined period of times on the same cell in subsequently digitized images and the movement vector between the first and the last images is stored for subsequent mathematical and statistical treatment.

4. Tracking algorithms

Spotting of an organism is based on a thresholding technique searching for pixels which exceed the background by a certain threshold. An organism is selected randomly from those within the field of view by scanning from a random position either horizontally or vertically (also randomly chosen, in order not to discriminate against elongated organisms swimming parallel to the scan direction). When a sufficiently bright pixel is found in the dark background (dark field microscopy) the outline of the cell is detected using standard edge detection algorithms (Grant and Reid, 1981; Berns and Berns, 1982). Once the outline has been analyzed it is good programming technique to determine whether the object falls into a predefined class of sizes in order to discriminate organisms from debris. Then the center of gravity (centroid) is calculated and the next image is requested from the digitizer (Fig. 2). Starting from the previously found centroid the new outline and centroid is determined and this process is repeated for a predefined period of time. The first and last centroids define a movement vector, the length of which can be used to determine the speed of movement by reading the elapsed time period from the computer's built-in hardware clock. Furthermore, the deviation angle from a stimulus direction (such as light or gravity) is determined and also stored in a disk file

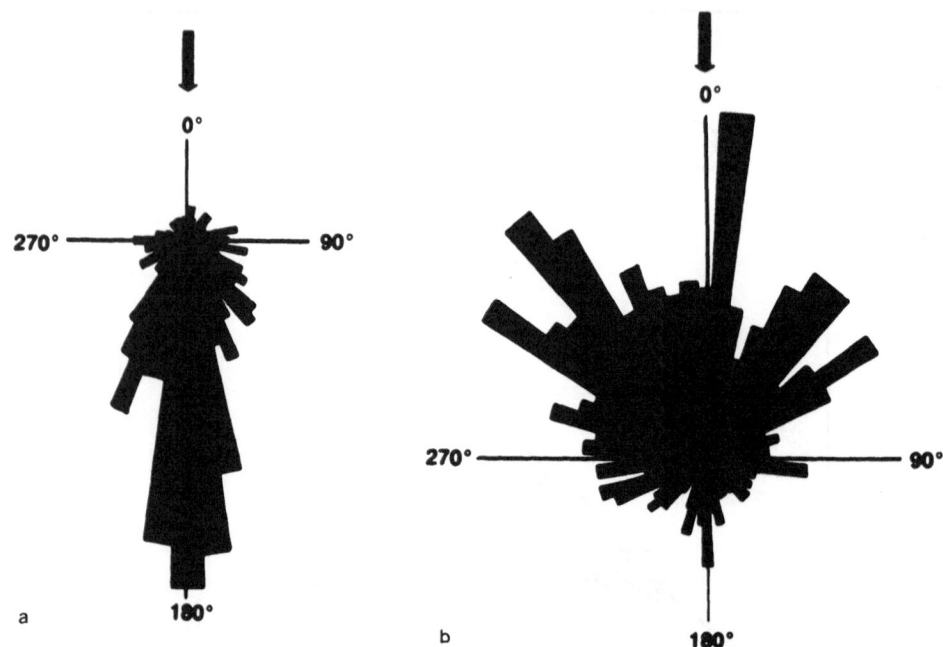

Fig. 3. Circular histograms of movement vectors of the green flagellate *Euglena gracilis* showing positive (a) and negative (b) phototaxis in weak and strong light, respectively. 1000 tracks have been binned in 64 5.6° sectors.

for subsequent mathematical and statistical analysis (Häder et al., 1981; Häder, 1985; Häder and Lebert, 1985). Additional safety checks determine whether the organism has left the field of view during analysis or has moved with an adequate speed in order to discriminate against nonmotile organisms or drift. This algorithm has proven to be rather robust and determine the directional movement of flagellated faithfully in real time at a rate of about 1000 tracks in 15 min.

The raw data can be binned in sectors to construct circular histograms (Fig. 3a,b). In many cases these histograms visualize the preferred direction of movement in the population under the stimulus used. In order to quantitate the degree of orientation, several

statistical methods have been employed such as the Rayleigh test (Batschelet, 1965, 1981; Mardia, 1972):

$$\bar{r} = \frac{\sqrt{(\Sigma \sin \alpha)^2 + (\Sigma \cos \alpha)^2}}{n}$$

An $\bar{r}$-value near 0 indicates a random distribution of the directions while an $\bar{r}$-value in the vicinity of 1 defines a high degree of orientation. However, the Rayleigh test is defined only for unimodal distributions and it does not yield the mean direction of movement. For this purpose Fast Fourier Transformation has been employed which calculates the amplitude and the angular phase of each Fourier coefficient. In addition it allows to discard either high frequency components or low amplitude coefficients in order to smooth the data set, which can be seen after IFFT (Häder and Lipson, 1986). Similar

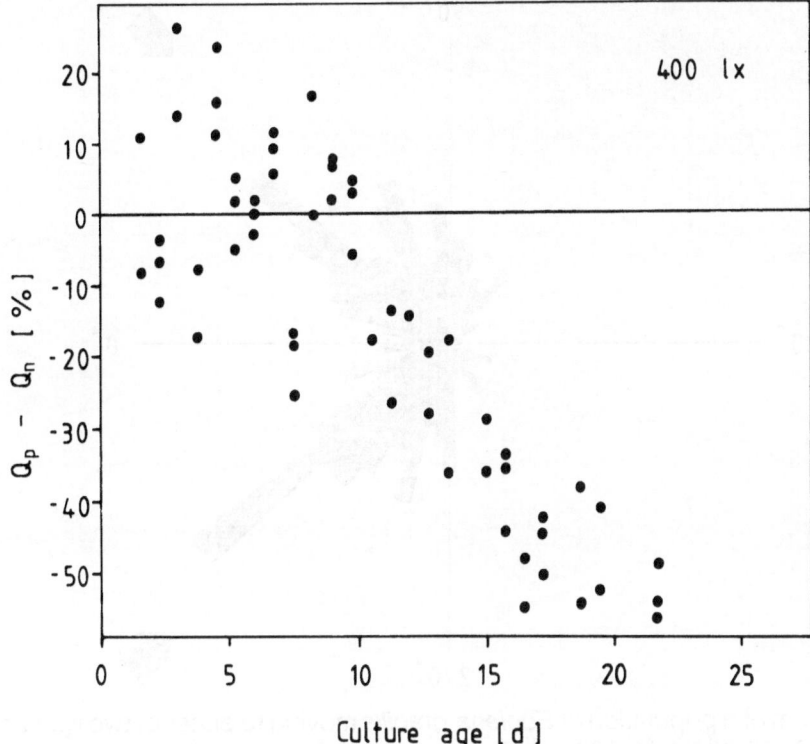

Fig. 4. Degree and dircetion of phototactic orientation in *Euglena gracilis* in a 400 lx test beam as quantified by substracting the percentage of cells moving away from the light source (Q_n) from the percentage of cells moving toward the light source (Q_p) in dependence of the culture age.

algorithms have been used to determine periodicities in electron micrographs (Harms et al., 1981; Squire et al., 1986) and to analyze eye and facial movements at high resolution (Hall, 1983; Dowideit et al., 1983).

5. Orientation of individual cells in flagellate populations

Histograms of the green flagellate *Euglena* moving with respect to a weak white light source indicate a not very pronounced positive phototaxis (movement toward the light source, Fig 3a). High irradiances (30 klx) result in a clear-cut negative phototaxis (away from the light source, Fig. 3b). The crossover point between the two modes of orientation depend on the culture conditions and age (Häder et al., 1987), e.g., the positive phototaxis at 400 lx found in a young culture reverses into a negative one in an older one (Fig. 4).

The strategy of light direction detection in *Euglena* has been described by a shading mechanism (Jennings, 1904; Bancroft, 1913). During forward locomotion the cell rotates around its long axis; thus, in lateral light, the stigma (an assembly of carotenoid colored lipid vesicles) periodically cast a shadow on the proposed photoreceptor, the paraflagellar body (PFB) which consists of a swelling near the basis of the one emer-

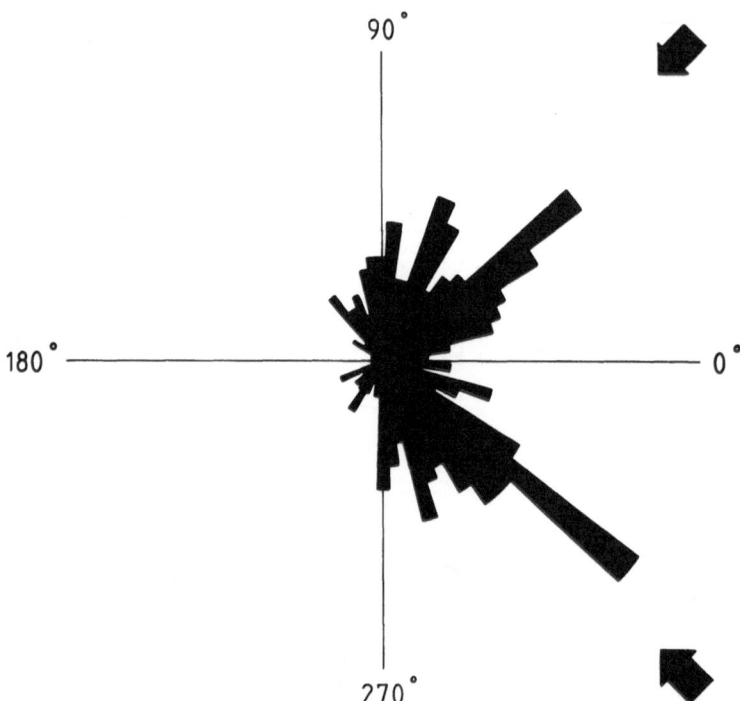

Fig. 5. Histogram of a population of *Euglena gracilis* moving to either of two light sources impinging perpendicularly to each other.

ging flagellum (Benedetti and Checcucci, 1975; Doughty and Diehn, 1980; Ghetti et al., 1985). According to the shading hypothesis the modulated signal at the photoreceptor causes the flagellum to swing out in synchrony with the shading which turns the cell's front end toward the light source until the long axis is aligned with the light direction in which case no modulated shading stimulates the PFB (Gössel, 1957; Diehn, 1969). There are two unstimulated positions with respect to the light source and the cell is sup-posded to distinguish between movement toward and away from the light source by the additional shading from the rear end of the cells with its chloroplasts or other cell organelles.

his shading hypothesis has been invalidated using different lines of evidence: First, stig-maless mutants orient in light; second, when irradiated by two beams of light perpen-dicular to each other the cells do not swim on the resultant as predicted by the shading hypothesis and described in earlier papers, but rather the population splits into two components moving to either light source (Fig. 5). The third argument is of biochemi-cal nature: Calcium channels and a potassium/sodium exchange pump have been found to be involved in the sensory transduction chain of photophobic responses in *Euglena* (Doughty and Diehn, 1982, 1983, 1984). If the phototactic orientation were based on stepwise reorientation of the cell, the same elements had to be involved in the transduction chain of phototaxis; however, inhibitor studies and the use of ionopho-res showed no such involvement (Häder et al., 1986b).

The alternative explanation for the direction detection mechanism is based on a dich-roic orientation of the absorbing vectors of the photoreceptor molecules (Häder, 1987a).

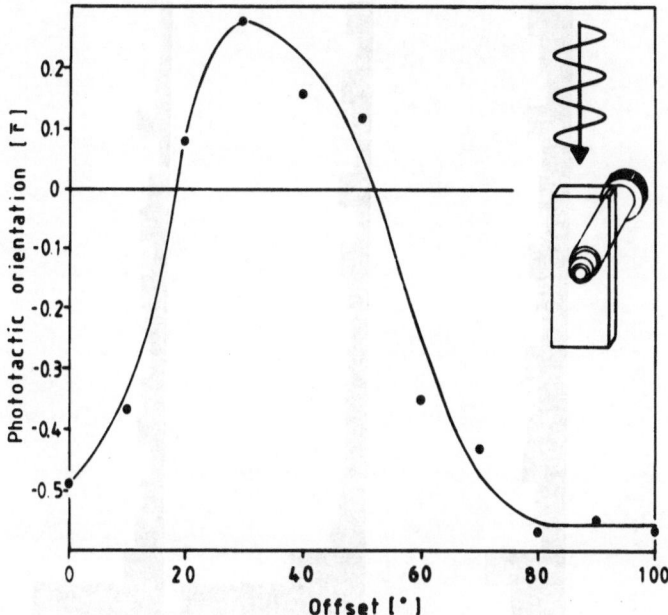

Fig. 6. Direction and degree of orientation of *Euglena gracilis* moving in a vertical cuvette under the antagonistical effects of negative gravitaxis and negative phototaxis to the light impinging from above in dependence of the fluence rate.

When irradiated from above with linearly polarized light the cells swimming in a horizontal cuvette orient at an angle 30° clockwise to the plane of the polarized light, indicating that the absorbing vectors are tilted at that angle with respect to the long axis of the cell.Upward movement is mechanically prevented by the cuvette wall. Using the same approach in the other two directions in space the dichroic orientation could be determined with respect to the three axis of the cell. Thus, the modulation of the photoreceptor signal is based on the dichroic orientation of the photoreceptor molecules. Electron microscopic analysis reveals that the PFB consists of a paracrystalline array which supports the above notion (Benedetti and Checcucci, 1975). Recently, flagella with the still attached PFB have been isolated from the cells (Gualtieri et al., 1986) and spectroscopic and biochemical analyses are under way.

In addition to phototactic orientation, several flagellates have been found to exhibit a remarkably precise gravitaxis. It is not clear, however, whether this orientation is based on a passive physical process (such as a heavy tail end so that the front end with the flagellum points upwards) or whether the cells have a sensor and orient actively (Brinkmann, 1968; Kessler, 1985, 1986). These questions will be tackled in space experiments which provide microgravity conditions. Since in *Euglena* negative gravitaxis is much more pronounced than positive phototaxis, this factor has to be taken into acount to characterize the behavior in the water column in the natural habitat. While the crossover point between positive and negative phototaxis was found at irradiances in the neighborhood of 1.5 W m^{-2}, far higher irradiances from above were necessary to compensate the gravitactic orientation (30 W m^{-2}; Fig. 6; Häder, 1987a).

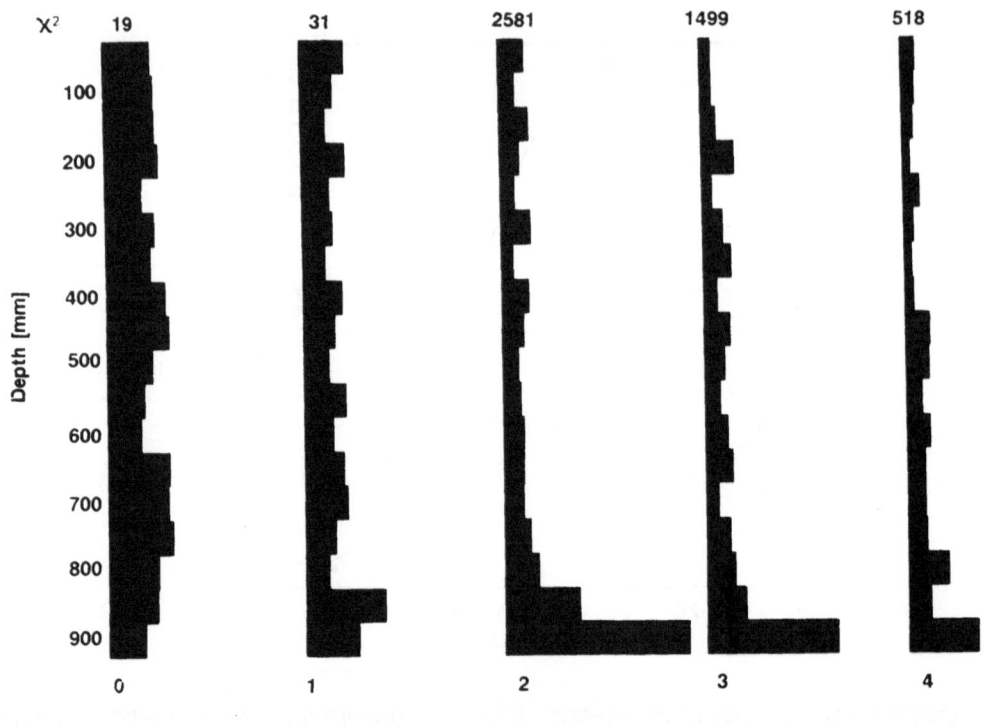

Fig. 7. Cell densities measured in samples taken from 18 outlets along the length of an submersed 1 m long plexiglass cuvette starting at 10 a.m. on July 12, 1987 at 1 h intervals. The chi square value for each data set is indicated above the density histogram.

Other flagellate species have been found to use a different strategy of orientation in their habitat. A freshwater *Cryptomonas* species has been found to show only positive phototaxis at all studied irradiances (Watanabe and Furuya, 1982); likewise two marine *Gyrodinium* species, which form red tides (algal blooms, Tangen, 1977; Spector, 1984) in the North and Baltic seas display exclusively positive phototaxis (Ekelund and Häder, 1988). It is not known by which mechanism these cells move downward in the water column, though they are known to perform massive vertical daily migrations (Burns and Rosa, 1980).

6. Orientation of the population in its habitat

In order to study the effects of the observed single cell orientation in a population in its natural habitat, a plexiglass column was constructed with 1 m length and 90 mm inner diameter which had 18 outlets spaced at 50 mm along the length of the column connected by silicon tubes with an inner diameter of 0.5 mm with a peristaltic pump which could draw 18 samples in parallel. The column was filled with a population of *Euglena* cells and inserted into a pond (Häder and Griebenow, 1988). After thermal equilibration samples were taken at regular time intervals and the cell densities determined using a modified image analysis system. The cell suspension of a sample is drawn through a quartz cuvette of known inner hight mounted on the stage of a dark field microscope by means of a peristaltic pump, taking care that no bubbles are trapped in the cuvette. The algorithm for cell counting is based on a successive scanning of the image in horizontal rows from the top until a cell is found (Häder and Griebenow, 1987). The size of the cell is determined by counting all connected pixels. Once the whole area of the cell has been covered the scan is resumed from that point until the whole image has been analysed. The cell density can easily be calculated from the objective magnification and the observed volume in the cuvette. Also, in mixed populations different cell types can be distinguished by separating them into size classes.

After thorough mixing of the cells at the beginning of the experiment, *Euglena* cultures have been found to move downward in the column when exposed to direct sunlight probably guided by negative phototaxis. In addition, there seems to be a hydrodynamic effect in the dense populations, since the cells move downward faster than their swimming speed allows (Fig. 7). During night the cells move upward, though this movement is much slower than in the opposite direction. These vertical movements of the population which have been described in a number of marine and freshwater phytoplankton species allow a fine-tuned adaptation of the organisms to the constantly fluctuating conditions and variations of the external factors in their habitat (Forward, 1975, 1976).

These movements are of vital importance for the populations, since on the one hand these photosynthetic organisms have to stay in the light for energetic reasons on the other hand, unlike higher plants, they are prone to photobleaching and photodamage when exposed to solar radiation of higher irradiances. The marine *Cryptomonas maculata* has been found to be bleached even at a fluence rate of 5 klx which corresponds to 5 % of full solar radiation (Häder et al. 1988); while the exclusive positive phototactically orienting *Gyrodinium* species show a higher resistance toward photobleaching (Ekelund and Häder, 1988).

7. Inhibition by UV-B radiation

In addition to white light photobleaching, phytoplankton organisms have been found to be affected by solar ultraviolet radiation in the range between 280 nm and 320 nm (UV-B) (Häder et al., 1986a; Häder, 1986a, 1987b). In addition to a direct damage of the cells by ultraviolet radiation, indirect effects have been found, which also affect the cells and impair growth and survival of the population.

Photoorientation has been found to be drastically affected in, e.g., *Euglena* even after short exposure to solar radiation (Fig. 8a-c; Häder, 1986b). This behavior is specifically due to the UV-B component of the radiation since reduction of specifically short wavelength by filtering through an ozon enriched cuvette increases the tolerated exposure time. This effect cannot be attributed to radiation absorbed by the photosynthetic pigments because dark bleached cells are affected likewise (Häder and Häder, 1988a). Similar UV-B inhibition of photoorientation has been found for a number of other flagellates (Häder and Häder, 1989).

In addition to photoorientation, also motility is drastically impaired by the solar UV-B radiation: The percentage of motile organisms decreases rapidly after prolonged exposure (Fig. 9). After a few hours all cells are imotile (Häder and Häder, 1988b). Likewise the average speed of movement decreases drastically as determined using the image analysis. The distances traversed by the organisms are extracted from the tracks during

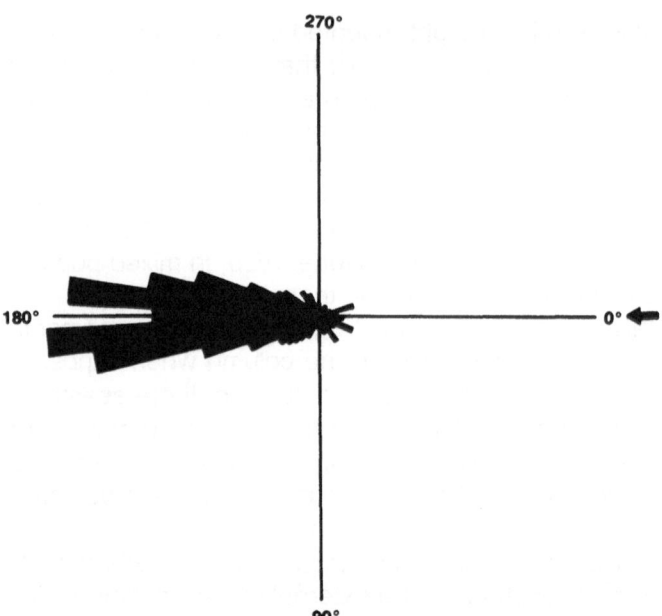

Abb. 8a

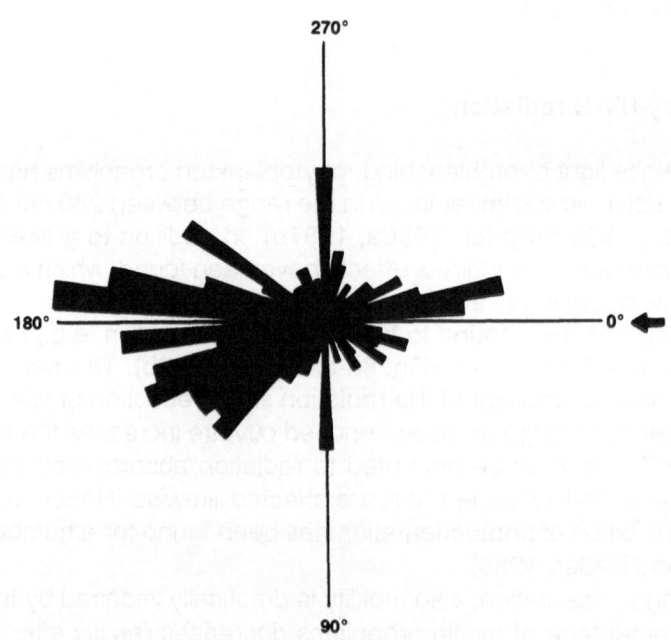

Abb. 8b

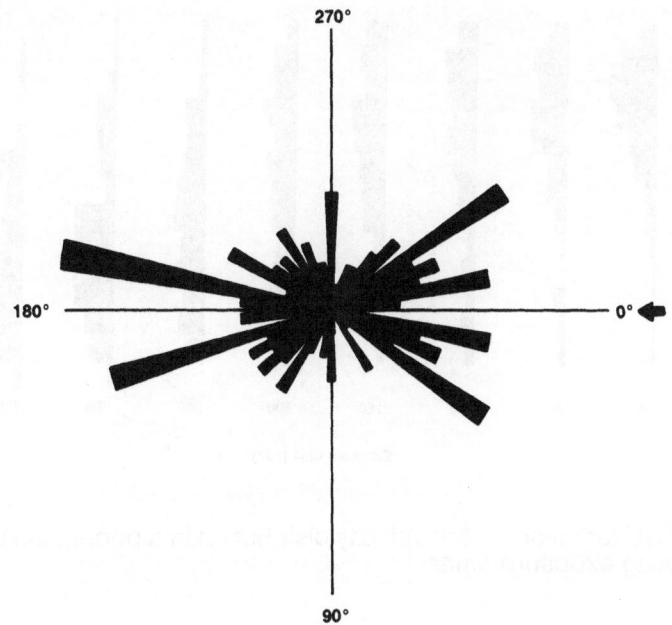

Abb. 8c

Fig. 8. Circular histograms of the negative phototactic orientation of *Euglena gracilis* to a 30 klx white test light after 90 min, 180 min and 300 min of solar radiation (1145 Wm⁻²) at a location south of Lisboa.

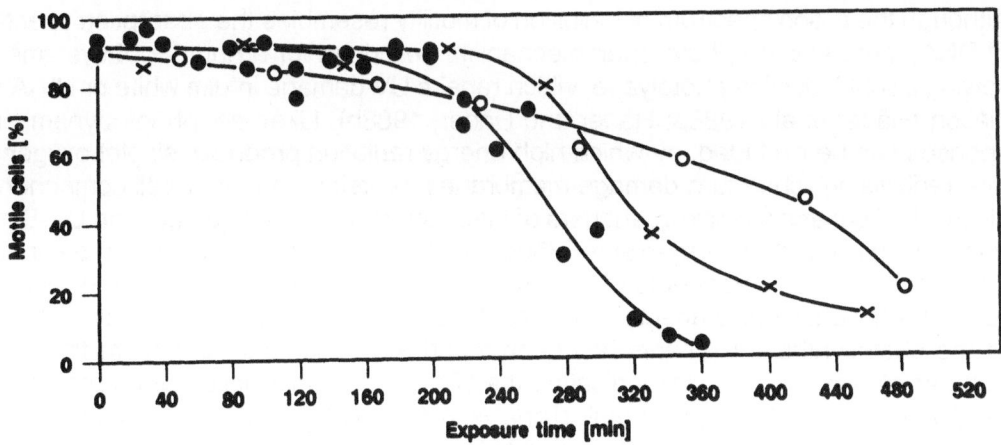

Fig. 9. Percentage of motile cells in a population of *Euglena gracilis* exposed to solar radiation in dependence of the exposure time to unfiltered solar radiaton (dark circles), covered with UV-B cut-off filter WG 320 (crosses), and radiation filtered by an ozone layer (open circles).

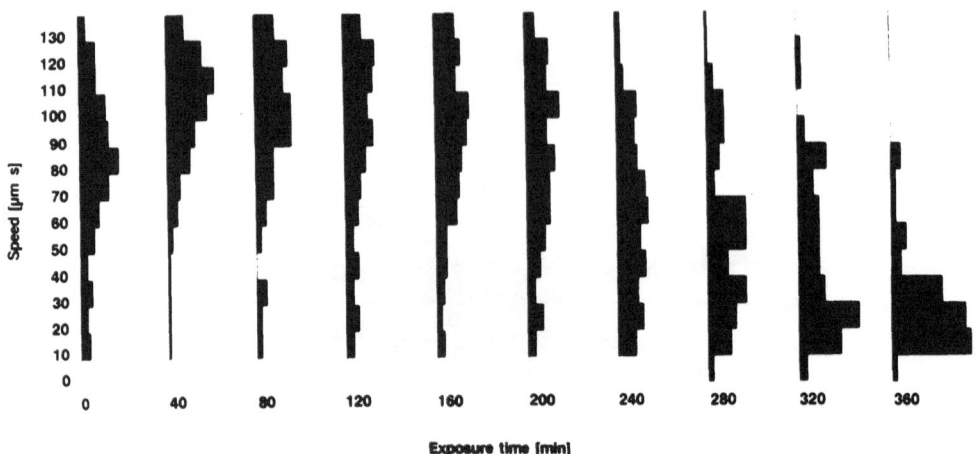

Fig. 10. Effect of UV radiation on the velocity distribution in a population of *Euglena gracilis* after increasing exposure times.

a predetermined period of time. During exposure to solar radiation the mean velocity of the cells increases which was initially thought to be due to the strong white light component (photokinesis; Wolken and Shin, 1958); however even artificial UV-B radiation produced from a transilluminator has a similar effect in both photosynthetic and colorless cells (Fig. 10). The effect of this response still needs to be determined. After the velocity has gone through an optimum, it quickly drops to lower values and after a few hours hardly a cell moves with a detectable speed.

Although the action spectrum of inhibition of motility resembles the absorption spection of DNA, there was no photorepair mechanism described for a number of systems involving a light inducible photolyase, which repairs UV damage in dim white or UV-A radiation (Häder et al., 1986a; Häder and Häder, 1988b). Likewise, phootodynamic responses can be excluded, by which high energy radiation produces singlet oxygen or free radicals, which in turn damage membranes, proteins and other cell components, since inhibitors and specific quenchers of these substances had no effect on UV-B mediated inhibition in these organisms (Häder and Häder, 1988a). Thus, it can be speculated that intrinsic components of the photoreceptor organelle and the motor apparatus of the cell are directly affected by solar UV-B radiation. In fact, a dramatic bleaching of the photosynthetic pigments has been found after even shor exposure times in a number of systems (Fig. 11) indicating the UV induced destruction of chromophores and/or protein structures. Biochemical analysis of the proteins involved using FPLC and gel electrophoresis are currently carried out.

In consequence, phytoplankton flagellates are under heavy UV-B stress at current radiation levels. Both motility and photoorientation are affected within even short exposure times. The organisms studied so far seem to lack a UV-B specific photoreceptor and orient exclusively using wavelengths in the visible or UV-A region of the solar spectrum. Thus, any increase in the UV-B level as expected to occur due to the partial destruction of the stratospheric ozon layer by manmade gaseous pollutants such as chlorinated fluorcarbohydrates (CFCs) is bound to have a detrimental effect on the population

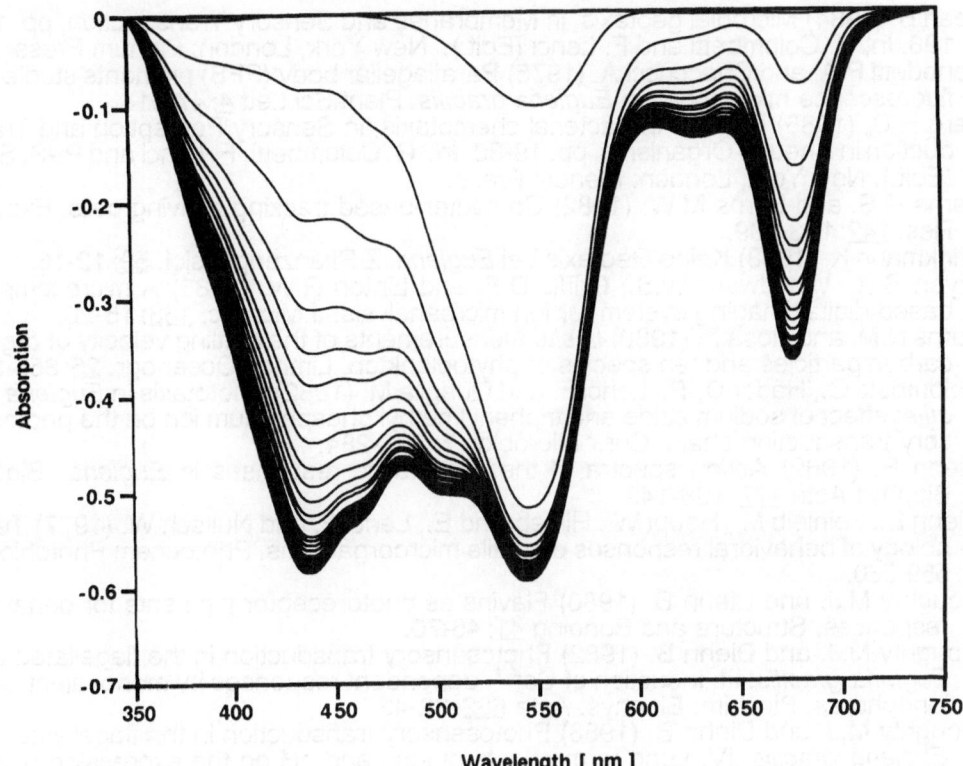

Fig. 11. Bleaching of the major photosynthetic pigments in *Cryptomonas maculata* as shown in absorption difference spectra after increasing exposure to a transilluminator UV radiation.

size of phytoplankton flagellates. This effect may adversely affect the biomass production which has been estimated to exceed that of terrestrial plants and in addition may augment the global warming due to the greenhouse effect, since the marine ecosystem is the major sink for the atmospheric CO_2.

Acknowledgements

This work was supported by the Bundesminister für Forschung und Technologie (KBF 57). The author gratefully acknowledges the skilfull technical assistance of H. Vieten and J. Schäfer.

REFERENCES

Bancroft F.W. (1913) Heliotropism, differential sensibility and galvanotropism in Euglena. J Exp Zool 15: 383-420.
Batschelet E. (1965) Statistical methods for the analysis of problems in animal orienta tion and certain biological rhythms, pp. 61-91. In: S.R. Galles , K. Schmidt-Koenig, G.J. Jacobs, R.F. Belleville (Edit) Animal orientation and navigation. Washington: NASA.
Batschelet E. (1981) Circular Statistics in Biology. London: Academic Press.

Bean B. (1984) Microbial geotaxis, in Membranes and Sensory Transduction, pp. 163-198. In: G. Colombetti and F. Lenci (Edit.). New York, London: Plenum Press

Benedetti P.A. and Checcucci A. (1975) Paraflagellar body (PFB) pigments studied by fluorescence microscopy in *Euglena gracilis*. Plant.Sci.Lett 4: 47-51.

Berg H.C. (1985) Physics of bacterial chemotaxis, in Sensory Perception and Transduction in Aneural Organisms, pp. 19-30. In: G. Colombetti, F. Lenci and P.-S. Song (Edit.). New York, London: Plenum Press.

Berns G.S. and Berns M.W. (1982) Computer-based tracking of living cells. Exp.Cell Res. 142: 103-109.

Brinkmann K. (1968) Keine Geotaxis bei *Euglena*. Z Pflanzenphysiol. 59: 12-16.

Bryan S.R., Woodward W.S., Griffis D.P. and Linton R.W. (1985) A microcomputer based digital imaging system for ion microanalysis. J.Microsc. 138: 15-28.

Burns N.M. and Rosa F. (1980) *In situ* measurements of the settling velocity of organic carbon particles and ten species of phytoplankton. Limnol. Oceanogr. 25: 855-864.

Colombetti G., Häder D.-P., Lenci F. and Quaglia M. (1982) Phototaxis in *Euglena gracilis*: effect of sodium azide and triphenylmethyl phosphonium ion on the photosensory transduction chain. Cur.r Microbiol. 7: 281-284.

Diehn B. (1969) Action spectra of the phototactic responses in *Euglena*. Biochim Biophys Acta 177: 136-143.

Diehn B., Feinleib M., Haupt W., Hildebrand E., Lenci F. and Nultsch W. (1977) Terminology of behavioral responses of motile microorganisms. Photochem Photobiol 26: 559-560.

Doughty M.J. and Diehn B. (1980) Flavins as photoreceptor pigments for behavioral responses. Structure and Bonding 41: 45-70.

Doughty M.J. and Diehn B. (1982) Photosensory transduction in the flagellated alga, *Euglena gracilis*. III. Induction of Ca^{2+}-dependent responses by monovalent cation ionophores. Biochim. Biophys. Acta 682: 32-43.

Doughty M.J. and Diehn B. (1983) Photosensory transduction in the flagellated alga, *Euglena gracilis*. IV. Long term effects of ions and pH on the expression of step-down photobehavior. Arch. Microbiol. 134: 204-207.

Doughty M.J. and Diehn B. (1984) Anion sensitivity of motility and step-down photophobic responses of *Euglena gracilis*. Arch. Microbiol. 138: 329-332.

Dowideit G.R., Newman D.G. and Young C.M. (1983) A new automated approach to high-density facial measurement. Part 1: The image capturing and processing hardware. Int.J.Bio-Medical Computing 14: 403-409.

Ekelund N. and Häder D.-P. (1988) Photomovement and photobleaching in two *Gyrodinium* species. Plant Cell Physiol. 29: 1109-1114.

Forward jr. R.B. (1975) Dinoflagellate phototaxis: Pigment system and circadian rhythm as related to diurnal migration, pp. 367-381. In: F. Vernberg (Edit.) Physiological Ecology of Estuarine Organisms. Columbia: Univ. South Carolina Press.

Forward jr. R.B. (1976) Light and diurnal vertical migration: Photobehavior and photophysiology of plankton, in Photochemical and Photobiological Reviews, Vol. 1 (Smith, K.C., Ed.), Plenum Pess, New York, London, pp. 157-209

Foster K.W. and Smyth R.D. (1980) Light antennas in phototactic algae. Microbiol.Rev. 44: 572-630.

Frankel R.B. (1984) Magnetic guidance of organisms. Ann.Rev. Biophys.Bioeng. 13: 85-103.

Ghetti F., Colombetti G., Lenci F., Campani E., Polacco E. and Quaglia M. (1985) Fluorescence of *Euglena gracilis* photoreceptor pigment: an in vivo microspectrofluorometric study. Photochem Photobiol 42: 29-33.

Gössel I. (1957) Über das Aktionsspektrum der Phototaxis chlorophyllfreier Euglenen und über die Absorption des Augenflecks. Arch. Mikrobiol. 27: 288-305.

Grant R. and Reid A.F. (1981) An efficient algorithm for boundary tracing and feature extraction. Comput. Graphics Image Process. 17: 225-237.

Gualtieri P., Barsanti L. and Rosati G. (1986) Isolation of the photoreeptor (paraflagellar body) of the phototactic flagellate *Euglena gracilis*. Arch.Microbiol. 145: 303-305.

Hall R.W. (1983) Image processing algorithms for eye movement monitoring. Comput.Biomed.Res. 16: 563-579.

Harms H., Boseck S., Aus H.M. and Lenz V. (1981) Untersuchungen der Abtastbedingungen bei Zellbildern mit einem Mikroskop-TV-System. Microscopica Acta 85: 69-82.

Haupt W. (1983) Photoperception and photomovement. Phil.Trans.R.Soc.Lond. B 303: 467-478.

Häder D.-P., Colombetti G., Lenci F. and Quaglia M. (1981) Phototaxis in the flagellates, *Euglena gracilis* and *Ochromonas danica*. Arch.Microbiol. 130: 78-82.

Häder D.-P. (1985) Computer-aided studies of photoinduced behaviors,pp. 75-91. In: G. Colombetti, F. Lenci, P.-S. Song (Edit.) Sensory perception and transduction in aneural organisms. New York, London: Plenum Press.

Häder D.-P.(1986a) The effect of enhanced solar UV-B radiation on motile microorganisms, pp. 223-233. In: R.C. Worrest and M.M. Caldwell (Edit.) Stratospheric ozone reduction, solar ultraviolet radiation and plant life. Berlin, Heidelberg, New York: Springer Verlag.

Häder D.-P. (1986b) Effects of solar and artificial UV irradiation on motility and phototaxis in the flagellate, *Euglena gracilis*. Photochem.Photobiol. 44: 651-656.

Häder D.-P. (1987a) Polarotaxis, gravitaxis and vertical phototaxis in the green flagellate, *Euglena gracilis*. Arch. Microbiol. 147: 179-183.

Häder D.-P. (1987b) Effects of UV-B irradiation on photomovement in the desmid, *Cosmarium cucumis*. Photochem.Photobiol. 46: 121-126.

Häder D.-P. (1988a) Computer-assisted image analysis in biological sciences. Proc.Indian Acad.Sci. (Plant Sci.) 98: 227-249.

Häder D.-P. (1988b) Ecological consequences of photomovement in microorganisms. J.Photochem.Photobiol.B: Biol. 1: 385-414.

Häder D.-P. and Griebenow K. (1987) Versatile digital image analysis by microcomputer to count microorganisms. EDV Med.Biol. 18: 37-42.

Häder D.-P. and Griebenow K. (1988) Orientation of the green flagellate, *Euglena gracilis*, in a vertical column of water. FEMS Microbiol.Ecol. 53: 159-167.

Häder D.-P. and Häder M. (1988a) Ultraviolet-B inhibition of motility in green and dark bleached *Euglena gracilis*. Current Microbiol. 17: 215-220.

Häder D.-P. and Häder M.A. (1988b) Inhibition of motility and phototaxis in the green flagellate, *Euglena gracilis*, by UV-B radiation. Arch.Microbiol. 150: 20-25.

Häder D.-P. and Häder M.A. (1989) Effects of solar UV-B irradiation on photomovement and motility in photosynthetic and colorless flagellates. Environ.Exp.Bot. 29: 273-282.

Häder D.-P. and Lebert M. (1985) Real time computer-controlled tracking of motile microorganisms. Photochem.Photobiol. 42: 509-514.

Häder D.-P. and Lipson E. (1986) Fourier analysis of angular distributions for motile microorganisms. Photochem.Photobiol. 44: 657-663.

Häder D.-P. and Tevini M. (1987) General photobiology. Pergamon Press.

Häder D.-P., Watanabe M. and Furuya M. (1986a) Inhibition of motility in the cyanobacterium, *Phormidium uncinatum*, by solar and monochromatic UV irradiation. Plant Cell Physiol. 27: 887-894.

Häder D.-P., Lebert M. and DiLena M. R. (1986b) New Evidence for the mechanism of phototactic orientation of *Euglena gracilis*. Curr.Microbiol. 14, 157-163.

Häder D.-P., Lebert M. and DiLena M. R. (1987) Effects of culture age and drugs on phototaxis in the green flagellate, *Euglena gracilis*. Plant Physiol. (Life Sci.Adv.) 6: 169-174.

Häder D.-P., Rhiel E. and Wehrmeyer W. (1988) Ecological consequences of photomovement and photobleaching in the marine flagellate *Cryptomonas maculata*. FEMS Microbiol.Ecol. 53: 9-18.

Jennings H.S. (1904) Reactions to light in ciliates and flagellates, pp 29-71. In: Contributions to the study of the behavior of microorganisms. Washington: Carnegie Inst Washington.

Julez B. and Harmon L.D. (1984) Noise and recognizability of coarse quantized images. Nature 308: 211-211.

Kessler J.O. (1985) Hydrodynamic focusing of motile algal cells. Nature (London) 313: 218-220.

Kessler J.O. (1986) The external dynamics of swimming microorganisms, pp. 258-307. In: Round anf Chapman (Edit.) Progress in Phycological Research. Biopress Ltd 4.

MacNab R.M. (1985) Biochemistry of sensory transduction in bacteria, pp. 31-46. In: G. Colombetti, F. Lenci and P.-S. Song (Edit.) Sensory Perception and Transduction in Aneural Organisms. New York, London: Plenum Press.

Mardia K.V. (1972) Statistics of Directional Data. London: Acad Press.

Mayfield C.I. (1984) A simple computer-based video image analysis system and potential applications to microbiology. J.Microbiol.Meth. 3: 61-67.

Mizuno T., Maeda K. and Imae Y. (1984) Thermosensory transduction in *Escherichia coli*, pp. 147-195. In: F. Oosawa, T. Yoshioka and H. Hayashi (Edit.) Transmembrane Signaling and Sensation. Japan Sci.Soc.Press, Tokyo and VNU Sci. Press BV, Netherlands.

Nultsch W. and Häder D.-P. (1988) Photomovement in motile microorganisms II. Photochem.Photobiol. 47: 837-869.

Ofer S., Nowik I., Bauminger E.R., Papaefthymiou G.C., Frankel R.B. and Blakemore R.P. (1984) Magnetosome dynamics in magnetotactic bacteria. Biophys.J. 46: 57-64.

Poff K.L. (1985) Temperature sensing in microorganisms, pp. 299-307. In: G. Colombetti, F. Lenci and P.-S. Song (Edit.) Sensory Perception and Transduction in Aneural Organisms. New York, London: Plenum Press.

Rhiel E., Häder D.-P. and Wehrmeyer W. (1988a) Photo-orientation in a freshwater *Cryptomonas* species. J.Photochem.Photobiol. B: Biol. 2: 123-132.

Rhiel E., Häder D.-P. and Wehrmeyer W. (1988b) Diaphototaxis and gravitaxis in a freshwater *Cryptomonas*. Plant Cell Physiol. 29: 755-760.

Spector D.L. (1984) Dinoflagellates. Orlando, Florida: Acad. Press Inc.

Squire J.M., Luther P.K. and Agnew G.D. (1986) Averaging of periodic images using a microcomputer. J.Microsc. 142: 289-300.

Stolz J.F., Chang S.-B.R. and Kirschvink J.L. (1986) Magnetotactic bacteria and single-domain magnetite in hemipelagic sediments. Nature 321: 849-851.

Tangen K. (1977) Blooms of *Gyrodinium aureolum* (Dinophyceae) in north European water, accompanied by mortality in marine organisms. Sarsia 63: 123-133.

Watanabe M. and Furuya M. (1982) Phototactic behavior of individual cells of *Cryptomonas sp.* in response to continuous and intermittent light stimuli. Photochem.Photobiol. 35: 559-563.

Wolken J.J., Shin E. (1958) Photomotion in *Euglena gracilis* I. Photokinesis II. Phototaxis. J. Protozool. 5: 39-46.

HELICAL ORIENTATION – A NOVEL MECHANISM FOR THE ORIENTATION OF MICROORGANISMS.

Hugh C. Crenshaw
Department of Zoology, Duke University
Durham, North Carolina 27706 U. S. A.

Abstract

This paper describes how helical motion can act both as a strategy for sampling a stimulus field and as a mechanism for orienting to that field. Namely, an organism can orient to a stimulus by pointing its rotational velocity vector towards the source of the stimulus. This is accomplished if the components of the rotational velocity are simple functions of the stimulus intensity. Evidence supporting this hypothesis is presented both from published observations of microorganisms and from experiments in which spermatazoa of the sea urchin, *Arbacia punctulata*, are tracked in three dimensions.

1. Introduction.

A surprising diversity of microorganisms swim in helices. Fungal spores, flagellates, ciliates, the motile spores of many plants, spermatozoa (of both invertebrates and vertebrates), many larval invertebrates, and rotifers all swim in helices.

It is not known why so many organisms move in helices. Some functional explanations have been proposed: (1) Jennings (1901) suggests that helical motion acts like the rifling of a rifle's barrel – it causes an otherwise meandering object (the bullet or the microorganism) to follow a straight path. (2) Theoretical analyses by Foster and Smyth (1980) suggest that the helical motion of phototactic algae is a sampling strategy to determine the direction of light. However, neither of these hypotheses has been tested.

Independent work by Brokaw (1958) and Crenshaw (1989a,b) raises a new hypothesis: Helical motion is an orientation mechanism. It enables an organism to sample and orient to a stimulus field. This hypothesis is addressed in several parts: Helical motion as a sampling strategy (Section 2). Helical motion as an orientation mechanism (Section 3). Evidence from 3D tracking of sea urchin spermatozoa (Section 4). Observations in the literature (Section 5).

The present discussion borrows heavily from the kinematics of helical motion. Box 6 describes the kinematics of helical motion and presents terms and conventions that are used in this discussion.

BOX   7

AN INTRODUCTION TO THE KINEMATICS
OF HELICAL MOTION

Hugh C. Crenshaw

This is a brief introduction to the kinematics of helical motion. Equations are developed that relate the translational and rotational velocities of an organism to the pitch, radius, and angular frequency of its helical path. Those preferring a more rigorous treatment are referred to Sugino and Naitoh (1988) and Crenshaw (1989).

The Helix:

A helix is a 3D curve. It can be generally described by three parameters: axis $\vec{z}$, radius r, and pitch p. Consider a helix that wraps a circular cylinder (fig. 1). The axis of the cylinder is $\vec{z}$; the radius of the cylinder's cross-section is r; and p is the distance travelled parallel to $\vec{z}$ for every turn around the cylinder. The helix in fig. 1 is a right-hand helix. This discussion uses right-hand helices, but the results apply equally well to left-hand helices.

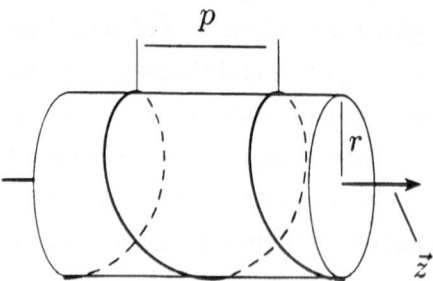

Figure 1. A right-hand helix that wraps a circular cylinder. The axis of the helix ($\vec{z}$) is the axis of the cylinder. The radius (r) of the helix is the radius of the cylinder's cross-section. The pitch (p) of the helix is the distance travelled parallel to $\vec{z}$ for each turn of the helix.

BOX 7

Helical Motion:

The right-hand helical motion of an object is described by the following vector function:

(1) $$\vec{H}(t) = r\cos(\gamma t)\vec{x} + r\sin(\gamma t)\vec{y} + \left(\frac{p\gamma t}{2\pi}\right)\vec{z}\,.$$

Here, $\overrightarrow{xyz}$ is a reference frame fixed to the helix such that $\vec{z}$ is the axis of the helix. γ is the angular frequency of the moving object, so the object completes one turn of the helix every $2\pi/\gamma$ units of time.

The motion of any rigid object, regardless of its trajectory, is described fully by its translation (change in position) and rotation (change in orientation), which can be determined from the object's translational and rotational velocities. It is easiest to describe an object's translational and rotational velocities relative to a reference frame fixed to the object. Let the object be some organism to which the reference frame $\overrightarrow{ijk}$ is fixed (fig. 2a) such that the origin of $\overrightarrow{ijk}$ is the organism's center of gravity; $\vec{i}$ is the organism's antero-posterior axis, pointing anteriorly; $\vec{j}$ is its left-right axis, pointing left; and $\vec{k}$ is its dorso-ventral axis pointing dorsally.

The organism's translational velocity $\vec{V}$ is then described relative to $\overrightarrow{ijk}$ by

(2) $$\vec{V}_{body} = V_1\vec{i} + V_2\vec{j} + V_3\vec{k}$$

where the subscript, *body*, denotes that $\vec{V}$ is described relative to $\overrightarrow{ijk}$ (fig. 2b). Similarly, the organism's rotational velocity, $\vec{\omega}$, is given by

(3) $$\vec{\omega}_{body} = \omega_1\vec{i} + \omega_2\vec{j} + \omega_3\vec{k}$$

where ω_1 is the rate of rotation around $\vec{i}$; ω_2 is the rate of rotation around $\vec{j}$; and ω_3 is the rate of rotation around $\vec{k}$ (fig. 2c). This organism completes one revolution every $2\pi/|\vec{\omega}|$ units of time.

If the three components of $\vec{V}_{body}$ and the three components of $\vec{\omega}_{body}$ all are variable, the organism is said to have six "degrees of freedom". Six degrees of freedom describe all possible motions of a rigid body. Unfortunately, the kinematics of helical motion of organisms with six degrees of freedom has not been described. This discussion considers motion with only one degree of translational freedom:

(4) $$\vec{V}_{body} = V_1\vec{i}, \quad V_1 = U$$

BOX 7

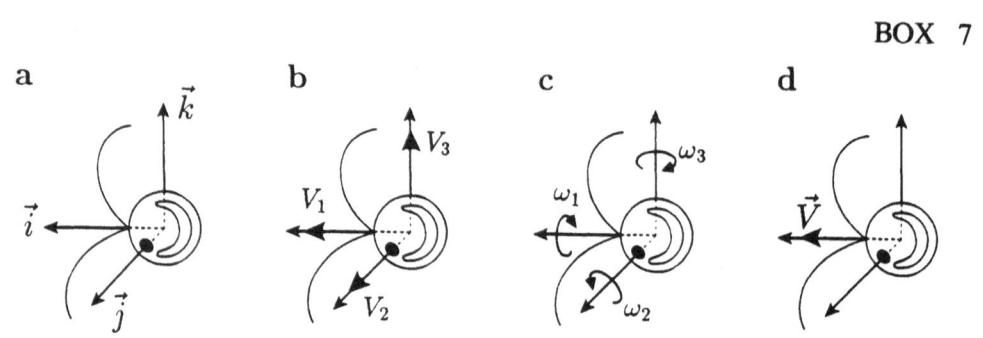

Figure 2. The orientation of $\overrightarrow{ijk}$ with respect to body axes and the components of translational and rotational velocities. (*a*) The reference frame $\overrightarrow{ijk}$ fixed to the body of a microorganism. (*b*) Three nonzero components of translational velocity. (*c*) Three nonzero components of rotational velocity. (*d*) Only the $\vec{i}$-component of $\overrightarrow{V}$ is nonzero, leaving the organism with four degrees of freedom.

where U is the organism's speed. V_2 and V_3 both equal zero (fig. 2d). This organism can only move forward and backward. It cannot move sideways, dorsally, *etc.* The only way it can change direction is by rotating its body such that $\vec{i}$ points in a new direction.

We now have an organism that translates in one direction, relative to $\overrightarrow{ijk}$, and rotates its body as it translates. This organism will move in a line, a circle, or a helix. First, consider an organism moving such that $\vec{\omega}$ is parallel to $\overrightarrow{V}$ ($\omega_1 \neq 0$, $\omega_2 = \omega_3 = 0$). This organism moves in a straight line parallel to the organism's axis of rotation (fig. 3a). The organism completes one revolution every $\frac{2\pi}{|\vec{\omega}|} = \frac{2\pi}{\omega_1}$ units of time and advances the distance $U(\frac{2\pi}{\omega_1})$ for every revolution. Second, consider an organism moving such that $\vec{\omega}$ is perpendicular to $\overrightarrow{V}$ ($\omega_1 = 0$, $\sqrt{\omega_2^2 + \omega_3^2} \neq 0$). This organism moves in a circle (fig. 3b), completing one revolution of the circle every $\frac{2\pi}{|\vec{\omega}|} = \frac{2\pi}{\sqrt{\omega_2^2 + \omega_3^2}}$ units of time. The circumference of the circle is the organism's speed times this amount of time (circumference $= \frac{2\pi U}{\sqrt{\omega_2^2 + \omega_3^2}}$), so the radius of the circle equals $\frac{U}{\sqrt{\omega_2^2 + \omega_3^2}}$.

The organism moves in a helix if $\vec{\omega}$ is neither parallel nor perpendicular to $\overrightarrow{V}$ ($\omega_1 \neq 0$, $\sqrt{\omega_2^2 + \omega_3^2} \neq 0$). Consider the cylinder formed by one revolution of the helix (fig. 4a). If this cylinder is split down the side, parallel to $\vec{z}$, it forms a rectangle when laid flat (fig. 4b). The length of the side parallel to $\vec{z}$ (side A) equals p. The length of the other side (B) equals the circumference of the cross-section of the cylinder ($= 2\pi r$). The diagonal is the flattened arc of the helix; its length

BOX 7

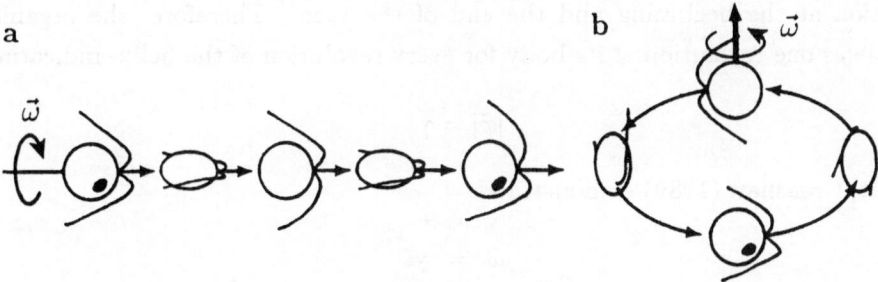

Figure 3. Circular and linear motions of an organism that swims forward while rotating its body. (a) If ω_1 is the only nonzero component of $\vec{\omega}$, the motion is linear. The organism advances the distance $U2\pi/\omega_1$ for every revolution of its body. (b) If ω_1 equals zero, the motion is circular. The radius of the circle equals $U/\sqrt{\omega_2^2 + \omega_3^2}$.

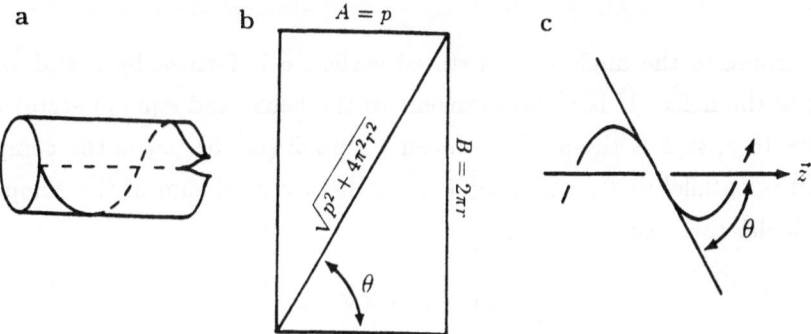

Figure 4. Geometry of the helix. (a, b) If the cylinder formed by one turn of the helix is cut down the side, parallel to $\vec{z}$, and laid flat, it forms a rectangle. The length of side A equals p, and the length of side B equals the circumference of the cylinder. The diagonal is the arc of one turn of the helix laid flat. The diagonal and side A form the pitch angle, θ. (c) θ is the angle formed by $\vec{z}$ and a line tangent to the 3D helix.

equals $\sqrt{p^2 + 4\pi^2 r^2}$. Also of interest is the pitch angle, θ, formed by the diagonal and side A. As stated earlier, A is parallel to $\vec{z}$, so for the unflattened helix θ is the angle formed by $\vec{z}$ and any line tangent to the arc of the helix (fig. 4c). It is evident in fig. 4b that

(5)
$$\tan \theta = \frac{2\pi r}{p} \ .$$

BOX   7

An organism that swims along this turn of the helix is oriented in the same direction at the beginning and the end of the turn.   Therefore, the organism completes one revolution of its body for every revolution of the helix, indicating

$$(6) \qquad\qquad\qquad |\vec{\omega}| = \gamma .$$

In fact, Crenshaw (1989) demonstrates

$$(7) \qquad\qquad\qquad \vec{\omega} = \gamma \vec{z}$$

(*i.e.* the rotation vector is parallel to the axis of the helix). This organism completes one turn of the helix every $\frac{2\pi}{|\vec{\omega}|}$ units of time, and its speed, U, is the length of the diagonal divided by this amount of time:

$$(8) \qquad\qquad\qquad U = \frac{|\vec{\omega}|}{2\pi}\sqrt{p^2 + 4\pi^2 r^2} .$$

This result is also obtained from $U = \left|\frac{d\vec{H}}{dt}\right|$, where $\vec{H}$ is given by equ.(1) of this Box.

Returning to the angle θ – as stated earlier, θ is formed by $\vec{z}$ and any line tangent to the helix. $\vec{V}$ is always tangent to the helix, and equ.(7) states that $\vec{\omega}$ is parallel to $\vec{z}$, so θ is the angle between $\vec{V}$ and $\vec{\omega}$ (fig. 5). ω_1 is the component of $\vec{\omega}$ that is parallel to $\vec{V}$, and $\sqrt{\omega_2^2 + \omega_3^2}$ is the vectorial sum of the components perpendicular to $\vec{V}$, so

$$(9a) \qquad\qquad\qquad \omega_1 = |\vec{\omega}| \cos \theta ,$$

$$(9b) \qquad\qquad\qquad \sqrt{\omega_2^2 + \omega_3^2} = |\vec{\omega}| \sin \theta .$$

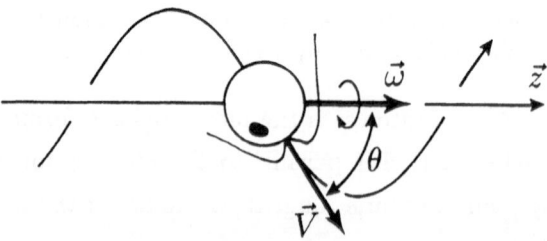

Figure 5. The rotational and translational velocities of an organism form the angle θ. Compare this figure with fig. 4c.

BOX 7

It is now possible to relate r, p, and γ of the helix to $\vec{V}$ and $\vec{\omega}$ of the organism. Combining equ.(5), (8), and (9) and rearranging yields

$$(10a) \qquad p = \frac{2\pi U \omega_1}{|\vec{\omega}|^2} \; ,$$

$$(10b) \qquad r = \frac{U \sqrt{\omega_2^2 + \omega_3^2}}{|\vec{\omega}|^2} \; .$$

Equ.(6) and (10) can then be rearranged, yielding

$$(11a) \qquad \omega_1 = \frac{\gamma p}{\sqrt{p^2 + 4\pi^2 r^2}} \; ,$$

$$(11b) \qquad \sqrt{\omega_2^2 + \omega_3^2} = \frac{2\pi \gamma r}{\sqrt{p^2 + 4\pi^2 r^2}} \; .$$

It is important to note that equ.(6)-(11) are only valid if the direction of the vectorial sum $\vec{\eta}_{body} = \omega_2 \vec{j} + \omega_3 \vec{k}$ is constant. This will occur if $\omega_2 = 0$, if $\omega_3 = 0$, or if $\omega_2 = \alpha \omega_3$ for constant α. Note that if the direction of $\vec{\eta}_{body}$ is constant, the organism has only two degrees of rotational freedom – one component (ω_1) parallel to $\vec{V}$ and one component $(\vec{\eta}_{body})$ perpendicular to $\vec{V}$. For reasons explained by Crenshaw (1989), the rotational acceleration must be considered if the direction of $\vec{\eta}_{body}$ is not constant. Development of the appropriate equations is outside the range of this discussion, but Crenshaw obtains

$$(12a) \qquad p = \frac{2\pi U (\omega_2^2 + \omega_3^2) \left[\omega_1 (\omega_2^2 + \omega_3^2) + \omega_2 \omega_3' - \omega_2' \omega_3 \right]}{(\omega_2^2 + \omega_3^2)^3 + \left[\omega_1 (\omega_2^2 + \omega_3^2) + \omega_2 \omega_3' - \omega_2' \omega_3 \right]^2} \; ,$$

$$(12b) \qquad r = \frac{U (\omega_2^2 + \omega_3^2)^{5/2}}{(\omega_2^2 + \omega_3^2)^3 + \left[\omega_1 (\omega_2^2 + \omega_3^2) + \omega_2 \omega_3' - \omega_2' \omega_3 \right]^2} \; ,$$

$$(12c) \qquad \gamma = \frac{\left[(\omega_2^2 + \omega_3^2)^3 + \left(\omega_1 (\omega_2^2 + \omega_3^2) + \omega_2 \omega_3' - \omega_2' \omega_3 \right)^2 \right]^{1/2}}{(\omega_2^2 + \omega_3^2)} \; .$$

where primes indicate derivatives with respect to time.

2. Helical Motion as a Sampling Strategy.

2.1. Organisms with directional receptors – photoreception.

Many unicellular algae orient to light. Foster and Smyth (1980) demonstrate that the helical motion of phototactic algae is an effective sampling strategy for determining the direction of a light ray. Readers are referred to their excellent paper for a comprehensive treatment. I will present only the simpler mathematical considerations.

Consider an organism with a photoreceptor located on the surface of its body. If this receptor's sensitivity to light is directional, due to shading by other body parts or to some other mechanism, then its direction of maximum sensitivity can be presented as a vector $\vec{S}$ pointing out from the body. As the organism swims in a helix, $\vec{S}$ changes direction in space according to

$$(1) \qquad \left.\frac{d\vec{S}}{dt}\right|_{space} = \left.\frac{d\vec{S}}{dt}\right|_{body} + \ \vec{\omega} \times \vec{S}$$

(Symon, 1971), where $\vec{\omega}$ is the organism's rotational velocity (see Box 6, equ.3). The subscript *space* indicates that the vector is described with respect to a reference frame fixed in space, and the subscript *body* indicates that the vector is described with respect to a reference frame fixed to the body of the organism (*i.e.* relative to $\overrightarrow{ijk}$, see Box 6, fig. 2). If the direction of $\vec{S}$ is constant with respect to $\overrightarrow{ijk}$, then eqn.(1) reduces to

$$(2) \qquad \left.\frac{d\vec{S}}{dt}\right|_{space} = \ \vec{\omega} \times \vec{S} \ .$$

The response R of the receptor (ignoring lag time, acclimation, differential output, *etc.*) when exposed to a ray of light $\vec{L}$ is

$$(3) \qquad R \propto \vec{S} \cdot \vec{L} \ .$$

If $\vec{L}$ is constant with respect to time, then

$$(4) \qquad \frac{dR}{dt} \propto (\vec{\omega} \times \vec{S}) \cdot \vec{L} \ .$$

Two important conclusions can be drawn from equ.(4):

$$(5) \qquad \frac{dR}{dt} = 0 \ \text{if} \ \vec{\omega} \parallel \vec{L} \quad \text{and} \quad \frac{dR}{dt} = \text{maximum if} \ \vec{\omega} \perp \vec{L} \ .$$

Equ.(3) can be rewritten as

$$(6) \qquad R \propto |\vec{S}||\vec{L}| \cos\phi$$

where ϕ is the angle between $\vec{S}$ and $\vec{L}$. ϕ varies with the rotation of the organism, so equ.(6) is an equation of the form

$$(7) \qquad |\vec{S}||\vec{L}| \cos(|\vec{\omega}|t)$$

when $\vec{\omega}$ is not parallel to $\vec{L}$. Therefore, the organism will experience a maximum receptor response when the receptor points towards the source of light (*i.e.* when $\vec{S}$ is anti-parallel $\uparrow\downarrow$ to $\vec{L}$) and a minimum receptor response when the receptor points away from the source (*i.e.* when $\vec{S}$ is parallel $\uparrow\uparrow$ to $\vec{L}$).

Foster and Smyth explain that this oscillating receptor response provides sufficient information to orient a cell. Namely, the phase angle of the peak R relative to $\vec{\omega}$ identifies when $\vec{S}$ is most nearly antiparallel to $\vec{L}$. If the cell turns, then from equ.(3) an increase in peak R indicates that $\vec{\omega}$ is turning more nearly perpendicular to $\vec{L}$, and a decrease in peak R indicates that $\vec{\omega}$ is turning more nearly parallel or antiparallel to $\vec{L}$. If $dR/dt = 0$ then $\vec{\omega}$ is parallel or antiparallel to $\vec{L}$, and the organism is oriented away from or toward the light source, as will be discussed later.

2.2. Organisms with Non-directional Receptors – Chemoreception.

The logic for chemoreception is similar to that for photoreception. Consider an organism that swims in a chemical concentration field $C(X,Y,Z)$ in $\overrightarrow{XYZ}$ space. A helically swimming organism follows the path given by Box 6, equ.(1). If the placement of the receptor on the organism is ignored, the organism's trajectory is the trajectory of the receptor. The receptor's output is then

$$(8a) \qquad R(t) \propto C(\vec{H}(t))$$

$$(8b) \qquad \frac{dR}{dt} \propto \vec{V}(t) \cdot \nabla C$$

where $\vec{V}(t) = d\vec{H}/dt$.

For simplicity, consider a concentration field that increases linearly with X:

$$(9) \qquad C = MX, \quad \nabla C = (M,0,0)$$

where M is a constant. R and dR/dt vary greatly with the orientation of $\overrightarrow{xyz}$ in $\overrightarrow{XYZ}$. For example, if $\vec{z} \uparrow\uparrow \vec{X}$ then $\vec{z} \uparrow\uparrow \nabla C$ and

$$(10a) \qquad R(t) \propto M\frac{p\gamma t}{2\pi}$$

$$(10b) \qquad \frac{dR}{dt} \propto M\frac{p\gamma}{2\pi}.$$

Therefore, if the axis of the helix points up the gradient, then R increases linearly.

Conversely, if $\vec{z} \uparrow\downarrow \nabla C$ then R decreases linearly. If $\vec{z} \perp \nabla C$, then

(11a) $$R(t) = Mr \cos \alpha \left[\cos(\gamma t)\right] + Mr \sin \alpha \left[\sin(\gamma t)\right]$$

(11b) $$\frac{dR}{dt} = Mr\gamma \cos \alpha \left[- \sin(\gamma t)\right] + Mr\gamma \sin \alpha \left[\cos(\gamma t)\right]$$

where α is the angle between $\vec{x}$ and the plane containing $\vec{z}$ and ∇C. Therefore, if the axis of the helix is perpendicular to ∇C, then $R(t)$ fluctuates with amplitude Mr. Finally if $\vec{z}$ is neither parallel, antiparallel, nor perpendicular to ∇C, then

(12a) $$R(t) = \cos \beta \left[M \frac{p\gamma t}{2\pi}\right] + \sin \beta \left[Mr \cos \alpha \left(\cos(\gamma t)\right) + Mr \sin \alpha \left(\sin(\gamma t)\right)\right]$$

(12b) $$\frac{dR}{dt} = \cos \beta \left[M \frac{p\gamma}{2\pi}\right] + \sin \beta \left[-Mr\gamma \cos \alpha \left(\sin(\gamma t)\right) + Mr\gamma \sin \alpha \left(\cos(\gamma t)\right)\right]$$

where β is the angle between $\vec{z}$ and ∇C. This states that R becomes some sum of equ.(10) and (11), *i.e.* a line with slope $\cos \beta \left(M \frac{p\gamma}{2\pi}\right)$ upon which is superimposed an oscillation with amplitude $\sin \beta (Mr)$ and period γt.

As with photoreceptors, there are unique signals if the helix is oriented parallel, antiparallel, perpendicularly, or obliquely to the chemical gradient. Similarly these signals provide the organism with sufficient directional information to orient to the gradient.

3. Helical Motion as an Orientation Mechanism.

3.1. Changes in the Direction of Motion.

Consider an organism that swims along a helix such that the radius, r, and pitch, p, are constant and $\vec{z}$ lies in the plane of the page (fig. 1a). From Box 6, equ.(5), the pitch angle, θ, also is constant. If the organism changes its motion such that θ changes (*i.e.* the ratio r/p changes) then either (1) the direction of the organism's path must change (*i.e.* the path is kinked at the point of transition – fig. 1b) or (2) the direction of the axis changes (*i.e.* the path is smooth at the transition point – fig. 1c).

If we assume that the path is smooth, then the axis of the helix changes direction whenever the organism changes θ. It is evident from fig. 1 that the axis of the helix is the organism's net direction of motion. If fact, the organism's net velocity (*i.e.* averaged over each turn of the helix) is

(13) $$U_{net} = U \cos \theta$$

where U is the organism's speed $|\vec{V}|$.

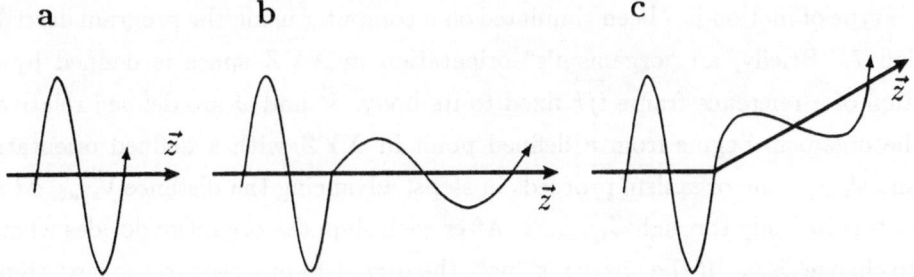

Figure 1. Two alternatives for accomodating changes in θ. (a) An organism swims along a helix of constant r and p, so θ is constant. If the organism changes θ then either (b) the path of the organism kinks to keep the axis straight or (c) the axis bends to keep the path smooth.

How can an organism change $\vec{z}$ while maintaining a smooth path? The answer comes from Box 6, equ.(7). The organism must change the direction of $\vec{\omega}_{space}$. Equ.(1) gives the operator for $\frac{d}{dt}|_{space}$, so

(14)
$$\left.\frac{d}{dt}\vec{\omega}\right|_{space} = \left.\frac{d}{dt}\vec{\omega}\right|_{body} + \vec{\omega} \times \vec{\omega} .$$

The last term equals zero, leaving

(15)
$$\left.\frac{d}{dt}\vec{\omega}\right|_{space} = \left.\frac{d}{dt}\vec{\omega}\right|_{body} .$$

This means that any change in $\vec{\omega}$, with respect to the organism's body, produces an equal change with respect to space.

This same result is obtained using Box 6, equ.(5). Substituting Box 6 equ.(10a,b) into this equation yields

(16)
$$\tan \theta = \frac{|\eta|}{\omega_1}$$

where $\vec{\eta} = \sqrt{\omega_2^2 + \omega_3^2}$. $|\vec{\eta}|$ and ω_1 are the magnitudes of the components of $\vec{\omega}$. The direction of $\vec{\omega}$ changes with respect to the body of an organism with two degrees of rotational freedom only if the ratio of the magnitudes of these two components changes. For an organism with three degrees of rotational freedom, substituting Box 6, equ.(12a,b) into Box 6, equ.(5) yields

(17)
$$\tan \theta = \frac{|\vec{\eta}|^3}{\omega_1|\vec{\eta}|^2 + \omega_2\omega_3' - \omega_2\omega_3'}$$

where the derivatives account for changes in the direction of $\vec{\eta}$ with respect to $\overrightarrow{ijk}$.

This type of motion has been simulated on a computer using the program described in section 7. Briefly, an "organism's" orientation in $\overrightarrow{XYZ}$ space is defined by the orientation of a reference frame $\overrightarrow{ijk}$ fixed to its body. $\vec{V}$ and $\vec{\omega}$ are defined relative to $\overrightarrow{ijk}$. The organism begins from a defined point in $\overrightarrow{XYZ}$ with a defined orientation, $\vec{V}_{body}$, and $\vec{\omega}_{body}$. The organism proceeds in steps, advancing the distance $\vec{V}_{space}\Delta t$ and then rotating its body through $\vec{\omega}_{space}\Delta t$. After each step, the organism decides whether or not to change $\vec{\omega}_{body}$. If the answer is "no", the organism proceeds to the next step. If the answer is "yes", the organism changes $\vec{\omega}_{body}$ according to some predefined formula and then proceeds to the next step. For simplicity, $\vec{V}_{body} = (1,0,0)$ in all simulations.

Fig. 2a presents the trajectory produced by a simulation in which $\vec{\omega}_{body}$ changes direction three times. Each change is marked by an dot (•). It is evident that at each change, the direction of the axis of the helical trajectory, $\vec{z}$, also changes.

Fig. 2b presents a trajectory in which the direction of $\vec{\omega}$ changes continuously. Again, $\vec{z}$ changes direction.

3.2. Orientation by Changing the Axis.

This change in the direction of the axis of the helix can be used to orient to stimulus gradients. The means by which this is accomplished is somewhat nonintuitive – the organism orients by pointing its rotational velocity vector $\vec{\omega}_{body}$, not its translational velocity vector $\vec{V}_{body}$, in the appropriate direction. When $\vec{\omega}_{body}$ is pointed correctly, then $\vec{z}$ (i.e. the net direction of motion) is pointed correctly.

The mechanism for pointing $\vec{\omega}_{body}$ up or down the gradient is quite simple. Consider the case of an organism swimming in a chemical concentration gradient. Unless $\vec{z}$, which is parallel to $\vec{\omega}$, is pointed exactly parallel or antiparallel to ∇C, $R(t)$ oscillates, as discussed in section 2.2. The oscillation is caused by the helical trajectory – as the organism swims around the helix, it swims up and then down ∇C. This means that when $R(t)$ is increasing, $\vec{V}$ is pointed up ∇C ($\vec{V} \cdot \nabla C > 0$). When $R(t)$ is decreasing, $\vec{V}$ is pointed down ∇C ($\vec{V} \cdot \nabla C < 0$).

The organism can point $\vec{\omega}_{body}$ up the gradient by making the angle θ between $\vec{V}_{body}$ and $\vec{\omega}_{body}$ smaller as $R(t)$ increases and by making θ larger as $R(t)$ decreases. Thus, an organism exhibits positive orientation if $\theta \propto \frac{1}{R}$. Conversely, an organism exhibits negative orientation if $\theta \propto R$.

Fig. 3 demonstrates positive orientation to a chemical concentration gradient. The gradient is a unidirectional gradient in $\overrightarrow{XYZ}$ space, increasing in the positive $\vec{X}$ direction. In fig. 3a the organism swims along a helix which has its axis parallel to the $\vec{Z}$ axis. When the organism reaches the point (1,0,0) it begins to change $\vec{\omega}$ such that θ is inversely proportional to chemical concentration. The organism proceeds from (1,0,0)

a   b

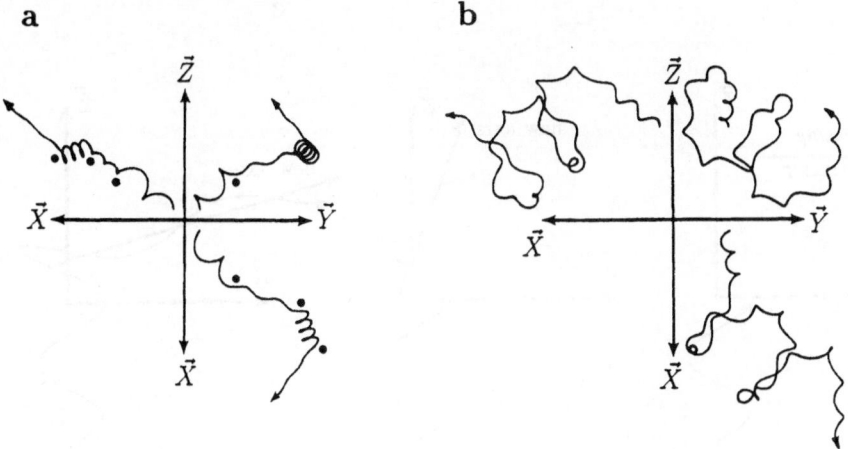

Figure 2. The axis of the helix changes direction when $\vec{\omega}$ changes direction – computer simulations. *(a)* Discrete changes of $\vec{\omega}$. $\vec{\omega}$ changes as a function of the arclength, s, of the trajectory. Total arclength of the trajectory is 12. Changes occur at the dots ($\bullet$).

$$
\begin{array}{llll}
s = 0 - 3: & \omega_1 = 2, & \omega_2 = 2, & \omega_3 = 2. \\
s = 3 - 6: & \omega_1 = 6, & \omega_2 = 2, & \omega_3 = 2. \\
s = 6 - 9: & \omega_1 = 1, & \omega_2 = 1, & \omega_3 = 5. \\
s = 9 - 12: & \omega_1 = 7, & \omega_2 = 1, & \omega_3 = 0.
\end{array}
$$

(b) Continuous change of $\vec{\omega}$. Total $s = 20$.

$$
\begin{array}{lll}
s = 0 - 2: & \omega_1 = 4, & \omega_2 = 5, \qquad\qquad \omega_3 = 1. \\
s = 2 - 18: & \omega_1 = 4 + \sin(2(s-2)), & \omega_2 = 3 + 2\cos(5(s-2)), \omega_3 = 4/\omega_1. \\
s = 18 - 20: & \omega_1 = 4.551, & \omega_2 = 2.779, \qquad\qquad \omega_3 = 0.879.
\end{array}
$$

The final values for ($s = 18 - 20$) for the components of $\vec{\omega}$ are the values for $s = 18$. Straight sections are placed at the beginning and end of the trajectory to better demonstrate the net change in the direction of $\vec{z}$. (*Note:* The trajectories are presented as projections on three different planes ($\overrightarrow{XY}$, $\overrightarrow{XZ}$, and $\overrightarrow{YZ}$) such that each figure folds to make three sides of a cube containing the trajectory. The axes are for a right-hand coordinate system.)

in the negative $\vec{X}$ direction, down the gradient. As the chemical concentration decreases, the organism increases θ, which is visible in fig. 3b as an increased r and a decreased p. The increase in θ causes $\vec{\omega}$, which is parallel to the axis of the helix, to rotate up the gradient. The organism then completes this turn of the helix and begins to move in the positive $\vec{X}$ direction, up the gradient. As the chemical concentration increases, the organism decreases θ, which is visible in fig. 3c as a decreased r and an increased p. The decrease in θ again causes the axis of the helix (and $\vec{\omega}$) to rotate up the gradient.

Box 6, equ.(5) demonstrates that changes in θ affect the appearance of the helical trajectory. As θ decreases for a positively orienting organism, the helix appears more

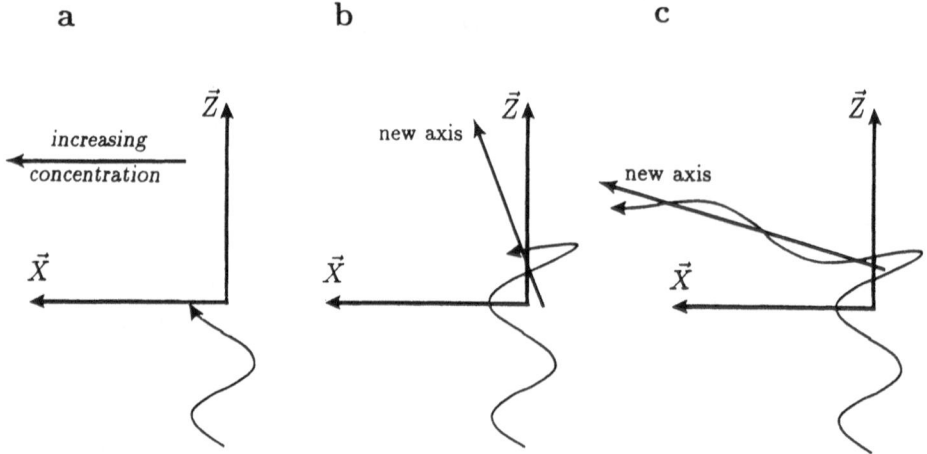

Figure 3. Positive orientation to a chemical concentration gradient by changes in θ. (a) A one-dimensional chemical concentration gradient with concentration increasing in the positive $\vec{X}$ direction. An organism swims in a helix with $\vec{z}$ parallel to $\vec{Z}$. The organism decreases θ with increasing concentration. (b) As the organism starts from (1,0,0) it first moves in the negative $\vec{X}$ direction, down the gradient. As the concentration decreases, the organism increases θ, rotating $\vec{z}$ up the gradient. (c) As the organism turns around the helix, it starts to move in the positive $\vec{X}$ direction. As the concentration increases, the organism decreases θ, again turning $\vec{z}$ up the gradient.

drawn out. Eqn.(13) demonstrates that this positively orienting organism has a larger net speed for smaller values of θ, even if U remains constant, a point that is widely recognized (Bullington, 1925; Ludwig, 1929; Brokaw, 1958; Párducz, 1964; Blake and Sleigh, 1974; Naitoh and Sugino, 1984; Fenchel and Jonsson, 1988).

It is also possible to simulate helical orientation using the program described in section 7. The simulations work as described earlier; however, now the organism moves through a chemical concentration field. As the organism moves through the field, a non-directional receptor elicits a response R according to the following function:

$$(18) \qquad\qquad R = m\left[0.5 + \frac{1}{\pi}\arctan(I^n - k^n)\right]$$

where R is the response; I is the chemical concentration; m is the maximum response; k is the threshold concentration; and n determines the steepness of the curve. The reasons for using this function are: (1) Its shape is that of a "typical" stimulus-response curve. (2) The response is bounded ($0 \leq R \leq m$). (3) The threshold stimulus, k, can be easily selected. (4) The strength of the response can be adjusted with n, higher values of n producing a steeper curve. Fig. 4 presents typical curves generated with equ.(18).

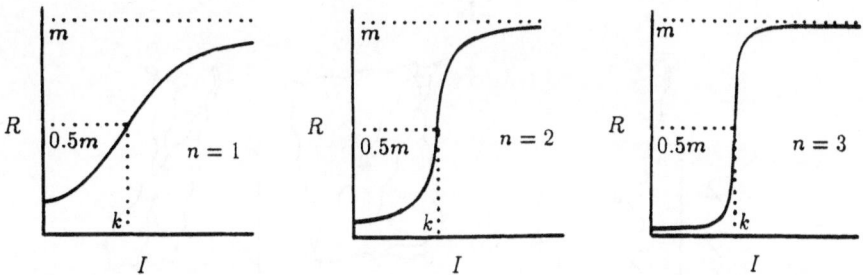

Figure 4. Stimulus-response function for three values of n. k is the threshold stimulus. m is the maximum response.

Fig. 5a models the situation presented in fig. 3 – positive orientation to a linear chemical concentration gradient. Initially, the "organism" does not respond to the chemical, then it begins to respond such that $\theta \propto \frac{1}{R}$. The axis of the helical trajectory orients up the chemical gradient.

Helical orientation also works for more complex gradients. Fig. 5b models orientation to a spherical gradient formed by diffusion of a chemical from a point source. This "organism" consistently turns back up the gradient, toward the source, when the chemical concentration drops below threshold.

4. Evidence from 3D Tracking of Sea Urchin Spermatozoa.

Preliminary experiments have investigated responses of the spermatozoa of the sea urchin, *Arbacia punctulata*, to an attractant released by eggs of the species. The trajectories of stimulated and unstimulated cells have been compared to determine if the rotational velocity of spermatozoa changes under stimulation and if these changes in the rotational velocity are sufficient to orient the cell to the stimulus.

4.1. Tracking Techniques.

Single spermatozoa have been tracked in three dimensions using the apparatus in fig. 6 (a more complete description of this apparatus is provided by Crenshaw, 1989b). Free-swimming cells are observed simultaneously by two video cameras that are attached to two microscopes that are perpendicular to each other. The raster scans of the two cameras are synchronized by a split-frame generator which also processes the two video signals such that the view from one camera occupies the left-hand side of the video monitor and the view from the other camera occupies the right-hand side. This combination of video images not only provides a convenient image for analysis, but it ensures that the separate images from the two cameras are made at exactly the same time.

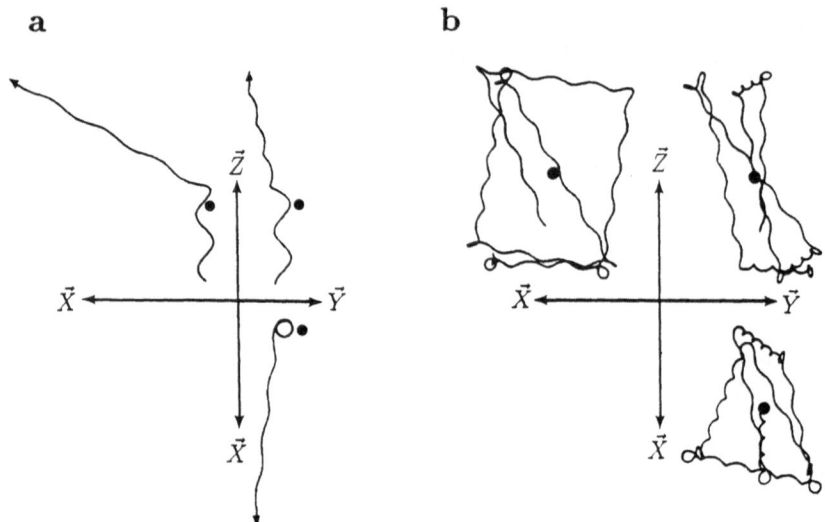

Figure 5. Orientation to chemical concentration gradients by changes in θ. *(a)* A one-dimensional gradient, similar to fig. 3. The chemical concentration, C is given by:

$$C(x) = 0 \qquad \{x < 10\}$$
$$C(x) = x + 10 \qquad \{-10 \le x \le 10\}$$
$$C(x) = 20 \qquad \{x > 10\}\ .$$

The "organism's" response is given by equ.(18): $m = 2,\ k = 10, n = 1.5$. For the straight portion of the helix, $\vec{\omega}$ is given by

$$\omega_1 = 2.5,\ \omega_2 = 2.5,\ \omega_3 = 0\ .$$

At the dot, X equals zero, and $\vec{\omega}$ begins to vary as

$$\omega_1 = 1.5 + R,\quad \omega_2 = 3.5 - R,\quad \omega_3 = 0 \quad (i.e.\ \theta \propto \frac{1}{R})\ .$$

(b) A three-dimensional gradient formed by diffusion from a point source. The source is marked with a •, and the gradient is given by:

$$C(X, Y, Z) = \frac{10}{((d/2)^2 + 1)}$$

where d is the distance from the source. The response is given by equ.(18): $m = 5, k = 3, n = 3$. The organism begins from (0,0,-1.5), and $\vec{\omega}_{body}$ varies as

$$\omega_1 = R,\quad \omega_2 = 7 - R,\quad \omega_3 = 0\ .$$

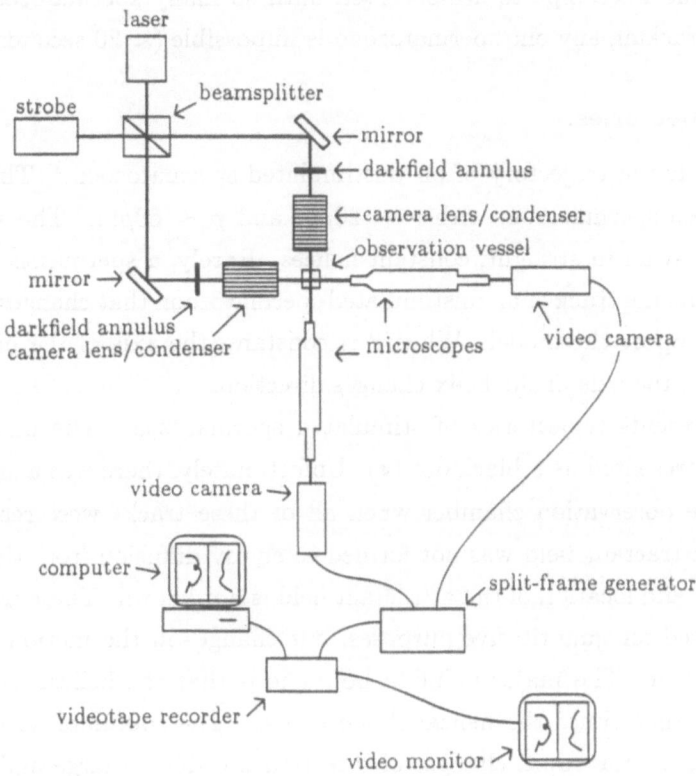

Figure 6. Basic design of the 3D tracking system. See text for explanation.

The videotapes are analyzed frame-by-frame to determine the screen coordinates of the cell. The image from one camera gives the X and Z coordinates of the cell; the image from the other camera gives the Y and Z coordinates. These coordinates are then used to construct the 3D trajectories of the spermatozoa.

It is important to note that the working distance is approximately 2.5 cm for both the objective and the condensor of the microscopes. These long working distances permit use of a large observation container (2.1 cm × 2.1 cm × 4.0 cm vertical). The microscopes are focused on the center of the container, so no organism within the field of view is closer than 1 cm to any surface, eliminating viscous wall effects.

Spermatozoa are added to the vessel and the videotape is started. No attempt is made to keep a spermatozoon in the view of the cameras; rather, spermatozoa are allowed to freely enter and exit the volume of water viewed by the two cameras. The unstimulated cells are observed for about 5 minutes. A glass micropipette that has been filled with agar and soaked in attractant is then placed in the field of view. The attractant diffuses out of the micropipette stimulating the spermatozoa. Spermatozoa attracted to the micropipette are observed until so many spermatozoa surround the pipette that tracking any one spermatozoon is impossible ($\approx$ 20 seconds).

4.2. 3D Trajectories.

Fig. 7a presents the trajectory of an unstimulated spermatozoon.[1] This trajectory is typical of these spermatozoa, with $r \sim 20\mu$m and $p \sim 60\mu$m. The spermatozoa of *A. punctulata* swim in straight, constant helices. Rarely, a spermatozoon will alter θ. Fig. 7b presents the track of an unstimulated spermatozoon that changes θ twice. These trajectories support the model. When θ is constant, the axis of the helix is straight. When θ varies, the axis of the helix changes direction.

Fig. 8 presents trajectories of stimulated spermatozoa. The mouth of the micropipette is presented as a black dot ($\bullet$). Unfortunately, there was a small convection current in the observation chamber when all of these tracks were recorded. Consequently, the attractant field was not formed solely by diffusion from the pipette's tip, and the shape and location of the attractant field is not known. These tracks, therefore, can not be used for quantitative purposes, but changes in the motion of the spermatozoa are evident. The major point to be made is that the helical motion is greatly changed. At some times the helical shape is lost. The spermatozoa in these figures never stopped, so the rapid changes of direction are due to large and rapid changes in the spermatozoa's rotational velocities. Presumably, these changes are elicited by exposure to the attractant. Whether $\vec{\omega}$ changes in the manner depicted in section 3.2 can not be determined from these trajectories.

5. Observations in the Literature.

There is considerable evidence for helical orientation in the literature. Although much of it is circumstantial, the evidence comes from a convincingly wide variety of techniques and organisms. There are two general types of evidence. The first comes from observations of free-swimming organisms, both unstimulated and stimulated. The second comes from studies of ciliary and flagellar beating.

---

[1] All trajectories of spermatozoa presented in this section have been smoothed by the following routine:

$$x'_n = 0.25x_{n-1} + 0.5x_n + 0.25x_{n+1}$$

where x'_n is the smoothed value of the coordinate in video frame number n, x_n is the unsmoothed value of the coordinate, x_{n-1} is the unsmoothed value from the preceeding video frame, and x_{n+1} is the unsmoothed value from the following video frame.

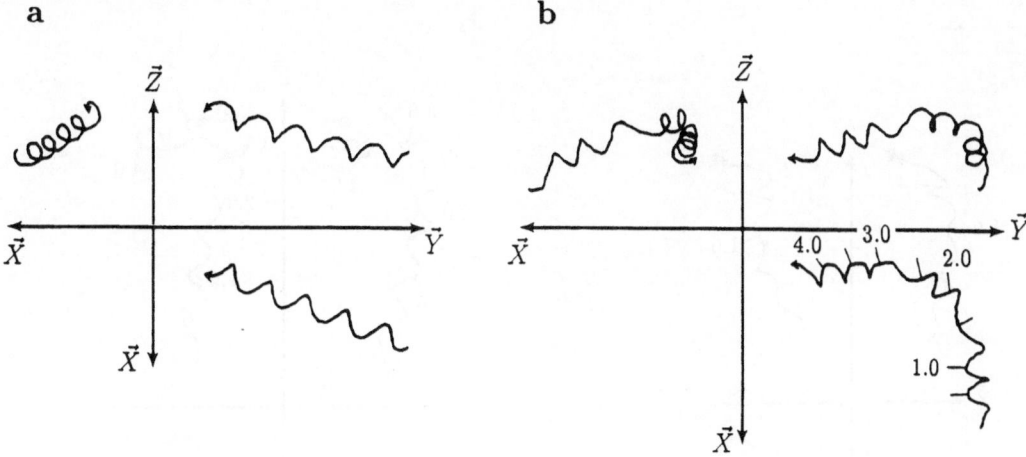

Figure 7. 3D trajectories of unstimulated sea urchin spermatozoa. *(a)* Typical trajectory of an unstimulated spermatozoon. $r \sim 20\mu m$, $p \sim 60\mu m$. r, p, and θ are constant and $\vec{z}$ is straight. *(b)* Trajectory in which θ changes twice. $\vec{z}$ also changes. Marks on the trajectory indicate 0.5 second time intervals. Magnification is the same for both trajectories.

5.1. Observations of Free-swimmming Organisms.

Jennings (1904) first reported that several microorganisms (*Chlamydomonas, Euglena,* rotifers) change the radius of their motion when they are stimulated. He also observed that these changes in radius alter the net direction of motion. However, Jennings was unable to fully explain his observation.

Brokaw (1958) first described how changes in $\vec{\omega}$, and thus r and p, can orient an organism to stimulus gradients. Brokaw examined the orientation of bracken fern spermatozoids to malate and presents some of the best evidence for helical orientation. Brokaw observed decreases in r and θ as spermatozoids moved up malate gradients. Concurrently, the axis of the helix turned up the gradient (see also Brokaw, 1974).

The large body of literature on the locomotion and chemotaxis of spermatozoa offers many examples of changes in radius, but interpretation of this information is difficult because many observations are of spermatozoa that swim next to a solid surface (the coverslip or the microscope slide). Viscous wall effects produced by the proximity of this surface alter the swimming behavior of spermatozoa – when a spermatozoon approaches a surface, its motion changes from helical to planar spiral in which ω_1 equals zero (Gray, 1955). The trajectories of these spermatozoa are concentric circles, a behavior called thigmotactic swimming (Miller, 1985). Nevertheless, qualitative evidence is available.

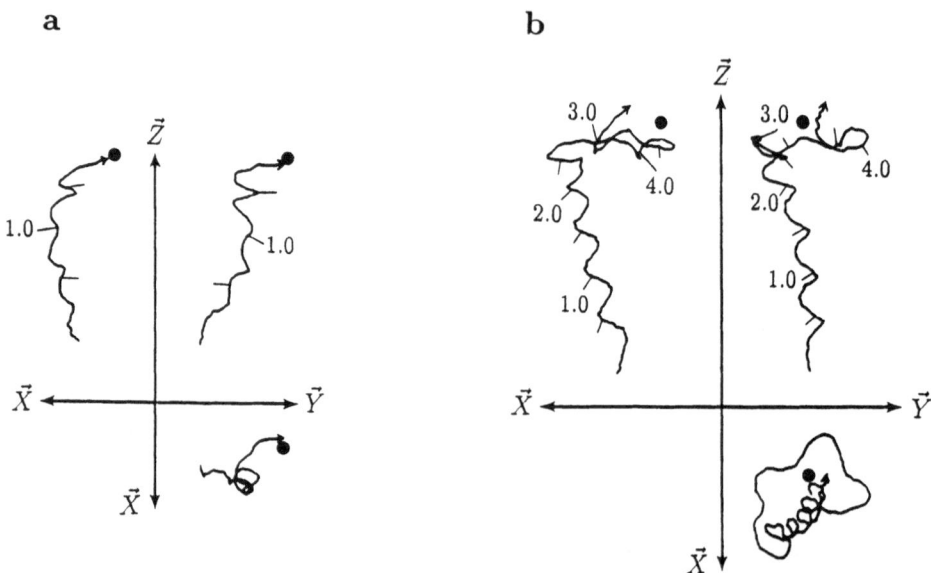

Figure 8. 3D trajectories of stimulated sea urchin spermatozoa. The stimulus is an egg attractant that diffuses from a micropipette. The location of the mouth of the pipette is marked with a •. Figures a and b are of different spermatozoa. Marks on the trajectories indicate 0.5 second time intervals. The magnification is the same as in fig. 7.

The spermatozoa of many invertebrates chemo-orient to a substance released by eggs of the same species (for review see Miller, 1985). Typically, a spermatozoon swimming thigmotactically changes the radius of its circular motion when exposed to the attractant, indicating that it has altered its rotational velocity (see Box 6, fig. 3). This response is displayed by the spermatozoa of many species (from at least four different phyla – Miller, 1985). For normal, helical motion Miller (1985) states that when exposed to an attractant, spermatozoa probably change the "loop" of their motion which orients the cell toward the attractant's source, probably with an increase in pitch. Interestingly, similar behavior is displayed by the motile spores of the brown alga, *Ectocarpus siliculosus* (Müller, 1978).

Some observations also suggest that positive photo-orientation by *Chlamydomonas* occurs by helical orientation. Negative orientation, which has been well-documented, occurs by a "stop-and-turn" response (Boscov and Feinleib, 1979), which violates this model's assumption of a smooth trajectory. However, positive orientation involves a smooth change in the direction of the axis of the helix; the cell never stops to turn. The mechanism for this smooth change of direction remains undescribed (Boscov and Fein-leib, 1979). Furthermore, Kamiya and Witman (1984) present multiple exposure photographs of demembranated *C. reinhardtii* swimming in solutions of Ca^{++}. At 10^{-9} M, $\theta = 0.66$; while at 10^{-7}M, $\theta = 0.44$ (personal measurements from their photographs).

Observations of the motion of ciliates also suggests that these organisms exhibit helical orientation. Párducz (1964) reports that theronts of *Ophyroglena* exhibit marked changes in θ, especially if small amounts of food are added to the culture media. These changes in θ cause the organism to change direction. Hildebrand and Dryl (1983) present numerous streak photographs of free-swimming *Paramecium caudatum* swimming in different concentrations of K^+, Na^+, and Ca^{++}. The radius and pitch of their motion change with changes in Ca^{++}. However, the strongest evidence is presented by Machemer (this volume and 1989 – see especially fig. 1) and by Machemer and Sugino (in press). They demonstrate that *Paramecium* and other ciliates exhibit graded changes of θ when exposed to solutions that change the electropotential of the cell membrane. Hyperpolarization, which occurs during positive stimulation, causes θ to decrease. Sub-threshold depolarization, which occurs during negative stimulation, causes θ to increase.

5.2. Observations of the Beating of Cilia and Flagella.

As discussed earlier, changes in θ are caused by changes in $\vec{\omega}$. Torque generated by propulsive appendages rotates the organism, and the propulsive appendages of many helical swimmers are cilia or flagella. Considerable evidence indicates that the beat of cilia and flagella changes during stimulation of the cell such that $\vec{\omega}$ is altered.

Observations of the ciliary beat of ciliates suggest these cells exhibit helical orientation. Naitoh and Sugino (1984), Sugino and Naitoh (1988), and Machemer and Sugino (in press), working with *Paramecium*, explain that ω_1 is generated by a ciliary beat that is oblique to the anterior-posterior (A–P) axis of the cell. When the cell's membrane hyperpolarizes the ciliary beat becomes more nearly parallel to the A–P axis, decreasing ω_1, and when the membrane depolarizes the ciliary beat becomes more nearly perpendicular, increasing ω_1. Curiously, hyperpolarization decreases ω_1, which according to eqn.(16) increases θ. However, as stated in section 5.1, θ decreases during hyperpolarization, so there must be an even greater decrease in $|\vec{\eta}|$. Unfortunately, aspects of the ciliary beat controlling $|\vec{\eta}|$ are not known (Naitoh and Sugino, 1984).

The flagellar beat of spermatozoa rotates these cells. Chwang and Wu (1971), Keller and Rubinow (1976), and Keller (1977) all explain how the helical waveform of the flagellum provides the torque generating ω_1. $|\vec{\eta}|$ is believed to be generated by asymmetries in the flagellar waveform. The circular path of thigmotactically swimming spermatozoa is due to the asymmetry of the flagellar beat; in fact, the radius of the circular motion is a good measure of the degree of asymmetry ($|\vec{\eta}| \propto \frac{1}{r}$: Rikmenspoel *et al.*, 1960; Brokaw, 1970; Goldstein, 1977; Okuno and Brokaw, 1981). The degree of asymmetry increases with increasing intracellular [Ca^{++}] (Brokaw *et al.*, 1974; Brokaw, 1979; Gibbons and Gibbons, 1980; Okuno and Brokaw, 1981). Furthermore, Ward *et al.* (1985) suggest that receptor-mediated regulation of intracellular [Ca^{++}] may be part of the intracellular signal transduction mechanism during orientation of spermatozoa of *A. puntulata* to egg attractant. Unfortunately, these results also contradict eqn.(16). Thigmotactically swimming spermatozoa increase $|\vec{\eta}|$, which should increase θ (see equ. 16), when exposed to attractant, which would not result in helical orientation. Effects of attractant on ω_1, however, are unknown. Furthermore, the viscous wall effects that induce thigmotactic swimming are also known to alter the flagellar beat (Katz and Blake, 1975), so the effects of attractants may be different for thigmotactically and free swimming spermatozoa.

Finally, Rüffer and Nultsch (1985, 1987) conclude from cinematographic studies that the helical path of *Chlamydomonas* is probably due to beat asymmetries between the two flagella. The degree of asymmetry in the flagellar beat of *Chlamydomonas* is affected by [Ca^{++}] (Kamiya and Witman, 1984; Omoto and Brokaw, 1985). Kamiya and Witman (1984) also report that this asymmetry is a graded response in demembranated cells when exposed to changes in [Ca^{++}]. The *trans*-axoneme is inactive at low [Ca^{++}]. As [Ca^{++}] increases, the *trans*-axoneme becomes increasingly active, and the *cis*-axoneme becomes less active, with the *cis*-axoneme being inactive at high [Ca^{++}]. These authors conclude that these changes in the beat of the flagella will result in changes in the radius and pitch of the organism's helical path.

6. Conclusions.

Helical orientation may be a mechanism by which many, diverse microorganisms orient to a wide variety of stimuli. In particular, helical orientation may be the mechanism by which spermatozoa orient to egg attractants, *Chlamydomonas* positively photo-orients, and ciliates orient to light and chemical gradients that do not elicit threshold depolarization of the membrane. Interestingly, helical orientation is a non-random mechanism of orientation – the rotational velocity vector always turns up the gradient during positive orientation and down the gradient during negative orientation. As such, helical orientation is one of the few non-random orientation mechanisms described for freeswimming microorganisms (see van Houten *et al.*, 1981). Even more remarkable, helical orientation permits non-random orientation to a stimulus when the organism possesses only the ability to temporally compare stimulus intensity. This is accomplished solely through the geometry of helical motion.

Helical orientation requires graded responses to stimulation, rather than all-or-nothing responses. It, therefore, provides a functional explanation for the graded responses of cilia and flagella to varying stimulation, to graded membrane electropotentials, and to varying concentrations of intracellular Ca^{++}. As such it provides a unifying paradigm in which to interpret sensory transduction, organelle (ciliary and flagellar) response, and cell locomotion in these organisms.

7. Computer Programs.

The kinematics used in this program are described in Crenshaw (1989a,b and briefly in Box 6). Two of the reference frames used in Box 6 are also used in this program – $\overrightarrow{XYZ}$, which is fixed in space, and $\overrightarrow{ijk}$, which is fixed to the organism. From here on I will refer to $\overrightarrow{XYZ}$ as Γ and to $\overrightarrow{ijk}$ as Φ. Vectors with the subscript Γ are described relative to Γ, and vectors with the subscript Φ are described relative to Φ.

The motion of the organism is described relative to Γ, so Φ is also described relative to Γ. For example, $\vec{i}$, which has the coordinates (1,0,0) in Φ, may have the coordinates $(1/\sqrt{3}, 1/\sqrt{3}, -1/\sqrt{3})$ in Γ. (Note: It would then be referred to as $\vec{i}_\Gamma$.) This permits the organism to assume any orientation in Γ.

The program proceeds as follows:

Step 1 – Determine Initial Location and Orientation.
The organism is placed at an initial point in Γ, and $\vec{i}_\Gamma, \vec{j}_\Gamma$, and $\vec{k}_\Gamma$ are given initial values.

Step 2 – Determine Stimulus Intensity.
The intensity of the stimulus is determined at that point according to the stimulus field defined for the simulation.

Step 3 – Determine $\vec{\omega}$ and $\vec{V}$.
$\vec{V}_\Phi$ and $\vec{\omega}_\Phi$ are calculated as functions of the stimulus intensity. Remember, $\vec{V}_\Phi$ has only one nonzero component – $V_{\Phi 1}$. Therefore,

$$\vec{V}_\Phi = V_{\Phi 1}\vec{i}$$

$$\vec{\omega}_\Phi = \omega_{\Phi 1}\vec{i} + \omega_{\Phi 2}\vec{j} + \omega_{\Phi 3}\vec{k} \, .$$

(Note: $\vec{i}, \vec{j}$, and $\vec{k}$ do not have subscripts. These equations are correct, regardless of the reference frame.)

Step 4 – Move to a New Position.
Now that the organism has a defined position, orientation, $\vec{V}_\Phi$, and $\vec{\omega}_\Phi$, the organism

moves forward at the designated velocity over a predefined period of time, Δt:

$$\vec{P}_\Gamma' = \vec{P}_\Gamma + \vec{V}_\Gamma \Delta t$$

where $\vec{P}_\Gamma'$ is the new position in Γ (at time $t + \Delta t$); $\vec{P}_\Gamma$ is the previous position (at time t); $\vec{V}_\Gamma \Delta t$ is the distance travelled; and $\vec{V}_\Gamma$ is given by

$$V_{\Gamma 1} = V_{\Phi 1} \vec{i}_{\Gamma 1} + V_{\Phi 2} \vec{j}_{\Gamma 1} + V_{\Phi 3} \vec{k}_{\Gamma 1} = V_{\Phi 1} \vec{i}_{\Gamma 1}$$
$$V_{\Gamma 2} = V_{\Phi 1} \vec{i}_{\Gamma 2} + V_{\Phi 2} \vec{j}_{\Gamma 2} + V_{\Phi 3} \vec{k}_{\Gamma 2} = V_{\Phi 1} \vec{i}_{\Gamma 2}$$
$$V_{\Gamma 3} = V_{\Phi 1} \vec{i}_{\Gamma 3} + V_{\Phi 2} \vec{j}_{\Gamma 3} + V_{\Phi 3} \vec{k}_{\Gamma 3} = V_{\Phi 1} \vec{i}_{\Gamma 3} \ .$$

Step 5 – Rotate to a New Orientation.

After moving forward, the organism rotates at the rotational velocity determined in **Step 3** over the period of time Δt. This is done as follows:

$$\vec{i}_\Gamma' = \vec{i}_\Gamma + \frac{d\vec{i}_\Gamma}{dt} \Delta t, \quad \vec{j}_\Gamma' = \vec{j}_\Gamma + \frac{d\vec{j}_\Gamma}{dt} \Delta t, \quad \vec{k}_\Gamma' = \vec{k}_\Gamma + \frac{d\vec{k}_\Gamma}{dt} \Delta t$$

where $\vec{i}_\Gamma' \vec{j}_\Gamma' \vec{k}_\Gamma'$ describes the new orientation,

$$\frac{d\vec{i}_\Gamma}{dt} = \vec{\omega}_\Gamma \times \vec{i}_\Gamma, \quad \frac{d\vec{j}_\Gamma}{dt} = \vec{\omega}_\Gamma \times \vec{j}_\Gamma, \quad \frac{d\vec{k}_\Gamma}{dt} = \vec{\omega}_\Gamma \times \vec{k}_\Gamma \ ,$$

$$\omega_{\Gamma 1} = \omega_{\Phi 1} i_{\Gamma 1} + \omega_{\Phi 2} j_{\Gamma 1} + \omega_{\Phi 3} k_{\Gamma 1}$$

$$\omega_{\Gamma 2} = \omega_{\Phi 1} i_{\Gamma 2} + \omega_{\Phi 2} j_{\Gamma 2} + \omega_{\Phi 3} k_{\Gamma 2}$$

$$\omega_{\Gamma 3} = \omega_{\Phi 1} i_{\Gamma 3} + \omega_{\Phi 2} j_{\Gamma 3} + \omega_{\Phi 3} k_{\Gamma 3} \ .$$

Step 6 – Repeat Steps 2-5.

Now that the organism's new position and orientation are known, the program returns to **Step 2** and repeats the process for as many times as required.

A Note About Approximation Errors.

This program is a finite approximation of a continuous process. Two types of error are intrinsic to this approach. First, the computer is a digital machine. Round-off in the digitization of numbers is a source of error. Use the highest precision available on your computer. Second, any finite approximation of a continuous process is unable to use infinitely small values of Δt, thus it is unable to approach the limit $dt \to 0$. This is known as an approximation error. To ensure that approximation error is negligible, the same simulation must be run repeatedly with successively smaller values of Δt. When the shape of the trajectory and the endpoint of the trajectory cease to change appreciably with increasingly smaller values of Δt, then approximation error is small. In the simulations presented in figs. 2 and 5 the following criterion was used: Let Δt_n

be the value used in the nth repetition; let $\vec{P}_n$ be the organism's position at the end of the nth repetition; and let s be the arclength of the simulation – s is the same for all repetitions. Approximation error was considered small if

$$\left|\vec{P}_n - \vec{P}_{n-1}\right| < 0.01s \text{ for } 10\Delta t_n = \Delta t_{n-1}.$$

Acknowledgments. Thanks to Leah Edelstein-Keshet and to the Duke Biomechanics Group for advice and assistance throughout this project. Thanks also to C.J. Brokaw for providing a copy of his thesis and to Donna Crenshaw and John Long for helpful comments on the manuscript. H.C.C. has been supported throughout this work by predoctoral fellowships from the U.S. Office of Naval Research and from Duke University and by grant DCB–8819271 from the U.S. National Science Foundation.

REFERENCES

Blake J.R. and Sleigh M.A. (1974) Mechanics of ciliary locomotion. Biol. Rev. 49: 85 – 125

Boscov J.S. and Feinleib M.E. (1979) Phototactic response of *Chlamydomonas* to light. II. Response of individual cells. Photochem. Photobiol. 30: 499 – 505

Brokaw C.J. (1958) Chemotaxis of bracken spermatozoids. Ph.D. thesis. Cambridge University, Cambridge

Brokaw C.J. (1970) Bending moments in free-swimming flagella. J. Exp. Biol. 53: 445 – 464

Brokaw C.J. (1974) Calcium and flagellar response during the chemotaxis of bracken spermatozoids. J. Cell. Physiol. 83: 151 – 158

Brokaw C.J. (1979) Calcium-induced asymmetrical beating of Triton-demembranated sea urchin sperm flagella. J. Cell Biol. 82: 401 – 411

Brokaw C.J., Josslin R., and Bobrow L. (1974) Calcium ion regulation of flagellar beat symmetry in reactivated sea urchin spermatozoa. Biochem. Biophys. Res. Commun. 58: 795 – 800

Chwang A.T. and Wu T.Y. (1971) A note on the helical movement of micro-organisms. Proc. Roy. Soc. Lond. B. 178: 327 – 346

Crenshaw H.C. (1989a) The kinematics of the helical motion of microorganisms capable of motion with four degrees of freedom. Biophys. J. 56: 1029 – 1035

Crenshaw H.C. (1989b) The helical motion of microorganisms: A novel orientation mechanism. Ph.D. Thesis. Duke University, Durham, North Carolina

Fenchel T. and Jonsson P.R. (1988) The functional biology of *Strombidium sulcatum*, a marine oligotrich ciliate (Ciliophora, Oligotrichina). Mar. Ecol. Prog. Ser. 48: 1 – 15

Foster K.W. and Smyth R.D. (1980) Light antennas in phototactic algae. Microbiol. Rev. 44: 572 – 630

Gibbons B.N. and Gibbons I.R. (1980) Calcium-induced quiescence in reactivated sea urchin sperm. J. Cell Biol. 84: 13 – 27

Goldstein S.F. (1977) Asymmetric waveforms in echinoderm sperm flagella. J. Exp. Biol. 71: 157 – 170

Gray J. (1955) The movement of sea-urchin spermatozoa. J. Exp. Biol. 32: 775 – 801

Hildebrand E. and Dryl S. (1983) Dependence of ciliary reversal in *Paramecium* on extracellular Ca^{++} concentration. J. Comp. Physiol. A. 152: 385 – 394

Jennings H.S. (1901) On the significance of spiral swimming in organisms. Amer. Nat. 35: 369 – 378

Jennings H.S. (1904) Contributions to the study of the behavior of lower organisms. Carnegie Inst. of Wash. Publ. No. 16

Kamiya R. and Witman G.B. (1984) Submicromolar levels of calcium control the balance of beating between the two flagella in demembranated models of *Chlamydomonas*. J. Cell Biol. 98: 97 – 107

Katz D.F. and Blake J.R. (1975) Flagellar motions near walls. In T.Y.-T. Wu, C.J. Brokaw, and C. Brennen (eds.) Swimming and Flying in Nature. vol. 1: 173 – 184. Plenum Press, New York

Keller J.B. and Rubinow S.I. (1976) Swimming of flagellated microorganisms. Biophys. J. 16: 151 – 170

Keller S.R. (1977) Mechanics of flagellar motion with an application to a conical spiral flagellate. J. Theor. Biol. 68: 73 – 94

Machemer H. (1989) Cellular behaviour modulated by ions: Electrophysiological implications. J. Protozool. 36: 463 – 487

Machemer H. and Sugino K. (In press) Electrophysiological control of ciliary beating: A basis of motile behaviour in ciliated Protozoa. Comp. Biochem. Physiol.

Mast G.O. (1911) Light and the Behavior of Organisms. John Wiley, New York

Miller R.L. (1985) Sperm chemo-orientation in the metazoa. In: C.B. Metz and A. Monroy (eds.) Biology of Fertilization. Vol. 2: 276 – 337. Academic Press, New York

Müller D.G. (1978) Locomotive responses of male gametes to the species specific sex attractant in *Ectocarpus siliculosus* (Phaeophyta). Arch. Protistenk. 120: 371 – 377

Naitoh Y. and Sugino K. (1984) Ciliary movement and its control in *Paramecium*. J. Protozool. 31: 31 – 40

Okuno M. and Brokaw C.J. (1981) Effects of Triton-extraction conditions on beat symmetry of sea urchin sperm flagella. Cell Motil. 1: 363 – 370

Omoto C.K. and Brokaw C.J. (1985) Bending patterns of Chlamydomonas flagella. II. Calcium effects on reactivated Chlamydomonas flagella. Cell Motil. 5: 53 – 60

Párducz B. (1964) Swimming and its ciliary mechanism in *Ophryoglena* sp. Acta Protozool. 2: 367 – 374

Rikmenspoel R., van Herpen G., and Eijkhout P. (1960) Cinematographic observations of the movements of bull sperm cells. Phys. Med. Biol. 5: 167 – 181

Rüffer U. and Nultsch W. (1985) High-speed cinematographic analysis of the movement of Chlamydomonas. Cell Motil. 5: 251 – 263

Rüffer U. and Nultsch W. (1987) Comparison of the beating of cis- and trans-flagella of *Chlamydomonas* cells held on micropipettes. Cell Motil. Cytoskel. 7: 87 – 93

Symon K.R. (1971) Mechanics. 3rd ed. Addison-Wesley, Reading, Massachusetts

Sugino K. and Naitoh Y. (1988) Swimming path measurement in *Paramecium* – Estimation of ciliary activity from the swimming path. Seitai Nō Kagaku. 39(5): 485 – 490

van Houten J., Hauser D.C.R., and Levandowsky M. (1981) Chemosensory behavior in protozoa. In: M. Levandowsky and S.H. Hutner (eds.) Biochemistry and Physiology of Protozoa. 2nd ed. vol. 4: 67 – 124. Academic Press, New York

Ward G.E., Brokaw C.J., Garbers D.L., and Vacquier V.D. (1985) Chemotaxis of *Arbacia punctulata* spermatozoa to Resact, a peptide from the egg jelly layer. J. Cell Biol. 101: 2324 – 2329

MOTILE BEHAVIOUR AND DETERMINATION OF THE MAGNETIC MOMENT
OF MAGNETIC BACTERIA IN ROTATING MAGNETIC FIELDS

H. Petermann (1), D. G. Weiss (2), L. Bachmann (3), N. Petersen (1)

(1) Institut für Geophysik, Universität München, Theresienstr 41, D-8000 München 2
(2) Institut für Zoologie, and
(3) Institut für Techn. Chemie, Technische Universität München,
Lichtenbergstr 4, D-8046 Garching, Fed. Rep. Germany

Abstract

Magnetic bacteria from limnic sediments of South-Bavarian lakes were investigated. With a special arrangement of Helmholtz coils, a rotating magnetic field, by which magnetic bacteria were forced to swim in circles, was applied to the stage of a light microscope.

This setup allows:
- observation of a single bacterium over a deliberately long time
- determination of the swimming speed
- determination of the magnetic moment of individual bacteria by motion analysis
- to force the bacterium to swim to locations chosen by the operator.

The magnetic moment of a given bacterium was also determined in a second way: The same bacterium, first studied by motion analysis, was then forced to swim onto a TEM-grid and observed under the Transmission Electron Microscope. The total magnetic moment of the bacterium was calculated from the size of the magnetite particles contained in the bacterium.

1. Introduction

Magnetic bacteria were discovered by Blakemore et al. (1975). The morphology of magnetic bacteria is very similar to other bacteria. We found cocci, spirillae, vibrios and rod-shaped forms in the Bavarian freshwater lakes. As far as investigated these bacteria are gram negative. Although anaerobic magnetic bacteria were described (Bazylinsky et al., 1989), most of the species seem to grow optimally under microaerophilic conditions.

In contrast to other bacteria, magnetic bacteria contain one or several chains of magnetite particles (Vali et al., 1987), so called magnetosomes (Balkwill et al., 1980). The size range of the magnetosomes is very restricted and according to theoretical calculations of Butler and Banerjee (1975), they lie in or close to the range of single-domain particles. That means the direction of magnetization is fairly uniform in the whole particle.

The magnetic bacteria studied here, live in the upper centimetres of lake sediments, where redox conditions are rapidly changing from oxidizing (in the overlying water column and in the upper millimetres of the sediment) to reducing. In that transition layer they appear in great numbers (up to 10^6 per cubic centimetre) .

2. Magnetotaxis

Why do bacteria synthesize magnetite particles?

For microaerophilic species high oxygen concentrations are toxic. Living in the sediment they don't need to move. Diffusion brings enough nutrients to them. But when they are removed from the microinvironment they are adapted to, for example by a storm swirling up the sediment, it will be of advantage to them to leave the oxidizing and therefore toxic region and reach the microaerophilic sediment layer as fast as possible.

The Earth's magnetic field exerts a torque on the bacteria until their magnetic moment is parallel to the field. On the northern hemisphere the magnetic field points to the north and downwards, with the downwards angle (inclination) depending on the geographical latitude. In Bavaria the inclination is about 64 degrees from the horizontal plane. That means that the bacteria swim straightforward down into the sediment without wasting energy or time for randomly orientated walk. When they reach microaerophilic sediment they can just stop.

This assumption is supported by our observation, that bacteria, brought under the light-microscope in dispersed sediment, are only activated to move, while a changing or rotating magnetic field is applied to them.

3. Experimental Setup: The Bacteriodrome

The Bacteriodrome (Fig. 1), first described by Petersen et al. (1989) consists of:
- Three orthogonal Helmholtz coils to compensate for the Earth's magnetic field,
- a light microscope with video camera and video recorder,
- two pairs of small coils around the microscope in the horizontal plane, and
- a current supply for the two pairs of coils, that alternatively provides three kinds of
    magnetic fields at the spot of observation under the light-microscope, namely:
    - a homogeneous field of constant amplitude rotating in the horizontal plane,
    - a homogeneous field that can be switched from one direction to the opposite, or
    - a homogeneous field where direction and intensity can be adjusted by a joy-stick.

Figure 1. Total view of the experimental setup – the 'Bacteriodrome'

Rotation frequency can be varied from 0.05 to 2 Hz and field intensity from 0 to 160 A/m. With this setup magnetic bacteria can be guided to move either in circles, to the left or the right, or in any desired direction, according to the applied magnetic field.

4. Swimming Speed

When the rotating magnetic field is applied, the magnetic bacteria swim in circles. For a given frequency of field rotation slow bacteria swim on small circles while fast ones swim on larger circles. By measuring the diameter of the circles it is possible to determine the swimming speed. Working with a video analysis system, it is possible to trace the trajectory of the swimming organism and to obtain a picture of the circular path, which can be recorded on video tape or video printer.

5. Magnetic Moment

It is also possible to determine the magnetic moment of individual bacteria under the light microscope although it is not possible to see single magnetic particles. With special image analysis systems, AVEC-micoscropy (Weiss et al., 1989) it is possible to visualize the chains of magnetic particles, but it is not possible to optically resolve them, i.e., get a good estimation of the size of individual magnetic particles. (They are about 100nm in size, that means usually not resolvable under the light microscope).

The method proposed here is based on motion analysis. Let us consider a dead bacterium to see how the determination of the magnetic moment works. To a chain of magnetic particles with their magnetic moments parallel to the chain an external magnetic field will exert a magnetic torque of:

(1) $T = M * B * sin(M, B)$

T = Magnetic torque
M = Magnetic moment
B = Magnetic induction

Since the magnetic chain is fixed in the bacterium the rotating magnetic field exerts a magnetic torque on the whole bacterium. There is also an opposing viscous torque, that increases with increasing rotational speed. For a sphere rotating with constant angular velocity the viscous torque is given by (Landau and Lifshitz 1987):

(2) $T = 8 * \pi * \eta * R^3 * \omega$

η = dynamic vicosity of the fluid (water),
R   = radius of the sphere-shaped bacterium,
ω = frequency of rotation.

Under stationary conditions the two opposing torques cancel each other and the bacterium rotates with constant angular velocity. For a given value of the magnetic field and the angular velocity there will be a defined angle between the direction of the external field and the orientation of the bacterium.

If the strength of the magnetic field is decreased (magnetic torque is decreased) or if the angular velocity is increased (viscous torque is increased) the angle between external field and magnetic moment will increase:

(3)
$$M * B * sin(M, B) = RCS * \omega$$

(RCS stands for the resistance coefficient of a sphere: $8 * \pi * \eta * R^3$)

The maximum value for sin(M,B) is 1. That means by reducing B or increasing ω one reaches a point where the magnetic torque needed for the bacterium to follow the external field becomes too small to compensate for the viscous torque.

As a consequence the bacterium tries to follow the external field but the angle (M,B) increases steadily (Fig. 2a–d). When the magnetic moment reaches 180° relative to the external field the direction of rotation flips to the opposite direction (Fig. 2e) until magnetic moment and external field are parallel again.

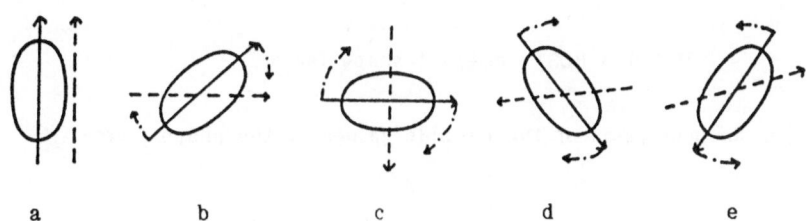

a   b   c   d   e

Figure 2. Magnetic torque exerted to a dead magnetic bacterium at various orientations by an external field, that is too small to compensate for the viscous torque.
--------→ magnetic moment; – – – – –▶ external B-field; –·–·–·–·▶ torque

By adjusting the critical values for B or ω , where M and B are just perpendicular to each other equation (3) becomes:

(4)
$$M * B = RCS * \omega$$

which can be used to determine M, the magnetic moment.

Our interest was to investigate a special type of bacterium with peculiar swimming behaviour. A morphologically similar type was reported by Vali et al. (1987). It occurred in great numbers in the Lake Chiemsee sediment samples. We tentatively named this type MB1 (for Magnetobacillus bavaricus). To describe this rod-shaped bacterium more exactly we have to alter the simple mathematics given above and calculate the magnetic moment of an ellipsoidal bacterium (still a simplifying assumption).

According to Jeffery (1922) is:

$$(5) \qquad\qquad RCE = 4 * \eta * \frac{r^2 + 1}{r^2 * K_3 + K_1} \cdot Vol$$

RCE   = resistance coefficient of an ellipsoid
Vol   = volume of the ellipsoid
r   = ratio of half-axes
K_1, K_3 = improper integrals describing the influence on the shape to the resistance
              coefficient (see Jeffrey 1922, Leal and Hinch 1971)

We can simplify this to : RCE = RCS' * shape factor, where RCS' means the resistance coefficient of a sphere with the same volume as the ellipsoid. The shape factor contains the geometric constants K_1, K_3, and r. So for an ellipsoidal body equation (4) is modified to :

(6)   M * B = RCS' * omega * shape factor

Figure 3. gives an impression of the possible values of the shape factor.

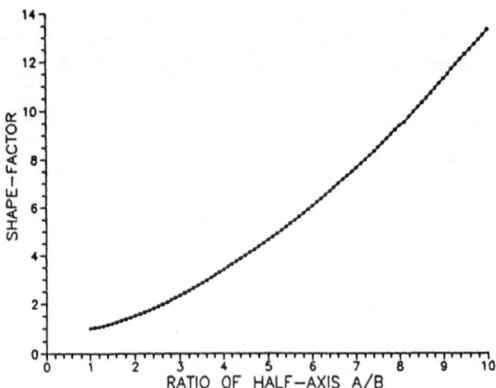

Figure 3. Calculated values of the shape factor versus r. Numerical examples: r = 1, shape factor = 1.0; r = 3, shape factor = 2.34; r = 5, shape factor = 4.64.

6. Assumptions and Uncertainties

In the calculation referring to an ellipsoidally shaped body in an infinite fluid several simplifications were made:
- Magnetic bacteria occur in shapes that are only approximately ellipsoidal.
- The surface is not necessarily smooth, there might be pili or other appendices.
- Flagella will contribute to the viscous torque.
- Bacteria do not move in infinite fluid but near the glass slide or the cover glass used for microscopic observation.

The magnetic moment determined from the measurement of the viscous torque is supposedly too small because all the above simplifications tend to decrease the viscous torque and consequently the magnetic moment.

7. Electron Microscopy on Selected Bacteria

To get a second independent value for the magnetic moment we guided the same bacterium that has previously been measured by motion analysis onto a specimen grid and then observed it under the TEM. The EM-grid was coated with collodium and a thin layer of silicon oxide to make it hydrophilic.

Using the joystick of the bacteriodrome it was possible to lead the observed magnetic bacterium out of the sediment through a region of pure water onto the grid. There we kept it by a rotating magnetic field until the water had evaporated and the bacterium adhered to the collodium film (Fig. 4).

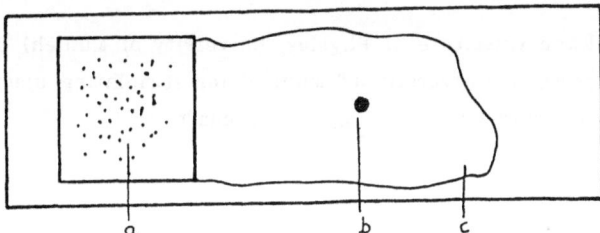

Figure 4. Microscope slide with TEM-grid. a, mud slurry; b, EM-grid; c, water.

8. Comparison of the Two Measurements

By estimating the volume of the magnetosomes from the two-dimensional EM-picture and assuming that the magnetite particles are magnetized up to saturation we calculated values for the magneteic moment of about three times the values measured by motion analysis.

Figure 5. EM-picture of an example of MB1. Several chains of projectile-shaped magnetosomes are visible.

The bacterium shown in Fig. 5 had a measured magnetic moment of 0.82 10^{-14}Am^2.
When the magnetosomes are assumed to be cuboids with identical lengths and widths,
the total volume of the magnetosomes in the sample shown in Fig. 5 gave a magnetic
moment of $2.9 \cdot 10^{-14}$Am^2.   Approximating the magnetosomes as ellipsoids of the same
lengths and widths, probably a better approximation for the irregularly shaped magnetosomes, led to $1.5 \cdot 10^{-14}$Am^2.   This is a quite good result taking into account the various simplifications made above.

We now estimate the errors still made and, in an attempt to reduce them, we are trying
to improve both our techniques and the calculations.

Acknowledgements

We want to thank Leo Schwab (Institute of Physics, University of Munich) and Jobst
Wippern (Institute of Geophysics, University of Munich) for stimulating discussions. This
work was supported by the Deutsche Forschungsgemeinschaft.

REFERENCES

Balkwill D.F., Maratea D., and Blakemore R.P. (1980) Ultrastructure of a magnetotactic spirillum. J. Bacteriology 141: 1399–1408.

Bazylinsky D.A., Frankel R.B. and Jannasch H.W. (1988) Anaerobic magnetite production by a marine magnetotactic bacterium. Nature 334: 518–519.

Blakemore R.P. (1975) Magnetotactic bacteria. Science 190: 377–379.

Butler D.F. and Banerjee S.K. (1975) Theoretical single domain size range in magnetite and titanomagnetite. J. Geophys. Res. 80: 4049–4058.

Jeffery G.B. (1922) Motion of ellipsoidal particles immersed in a viscous fluid. Proc. Roy. Soc. Lond. A102: 161–179.

Landau L.D. and Lifshitz E.M. (1987) Course of Theoretical Physics Vol. 6 (Fluid Mechanics) p. 65.

Leal L.G. and Hinch E.J. (1971) The effect of weak Brownian rotation on particles in shear flow. J. Fluid Mech. 46: 685–703.

Petersen N., Weiss D.G. and Vali H. (1989) Magnetic bacteria in lake sediments, pp. 231–241. In: Geomagnetism and Palaeomagnetism, Kluwer Academic Publishers.

Vali H., Förster O., Amarantidis G. and Petersen N. (1987) Magnetotactic bacteria and their magnetofossils in sediments. Earth Planet. Sci. Lett. 86: 389–400.

Weiss D.G., Maile W. and Wick R.A. (1989) Video Microscopy, pp. 221–278. In: Lacey A.J. (Edit.) Light Microscopy in Biology. A Practical Approach. Oxford :IRL Press.

CHEMOKINESIS, CHEMOTAXIS AND GALVANOTAXIS DOSE-RESPONSE CURVES AND SIGNAL CHAINS

Hans Gruler
Biophysics Department,
University of Ulm
D 7900 Ulm, Germany

Abstract

The translational kinematics of cells in a polar field as electric field, concentration gradient, etc., can be described by two independent cellular responses: the speed and the direction of migration. It is shown that the speed can be described by a steerer (=controller without feedback) and the direction of migration by an automatic controller (=controller with feedback). The steerer and automatic controller can be regarded as the framework for the directed movement or growth and it can be applied even when the physico-chemical signals of the cell are unknown. The models are verified by data of human granulocytes.

1. Introduction

Polymorphonuclear leukocytes (=granulocytes) are attracted to sites of inflammation to destroy microorganisms. Different mechanisms exist which cause granulocytes to be attracted to and remain in the general vicinity of the infection. The signal can be chemical in nature; the resulting directed movement is then called chemotaxis. The initial part of the peptides responsible for chemotaxis is a formylmethionyl moiety. These peptides allow the granulocytes to distinguish between prokaryote and eukaryote cells. The signal may also be electrical in nature; the directed migration is then called galvanotaxis (Wilkinson 1974). For example, when a cell is lysed, ions inside and outside the cell diffuse to reestablish a concentration equilibrium. The different diffusion constants of the ions involved result in the separation of small and large ions, and a diffusion potential is created (Rapp et al. 1988, Atkins 1986).

An important concept in understanding biological phenomena is the cybernetics also known as the theory of the automatic control (Wiener 1961). The theory of automatic controls is widely used in physiology and neurophysiology. Here it is shown that chemotaxis, galvanotaxis, galvanotropism, etc. are functions of cells having an automatic controller as goal-seeking system. The model is veryfied by means of galvanotaxis and chemotaxis data of granulocytes. The chemokinetic response is explained by a steerer of the cells (=controller without feedback). The description of the cellular functions via automatic controller and steerer allows the incorporation of effects of secondary factors for which a detailed knowledge is missing. Due to the noise in the steerer and automatic controller, the cellular systems are analysed via probabilistic models rather than deterministic ones.

2. Chemokinesis

The migration of granulocytes can be induced by exogenous signals, such as peptides with formylmethionyl as the initial part of the molecule. The cell has the membrane-bound receptors that recognize these molecules and the signal transduction chain is initiated. The steerer as a model for the chemokinesis is based on the observation that the mean speed is proportional to mean concentration of membrane-bound receptors loaded with chemokinetic molecules.

2.1 Steerer - a Controller Without Feedback

The steerer is explained in the following way: The increase of the mean concentration of the second messenger, M, is proportional to the primary cellular signal, S and the cellular response is proportional to the second messenger ($v_c = k_t \cdot S$). k_t is a signal transduction coefficient (Tranquillo et al. 1987, 1988a). The decrease of the second messenger is assumed to be proportional to its concentration, or $-k_d \cdot v_c$, where k_d is a decay coefficient. The rate equation for the track velocity is

$$(1) \qquad \frac{dv_c}{dt} = k_t \cdot S - k_d \cdot v_c + \Gamma(t)$$

The deterministic part in the signal transduction/response mechanism is described by $k_t \cdot S$ and $k_d \cdot v_c$. The noise induced in the signal transduction/response system is described by $\Gamma(t)$. Here we assume δ-correlated white noise with the strength q ($<\Gamma(t)>=0$).

$$(2) \qquad <\Gamma(t) \cdot \Gamma(t')> = q \ \delta(t-t')$$

In the next step the predictions of the model are compared with experimental findings as chemokinetic dose-response curve and track velocity distribution function.

2.2 Chemokinetic Dose-Response Curve

If the primary cellular signal in chemokinesis is the number of occupied receptors as assumed by Tranquillo et al. (1987, 1988a) then the concentration dependence of the track velocity derived from equ. 1 is

$$(3) \qquad v_c = R_O \cdot \frac{k_t}{k_d} \cdot \frac{[c]}{[c] + K_R}$$

where R_O is the total number of receptors exposed to the cell surface, and K_R is the equilibrium binding constant of the chemokinetic molecule to the receptor. The chemokinetic dose-response curve (= the mean track velocity as a function of the concentration of the chemokinetic molecule, [c]) has the predicted concentration dependence (Fig. 1). The receptor occupancy as the primary cellular signal as well as the linear response of the steerer are justified.

2.3 Track Velocity Distribution Density

The deterministic signal in equ. 1 determines the chemokinetic dose-response curve. But the distribution of the track velocity around the expected velocity is a consequence of the noise in the signal. (Granulocytes seem to be a homogeneous population with respect to their chemokinesis since there are no significant differences between the

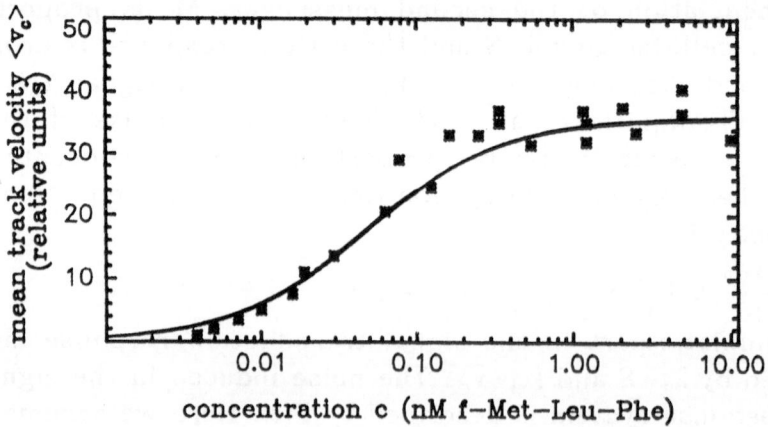

Figure 1: Chemokinetic dose-response curve of granulocytes. Solide line (equ. 3). (Fit: K_R= 0.04 nM, measured high affinity site of the receptor K_R =0.05 to 0.5 nM)

distribution obtained from a single cell observed over a long period of time and the distribution obtained from multiple cells (Gruler 1984, 1989)). The track velocity distribution density, $f(v_c)$, predicted from the stochastic differential equation (equ. 1), can be obtained from the Fokker–Planck equation (see Box 9, $\langle$equ. 20$\rangle$, Risken 1985):

$$(4) \qquad f(v_c) = f_0 \cdot e^{ -\dfrac{(v_c - v_c^{det})^2}{2\sigma^2} }$$

The maximum and the width of the gaussian are described by the deterministic and stochastic signal. f_0 is a calibration constant.

$$(5) \qquad v_c^{det} = \frac{k_t}{k_d} \cdot S$$

$$(6) \qquad 2\sigma^2 = \frac{q}{k_d}$$

The measured track velocity density, $N(v_c) \cdot (\sum N(v_c))^{-1} \cdot (2\pi v_c)^{-1}$, is approximately a gaussian as predicted by the model (Fig. 2a). (The measured histogram of the track velocity, $N(v_c) / \sum N(v_c)$, has to be divided by the "area" of the segment, $2\pi v_c \cdot dv_c$ (see Box 8) before it can be compared with the predicted distribution density, $f(v_c)$).

We expect that the basic noise source is at the initiation of the signal chain. The noise source cannot originate from receptor binding fluctuations since it is a fast process but a slow varying process is observed: The track velocity autocorrelation function, $\langle v_c(t) \cdot v_c(t')\rangle$, is nearly a horizontal line within a few minutes ($t-t' \neq 0$) (Fig. 2b, Alt, this volume, section II.1, Franke & Gruler, will be published). Possible main noise sources might be (i) the fluctuations of the total number of receptors exposed to the cell surface, or (ii) the fluctua-

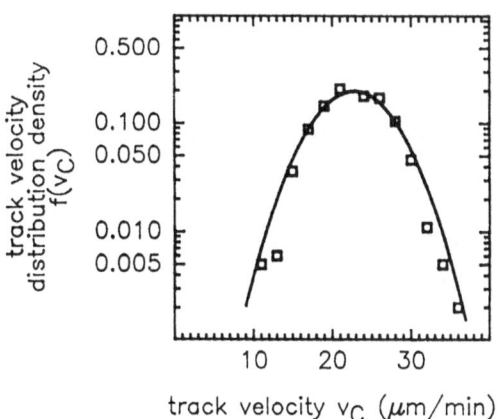
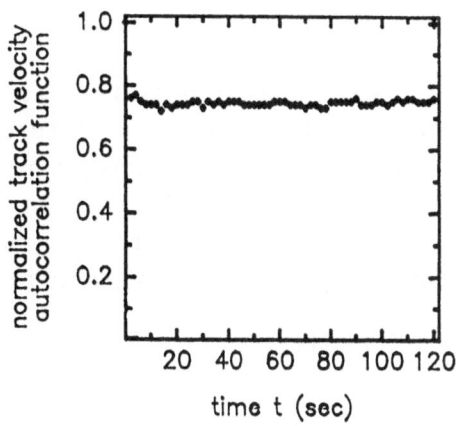

Figure 2: a) Track velocity distribution density (granulocytes). b) Normalized track velocity autocorrelation function of granulocytes exposed to 0.8 $V \cdot mm^{-1}$.

tions in the concentration of regulatory proteins in the signal chain.

In summary the steerer model for the chemokinetic response is obviously correct. It predicts the chemokinetic dose-response curve and the distribution of the track velocity.

BOX 8

HISTOGRAM, DISTRIBUTION DENSITY, AND GENERATING FUNCTION

Hans Gruler

Histogram: Experimental results are very often represented in histograms, for example the histogram of the cells moving on a plane. The angle Φ (=direction of migration) is a continuous random variable since it can have every value between 0 and 2π or between 0 and 360°. In the histogram, N_i, the cells are counted which migrate in a certain direction in the interval between Φ_i and $\Phi_i + \Delta\Phi$. (The circle is divided into n segments). The probability of finding a cell migrating in the interval between Φ_i and $\Phi_i + \Delta\Phi$ is $N_i/\Sigma_i N_i$.

Density Distribution: The histogram is a discrete function which can be approximated by a continous function - the distribution density, $f(\Phi)$. The distribution density times the interval, $f(\Phi_i) \cdot \Delta\Phi$, describes the probability to find a cell migrating in the interval between Φ_i and $\Phi_i + \Delta\Phi$. Thus we have

$$(1) \qquad f(\Phi) \cdot \Delta\Phi \approx \frac{N_i}{\Sigma_i N_i}$$

Size of Segment: The interval of the random variable has not always the simple structure as shown in equ. 1. For example the track velocity histogram where the number of cells migrating between v_{ci} and $v_{ci} + \Delta v_c$ is counted. If the cells migrate on a plane the "size" of the segment is the "area" between the circles with radius $v_{ci} + \Delta v_c$ and the radius v_{ci}. One obtains thus

$$(2) \quad f(v_c) \cdot 2\pi \cdot v_c \cdot \Delta v_c \approx \frac{N_i}{\Sigma_i N_i}$$

Generating Function: A histogram consists of positive or zero numbers and therefore the distribution density is also a function with only positive or zero values. The distribution density can be described by another function – the generating function $V(\Phi)$ – without loosing any information. (Φ is the continuous random variable).

(3) $f(\Phi) = e^{V(\Phi)}$

If a stochastic differential equation is known for the problem then the generating functions can be predicted from the corresponding Fokker-Planck equation. A few examples are shown in BOX 9. But if there is no theory available, the generating function can be approximated in the following way where the symmetry of the experimental setup is used.

If the continuous random variable, Φ, varies from 0 to 2π, then the generating function can be expressed by a Fourier Series.

(4) $V(\Phi) = a_0 + \sum_i a_i \cos(i\cdot\Phi) + \sum_i b_i \sin(i\cdot\Phi)$

The coefficients a_i and b_i are fitting parameters where a_0 can be used to calibrate the distribution ($\int f(\Phi)\cdot d\Phi = 1$). The number of unknown coefficients can be reduced by applying the symmetry of the experimental setup.

If the random variable varies from $-\infty$ to $+\infty$ or from 0 to $+\infty$, then the generating function, $V(x)$, can be expressed by a Taylor series. (x is the continuous variable and the coefficients A_i are fitting parameters).

(5) $V(x) = A_0 + \sum_i A_i \cdot x^i$

Why is it so important to use the generating function? (i) First a practical reason: A large number of data points contribute to a small number of fitting parameters. (ii) Second, the basic principles of special biological functions can be uncovered by means of the generating functions (see BOX 9). (iii) Third, complex phenomena may be separated by means of the generating function. E.g. the joint probability of two independent random variables, x and y, is the product of their density functions, $f(x)$ and $f(y)$. Or, more simple, the joint probability is obtained by adding the two generating functions, $V(x)$ and $V(y)$.

3. Directed Movement

One important biological function of cells is their capacity for directed movement. The automatic controller as a model for the directed movement is based on the observation that the cell orientation fluctuates around the desired direction.

The mean displacement, $\langle x \rangle$, of cells is quantified by the mean velocity parallel to the polar field, $\langle v_{||} \rangle$, (= drift velocity) times the observation time, t.

(7) $\langle x \rangle = \langle v_{||} \rangle \cdot t$

The translational movement in a polar field requires two components of the cellular response: the track velocity and the direction of migration $\Phi(t)$.

(8) $v_{||}(t) = v_c(t) \cdot \cos \Phi(t)$

One might well expect that these two parameters, $v_c(t)$ and $\Phi(t)$, would depend on each other. But they are independent of each other since the average drift velocity, $\langle v_{||} \rangle$, is equal the average track velocity, $\langle v_c \rangle$, times the average of $\cos \Phi$. This holds at least for human granulocytes (Rapp et al. 1988, Gruler 1984), somitic fibroblasts (Gruler & Nuccitelli 1986), and neural crest cells (Gruler & Nuccitelli, will be published).

Now it will be shown that the direction of migration can be described by an automatic controller.

3.1 Automatic Controller

The basic elements of an automatic controller are: first, an element which measures the output of the biological system; second, a means of comparing that output with the desired one; third, a means of feeding back this information into the input in such a way as to minimize the deviation of the output from the desired level.

The cell must have the ability to measure its orientation with respect to the applied polar field (e.g. electric field or concentration gradient). This orientation is compared with the desired one - e.g. to be parallel to the polar field. The created intracellular signal is such that the cell rotates to approach the desired orientation. The rate equation for the angle of migration is thus

(9) $\dfrac{d\Phi}{dt} = -\Gamma_D(E,\Phi) + \Gamma_N(t)$

The deterministic virtual torque, $-\Gamma_D(E,\Phi)$, tries to render the movement parallel to the applied polar field. The noise in the signal transduction/response system induces a stochastic virtual torque, $\Gamma_N(t)$, and we assume here a δ-correlated white noise with the strength Q ($\langle \Gamma_N(t) \rangle = 0$).

(10) $\langle \Gamma_N(t) \cdot \Gamma_N(t') \rangle = Q \cdot \delta(t-t')$

We will show that the deterministic virtual torque of granulocytes has two contributions originating from (i) a proportional controller and (ii) an integral controller. The discussion is focussed to the proportional controller since it is the important one. At the end the integral controller is discussed.

In order to find a function for the deterministic virtual torque, $\Gamma_D(E,\Phi)$, it is common to begin with a very general equation such as an infinite series. The symmetry requirements for the directed movement restrict the possible series and the number of terms in the infinite series.

The physical state is unchanged if the coordinate system is rotated by $n \cdot 360$ degrees (n=integer) and the torque changes sign if the co-ordinate system is reflected at a mirror containing the polar field vector.

(11) $\Gamma_D (E, \Phi) = \Gamma_D (E, \Phi \pm n \cdot 360 \text{ degrees })$

(12) $\Gamma_D (E, \Phi) = - \Gamma_D (E, -\Phi)$

One well known series that meets these symmetry requirements is the Fourier Series. One can immediately eliminate half of the series by applying the symmetry restrictions (equ. 12). Thus one finds

(13) $\Gamma_D(E, \Phi) = c_1 \cdot \sin\Phi + c_2 \cdot \sin 2\Phi + \cdots$

c_1 describes the directed (or polar) movement or growth, and c_2, the bidirectional or apolar movement or growth (and a constant c_0 would describe a spiral movement).

In the following steps it will be shown how the virtual deterministic torque, Γ_D, can be determined from experimental data.

3.2 Angle Distribution Density and Generating Function

To do this, the generating function, $V(\Phi)$, of the directed movement is determined from the measured steady state angle distribution (see BOX 8) and compared with the predicted one. The angle distribution density, $f(\Phi)$, predicted from the stochastic differential equation (equ. 9), can be obtained from the Fokker–Planck equation (see BOX 9, $\langle$equ. 12$\rangle$). The predicted generating function, $V(\Phi)$, is the integrated deterministic virtual torque divided by half of the white noise strength.

(14) $V(\Phi) = \dfrac{2}{Q} \displaystyle\int_0^{\Phi} \Gamma_D (E, \Phi') \cdot d\Phi'$

(15) $f(\Phi) = e^{V(\Phi)}$

A plot of the data, $\ln f(\Phi)$ vs $\cos\Phi$, as shown in Fig. 3a yields a straight line. It proves that the only important torque in directed movement is $c_1 \cdot \sin\Phi$. (The slope of the line is a_1). Such straight line behaviour is found for galvanotaxis of granulocytes (Fig. 3a), of neural crest cells (Gruler & Nuccitelli, will be published), and of somi-

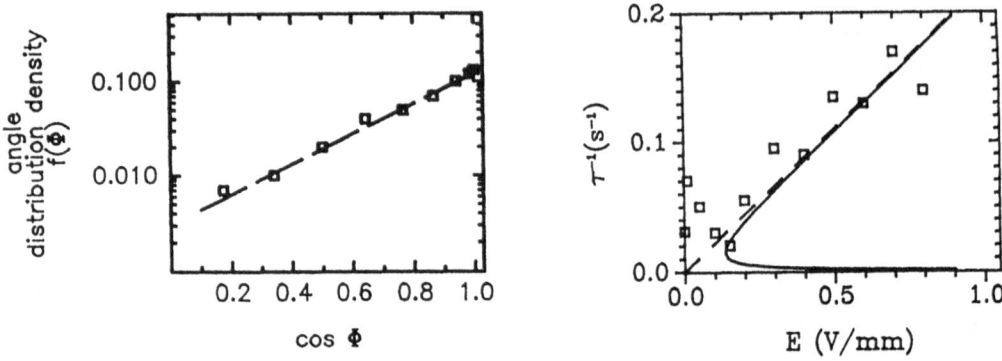

Figure 3: Galvanotaxis of human granulocytes. a) Logarithm of the measured angle distribution vs $\cos\Phi$. b) Inverse of the characteristic time vs E. Proportional controller (dashed line, equ.16, k_P =-0.22 mm $\cdot$ V^{-1} $\cdot$ s^{-1}), proportional-integral controller (thick line, BOX 9, k_{IC} = 0.00022 s^{-2}).

tic fibroblasts (Gruler & Nuccitelli 1986), for chemotaxis of granulocytes (Gruler 1988, 1989), for necrotaxis of granulocytes and monocytes (deBoisfleury-Chevance et al. 1989), for galvanotropism of Fungal Hyphae (Gruler & Gow 1990).

In summary, the basic term in describing directed movements or growth is the term $c_1 \cdot \sin\Phi$ of the infinite Fourier Series (equ. 13).

Next the coefficient c_1 will be investigated, first for galvanotaxis and then for chemotaxis.

3.3 Galvanotaxis

The coefficient, c_1, as a function of the applied electric field, E, can be obtained from the angle autocorrelation function, $\langle\Phi(t_1) \cdot \Phi(t_2)\rangle$. The predicted function derived from the stochastic differential equation (equ. 9), decays exponentially in time, $|t_1-t_2|$, with a characteristic time τ (= c_1^{-1}). One expects in the case of a linear response
$$(16) \qquad \tau^{-1} = c_1 = k_P \cdot E$$
The assumed linear behaviour (τ^{-1} vs E) is actually found (E>0.1 V $\cdot$ mm^{-1}) (Fig. 3b). The deterministic part of the controller is determined by the slope and yields k_P=-0.22 mm $\cdot$ V^{-1} $\cdot$ s^{-1}.

The field dependence of c_1 is also obtained by investigating the measured generating function $V(\Phi)$ (=$a_1 \cdot \cos\Phi$). A plot of the experimentally obtained data (a_1 vs E) yields a straight line (granulocytes). The prediction from equ. 9 is

$$(17) \qquad a_1 = \frac{2}{Q} \cdot c_1 = K_G \cdot E$$

$$(18) \qquad K_G = \frac{2 \cdot k_P}{Q}$$

where K_G is the galvanotaxis coefficient. The found straight line behaviour proves the assumed linear response.

The linear response can also be shown by means of the galvanotaxis dose-response curve (= average of $\cos\Phi$ vs E). The prediction is if the deterministic virtual torque can be expressed by a single term (Gruler 1988, 1989, Gruler & Nuccitelli 1986)

$$(19) \qquad \langle\cos\Phi\rangle = \frac{I_1(a_1)}{I_0(a_1)}$$

$I_1(a_1)$ and $I_0(a_1)$ are hyperbolic Bessel functions. The assumed linear response is again proved since the polar order parameter as a function of the electric field strength has the predicted dependence with only one fitting parameter, K_G (Fig. 4, Rapp et al. 1988, Franke & Gruler 1990) ($E < 0.8$ V $\cdot$ mm^{-1}).

In summary, the automatic controller for the galvanotactic response of granulocytes is obviously correct. The linear response of the controller is justified. Similar results were obtained for the galvanotaxis of spermatozoids (Gruler 1988, Brakow 1958) and for the directed growth of spores in an electric field (Gruler & Gow 1990).

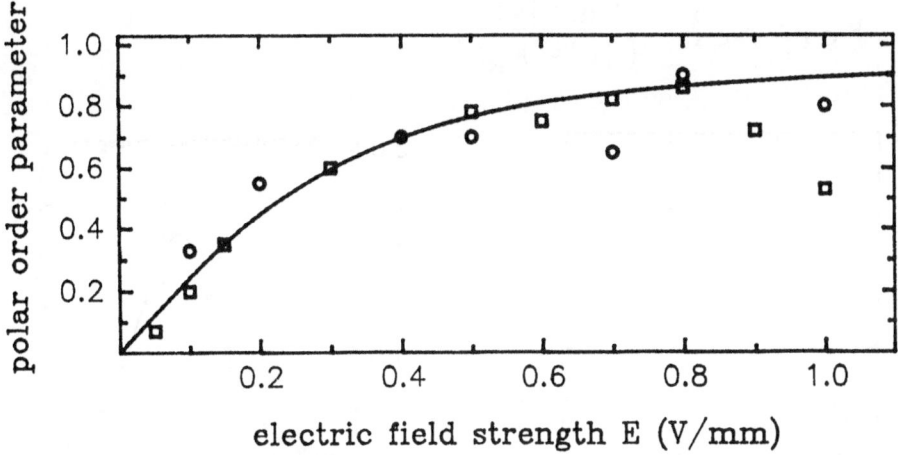

Figure 4: Galvanotaxis dose-response curve of granulocytes. Line: equs. 17 and 19 with $K_G^{-1} = -0.2$ V $\cdot$ mm^{-1} (open circles: Rapp et al. 1988, open squares: Franke & Gruler 1990).

3.4 Chemotaxis

The chemotaxis can be treaded in the same way as galvanotaxis. E has to be replaced by the gradient of the chemical activity of the chemotactic molecule. In the case of a linear relation, the torque is proportional to grad $\ln[c]$ and hence

$$(20) \qquad a_1 = K_{CT} \cdot grad\ \ln[c]$$

The expected linear response is actually found (Fig 5a, Gruler 1988, 1989). But there is a problem since the chemotaxis coefficient K_{CT} is a function of the mean concentration of the chemotactic molecule (Fig. 6b, Gruler 1988, Rapp et al. 1988).

The concentration dependence of K_{CT} can be predicted if the molecular model of the chemokinetic response of granulocytes is enlarged (Tranquillo et al. 1987, 1988a): It is assumed that the cell measures the concentration, $[c_L]$ and $[c_R]$, at two separate parts of the membrane and the created deterministic virtual torque is proportional to the difference between these primary cellular signals, S_L and S_R

$$(21) \quad S_L - S_R = [R \cdot c]_L - [R \cdot c]_R = \frac{R_O}{2} \ \frac{K_R}{([c] + K_R)^2} \ \frac{d[c]}{dx} \ l_O \sin\Phi$$

l_O is the mean distance between the two membrane patches. The model predicts a concentration dependence of the chemotaxis coefficient, K_{CT}, as actually measured (Fig. 5b, Gruler 1988, Rapp et al 1989, Tranquillo et al 1988b).

$$(22) \qquad K_{CT} = K_{CT}^O \ \cdot \frac{K_R \cdot [c]}{([c] + K_R)^2}$$

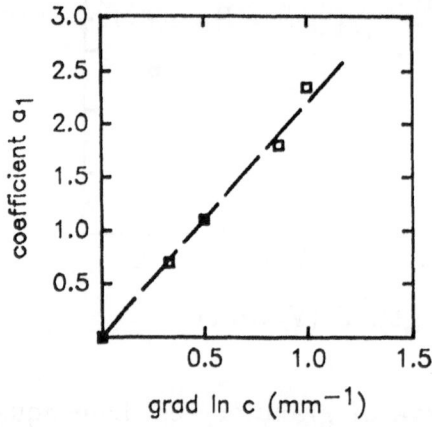
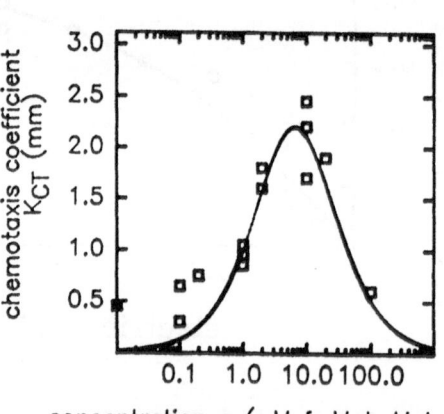

Figure 5: a) The coefficient a_1 as a function of grad $\ln [c]$ with constant mean concentration (f-Met-Met-Met). b) Chemotaxis coefficient, K_{CT} as a function of the mean concentration of the chemotactic molecule. Solid line (equ. 22, $K_{CT}^O = 9$ mm, $K_R = 6.6$ µM)

In summary, the automatic controller for galvanotaxis is obviously correct. The linear response of the controller is justified. The hypothesis of a spatial recognition system is consistent with the recorded data.

3.5 Patlak-Keller-Segel Model

A few words to the Patlak-Keller-Segel model of the directed movement (Patlak 1953, Keller & Segel 1971): There the basic assumption is that the mean drift velocity, $\langle v_{||} \rangle$, is proportional to $d[c]/dx$

$$(23) \qquad \langle v_{||} \rangle = \chi_{CT} \cdot \frac{d[c]}{dx}$$

χ_{CT} is a coefficient characterizing the chemotactic response. This model is only an approximation which holds for small concentration gradients since $\langle v_{||} \rangle$ increases in the same way as $d[c]/dx$. This model cannot predict the observed saturation in the chemotactic response ($\langle v_{||} \rangle \rightarrow \langle v_c \rangle$ for $d[c]/dx \rightarrow \infty$). Eq. 23 can be compared with our calculation.

$$(24) \qquad \chi_{CT} = \frac{1}{2} \cdot \langle v_c \rangle \cdot K_{CT}^0 \cdot \frac{K_R}{([c] + K_R)^2}$$

3.6 Galvanotaxis and Chemotaxis Coefficient

The galvanotaxis and the chemotaxis coefficients quantify the cellular sensitivity to an electric field or to a concentration gradient and their value can be determined from the dose-respond curves. K_G or K_{CT} is the ratio of the coefficient, k_P, which characterizes the deterministic part of the controller, and of $Q/2$, which characterizes the noise in the controller (see equ. 18). K_G or K_{CT} can be determined from the angle autocorrelation function. The mean square angle, $\langle \Phi(t)^2 \rangle$, predicted from the stochastic differential equ. 9 is

$$(25) \qquad \langle \Phi(t)^2 \rangle = \tau \cdot \frac{Q}{2} = (K_G \cdot E)$$

The inverse of the galvanotaxis coefficients derived directly from trajectories are -0.19 and -0.18 $V \cdot mm^{-1}$ for E=0.5 and 0.8V $\cdot mm^{-1}$, respectively. K_G^{-1} determined from the dose-response curve is -0.2 $V \cdot mm^{-1}$. This accordance demonstrates the quality of the model.

3.7 Limitations of the Model of the Automatic Proportional Control

The characteristic time of the controller is a function of the applied electric field: τ is small at large field and large at small field strength. One expects for $E \to 0$, $\tau \to \infty$. But τ approaches a finite value τ_O ($\sim 30s$) which is identical with that one obtained from the mean square displacement (Fig. 6a). (τ_O can be regarded as the internal clock of the cell.) Obviously the cell regulates continously its direction of migration at large field strength whereas at low field strength the cell choses a new direction of migration after the time τ_O. If the goal-seeking system of the cell is a proportional-integral controller (see BOX 9, $k_{PC}=k_P \cdot E$), then the system exhibts an oscillatory state when the deterministic signal of the proportional controller is too small.

$$(26) \qquad k_P{}^2 \cdot E^2 \le 4 \cdot k_{IC}$$

The proportional-integral controller exhibits two characteristic times τ_1 and τ_2, if $k_P{}^2 \cdot E^2 \ge 4 \cdot k_I$ (Fig. 3b). At large field strength a characteristic time of several seconds and several minutes is predicted e.g. $\tau_1 = 9.2$ s and $\tau_2 = 8.2$ min with $k_P=0.22$mm $\cdot$ V$^{-1} \cdot$ s^{-1} and $k_{IC}=0.00022$s^{-2} for $E=0.5$ V $\cdot$ mm^{-1} . The characteristic time of the fast process was experimentally found as already discussed. The slow process with a characteristic time of more then several minutes was observed in granulocytes (Franke & Gruler 1990) and in monocytes (deBoisfleury-Chevance et al. 1989). The situation at low field strength is not yet

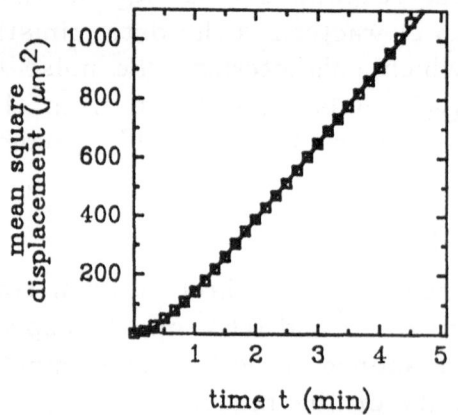
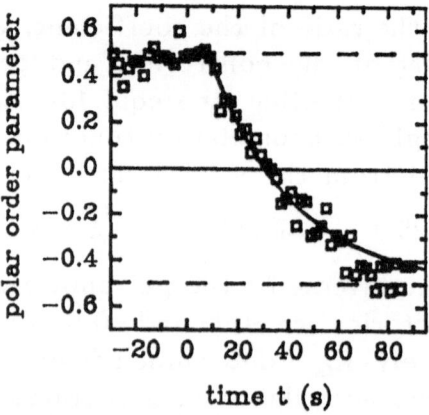

<u>Figure 6:</u> a) Mean square displacement as a function of time of granulocytes in a homogeneous environment. b) Polar order parameter as a function of time. At time t=0, the electric field was changed from –E to +E (=1 V $\cdot$ mm^{-1}) (characteristic time =32s).

solved. More experimental material is necessary.

The model of the automatic controller needs a further improvement since the characteristic time of the signal transduction/response system is neglected. The delay of the system can be determined by an adaptation experiment (Franke & Gruler 1990): a constant electric field, E, is applied for a longer period to equilibrate the cells to their environment. Then at t=0 the electric field is altered from -E to +E. (Fig. 6b). If the cells could react immediately to the altered environment one would obtain a step function for the polar order parameter. However, the cells need some time to react: There exists a lag-time of 8.3 s. This time has to be incorporated into the model of the automatic controller.

The mean turning behaviour for large angles can be obtained by the above mentioned adaptation experiment. The whole cellular response can be described by a single exponential with a characteristic time of 32 s. For $E=1V \cdot mm^{-1}$ a much smaller value is expected. The model of the automatic controller has to be improved.

In summary, the model of the automatic proportional-integral control explains basic features of the directed movement but the model has to be improved.

Acknowledgements

This work was supported by "Fond der chemischen Industrie" and by a NATO travel grant.

REFERENCES

Alt W. (1980) Biased random walk models for chemotaxis and related diffusion approximations, J. Math.Biol. 9, 147-177

Atkins P.W. (1986) Physical Chemistry. Oxford University Press, Oxford, 260

Brokaw C.J. (1958) Chemotaxis of bracken spermatozoids. Implications of electrochemical orientation. J.Exp.Biol. 35, 197-212

deBoisfleury-Chevance A., Rapp B., Gruler H. (1989) Locomotion of white blood cells: A biophysical analysis. Blood Cells 15, 315-333

Franke K. and Gruler H. (1990) Galvanotaxis of human granulocytes: electric field jump studies. Eur. Biophys.J. 18, 335-346

Gruler H. (1984) Cell Movement analysis in a necrotactic assay. Blood Cells 10, 107-121

Gruler H., and Nuccitelli R. (1986) New insights into galvanotaxis and other directed cell movements: an analysis of the translocation distribution function. In: Ionic Currents in Development. Nuccitelli R., Ed., Alan R. Liss, New York, 337- 347

Gruler H. (1988) Cell movement and symmetry of the cellular environment. Z.Naturforsch. 43c, 754-764

Gruler H. (1989) Biophysics of leukocytes: neutrophil chemotaxis, characteristics and mechanisms. In: The Cellular Biochemistry and Physiology of Neutrophil (M.B. Hallett, Ed.) CRC-Press UNISCIENCE, 63-95.

Gruler H. and Nuccitelli R., Neural crest cell galvanotaxis: new data and a novel approach to the analysis of both galvanotaxis and chemotaxis. Will be published.

Gruler H. and Gow N.A.R. (1990) Directed growth of fungal hyphae in an electric field. Z.Naturforsch. 45c, 86-93

Keller E.F., and Segel L.A.(1972) Travelling bands of chemotactic bacteria: a theoretical analysis. J. theor. Biol. 30, 235-248

Patlak C.S. (1953) Random walk with persistence and external bias. Bull of Math. Biophys. 15, 311-338

Rapp B., de Boisfleury-Chevance A., and Gruler H. (1988) Galvanotaxis of human granulocytes. dose-response curve. Eur.Biophys.J. 16, 313-319

Risken H. (1985) Fokker-Planck-Equation, Springer Verlag, Berlin, Heidelberg

Tranquillo R.T. and Lauffenburger D.A. (1987) Stochastic model of leukocyte chemosensory movement. J.Math.Biol. 25, 229-262

Tranquillo R.T., Lauffenburger D.A., and Zigmond S.H. (1988a) A stochastic model for leukocyte random mobility and chemotaxis based on receptorbinding fluctuations. J.Cell Biol. 106, 303-309

Tranquillo R.T., Zigmond S.H., and Lauffenburger D.A. (1988b) Measurement of the chemotaxis coefficient for human neutrophils in the under-agarose migration assay. Cell Motility & Cytoskeleton 11, 1-15

Wiener N. (1961) Cybernetics: or Control and Communication in Animal and the Machine, M.I.T. Press, Cambridge

Wilkinson P.C. (1974) Chemotaxis and Inflammation, J.& A. Churchill London, chap. 1 .

BOX 9

FOKKER–PLANCK EQUATION

Hans Gruler

The system is described by the continous state variable s. The temporal change of the variable is assumed to be only a function, $K(s)$, of the state variable.

$$(1) \qquad \frac{\partial s}{\partial t} = K(s)$$

The solution of this equation leads to $s(t)$ which describes in a deterministic way the state variable. However, if noise, described by the stochastic function $\Gamma(t)$, is involved, the state variable, $s(t)$, becomes a stochastic quantity.

$$(2) \qquad \frac{\partial s}{\partial t} = K(s) + \Gamma(t)$$

Now the state variable has to be described by a probability distribution, $f(s) \cdot \Delta s$ where $f(s)$ is the probability density in respect to the continous state variable s. A biological example is the migration direction of cells exposed to a concentration gradient. In this case the variable is the angle Φ between the direction of migration and the direction of the concentration gradient of the chemotactic molecules. The probability that the angle of migration falls between Φ and $\Phi+\Delta\Phi$ is described by $f(\Phi) \cdot \Delta\Phi$.

The temporal change of the probability density is described by the transition probability $P(s,t+\tau \,|\, s,t)$.

$$(3) \qquad f(s,t+\tau \,|\, s',t) = \int P(s,t+\tau \,|\, s',t) \cdot f(s',t) \cdot ds'$$

To derive an expression for the differential $\partial f/\partial t$ we must know the transition probability $P(s,t+\tau \,|\, s',t)$. The unknown transition probability is approximated by the transition moments (equ. 5) as

$$(4) \qquad P(s,t+\tau \,|\, s',t) = \left[1 + \sum (n!)^{-1} \cdot \left(-\frac{\partial^n}{\partial s}\right) M_n(s,t,\tau) \right] \delta(s-s')$$

$$(5) \qquad M_n(s',t,\tau) = \int (s-s') \cdot P(s,t+\tau \,|\, s',t) \cdot ds$$

We now assume that the moments M_n can be expanded into a Taylor series with respect to τ.

$$(6) \qquad M_n(s,t,\tau) = n! \cdot D^{(n)}(s,t) \cdot \tau + \cdots O(\tau^2)$$

BOX   9

The final result is the Fokker-Planck equation

$$(7) \qquad \frac{\partial f}{\partial t} = \frac{\partial}{\partial s} \left[- D^{(1)}(s,t) + \frac{\partial}{\partial s} D^{(2)}(s,t) \right] \cdot f(s,t)$$

The drift term, $D^{(1)}$, is given by the deterministic part of the differential equation ($D^{(1)} = K(s)$). The diffusion term, $D^{(2)}$, equals

$$(8) \qquad D^{(2)} = \frac{q}{2}$$

where q is the strength of the assumed white noise source, $\Gamma(t)$ ($<\Gamma(t)> = 0$)

$$(10) \qquad <\Gamma(t) \cdot \Gamma(t')> = q \cdot \delta(t-t')$$

The steady state solution of the Fokker-Planck solution is simply

$$(11) \qquad f(s) = f_0 \cdot e^{V(s)}$$

$$(12) \qquad V(s) = \frac{2}{q} \int^s K(s') \cdot ds'$$

where f_0 is a constant and V(s) the generating function.

A few examples will be discussed.

1. Torque Balance Equation and Dipol Orientation

The angle, Φ, between the applied electric field, E, and the dipole, p, is the state variable and eq. 2 is the torque balance equation. The deterministic torque $K(\Phi)$ is then

$$(13) \qquad K(\Phi) = - |p| \cdot |E| \cdot \sin\Phi$$

The steady state angle distribution density is the Boltzmann statistics

$$(14) \qquad f(\Phi) = f_0 \cdot e^{\frac{|p| \cdot |E|}{kT} \cdot \cos\Phi}$$

where the noise source strength is expressed by the thermal energy kT ($= q/2$). The bell-shaped curve has its maximum in the direction of the applied field.

2. Force Balance Equation and Motion

The velocity, v, is the state variable and eq. 2 is the force balance equation of a particle driven through a viscous liquid. The deterministic force is then

$$(15) \qquad K(v) = m^{-1} \cdot (F_{ext} - \gamma \cdot v)$$

where m, F_{ext}, and γ are the mass of the particle, the external

BOX   9

force, and the friction coefficient, respectively. The deterministic steady state velocity, v_{det}, is

(16) $v_{det} = \gamma^{-1} \cdot F_{ext}$

The steady state velocity distribution density is a gaussian if a stochastic force is involved

(17) $f(v) = f_0 \cdot e^{-\frac{(v-v_{det})^2}{2 \cdot kT}}$

with $q/2 = \gamma \cdot kT \cdot m^{-1}$. The gaussian has its maximum at the deterministic value.

3. Proportional Steerer

The state variable, v, is the output of a steerer device. The rate of the state variable originates from two contribution. (i) The external signal S alters the state variable where the rate is proportional to the external signal, S, (k_{PS}=constant of proportionality) and (ii) the state variable depends on intrinsic properties of the steerer. The decrease of the rate is proportional to the state variable (k_d^{-1} = memory time of the steerer).

(18) $K(v) = k_{PS} \cdot S - k_d \cdot v$

The expected steady state value of v is then

(19) $v_{det} = \frac{k_{PS}}{k_d} \cdot S$

The steady state distribution density of the state variable is a gaussian if a white noise source is involved.

(20) $f(v) = f_0 \cdot e^{-\frac{(v-v_{det})^2 k_d}{Q}}$

The gaussian has its maximum at the expected value, v_{det}.

4. Proportional Controller

The state variable, Φ, is the output of an automatic controller. The measured state variable is compared with the desired one and this information is feeded back in such a way as to minimize the deviation of the output from the desired level. At a proportional controller, the state variable is altered with a rate which is proportional to ($\Phi - \Phi_{set}$) where Φ_{set} is the set point for the

BOX   9

state variable.

(21) $K(\Phi) = - k_{PC} \cdot (\Phi - \Phi_{set})$

In the case of no noise source in the controller, the state variable approaches the set point with the characteristic time, k_{PC}^{-1},

In the case of a noise source, the density distribution of the state variable is a gaussian

(22) $f(\Phi) = f_0 \cdot e^{-\dfrac{k_{PC}}{q} \cdot (\Phi - \Phi_{set})^2}$

where the maximum of the distribution is determined by the set point, Φ_{set}.

5. Proportional-Integral Controller

The state variable and the set point are Φ and $\Phi_{set}(=0)$, respectively. At an integral controller the state variable is altered with a rate which is proportional to $\int_\tau (\Phi(t) - \Phi_{set}) \cdot dt$ (k_{IC}=constant of proportionality). The function $K(\Phi)$ is then

(21) $K(\Phi) = - k_{PC} \cdot \Phi - k_{IC} \cdot \int^\tau \Phi(t) \cdot dt$

The rate equation for the state variable reads then

(22) $\dfrac{\partial^2 \Phi}{\partial t^2} = - k_{PC} \dfrac{\partial \Phi}{\partial t} - k_{IC} \Phi + \Gamma(t)$

This equation is identical with that one of a damped oscillator. For $k_{PC}^2 < 4 k_{IC}$ the controller is instable and oscillates ($\omega = \sqrt{(k_{IC} - \tau^{-2})}$, $\tau^{-1} = k_{PC}/2$). For $k_{PC}^2 > 4 k_{IC}$ the controller is stable ($2 \cdot \tau^{-1} = k_{PC} \overset{+}{-} \sqrt{(k_{PC}^2 - 4k_{IC})}$). The Fokker-Planck equation for this case is more complex but a solution can be found in the book of Risken (1985).

MODELS OF CHEMICAL GRADIENT SENSING BY CELLS

Robert T. Tranquillo

Department of Chemical Engineering and Materials Science, University of Minnesota
Minneapolis MN 55455, USA

Abstract

This paper first provides a review of probabilistic models which have been proposed to understand various aspects of how receptor-sensing of chemoattractant influences directional movement. These all share a common premise, that directional *orientation* is determined by a spatial or temporal difference in receptor-measured concentrations, which fluctuate in time because receptor binding is inherently a stochastic process. After discussing the limitations of these models, a stochastic model recently proposed by the author is reviewed in detail. It is again based on fluctuations in receptor binding, but, in contrast to the simpler premise above, directional *movement* is regulated by the spatiotemporal pattern of a receptor signal and integrated signal-response coupling. The formulation of the modeling stochastic differential system and its analytical and numerical analysis are described in detail.

1. Introduction

Cells use specialized surface macromolecules, termed receptors, to sense chemical stimuli in the environment. Sensing occurs when some event related to the binding of stimulus molecules to receptors triggers a complex signal transduction process involving intracellular biochemical pathways and biophysical events. A cell expresses several populations of receptors each of which binds complementary stimulus molecules in a highly specific fashion. Thus, each receptor population essentially acts as an instrument to measure the extracellular concentration of a chemical stimulus, and, as with any instrument, will have an associated measurement error (Berg and Purcell, 1977). Receptor-sensing for the purpose of initiating and regulating a specific response or function appears to be ubiquitous. In particular, virtually all motile cells use information in the form of signals generated by receptor binding to regulate their speed and/or directionality of movement. "Receptor-signal noise" will be seen to be crucial in understanding the directional characteristics of cell movement in a chemical stimulus gradient, where *gradient* will mean a steady-state, linear, one-dimensional concentration gradient hereafter. Unlike most other models of biological motion, the noise characteristics of the ones discussed herein are not chosen *ad hoc* to yield agreement with movement data but are completely defined mechanistically *a priori*, in this case by the receptor binding characteristics.

The white blood cell known as the polymorphonuclear neutrophil leukocyte (PMN) is a well-studied prototype of the class of eukaryotic cells which translocate by "crawling" in ameoboid-like fashion and exhibit receptor-mediated movement responses. These cells express a population of $10^3 - 10^5$ receptors distributed over the cell surface for each of several known chemical stimuli which regulate movement, generically termed chemoattractant (CA). Receptor signals, which will refer to intracellular biochemical species generated by receptor binding, locally activate intracellular proteins involved in the generation of forces for movement (Singer and Kupfer, 1986; Omann *et al.*, 1987; Devreotes and Zigmond, 1988; see Tranquillo (1990) for a review of the cell biology relevant to the modeling herein). How the cell coordinates these signals distributed in space across its dimension so as to exhibit the observed movement responses, especially when the signals also vary with time when a cell moves in a CA gradient, will be the focus of this paper.

The different movement responses exhibited by PMN depend upon whether the CA is present in uniform concentration or as a gradient (Wilkinson, 1982). When placed in a uniform CA concentration, PMN are observed to translocate in the same general direction on a short time scale (i.e., minutes), a phenomenon termed "directional persistence" (see Box 14, this volume, section II/3 for definitions of cell migration terms). On a slightly longer time scale, cells exhibit directional changes during translocation, which can vary from a continuous turn, the cell maintaining its polarized morphology during the directional change, to an abrupt redirection, the directional change being due to a morphological repolarization. On a long time scale (i.e., hours), the cumulative result of many directional changes is a cell trajectory with a meandering appearance that can be adequately described as a random walk (Gruler and Bultmann, 1984), illustrated in Fig. 1a. For PMN exhibiting random motility, the mean speed is known to depend on the CA concentration (Maher *et al.*, 1984). This concentration dependence is termed orthokinesis and can cause a nonzero mean displacement of a cell moving in a spatial gradient (Doucet and Dunn, this volume, section II/3). A concentration dependence of the random turning behavior is termed klinokinesis. It can also result in a net displacement of a cell in a CA gradient and, in certain cases, result in an observable directional orientation bias for the cell, defined by the alignment of the cell polarity axis (and, hence, the direction of movement) with the direction of the gradient. That PMN exhibit klinokinesis is not as well-established because only recently has turning behavior been characterized with objective indices, such as the directional persistence time (Dunn, 1983; Dunn and Brown, 1987; Alt, this volume, section II/1). However, initial investigations indicate that leukocytes may exhibit klinokinesis (Farrell *et al.*, 1990).

When placed in a CA gradient, PMN still exhibit a random directional component of movement, but the gradient evidently presents directional information to the cell which results in biased movement up the gradient (Fig. 1b). Orientation bias associated with directed turning, the tendency of a cell to turn towards the direction of a gradient (Nossal and Zigmond, 1976), underlies the biased random walk resulting in a mean net displacement up-gradient. Quantitative studies reveal that PMN orientation bias increases with the gradient steepness (Zigmond, 1977, 1982).

a.   b.

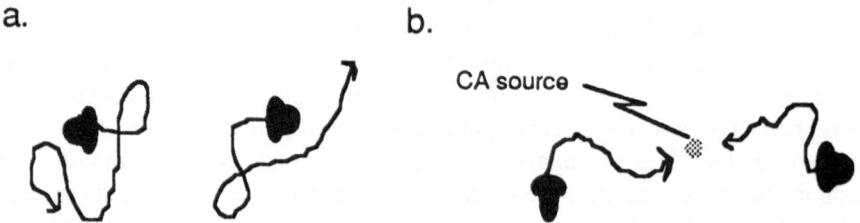

<u>Figure 1</u>. PMN chemosensory movement responses: (a) random motility, (b) chemotaxis.

Microscopic examination of the cells, though, indicates that chemotactic movement is not fundamentally different from random motility (Zigmond *et al.*, 1981). The cells exhibit the same polarized morphology in both cases, with the motile activity predominately at the broad leading front. This suggests that the directed turning may derive mainly from a difference in the the spatiotemporal pattern of receptor signals which activate the motile mechanisms and not from an alteration in the intrinsic motile mechanisms.

2. Theory of Receptor-Sensing and Probabilistic Models

It is appropriate to first consider sources of stochasticity that exist and could contribute to the random directional component of cell movement. These include thermal fluctuations in the CA concentration local to the cell, fluctuations in the number of receptor binding events, and random intracellular responses by the cell to a receptor binding event. Using the formalism of the cell as a sensory system, where here signal refers to the local CA concentration, the stochastic nature of chemosensory movement might arise then from true signal noise, noise in the signal measurement process, and/or a noisy transformation of the measured signal.

Although the possibility of a stochastic transformation of a receptor binding event is entirely possible, assessing its quantitative significance is difficult given incomplete knowledge of the molecular basis of the transformation (i.e., the details of the biochemical transduction pathways and motile force generation events). For the two remaining sources, the following analysis reveals that for PMN, noise associated with fluctuations in receptor binding predominates over noise associated with fluctuations in the local CA concentration. Consider a hypothetical sampling volume (volume V) associated with a monovalent receptor reversibly binding CA molecules present at true uniform concentration C_∞ (*true* refers to the macroscopic concentration associated with averaging over an infinite volume). At any instant, the number of CA molecules the sampling volume contains will deviate from the expected number (VC_∞) due to thermal fluctuations (i.e., molecules diffuse into and out of the sampling volume due to thermal motion). Assuming the sampling volume is spherical with radius R determined by the root mean square distance that a CA molecule will diffuse during the mean time that a receptor is bound (to ensure statistical independence of binding events), then $R \propto (D_{CA}/k_r)^{1/2}$, where D_{CA} is the CA diffusion coefficient in the extracellular medium and k_r is the reverse (dissociation) rate constant for the CA-receptor complex. For rabbit PMN binding N-formyl-norleucyl- leucyl-phenylalanine (FNLLP), a bacterial metabolite that elicits PMN chemotaxis, which serves as the case study in this paper, $D_{CA} = 10^{-5}$ cm^2/s (Stickle et al., 1985), $k_r = 0.4$ min^{-1} and $K_D = 2 \times 10^{-8}$ M (Sullivan and Zigmond, 1980), where K_D is the equilibrium dissociation constant for the CA-receptor complex ($K_D = k_r/k_f$, where k_f is the forward (binding) rate constant). Choosing $C_\infty = K_D$, since most movement responses are optimal at this CA concentration, the relative noise due to thermal fluctuations is $\approx 10^{-6}$ %, defined as $\sigma/C_\infty \cdot 100$ where σ is the standard deviation for the concentration in the sampling volume relative to C_∞, based on the result

$$(1) \qquad \left(\frac{\sigma}{C_\infty}\right)^2 = \frac{1}{C_\infty V N_{Av}}$$

which describes the Poisson counting process in the sampling volume.

Consider now the relative noise due to fluctuations in receptor binding for a population of N_T monovalent receptors reversibly binding CA molecules present at true uniform concentration C_∞ (neglecting thermal fluctuations). Because receptor binding is essentially a reversible bimolecular reaction, it is inherently a stochastic process (Del Grosso and Marchetti, 1983): there will be a distribution of times a receptor remains bound around some mean time and likewise for the time a receptor remains free. The consequence of these times being random variables along with N_T being a finite number for the cell is that the instantaneous fraction of bound receptors, I, will deviate at almost every instant in time from the mean, P, associated with an ensemble average ($N_T \to \infty$). Since C_∞ is constant and binding is presumed to have attained equilibrium, the equilibrium binding relationship for these receptors is simply $P = C_\infty/(K_D + C_\infty)$. If the cell had the information P then it could "measure" C_∞ by the equivalent relationship $C_\infty = K_D \cdot P/(1-P)$. However, since the cell only has the information I, which deviates from P, the cell's "concentration measurement", C, necessarily deviates from C_∞. Choosing $N_T = 12,500$, which is one-half (see below) the measured value of N_T at $C_\infty = K_D$ (Zigmond and Sullivan, 1979), the relative noise due to fluctuations in the fraction of bound receptors is $\approx 2\,\%$ based on the result

$$(2) \qquad \left(\frac{\sigma_C}{C_\infty} \right)^2 = \frac{P}{N_T(1-P)^3}$$

which follows from the stochastic description of receptor binding (see Box 10, this section). This shows that the noise associated with fluctuations in the local CA concentration is insignificant compared to that associated with fluctuations in receptor binding.

It remains to be demonstrated that receptor-signal noise is crucial to understanding the directional characteristics of cell movement in CA gradients, as postulated at the outset. That, in fact, is quite easy to do given the observation that PMN exhibit significant orientation bias in FNLLP gradients corresponding to a relative concentration difference across their dimension of $< 1\,\%$ at $C_\infty = K_D$ (Zigmond, 1977), defined as $\Delta/C_\infty \cdot 100$ where Δ is the true concentration difference and C_∞ here is the CA concentration at the cell midpoint (significant orientation bias means that $\approx 75\,\%$ of the cell population is oriented up-gradient at each instant). However this true concentration difference ($< 1\,\%$) that the cells can sense quite accurately is smaller than the average fluctuation in receptor-measured concentration (2 %) that each of two receptor subpopulations on either side of the cell would perceive ($N_T = 12,500$ for each). Thus, the paradox arises: How can PMN sense these gradients so accurately?

This same paradox, first identified for bacteria and the cellular slime mold by Berg and Purcell (1977), is due to the assumption thus far that the cell bases its concentration measurement on the *instantaneous* fraction of bound receptors. The paradox is resolvable with the "time-averaging hypothesis": the cell has some means by which it can time-average the fraction of bound receptors so as to reduce the noise in the associated receptor-measured concentration. The advantage of time-averaging for measurement accuracy follows from the following expression for σ_{CT} (DeLisi *et al.*, 1982), where the subscript "T" denotes time-averaging, again for the case of a population of N_T monovalent receptors reversibly binding CA molecules present at true uniform concentration C_∞ with binding at equilibrium (see Box 10, this section):

$$(3) \qquad \left(\frac{\sigma_{CT}}{C_\infty} \right)^2 = \frac{2}{N_T P k_r T} \left[1 - \frac{1-P}{k_r T} \left(1 - \exp\left[-\frac{1-P}{k_r T} \right] \right) \right]$$

T is the time-averaging period. k_r is a function of P in the general case where receptor binding is limited both by diffusion of the CA molecule to form an "encounter complex" with the receptor and the binding reaction once in the encounter complex (DeLisi and Wiegel, 1981; Wiegel, 1983). Dependencies on the parameters are more clearly seen for σ_I^2 and σ_{IT}^2, the related variances of the instantaneous and time-averaged fraction of bound receptors relative to P, respectively (taking the limiting case of $T \gg \tau$ for σ_{IT}^2):

(4a) $\quad \sigma_I^2 = \dfrac{P(1-P)}{N_T}$

(4b) $\quad \sigma_{IT}^2 = \dfrac{P(1-P)}{N_T} \dfrac{\tau}{T}\quad$ where $\tau^{-1} = \tau_f^{-1} + \tau_r^{-1} = k_r(C_\infty/K_d + 1)$

τ is a time constant related to the mean times for a free receptor to become bound ($\tau_f = (k_f C)^{-1}$) and a bound receptor to become free ($\tau_r = k_r^{-1}$). In both cases, the mean square deviations decrease with increasing N_T. For the time-averaged case, they also decrease with increasing T and decreasing τ (e.g., greater k_r for fixed relative concentration C_∞/K_D). These dependencies can be thought of as improvement in measurement accuracy resulting from either more sensors (N_T) or from a greater sampling frequency per sensor ($k_r T$) at a fixed relative concentration.

A series of investigations have used the time-averaging hypothesis to distinguish whether a particular cell type senses a gradient with either a "spatial" or a "temporal" mechanism, i.e., whether the direction of movement is determined by a receptor-measured concentration difference associated with two positions on the cell at one point in time, or with two points in time at one position (Zigmond, 1982). As discussed later, these alternative mechanisms are likely inappropriate for understanding the chemosensory movement of PMN and related cell types. However, a brief summary and critique of the approaches used in these investigations is worthwhile to highlight their limitations and to justify the more complicated stochastic model discussed at length subsequently.

The seminal work on this problem is due to Berg and Purcell (1977). Besides proposing the time-averaging hypothesis, they also first derived an expression for σ_{CT} (restricted to diffusion-limited binding, k_r constant, and $T \gg \tau$) and postulated feasibility criteria consistent with spatial and temporal mechanisms. They assume that a signal to noise (S/N) ratio must exceed unity for a mechanism to be feasible. The signal is defined here by the true concentration difference that the cell is attempting to measure, as consistent with the spatial or temporal mechanism. The noise is defined by the root-mean-square deviation associated with the two receptor-measured concentrations, given by $\sqrt{2}\sigma_{CT}$ upon assumption that the measurements are statistically independent. Thus the feasibility criteria can be stated as:

(5) spatial mechanism: $\qquad DdC_\infty/dx > \sqrt{2}\sigma_{CT}$
(6) temporal mechanism: $\qquad vTdC_\infty/dx > \sqrt{2}\sigma_{CT}$

where D is the distance across the cell surface between two hypothetical independently-sensing subpopulations of the receptors and v is the component of cell velocity in the direction of the true CA gradient, dC_∞/dx.

Although Berg and Purcell did not consider PMN, DeLisi et al. (1982) did using their more general expression for σ_{CT}, eq. (3). Using independent estimates for all other parameters in order to predict T, it was concluded that PMN would require days for a spatial mechanism to be feasible but only minutes for a temporal mechanism. In subsequent work, DeLisi and Marchetti (1983) further relaxed the usual assumption of equilibrium binding for the particular case where at some initial time a cell not previously exposed to CA is subject to a CA gradient. This required the introduction of another parameter related to time-averaging for the temporal mechanism case, the time delay t_1 between two successive time-averaging periods, with an optimum t_1 predicted for minimizing T. The original conclusion that a temporal mechanism must be operative based on a S/N ratio criterion (generalized for nonequilibrium binding) was unchanged.

BOX 10

STOCHASTIC ANALYSIS OF RECEPTOR BINDING:
THE DIFFUSION APPROXIMATION OF A POPULATION PROCESS AND
ASSOCIATED STOCHASTIC DIFFERENTIAL EQUATION

Robert T. Tranquillo

A rigorous description of the stochastic receptor binding process was first provided by Del Grosso and Marchetti (1983). They show how a limit theorem for a population birth-death process due to Kurtz (1981) can be applied to describe the global statistical properties of a large but finite receptor population reversibly binding chemoattractant (CA) at uniform concentration C_∞. The limit theorem states that under certain regularity conditions involving the generator associated with the Markov processes describing receptor binding and dissociation (i.e., the birth and death of a bound receptor), the statistical behavior of the population can be described by a continuous process. However, their result only applies for the case of random motility. It was extended to the more general case where the concentration local to the receptor population is variable, as determined by the system dynamics (i.e., cell movement in a CA gradient). A sketch of the application of the theorem is given below. The reader is referred to Tranquillo and Lauffenburger (1987) for a statement of Kurtz' theorem and further details.

The quantity of interest and to be derived is the instantaneous fraction of bound receptors on a cell with a large receptor population (large enough so as to justify use of the theorem). Following Del Grosso and Marchetti, this begins with the definition of a normalized deviation from the mean value, P:

(1) $\qquad \eta(\underline{w}) = \dfrac{1}{N_T^{1/2}} \displaystyle\sum_{i=1}^{N_T} (w_i - P) = z$

where $\underline{w}$ represents the vector defining the binding state of each of the N_T receptors and P is the mean fraction of bound receptors satisfying the familiar kinetic binding equation:

(2) $\qquad \dfrac{dP}{dt} = \lambda_+(P)\,(1-P) - \lambda_-(P)\,P$

The parameters $\lambda_+(P)$ and $\lambda_-(P)$ are the transition intensities associated with the Markov processes of binding and dissociation, respectively, which in general for cell surface receptor binding are dependent on P (DeLisi and Wiegel, 1981). Kurtz' theorem can be applied to derive eq. (2), as Del Grosso and Marchetti show.

Evaluating the generator of the Markov process for receptor binding and dissociation, A_N, and using a Taylor series expansion of $f_N(\underline{w} \pm \underline{e}_i)$ up to second order with eq. (1) yields:

(3) $\quad A_N f_N(\underline{w}) \approx [\lambda_+(P')\,(1-P) + \lambda_-(P')\,P]\left(\dfrac{1}{2}\dfrac{\partial^2 f}{\partial z^2}\right) + [\lambda_+(P') + \lambda_-(P')]\left(-z\dfrac{\partial f}{\partial z}\right)$

$\qquad\qquad\qquad\qquad\qquad\qquad\qquad$ (i) $\qquad\qquad\qquad\qquad\qquad\qquad\qquad$ (ii)

$\qquad\quad + [\lambda_+(P')\,(1-P) - \lambda_-(P')\,P]\left(N^{1/2}\dfrac{\partial f}{\partial z}\right) + [\lambda_+(P') - \lambda_-(P')]\left(\dfrac{-z}{2N^{1/2}}\dfrac{\partial^2 f}{\partial z^2}\right)$

$\qquad\qquad\qquad\qquad\qquad\qquad\qquad$ (iii) $\qquad\qquad\qquad\qquad\qquad\qquad\qquad$ (iv) where $f_N \in D(A_N)$, $f \in D(A)$, and $P' = P + \dfrac{z}{N^{1/2}}$ BOX 10

A is the infinitesimal generator of the limiting process in the limit $N \rightarrow \infty$. The following remarks apply to eq. (3), where $N \equiv N_T$:

(i)   in the limit $N_T \rightarrow \infty$, $\lambda_+(P') \rightarrow \lambda_+(P)$ and $\lambda_-(P') \rightarrow \lambda_-(P)$
(ii) in the limit $N_T \rightarrow \infty$, term (iv) vanishes
(iii) for the case where the CA concentration is constant and binding equilibrium conditions exist (dP/dt = 0), the coefficient of term (iii) is zero and the term vanishes

The *continuous* process $x(t)$ whose generator A is defined by:

$$(4) \qquad A f(x(t)) = a(x(t),t) \frac{\partial f}{\partial x} + \frac{1}{2} \sigma(x(t) \ t)^2 \frac{\partial^2 f}{\partial x^2}$$

is known to satisfy the Ito stochastic differential equation (SDE)

$$(5) \qquad dx(t) = a(x(t),t)dt + \sigma(x(t),t)dW(t)$$

where $W(t)$ denotes the Wiener process. A comparison of eqs. (3)-(5) in the limit $N_T \rightarrow \infty$ yields the following Ito SDE by Kurtz' theorem (after algebraic manipulation), which defines the instantaneous fraction of bound receptors, I, for a large receptor population of total N_T:

$$(6) \qquad dI(t) = [(\lambda_+(P) \ (1-P) - \lambda_-(P) \ P) - (\lambda_+(P) + \lambda_-(P)) \ (I(t)-P)] \ dt$$

$$+ \frac{1}{N_T^{1/2}} \ [\lambda_+(P) \ (1-P) + \lambda_-(P) \ P]^{1/2} dW(t)$$

Substituting for the transition intensities $\lambda_+(P) = k_f C$ and $\lambda_-(P) = k_r$ in eq. (6), which assumes reaction-limited binding, yields eqs. (10) of §3. The reader is referred to van Kampen (1981) and Gardiner (1985) for alternative diffusion approximation methods and to Kurtz (1986) for the complete theoretical foundation of this method.

   The condition that C and P are constant in remark (iii) is a simplification applying for random motility, and when invoked the result of Del Grosso and Marchetti is recovered from eq. (6) (the authors also demonstrate the remaining conditions of the theorem that need to be satisfied). The simplified SDE is recognized to be an Ornstein-Uhlenbeck process for the variable $y(t) = I(t) - P$:

$$(7) \qquad dy(t) = -ay(t)dt + \sigma dW(t), \text{ where } a = k_f C + k_r, \ \sigma = \frac{1}{N_T^{1/2}}[k_f C(1-P) + k_r P]^{1/2}$$

The Ornstein-Uhlenbeck process is related to the celebrated Langevin equation, which in its simplest form is given by (Gardiner, 1985):

$$(8) \qquad \frac{d \ y(t)}{d \ t} = -ay(t) + \xi(t)$$

Langevin proposed this equation as a simpler treatment of Brownian motion over Einstein's. In this context, the equation represents a force balance on a particle in a viscous fluid, where $y(t)$ is a component of the particle velocity. The first term on the right, generally referred to as the drift term, is associated with a viscous drag force on the moving particle and the second term on the right, generally referred to as the diffusion term, represents a rapidly fluctuating force on the particle due to incessant randomly-directed collisions of the particle with fluid molecules. The drift term is associated with the deterministic motion of the particle (i.e., in the absence of the collisions) with time constant $\tau = a^{-1}$ and the diffusion term with the stochastic motion. A common mathematical idealization of the rapidly varying, highly irregular function $\xi(t)$ is that for $t \neq t'$, $\xi(t)$ and $\xi(t')$ are statistically independent (or "delta-correlated") with zero correlation time,

$$(9) \qquad <\xi(t)\xi(t')> = \delta(t-t')$$

and $\xi(t)$ has zero mean. In this case $\xi(t)$ is referred to as "white noise" and it can be formally shown that $\xi(t)dt = dW(t)$, thus relating the Ornstein-Uhlenbeck process to the Langevin equation.

The solution to the SDE defined by eq. (7) is (Gardiner, 1985):

$$(10) \qquad y(t) = e^{-at}y(0) + \sigma \int_0^t e^{-a(t-s)}dW(s)$$

where stochastic integration is in the sense of Ito. Using the Ito calculus, the autocorrelation function of $y(t)$ is shown to be:

$$(11) \qquad < y(t)y(s) > \; = e^{-a(t-s)}< y(0)^2 > + \sigma^2 \int_0^{\min(t,s)} e^{-a(t+s-2t')}\,dt'$$

$$= \left[var(y(0)) - \frac{\sigma^2}{2a} \right] e^{-a(t+s)} + \frac{\sigma^2}{2a}e^{-a|t-s|}$$

These results can be used to derive eqs. (2)-(4) in §2 which apply for uniform CA concentrations, as follows. Upon assumption that the process is stationary (i.e., t, s $\rightarrow$ ∞), eq.(11) directly gives:

$$(12) \qquad \sigma_I^2 = < y(t)^2 > \; = \frac{\sigma^2}{2a} = \frac{P(1-P)}{N_T}$$

The result for σ_{IT}^2 follows from calculation of the defining integral using eq. (11):

$$(13) \qquad \sigma_{IT}^2 = \frac{1}{T^2}\int_0^T dt \int_0^T < y(t)y(s) > ds = \frac{P(1-P)}{N_T} \frac{2\tau}{T}\left[1 - \frac{\tau}{T}(1-e^{-\frac{T}{\tau}}) \right]$$

These results can then be used to directly derive the associated expressions for σ_C^2 and σ_{CT}^2 using the following relationship for a sufficiently smooth nonlinear function $g(x)$ of a random variable X (Papoulis, 1965):

$$(14) \qquad \sigma_{g(x)}^2 \approx \left(\frac{dg}{dx}\Big|_{<x>} \right)^2 \sigma_x^2$$

Making the correspondences: $X = I$ (or I_T) and $g(X) = C = K_D \cdot I/(1-I)$ (or $C_T = K_D \cdot I_T/(1-I_T)$), σ_C^2 (or σ_{CT}^2) follows.

Lauffenburger (1982) postulated a different feasibility criterion consistent with a spatial mechanism: a signal to threshold (S/T) ratio must exceed unity for a cell to exhibit orientation bias in a gradient. The signal is defined here by the (fluctuating) difference in the number of bound receptors across the cell dimension. The threshold is some minimum difference that elicits orientation bias. Probabilities were computed for all possible signals for a specified gradient with independent estimates for all other parameters besides an assumed value of T. Comparing each signal to an assumed threshold value, the total probability for orientation bias up-gradient versus down-gradient was computed (the model cell has only two orientation states), where random orientation was assumed whenever the signal was sub-threshold. For sufficiently small thresholds of order ten bound receptors, it was found that T of order minutes yielded predictions of orientation bias consistent with PMN data. Perhaps more significantly, the validity of the S/N criterion was challenged, as the model predicted perfect orientation in some cases where S/N < 1 and did not predict perfect orientation in all cases where S/N > 1 (perfect orientation $\equiv$ Pr[orientation up-gradient] $\approx$ 1) This probabilistic model was extended to account for more complicated receptor characteristics, including down- regulation, asymmetric distribution, and multiple affinity subpopulations (Tranquillo and Lauffenburger, 1986). Also, the nature of the signal used in the S/T criterion was changed to a difference in the *fraction* of bound receptors, consistent with a scheme for sensory adaptation (see Discussion).

Several significant limitations in these models can be identified: (1) They all implicitly assume unrealistic cell movement in order to perform the calculations: the spatial mechanism criterion only rigorously holds if during the time-averaging period the cell is stationary; for the temporal mechanism criterion, the cell must move with constant v, i.e., in the same direction. Consequently, there is no account for changes in sensing associated with turning away from (or towards) the direction of the gradient, which seems intrinsic to the directional response, (2) Increasing understanding of the cell biology of PMN chemosensory movement supports a more complicated "pseudospatial" mechanism (Zigmond, 1982; Lackie 1986), where sensing results from the "temporal signal" generated within an extending cell process (i.e., a lamellipodium). This raises the question of global spatial regulation of signals and coordination of lamellipodial activity, (3) None of the models address random motility, which seems to be inextricably related to chemotaxis and for which considerable data exists, and (4) Although estimates for T are interesting for comparison to cell movement time scales, the time-averaging period *per se* offers no mechanistic insight.

Despite this last criticism, that the concept of a time-averaging period is somewhat artificial, the concept of time-averaging definitely is not: a receptor signal has a finite lifetime due to its propagation along the transduction and response pathways, and this propagation lifetime is effectively a time-averaging period. Thus, a new model for understanding the consequences of receptor-signal noise was proposed, one which confers the cell with intracellular mechanisms through which receptor signals are generated and from which the directional change of cell movement is determined (Tranquillo and Lauffenburger, 1987). There is no explicit time-averaging period upon which the earlier models are based, the parameters associated with a particular set of mechanisms determining implicitly the extent of time-averaging. Indeed, the functional form of the mechanisms and their interrelationships determine the characteristics of the implicit time-averaging, such as parametric sensitivity. In addition, the new model cell is capable of autonomous movement in two-dimensional space given a specified CA concentration field (uniform or gradient), the turning behavior being determined by a simple pseudospatial mechanism. As seen from these features, the limitations of the earlier models are rectified in the new stochastic model while maintaining the focus on the certain occurrence of statistical fluctuations in receptor binding.

3. Stochastic Model Description

The model to be analyzed is based on the idealized cell illustrated in Fig. 2. The biological justification for its general features has been presented elsewhere (Tranquillo and Lauffenburger, 1987; Tranquillo *et al.*, 1988) and only the assumptions underlying the model equations are presented here. The cell remains morphologically and behaviorally polarized so that directional change occurs via a continuous "smooth turn" governed by a stable leading front. Since the turning behavior is of primary interest, it is assumed that translocation is uncoupled from turning and is described with a constant speed. A pseudospatial-type mechanism which governs turning is accomplished by modeling the leading front as two interacting compartments. Assuming a receptor signal is the critical regulator of the motility system, the turning rate of the cell can be related to an imbalance of the receptor signal between the two compartments. The generation of the receptor signal in each compartment is related to the stochastic receptor binding process on the cell surface associated with each compartment. In the case of movement in a CA gradient, a coupling then arises between the movement response (translocation with turning) and the receptor signal since receptor binding depends on CA concentration, which depends on the cell position and orientation in the CA gradient. The model can be formalized as a stochastic system with a deterministic transformation (a) of and response (b) to a stochastic signal (c) which is feedback coupled to the response (d), where a = transduction of receptor binding to a receptor signal, b = cell translocation with turning, c = receptor binding process, d = movement in the CA gradient. For random motility conditions (uniform CA concentration), the stochastic signal is not coupled to the response, which affords considerable simplification as will become evident.

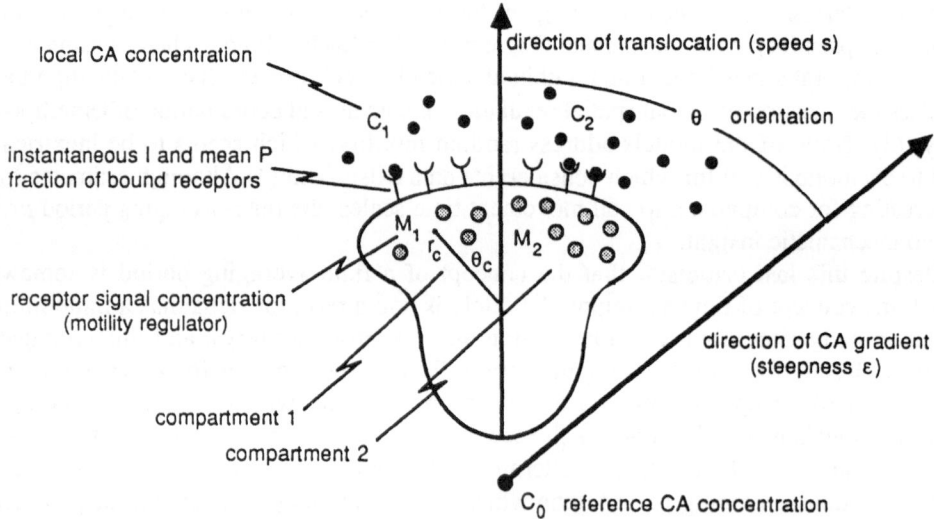

<u>Figure 2</u>.   Model chemosensory cell.   See text for description.

The model is not entirely consistent with the pseudospatial mechanism as there is no account for lamellipodium extension, but there is an account for a temporal signal associated with each compartment, both of which are moving through space. Moreover, the issue of global regulation and coordination is addressed, albeit simplistically, by the dual compartment structure. However, the model provides a single unifying mechanism to qualitatively explain both random motility and chemotaxis as seen in Fig. 3, consistent with their inextricable relationship. In the case of random motility, each receptor subpopulation is constantly subject to the same CA concentration and perceives statistically equivalent fluctuations. Thus, the cell perceives fluctuating gradients without a reference direction. In the case of chemotaxis, the cell is subject to a CA gradient with each subpopulation perceiving, in general, statistically different fluctuations from the true local concentrations. At any instant, the cell perceives some deviation from the true gradient and may even transiently perceive a gradient in the reverse direction. Note that for greater (lesser) receptor noise as represented by larger (smaller) error bars, directional persistence in random motility and orientation bias in chemotaxis should both decrease (increase).

a.   b.

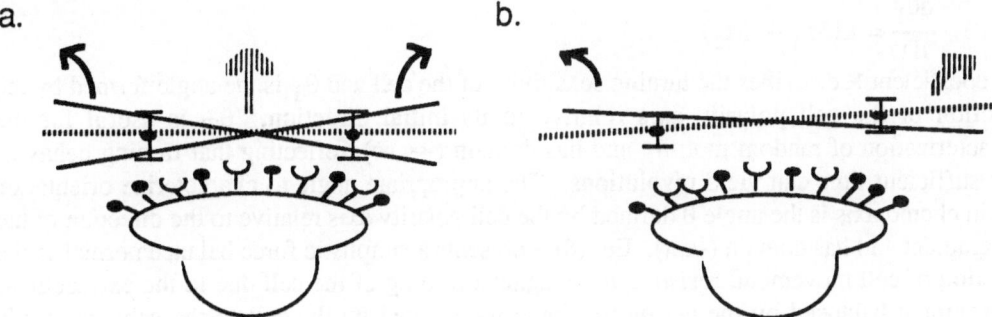

Figure 3. Consequences of receptor noise on (a) random motility and (b) chemotaxis. The striped lines represent the true CA gradients and the solid lines represent possible transiently *perceived* gradients. Both receptor subpopulations *perceive* fluctuating CA concentrations, represented by the error bars, around the true local concentration, indicated by the closed circles in their receptor measurement of concentration. Striped arrows represent the expected directional responses in the absence of receptor noise and solid arrows the possible responses in the presence of noise.

The specific mechanisms of receptor binding, signal transduction, and turning response to be analyzed for the stochastic pseudospatial model based on Fig. 2 are now considered. Although many details of these processes have recently been elucidated, mechanisms representing a high level of abstraction were chosen since the available data at the time this model was formulated were largely preliminary and incomplete (see Discussion). However, this facilitates elucidating the fundamental characteristics of a stochastic dynamical system with signal-response structure consistent with the stochastic pseudospatial model. The following mechanisms represent a set which incorporates the minimal complexity and number of components.

Receptor signal

A single intracellular messenger (i.e., the receptor signal), considered to be the critical regulator of the motility system, is defined with concentration M and its dynamics in the two compartments are modeled by the following conservation equations:

$$(7) \qquad \frac{dM_1}{dt} = k_t R_1 - k_d M_1 - D(M_1 - M_2) \qquad \frac{dM_2}{dt} = k_t R_2 - k_d M_2 - D(M_2 - M_1)$$

where the numeric subscript designates the compartment. In each compartment M is generated at a rate proportional to the magnitude of some as yet undefined receptor signal R with first-order transduction rate constant k_t, decays according to first-order kinetics with decay rate constant k_d, and is transported between compartments at a rate proportional to the difference in M with diffusive rate constant D.

Cell turning

The turning rate is assumed proportional to the difference in receptor signals between the compartments and given by:

$$(8) \qquad \frac{d\theta_T}{dt} = \kappa(M_1 - M_2)$$

The coefficient κ describes the turning sensitivity of the cell and θ_T is the angle formed by the direction of the cell polarity axis relative to its initial direction. θ_T is useful for the characterization of random motility and has domain $(-\infty,\infty)$, reflecting that turning behavior over sufficient time can yield revolutions. The appropriate angle to characterize orientation bias in chemotaxis is the angle θ defined by the cell polarity axis relative to the direction of the CA gradient and has domain $(-\pi,\pi)$. Eq. (8) represents a simplistic force balance normal to the direction of cell movement: resistive force against turning of the cell due to the extracellular environment balanced by the net motile force transmitted by the cell to the substratum via adhesive bonds with the leading front. It is implicitly assumed that the rate limiting step in activation of the motility system is the generation of M and that the motility components and adhesion sites are are uniformly distributed between the compartments.

Eqs. (7,8) comprise the deterministic signal transformation and response phases of the stochastic model. Given the complex nonlinearities undoubtedly associated with a true structured model, the linearity of the equations implies the model may only apply to pseudo-steady state cell behavior, i.e., responses to relatively small receptor noise and small CA gradients. However, when analyzed in conjunction with the receptor binding equations consistent with random motility, the resultant linear system will be seen to yield a valuable analytical result. In this case, only the small receptor noise requirement must be met to justify the linear description.

CA concentration

The following equations govern the CA concentration, C, local to the receptor population of each compartment, accounting for translocation and turning of the cell in the gradient:

$$(9) \qquad \frac{dC_1}{dt} = \left[s\cdot\cos\theta - r_c\frac{d\theta}{dt}\sin(\theta+\theta_c) \right]\varepsilon C_0 \qquad \frac{dC_2}{dt} = \left[s\cdot\cos\theta - r_c\frac{d\theta}{dt}\sin(\theta-\theta_c) \right]\varepsilon C_0$$

where r_c and $\pm\theta_c$ define the characteristic positions for receptor binding for the two compartments (Fig. 2). ε is the fractional CA concentration difference across one-half the effective cell diameter, $\Delta C/C_0$, defining the gradient steepness.

Receptor binding

The following stochastic differential equations (SDEs) describe the evolution of the instantaneous fraction of bound receptors, I, for a constant population of number N_T

monovalent receptors on each compartment reversibly binding CA molecules present at true concentration C (see Box 10, this section):

$$d\,\mathbf{I_1} = [(k_fC_1(1-\mathbf{P_1}) - k_r\mathbf{P_1}) - (k_fC_1 + k_r)(\mathbf{I_1} - \mathbf{P_1})]dt + \frac{1}{N_T^{1/2}}[k_fC_1(1-\mathbf{P_1}) + k_r\mathbf{P_1}]^{1/2}d\mathbf{W_1}$$

(10)

$$d\,\mathbf{I_2} = [(k_fC_2(1-\mathbf{P_2}) - k_r\mathbf{P_2}) - (k_fC_2 + k_r)(\mathbf{I_2} - \mathbf{P_2})]dt + \frac{1}{N_T^{1/2}}[k_fC_2(1-\mathbf{P_2}) + k_r\mathbf{P_2}]^{1/2}d\mathbf{W_2}$$

$\mathbf{I}$ is formally a stochastic process, as are all variables defined in the model formulation for chemotaxis conditions (stochastic processes are denoted by bold type). $\mathbf{P}$ is the associated mean fraction of bound receptors. $\mathbf{W}$ is the normal Wiener process and the integrations implied by eqs. (10) are in the sense of Ito (Gihman and Skorohod, 1969). The equations which govern $\mathbf{P}$ are given by:

$$(11) \qquad \frac{d\mathbf{P_1}}{dt} = k_fC_1(1-\mathbf{P_1}) - k_r\mathbf{P_1} \qquad \frac{d\mathbf{P_2}}{dt} = k_fC_2(1-\mathbf{P_2}) - k_r\mathbf{P_2}$$

For PMN binding FNLLP, it is believed that the global binding reaction is limited by intrinsic reaction of the encounter complex (Lauffenburger, 1982). Therefore, k_f and k_r are true constants (independent of $\mathbf{P}$), which is already incorporated into eqs. (10,11).

The receptor event $\mathbf{R}$ of eqs. (7) involved in receptor-signal transduction in random motility and chemotaxis needs to be specified. One use of this model is to determine what receptor event(s) yield predictions consistent with the available data. The case where the event is a bound receptor is considered in detail here. This is the simplest choice and historically has been the most common assumption in the interpretation of the data. Making the substitution $\mathbf{R} = \mathbf{I} \cdot N_T$ into eqs. (7) completely specifies the binding dynamics of each receptor population in conjunction with eqs. (8-11) for the case of chemotaxis. The case where $\mathbf{R} = \mathbf{I}$, consistent with a simple scheme of sensory adaptation, has also been analyzed in parallel. The motivation and results for that case are deferred to the Discussion.

It is instructive to illustrate the model behavior at this point with some sample paths, i.e., simulated paths of cell movement in the x-y plane for both random motility and chemotaxis obtained by numerical integration of the governing SDE system (see Box 11, this section). Simulations involve generating cell paths in the x-y plane for a given set of cell parameter values and specified gradient. Each path is an approximation to one realization of the SDE system corresponding to one realization of random initial orientation at the origin and one realization of the driving Wiener process for each of the two compartments. The cell is assumed to have been subject to a constant uniform concentration, C_0, prior to the instantaneous establishment of the gradient, defined by the gradient steepness, ε, and the absolute concentration at the origin, C_0. The expected values for $\mathbf{P}, \mathbf{I}$, and $\mathbf{M}$ determined by C_0 ($= K_D$) are assumed for both compartments initially, i.e., deterministic initial conditions except for the initial orientation with respect to the gradient, $\theta(0)$.

Each simulated path of random motility in Fig. 4a represents the migration response of the model cell to one of the infinitely many possible realizations of receptor binding fluctuations occurring in the receptor population of each compartment. The characteristics of the persistent random walk observed for PMN behavior are evident. The influence of a moderate gradient on each of these paths indicates a directed turning response in the direction of the gradient. These chemotactic responses correspond to the identical driving noise and same cell parameter values used for the random motility simulation. The gradient is of magnitude

a.

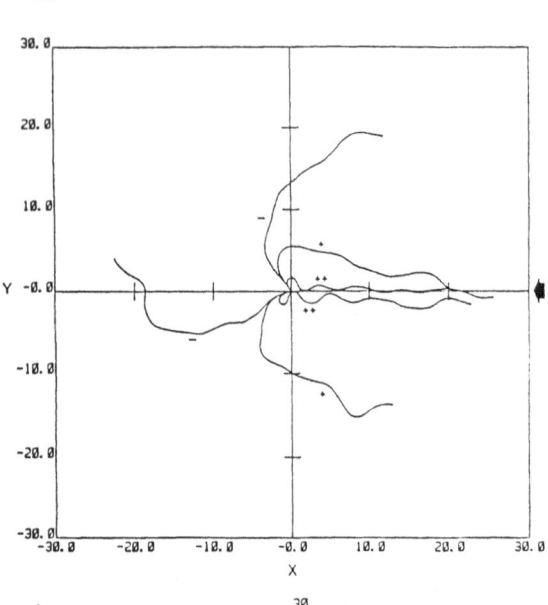

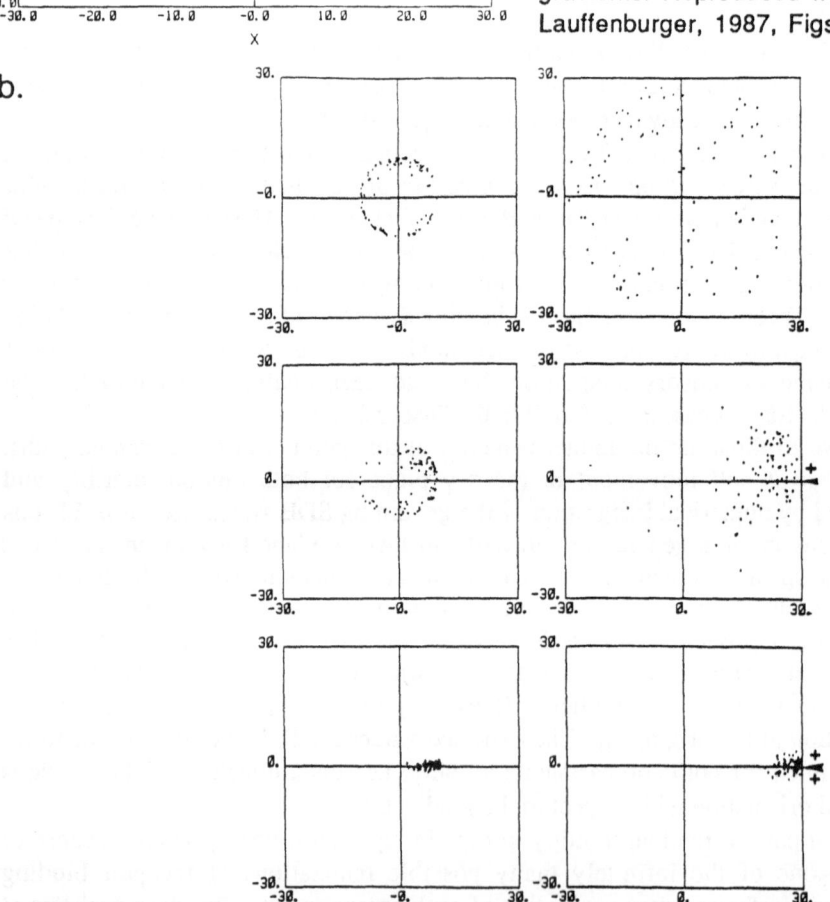

Figure 4. Stochastic model simulations of random motility and chemotaxis. (a) Cell paths: two sets of results are indicated. In each set, the cell has the same initial orientation at the origin. For each set, one path is obtained for the cases of random motility (-) and chemotaxis in moderate (+) and steep (++) gradients. (b) Scatter diagrams: coordinates of the paths of 100 cells at a relatively early time (left column) and later time (right column) in the simulations are plotted. As in (a), the individual cells have the same initial orientation at the origin and experience the same driving noise for the cases of random motility and chemotaxis in the moderate and steep gradients. Reproduced from Tranquillo and Lauffenburger, 1987, Figs. 5 & 6.

b.

typically established in the visual assay system of Zigmond ($\varepsilon = 0.008$) from which orientation bias data are obtained (Zigmond, 1977). Also included is the influence of a ten times steeper gradient on each of the random motility paths. A faster turning response with initial oscillatory behavior occurs. Greater bias along the gradient later in the simulation is also evident. The degree of oscillatory behavior for this case may be related to the dual compartmental structure of the model; however, a gradient of this magnitude has yet to be established experimentally, so the nature of the response of PMN to such steep gradients is unknown. Using a semi-empirical characterization of the chemotactic response based on conditional turn angle probabilities (related to orientation bias), oscillatory behavior of the cell path in steep gradients has been predicted (Nossal and Zigmond, 1976). The scatter diagrams of Fig. 4b give a different perspective of the model behavior. In the random motility case, directional persistence is obvious at the early time and the dispersion characteristic of an unbiased random walk is developing at the later time. For the moderate gradient simulation, the response is consistent with the biased random walk underlying chemotaxis. In the steep gradient simulation, the strongly biased random walk yields a chemotactic wave of migrating cells. These results suggest that the model behavior is at least qualitatively consistent with the true cell behavior for chemotactic cells like PMN and motivates the following quantitative analyses.

BOX 11

NUMERICAL INTEGRATION OF SDEs

Robert T. Tranquillo

Simulation of random motility and chemotaxis involves numerical integration of the respective SDE systems. A key distinction to be made in choosing a numerical integration algorithm is the type of stochastic integration associated with the SDEs. Ito SDEs necessitate algorithms in which the Wiener increment dW is statistically independent of the current solution. Thus, stochastic versions of explicit methods used in numerical integration of deterministic ordinary differential equations are consistent with Ito integration whereas implicit methods are not. The algorithm employed here is the stochastic version of the explicit Euler method known as the Cauchy-Euler method (Gardiner, 1985; Gard, 1988). It is known that this yields a numerically realizable mean square approximation to an Ito stochastic differential system with a first-order global rate of convergence:

$$(1) \quad < ||\underline{x}(T) - \underline{x}^*(T)||^2 > = O(h)$$

where $\underline{x}(t)$ is the vector of stochastic processes in the general Ito system defined by eq. (2) and $\underline{x}^*(t)$ denotes the mean square approximation defined by eq. (3).

$$(2) \qquad d\underline{x}(t) = \underline{a}(\underline{x}(t), t)\, dt + \sum_{r=1}^{M} \underline{\sigma}_r(\underline{x}(t), t) dW_r$$

The recursion formula for the approximation follows from dividing the integration interval $[0,T]$ into K equal partitions, defining $h = T/K$ and $t_{k+1} = t_k + h$, $k=0,1,...,K-1$:

$$(3) \qquad \underline{x}^*(t_{k+1}) = \underline{x}^*(t_k) + \underline{a}(\underline{x}^*(t_k), t_k)h + \sum_{r=1}^{M} \underline{\sigma}_r(\underline{x}^*(t_k), t_k) N_r(0,h), \quad \underline{x}^*(t_0) = \underline{x}_0$$

where $\underline{x}_0$ denotes a given initial value and $N(0,h) = \sqrt{h} \cdot N(0,1)$ denotes a normally distributed random variable with zero mean and variance h. A schematic representation of this algorithm for a one-dimensional SDE is shown in Fig. 1

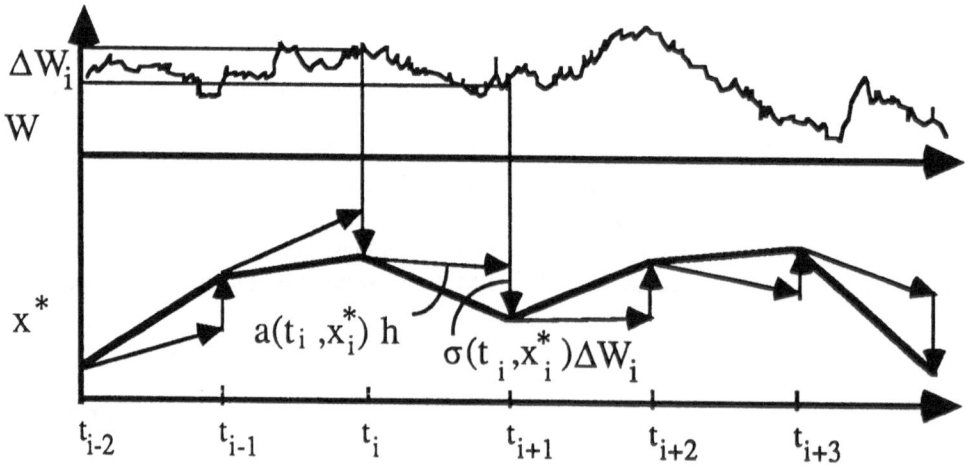

Figure 1. Numerical integration of a one-dimensional SDE with the Cauchy-Euler method. The upper panel is one realization of the driving Wiener process approximated by a series of normal random deviates, ΔW. The thick line of the lower panel shows the evolution of the corresponding sample path.

Mil'shtein (1974) proposed a method for systematically increasing the rate of convergence of the approximating process for one-dimensional (M=1) Ito systems. This method cannot be directly extended to multi-dimensional (M>1) systems. However, when the following commutativity condition is satisfied:

(4) $(J\underline{\sigma}_r)\underline{\sigma}_s = (J\underline{\sigma}_s)\underline{\sigma}_r$

where $J\underline{\sigma}$ denotes the Jacobian matrix associated with $\underline{\sigma}$, then $O(h^2)$ convergence can be obtained for multi-dimensional systems with the following formula (Pardoux and Talay, 1985):

(5) $$\underline{x}^*(t_{k+1}) = \underline{x}^*(t_k) + \left(\underline{a} - \frac{1}{2}\sum_{r=1}^{M}(J\underline{\sigma}_r)\underline{\sigma}_r \right)h$$

$$+ \sum_{r=1}^{M}\underline{\sigma}_r N_r(0,h) + \frac{1}{2}\sum_{r=1}^{M}\sum_{s=1}^{M}(J\underline{\sigma}_r)\underline{\sigma}_s N_r(0,h)N_s(0,h)$$

where $\underline{a}$ and the $\{\underline{\sigma}_r\}$ are evaluated at $(\underline{x}^*(t_k), t_k)$. This commutativity condition is a statement that the Lie algebra generated from the columns $\{\underline{\sigma}_r\}$ of the diffusion matrix of eq. (3) is abelian. In this case the stochastic differential system can be decomposed into

BOX   11

two systems, one containing all of the driving Wiener processes which are now uncoupled (Krenner and Lobry, 1981). For the diffusion matrix defined for the SDE system applicable to chemotaxis, the commutativity condition is satisfied. In fact, all of the terms $(J\underline{\sigma}_r)\underline{\sigma}_S$ are identically zero so that at least second order convergence is obtained directly from the Cauchy-Euler algorithm. The coefficients for higher rates of convergence have yet to be determined for the general multi-dimensional Ito system; however, additional restrictions apply on the drift and diffusion coefficients beginning with third-order rate of convergence for the one-dimensional case in the sense of numerically realizable accuracy (Rumelin, 1982).

The method employed for assessment of accuracy in the numerical integrations is that described by Wright (1974). Briefly, this requires first generating a finest approximation to a sample path by numerical integration using some smallest partition, h^*. Then the partition is successively doubled with the Wiener increments defined by the series of $N(0,h^*)$ pseudo-random deviates of the finest approximation being appropriately summed to generate successively less accurate approximations to the same sample path, as illustrated in Fig. 2. This procedure ensures that accuracy is being assessed for the same realization of noise (i.e., the same realization of the driving Wiener process).

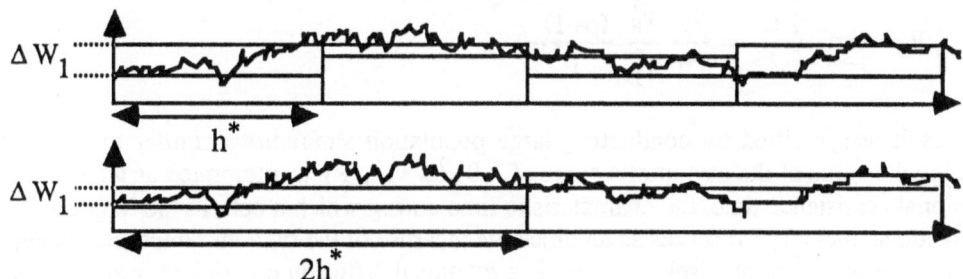

Figure 2. Accuracy assessment in numerical integration of a scalar SDE.
The upper panel illustrates a finest approximation to a particular realization of W(t) which yields a most accurate solution, x^*. The lower panel illustrates a coarser approximation to the same realization by summing Wiener increments pairwise (doubling the effective sampling time interval of W(t)) which yields a solution less accurate than x^*.

4. Stochastic Model Analysis and Results

A. Random Motility

Discussion of random motility implies the cell has been subject to a uniform CA concentration, C, for effectively an infinite time period, so that conditions of binding equilibrium have been established. Then the mean fraction of bound receptors, **P**, is constant and given by $P = C/(K_D+C)$. The Ito SDE system affords a considerable simplification in this case. The first term of the drift coefficient of eqs. (10) vanishes as a result of the assumed binding equilibrium, and the resulting drift and diffusion coefficients are also constants:

$$d\,I_1 = -(k_fC + k_r)(I_1 - P)dt + \frac{1}{N_T^{1/2}}[k_fC(1-P) + k_rP]^{1/2}dW_1$$

(10')

$$d\,I_2 = -(k_fC + k_r)(I_2 - P)dt + \frac{1}{N_T^{1/2}}[k_fC(1-P) + k_rP]^{1/2}dW_2$$

The simplified Ito SDE system for random motility described by eqs. (7-9,10') is linear and constant coefficient. The system can be written in vector-matrix form

(12) $d\underline{x}(t) = A\underline{x}(t)dt + \sigma d\underline{W}(t)$

and has analytical solutions (Gardiner, 1985). A zero eigenvalue is associated with **A**, which has the simple interpretation that any steady-state mean direction, $< \theta_T >_s$, saisifies the algebraic system for $< \underline{x} >_s$ because $< I_1 >_s = < I_2 >_s$ $(= P)$ and $< M_1 >_s = < M_2 >_s$. So the steady-state mean directional response for random motility is straight line movement along the initial direction of movement. This is an expected result because of the symmetry of the intracellular mechanisms and the identical stationary receptor signal processes for both compartments in this case.

For the linear Ito matrix equation defined by eq. (12), it is possible, in principle, to solve analytically for any desired moment. Since characterizing the directional randomness of movement is of primary interest, the second moment of the orientation process $< \theta_T^2 >$ was solved for explicitly (see Box 12, this section). The main result of this analysis is given by:

(13) $$P_T = \lim_{t \to \infty} \frac{2\,t}{<\theta_T^2>} = \frac{f_s}{N_T} \frac{\tau_R^4}{\tau_D^2} \frac{(\rho+1)^3}{\rho}$$

This result was verified by conducting large population simulations similar to Fig. 4 and checking the value of the asymptotic slope of $< \theta_T^2 >$ vs t. P_T is interpreted as an asymptotic directional persistence time, the characteristic time during which a cell, in the long run, loses its directional memory. It serves as an objective measure of the random directional change in random motility. More precisely, $1/(2P_T)$ is a rotational diffusion coefficient, where P_T is the final exponential decay rate of the autocorrelation function for the directional process, $(\cos\theta, \sin\theta)$ (Alt, this volume, section II/1). f_s is considered to be the sampling frequency for a receptor and given simply by $f_s = k_r$, and N_T is the total number of receptors associated with each compartment as before. ρ is the dimensionless uniform concentration, $\rho = C/K_D$. However, the model only applies for $\rho \approx 1$ which is the CA concentration eliciting stable cell polarity (Shields and Haston, 1985), a model assumption (see Fig. 2). τ_R and τ_D are interpreted as time constants for turning and turning signal decay, respectively, and are defined in terms of the four parameters associated with the intracellular mechanisms:

(14) $\tau_R = \dfrac{1}{(\kappa k_t)^{1/2}}$ $\tau_D = \dfrac{1}{2D + k_d}$

Considering $\Delta M = M_1 - M_2$ as a turning signal, the random turning behavior is a consequence of ΔM fluctuating about zero due to I_1 and I_2 fluctuating about P. The result for P_T indicates how, for example, the model cell can be conferred greater directional persistence by appropriately altering f_s, N_T, τ_R, τ_D: the larger τ_R, the slower the cell responds to a receptor binding fluctuation; the smaller τ_D, the faster the cell eliminates a turning signal created by a receptor binding fluctuation. The dependence of P_T on f_s might be anticipated from eq. (4b). As f_s increases, the magnitude of the receptor binding fluctuations decreases. The dependence of P_T on N_T is seemingly counterintuitive given that eq. (4b) also states that as N_T increases, the magnitude of the receptor binding fluctuations decreases. However, the rate of receptor signal transduction has been assumed proportional to the total number of bound receptors, $N_T \cdot I$, not to just the fractional number of bound receptors, I (see Discussion). Even though the receptor binding fluctuations decrease with increasing N_T, the transduction process is amplifying them in proportion to N_T. The net effect of this trade-off from increasing N_T is a decrease in P_T. Recall that the analysis is based on eqs. (10') which are valid in the large N_T limit; thus, the conclusion that P_T increases with decreasing N_T may be invalid when N_T becomes sufficiently small. Finally, note that P_T is independent of cell speed, s, consistent with the model assumption that turning behavior is uncoupled from translocation.

BOX 12

DETERMINATION OF MOMENTS FOR LINEAR SDE SYSTEMS

Robert T. Tranquillo

For the general Ito system described by:

(1) $d\underline{x}(t) = \underline{a}(\underline{x}(t),t)dt + \sigma(\underline{x}(t),t)d\underline{W}(t)$

the "moment equation" which describes the time evolution of any desired moment $< h >$, where $h = x_1^{k_1} \cdots x_n^{k_n}$, is given by (Soong, 1973):

(2) $$\frac{d <h(t)>}{dt} = \sum_{j=1}^{n} <a_j \frac{\partial h}{\partial x_j}> + \frac{1}{2} \sum_{i,j=1}^{n} <(\sigma\sigma^T)_{ij} \frac{\partial^2 h}{\partial x_i x_j}> + <\frac{\partial h}{\partial t}>$$

This was used to generate a vector differential equation describing the time evolution of the vector $\underline{u}(t)$ of all the second moments associated with the SDE system simplified for random motility:

(3) $\frac{d\underline{u}}{dt} = B\underline{u} + \underline{b}$

which has a solution in terms of the state transition matrix $\Phi(t,t_0) = \exp[B(t-t_0)]$:

(4) $$\underline{y}(t) = \Phi(t,t_0)\underline{y}_0 + \int_{t_0}^{t} \Phi(t,s)\underline{b}ds$$

The problem was solved analytically in symbolic form using MACSYMA[1] by finding the set of eigenvalues and eigenvectors $(\lambda,\underline{v})$ associated with B, then applying the similarity transformation $\Phi(t,t_0) = V\exp[\Lambda(t-t_0)]V^{-1}$, where $\Lambda = \text{diag}[\lambda_1,...\lambda_n]$ and $V = [\underline{v}_1,...\underline{v}_n]$, and performing the integration.

[1]EUNICE MACSYMA Release 305, Massachusetts Institute of Technology, 1983.

B. Chemotaxis

Having the characterization of the random turning behavior in a uniform CA concentration in terms of P_T, the biased turning behavior in a CA gradient underlying chemotaxis is examined in this section. The goal is not only to demonstrate that the model cell exhibits chemotactic behavior consistent with observations of PMN, but to elucidate the relationship between the two types of turning behavior as suggested by Fig. 3.

The full system of equations presented in the Model Description section, eqs. (7-11), applies in this case. It is nonlinear and stochastically coupled reflecting feedback on receptor-sensing associated with translocation and turning in the gradient. Although perturbation techniques could be employed for the limiting case of small receptor noise and small CA gradient (Gardiner, 1985), simulation results are presented here for the general case. Orientation bias was chosen to characterize chemotaxis for the purpose of convenient comparison to available data for PMN, although measurable orientation bias alone is not always unique to chemotaxis (Doucet and Dunn, this volume, section II/3). The Von Mises distribution, which is the analog of the normal distribution on the line for directional statistics on the circle (Mardia, 1972), was found to satisfactorily describe the distributions of orientations in the simulations and for PMN data. Results presented here are in the form of "percent correct orientation", the percentage of cells with orientation towards higher CA concentration, which is derived from the value of the Von Mises distribution concentration parameter.

The first finding of significance is that simulated paths depend only on the parameter groups used to characterize the random motility response (f_s, N_T, τ_R, τ_D) rather than on the individual cell parameters ($k_r, N_T, k_t, k_d, D, \kappa$) for a specified gradient steepness, ε, i.e., if identical initial directions and realizations of the Wiener process are used, identical paths are obtained for any combination of the cell parameters that yield the same values for f_s, N_T, τ_R, and τ_D. Evidently, τ_R and τ_D completely characterize the dynamical characteristics of the intracellular mechanisms. The parameter space can be further reduced using measured values for P of 1-5 min (Lackie and Wilkinson, 1984; Zigmond et al., 1985) for $\rho = 1$ and assuming $P_T \approx P$ in eq. (13), which yields the constraint $\tau_R^4/\tau_D^2 = 0.312\text{-}1.56 \times 10^4$ min^2 (FNLLP receptors appear to be anterior-posterior asymmetrically distributed (Sullivan et al., 1984) - estimating that 40% would be associated with each compartment, then $N_T = 10,000$).

The next finding of significance is that the orientation bias for the population approaches a quasi-steady state value despite population drift to higher CA concentration. This allows the dependence of orientation bias on ε to be compared directly to that observed for PMN, notwithstanding the complication that the model results do not apply exactly for $\rho = 1$, as do the data, due to drift of the population up-gradient. It is striking from Fig. 5 that the agreement is quantitative as well as qualitative given the variability of a PMN radius upon which the experimental value of ε is calculated.

Given that validation, the model can be used to examine an intriguing question: what is the relationship between directional persistence (in the absence of a CA gradient) and orientation bias (in the presence of a CA gradient) for a cell? The answer was qualitatively suggested in Fig. 3 and is quantitatively provided here in part by the results summarized in Fig. 6. These plots show the dependence of orientation bias as a function of τ_R^4 and τ_D^2. Lines of constant P_T are indicated. In crossing lines from small to large P_T, the orientation bias passes through a relatively shallow maximum, corresponding to an optimal $P_T \approx 3$ min, in agreement with the experimental data. Following along the line of optimal P_T to smaller values of τ_R^4 and τ_D^2, the orientation bias is seen to increase slightly. So at least for the values of f_s and N_T applicable to the case study, an optimal P_T is suggested; furthermore, one which reflects small time constants, i.e. a cell that rapidly responds to a turning signal and rapidly eliminates it.

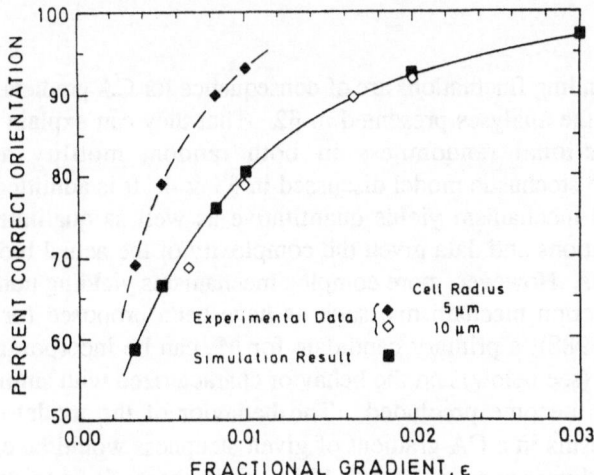

<u>Figure 5.</u> Dependence of orientation bias on gradient steepness.
Experimental data for PMN are from Zigmond (1977). Simulation results incorporate the previously stated estimates for N_T and f_s and values for k_t, k_d, D, and κ which yield $\tau_R = 3.54$ min and $\tau_D = 0.114$ min, determining $P_T = 3.88$ min and $\tau_R{}^4/\tau_D{}^2 = 1.21 \times 10^4$ min^2. A population size of 300 with the same initial orientations and realizations of the Wiener process apply to the simulations for the various ε. Other parameter values needed to define the simulation conditions are $r_c = 5$ μm, $\theta_c = \pi/4$, s = 20 μm/min.

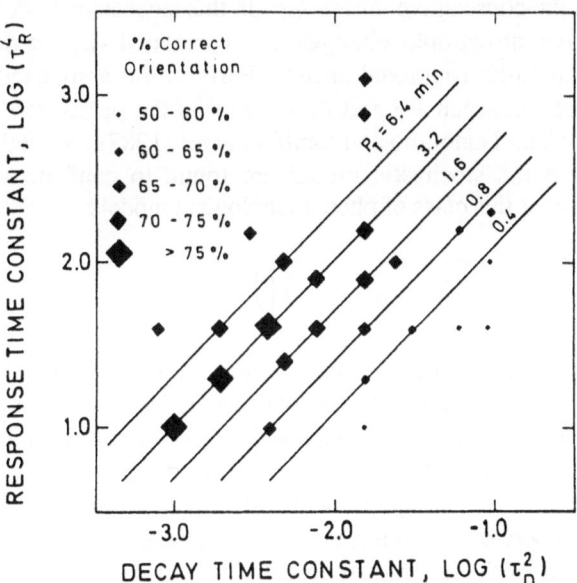

<u>Figure 6.</u> Relationship between orientation bias and directional persistence.
Simulation parameters are the same as Fig. 5 except the population size is 100. $\varepsilon = 0.008$ applies to all simulations here.

5. Discussion

That receptor binding fluctuations are of consequence for CA gradient sensing by PMN is well-supported from the analyses presented in §2. That they can explain simultaneously the component of directional randomness in both random motility and chemotaxis is well-supported by the stochastic model discussed in §3 & 4. It is admittedly surprising that a simple pseudospatial mechanism yields quantitative as well as qualitative agreement with experimental observations and data given the complexity of the actual biochemical pathways and biophysical events. However, more complex mechanisms yielding nonlinear functionality in the signal transduction mechanisms, such as have been proposed for intracellular Ca^{2+} (Meyer and Stryer, 1988), a primary candidate for M, can be incorporated into the general model as appropriate (see below) and the behavior characterized with simulations even though analytical treatments become precluded. The behavior of the model cell with nonlinear transduction mechanisms in a CA gradient of given steepness would be expected to be more complex than for the linear ones investigated here given the positive feedback inherent to the model. Whether this improves the gradient-sensing characteristics of the idealized cell is under investigation.

It is interesting to compare the results for the stochastic model with those from phenomenological models, where the stochasticity is prescribed, not having a mechanistic basis. For random motility, the result for the directional persistence time defined herein, P_T, can be compared to that associated with the class of models known as the "persistent" or "correlated" random walk (Hall, 1977; Dunn, 1983). These models assume that consecutive *directional changes* (i.e., left *vs* right turn) between straight-line runs are uncorrelated; however, since directional changes are assumed nonuniformly distributed about the current direction, there is correlation between the consecutive *directions*. In this stochastic model, however, it can be shown that consecutive directional changes *are* correlated (i.e., $< d\theta_T(t)d\theta_T(s) > \neq 0$), constraining the cell to finite rotational speed. Further, the associated correlation time is related to the difference between P_T and P, where P is the persistence time defined in the persistent random walk (see Tranquillo and Lauffenburger (1987) for details). Nonetheless, the simulation results from the stochastic model are found to conform with the $< d^2 > vs\ t$ expression characteristic of this class of phenomenological models:

$$(15) \quad < d^2 > = 2s^2 \left(Pt - P^2 \left[1 - \exp\left(-\frac{t}{P} \right) \right] \right)$$

where d is the displacement of a cell at time t from its initial position, as seen in Fig. 7a. If the correlation time was small relative to the total simulation time, then it would be expected that eq. (15) would apply. However, it applies well even when the correlation time is $\approx 1/3$ of the simulation time. Given that PMN data are also well-described by this expression as seen in Fig. 7b, it is clear that a more stringent statistical examination of PMN movement paths is necessary to distinguish between alternative stochastic descriptions of random motility, such as use of the autocorrelation of the directional process (Alt, this volume, section II/1) .

Alt has analyzed the "velocity jump" process, which is equivalent to the persistent random walk model discussed above in uniform CA concentrations, for the case where the turn angle distribution is affected by a CA gradient but the turning frequency and speed are constant (Othmer *et al.*, 1988). Assuming that the turn angle distribution is the appropriately normalized sum of two symmetric distributions, one dependent on the turn angle with respect to the current direction, representing the random component, and the other dependent on the

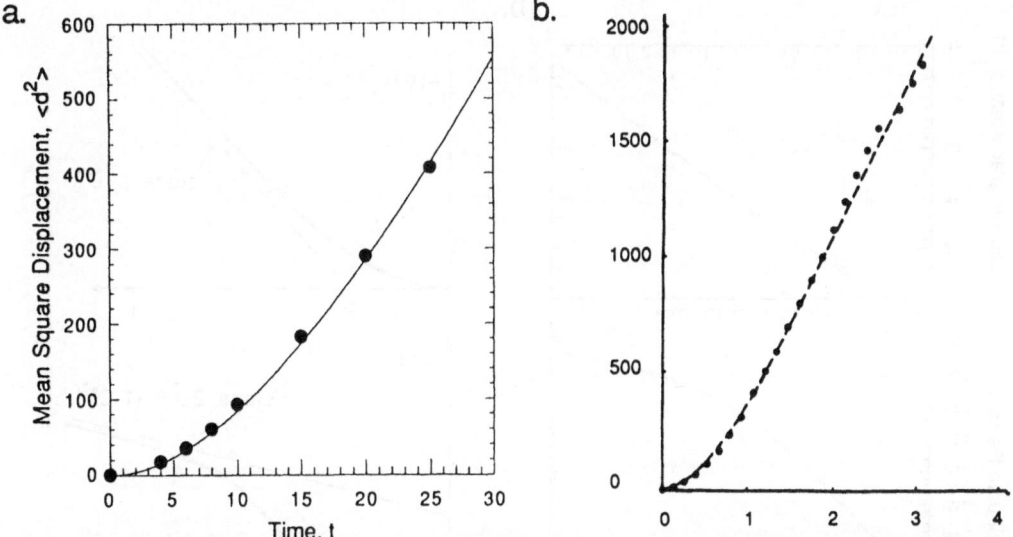

Figure 7. Comparison of stochastic model with phenomenological model of random motility. (a) Theoretical predictions: the dependence of mean square displacement $<d^2>$ on time t from the simulation of random motility conforms with the persistent random walk model. The points are simulation data and the solid line is the nonlinear least squares fit of the data to eq. (15). (b) Experimental data: $<d^2>$ (square microns) is plotted *vs.* t (min) for PMN responding to an analog of FNLLP. The dashed line is the fit of eq. (15). Fig. 7b reproduced from Gruler and de Boisfleury Chevance (1987), Fig. 2a.

turn angle with respect to the gradient direction, representing the biased component, analytic expressions were derived for the expected displacement up-gradient and the mean squared displacement relative to the expected displacement up-gradient. As seen from Fig. 8, the stochastic model also conforms to the velocity jump process extended to chemotaxis by these criteria, underscoring again the need for careful discrimination between alternative phenomenological models.

One unsatisfying feature of the stochastic model is with respect to evaluating candidates for the receptor signal: directional persistence and orientation bias depend on the time constants and not on the individual intracellular parameters. Evidently, this is a natural consequence of modeling the cell as an integrated dynamical system. However, there is at least one direct test of the particular model mechanisms analyzed here: the prediction that P_T is inversely proportional to N_T. N_T would have to be modulated in a manner which does not perturb the original state or function of the cell. A competitive inhibitor which binds to the same receptor but neither elicits a signal nor perturbs the cell otherwise would be ideal. Of course, failure of this test would not necessarily invalidate the model structure combining receptor binding fluctuations with a pseudospatial mechanism. It might indicate that the particular mechanisms used to illustrate the general structure are inadequate or incorrect. Note that if, in fact, the rate of receptor signal transduction was proportional to only the *fraction* of bound receptors (i.e., $\mathbf{R} = \mathbf{I}$), which is consistent with a simple adaptation mechanism wherein a bound receptor both rapidly produces a nondiffusible generator of $\mathbf{M}$ and slowly produces a rapidly diffusible degrader of $\mathbf{M}$ (Tranquillo and Lauffenburger, 1986), then the random motility analysis yields the same result except that P_T is now directly rather than inversely proportional to N_T. In addition, chemotaxis simulations reveal that a maximum in orientation bias corresponding to

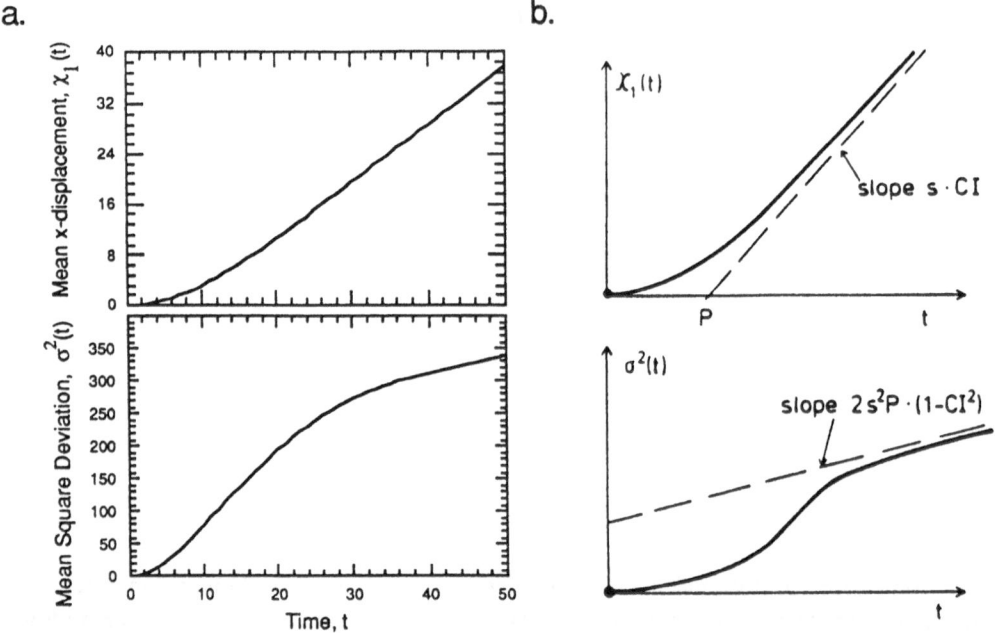

Figure 8. Comparison of stochastic model with phenomenological model of chemotaxis. Predictions from (a) the stochastic model of expected displacement up-gradient (upper panels) and the mean squared displacement relative to the expected displacement up-gradient (lower panels) conform to the (b) extended "velocity jump" process. Fig. 8b reproduced from Othmer *et al* (1988), Fig. 5.

some optimal N_T does not exist for this "adapting signal" case, as is true for the case of $R = I \cdot N_T$ presented herein; rather, orientation bias increases monotonically with N_T over the relevant range for N_T. Thus, the model predicts that these alternative transduction mechanisms can be distinguished based on the qualitative dependence of directional persistence and orientation bias on N_T.

Recently, many details of chemotactic peptide receptor signaling have been elucidated (see Tranquillo (1990) for a review). The stochastic description of receptor binding used herein applies only for the simplest case of a constant population of receptors on each compartment (i.e., receptors *only* reversibly bind CA molecules). However, the receptor distribution on the cell surface, critical to the pseudospatial mechanism, is believed to represent a dynamic steady-state involving internalization of some bound and perhaps free receptors into the cell which after processing are reinserted into the cell surface, where they are likely differentially diffusible. The intriguing possibility that the cell optimizes its orientation bias in a gradient by controlling the spatial pattern of reinsertion is being investigated as are more complex transduction schemes based on the requisite interaction of bound receptors with membrane-bound GTP-binding proteins for activation of the enzyme triggering the transduction cascade.

One major limitation of the stochastic model is that it applies only for CA concentrations where the cell retains its morphological polarity (i.e., $C \approx K_D$). There is considerable data for indices of PMN random motility and chemotaxis at concentrations away from K_D. Modeling the cell behavior when the morphology is unstable (i.e., where the assumption that the cell turning is governed by a single stable leading front does not apply) requires a considerably more complex mechanical model for cell motility than is incorporated at present (Dembo *et al.*, 1984; Oster and Perelson, 1985; Zhu *et al.*, 1989; Evans and Dembo, 1990). A model which replaces the dual compartment turning mechanism with a contractive cortical fluid (Alt, 1990) forming lamellipodia in response to stochastic binding of diffusible surface receptors is under development in collaboration with Wolfgang Alt. Preliminary results can be found in Tranquillo (1990). Ultimately, such an extended stochastic model should give insight into the relationship between cell speed and directional movement behavior. However, the present stochastic model based on the high polarity limit analyzed here yields significant insight into the mechanistic relationships between random motility and chemotaxis.

Finally, it is instructive to make a connection between the stochastic model and the phenomenological description of movement of a chemotactic cell population (Tranquillo and Lauffenburger, this volume, section II/3). Two motility coefficients appear in the constitutive cell flux expression of this description: the random motility coefficient, $\mu = 1/2Ps^2$, and the chemotaxis coefficient, $\chi = \phi \cdot s$, where ϕ is the orientation bias per unit CA gradient. Although s is assumed constant in the model, the characterization of P and ϕ (P_T and the Von Mises distribution concentration parameter, respectively, in the model) in terms of cell parameters provides a link between indices of population migration and parameters of cellular mechanisms.

Acknowledgements

The author is indebted to Doug Lauffenburger and Sally Zigmond for their many contributions to these ideas and their development. A travel grant from the Office for International Education for attending the Workshop is also gratefully acknowledged.

REFERENCES

Alt, W. (1990) Mathematical models and analysing methods for the lamellipodial activity of leukocytes. In: Biomechanics of Active Movement and Deformation of Cells: 403-422. Springer (NATO ASI Ser. H).

Berg H.C. and Purcell E.M. (1977) Physics of chemoreception. Biophys. J. **20**: 193-219

Del Grosso G. and Marchetti F. (1983) Limit theorems in stochastic biochemical modeling. Math. Biosci. **66**: 157-165

DeLisi C. and Wiegel F.W. (1981) Effect of nonspecific forces and finite receptor number on rate constants of ligand-cell bound-receptor interactions. Proc. Natl. Acad. Sci. USA **78**: 5569-5572

DeLisi C., Marchetti F. and Grosso G.D. (1982) A theory of measurement error and its implications for spatial and temporal gradient sensing during chemotaxis. Cell Biophysics **4**: 211-229

DeLisi C. and Marchetti F. (1983) A theory of measurement error and its implications for spatial and temporal gradient sensing during chemotaxis - II. The effects of non-equilibrated ligand binding. Cell Biophysics **5**: 237-253

Dembo M., Harlow F.J. and Alt W. (1984) The biophysics of cell surface motility. In: A. Perelson, C. DeLisi and F. Wiegel (eds.) Cell surface dynamics: Concepts and models: 495-543. Marcel Dekker, New York

Devreotes P.N. and Zigmond S.H. (1988) Chemotaxis in eukaryotic cells: A focus on leukocytes and *Dictyostelium*. Ann. Rev. Cell Biol. **4**: 649-686

Dunn G.A. (1983) Characterizing a kinesis response: time averaged measures of cell speed and directional persistence. Agents and Actions Suppl. **12**: 14-33

Dunn G.A. and Brown A.F. (1987) A unified approach to characterizing cell motility. J Cell Sci. Suppl. **8**: 81-102.

Evans E. and Dembo M. (1990) Physical model for phagocyte motility: Local growth of a contractile network from a passive body. In: Biomechanics of Active Movement and Deformation of Cells. Springer (NATO ASI Ser. H).

Farrell B.E., Daniele R.P. and Lauffenburger D.A. (1990) Quantitative relationships between single-cell and cell-population model parameters for chemosensory migration responses of alveolar macrophages to C5a. Cell Motility Cytoskel. (in press)

Gard T.C. (1988) Introduction to stochastic differential equations. M. Dekker, New York

Gardiner C.W. (1985) Handbook of stochastic methods for physics, chemistry and the natural sciences. Springer-Verlag, New York

Gihman I.I. and Skorohod A.V. (1969) Introduction to the theory of random processes, Saunders, Philadelphia

Gruler H. and de Boisfleury Chevance A. (1987) Chemokinesis and necrotaxis of human granulocytes: The important cellular organelles. Z. Naturforsch **42c**: 1126-1134

Gruler H. and Bultmann B.D. (1984) Analysis of cell movement. Blood Cells **10**: 61-77

Hall R.L. (1977) Amoeboid movement as a correlated walk. J. Math. Biol. **4**: 327-335

Krenner A.J. and Lobry C. (1981) The complexity of stochastic differential equations. Stochastics **4**: 193-203

Kurtz T.G. (1981) Approximation of population processes. SIAM, Philadelphia

Kurtz T.G. (1986) Markov processes characterization and convergence, Wiley, New York

Lackie J.M. (1986) Cell movement and cell behavior, Allen & Unwin, London

Lackie J.M. and Wilkinson P.C. (1984) Adhesion and locomotion of neutrophil leukocytes on 2-D substrata and in 3-D matrices. In: White cell mechanics: Basic science and clinical aspects: 237-254. Alan R. Liss, New York

Lauffenburger D.A. (1982) Influence of external concentration fluctuations on leukocyte chemotactic orientation. Cell Biophysics **4**: 177-209

Maher J., Martell J.V., Brantley B.A., Cox; E.B., Neidel J.E. and Rosse W.F. (1984) The response of human neutrophils to a chemotactic tripeptide (N-formyl-methionyl-leucyl-phenylalanine) studied by microcinematography. Blood **64**: 221-228

Mardia K.V. (1972) Statistics of directional data. Academic Press, New York

Meyer T. and Stryer L. (1988) Molecular model for receptor-stimulated calcium spiking. Proc. Natl. Acad. Sci. **85**: 5051-5055

Mil'shtein G.N. (1974) Approximate integration of stochastic differential equations. Theor. Prob. **19**: 557-562

Nossal R. and Zigmond S.H. (1976) Chemotropism indices for polymorphonuclear leukocytes. Biophys. J. **16**: 1171-1182

Omann G.M., Allen R.A., Bokoch G.M., Painter R.G., Traynor A.E. and Sklar L.A. (1987) Signal transduction and cytoskeletal activation. Physiol. Rev. **67**: 285-321

Oster G.F. and Perelson A.S. (1985) Cell spreading and motility. J. Math. Biol. **21**: 383-388

Othmer H.G., Dunbar S.R. and Alt W. (1988) Models of dispersal in biological systems. J. Math Biol. **26**: 263-298

Papoulis A. (1965) Probability, random variables, and stochastic processes, McGraw-Hill

Pardoux E. and Talay D. (1985) Discretization and simulation of stochastic differential equations. Acta Applicandae Math. **3**: 23-47

Rumelin W. (1982) Numerical treatment of stochastic differential equations. SIAM J. Numer. Anal. **19**: 604-613

Shields J.M. and Haston W.S. (1985) Behaviour of neutrophil leucocytes in uniform concentrations of chemotactic factors: Contraction waves, cell polarity, and persistence. J. Cell Sci. **74**: 75-93

Singer S.J. and Kupfer A. (1986) The directed migration of eukaryotic cells. Ann. Rev. Cell Biol. **2**: 337-365

Soong T.T. (1973) Random differential equations in science and engineering. Academic Press, New York

Stickle D.F., Lauffenburger D.A. and Zigmond, S.H. (1984) Measurement of chemoattractant concentration profiles and diffusion coefficient in agarose. J. Immunol. Meth. **70**: 65-74

Sullivan S.J., Daukas G. and Zigmond S.H. (1984) Asymmetric distribution of the chemotactic receptor on polymorphonuclear leukocytes. J. Cell Biol. **99**: 1461-1467

Sullivan S.J. and Zigmond S.H. (1980) Chemotactic peptide receptor modulation in polymorphonuclear leukocytes. J. Cell Biol. **85**: 703-711

Tranquillo R.T. (1990) Theory and models of gradient perception. In: J.M. Lackie and J. Armitage (eds.) Motility & Taxis (in press). Cambridge University Press, Cambridge

Tranquillo R.T. and Lauffenburger D.A. (1986) Consequences of chemosensory phenomena for leukocyte chemotactic orientation. Cell Biophysics **8**: 1-46

Tranquillo R.T., Lauffenburger D.A. and Zigmond S.H. (1988) Stochastic model for leukocyte random motility and chemotaxis based on receptor binding fluctuations. J. Cell Biol. **106**: 303-309

van Kampen N.G. (1981) Stochastic processes in physics and chemistry. North-Holland, Amsterdam

Wiegel F.W. (1983) Diffusion and the physics of chemoreception. Phys. Rep. **95**: 283-319

Wilkinson P.C. (1982) Leukocyte chemotaxis. Churchill Livingstone, Edinburgh

Wright D.J. (1974) The digital simulation of stochastic differential equations. IEEE Trans. on Auto. Control **19**: 75-76

Zhu C., Skalak R. and Schmid-Schonbein G.W. (1989) One-dimensional steady continuum model of retraction of pseudopod in leukocytes. J. Biomech. Eng. **111**: 69-77

Zigmond S.H. (1977) Ability of polymorphonuclear leukocytes to orient in gradients of chemotactic factors. J. Cell Biol. **75**: 606-616

Zigmond S.H. (1982) Polymorphonycgear leucocyte response to chemotactic gradients. In: Curtis and Dunn (eds.) Cell behavior, Bellairs.

Zigmond S.H. and Sullivan S.J. (1979) Sensory adaptation of leukocytes to chemotactic peptides. J. Cell Biol. **82**: 517-527

Zigmond S.H., Levitsky H.I. and Kreel B.J. (1981) Cell polarity: An examination of its behavioral expression and its consequences for polymorphonuclear leukocyte chemotaxis. J. Cell Biol. **89**: 585-592

Zigmond S.H., Klausner R., Tranquillo R.T. and Lauffenburger D.A. (1985) Analysis of the requirements for time-averaging of the receptor occupancy for gradient detection by polymorphonuclear leukocytes. In: Membrane receptors and cellular regulation: 347-356. Alan R. Liss, New York

ENDOTHELIAL CELL CHEMOTAXIS IN ANGIOGENESIS

C. L. Stokes   D. A. Lauffenburger
Department of Chemical Engineering
University of Pennsylvania
Philadelphia, Pennsylvania 19104

S. K. Williams
Department of Surgery
Thomas Jefferson University
Philadelphia, Pennsylvania 19107

Abstract

A probabilistic model for angiogenesis was developed to investigate the possible role of microvessel endothelial cell (MEC) chemotaxis in determining microvessel network morphology and growth rate. The model simulates developing microvessels, providing theoretical pictures of the networks. The cell at the tip of a growing capillary is hypothesized to guide the path of the capillary according to a model of single cell migration, using experimentally measured values for MEC speed and persistence time. The simulations demonstrate that random motility alone cannot account for the directional growth of vessels observed *in vivo*. A moderate chemotactic response, like that we have measured in acidic fibroblast growth factor (Stokes *et al.*, 1990), is necessary to provide directional growth of vessels similar to that observed *in vivo*.

1. Introduction

Angiogenesis, the growth and development of capillary blood vessels, is common in numerous physiological and pathological conditions (Folkman and Klagsbrun, 1987). Several functions of the MEC, the cells which comprise the walls of the capillaries, make up the process of angiogenesis. These functions include cell proliferation, migration, protease production, and basement membrane production. A central question is: what is the role of MEC chemotaxis in this process? To help answer this question, we have developed a probabilistic mathematical model of microvessel network development. The model has been designed to explore specifically how the cell migration properties of speed, persistence time, and chemotactic sensitivity affect the rate of growth and morphology of developing microvessel networks. The model results consist of theoretical pictures of microvessel networks which reveal the predicted network morphology. The results can also be analyzed for quantitative information, such as vessel lengths and mean network expansion rates.

In this communication, we outline the model and demonstrate how it can be used to explore several questions regarding the role of MEC chemotaxis in angiogenesis. For instance, is a MEC chemotactic response necessary to result in the directional development of microvessel networks which is commonly observed *in vivo*? If so, is the magnitude of the chemotactic response necessary for that directional growth on the same order as that we have observed in the laboratory? On the other hand, if the vessels begin growing towards an attractant source, which is observed experimentally (Clark and Clark, 1939; Eddy and Casarett, 1973), and are persistent for long enough, will the vascular network develop in the direction of the stimulus without a chemotactic response?

2. Structure of the Model

The model is confined to two dimensions. This should be a good approximation of the vessel growth which takes place in certain experimental situations such as thin (< 50 μm) rabbit ear chambers (Clark and Clark, 1939; Zawicki *et al.*, 1981). It is probably a reasonable first approximation for tissues which are thicker, such as thicker ear chambers and the cornea assay (about 100 μm). Capillaries are typically 10 to 20 μm in diameter.

The geometry of the model is illustrated in Figure 1. We model the existing vasculature as a single, straight, parent vessel located at y=0. New buds start to grow off the parent vessel at time t=0. They are placed randomly, with a uniform probability density p_d. All

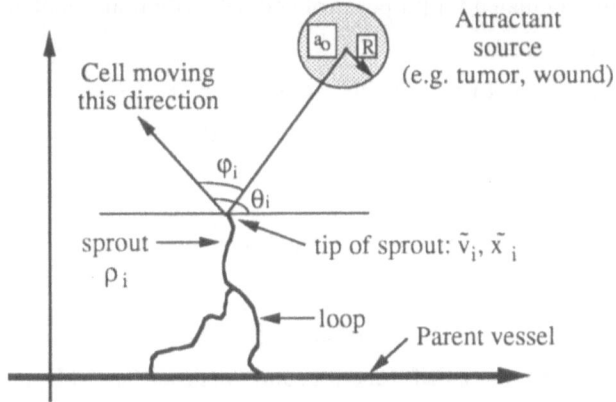

Figure 1. Geometry of angiogenesis model. Subscript i denotes the ith sprout. Each sprout has cell density ρ, and its tip cell has position $\tilde{x}$, velocity $\tilde{v}$, and moves in direction θ. ϕ is the angle between the direction the cell is moving and that towards the attractant source. R is the radius of the attractant source, and a_0 is its concentration.

buds begin to grow off the parent vessel at the same time (t=0), and no budding off the parent vessel occurs at later times.

We hypothesize that the path of each capillary is determined by the trajectory of an actively migrating cell located at its tip. This cell migrates with velocity $\tilde{v}$. All other cells in the sprout follow directly behind the tip cell. The velocity of the tip cell is assumed to be unaffected by mechanical attachment to the rest of the sprout, so that its velocity can be calculated using a model of single cell migration. The cells within the sprout proliferate to provide additional cells for the elongation of the sprout. The linear cell density within the sprout, ρ, is the average number of cells per unit length. It is not a distributed variable in length along the sprout. Two constraints are placed on the value of ρ: First, there is a minimum density, ρ_{min}, required which corresponds to the lowest density needed to maintain a contiguous vessel. Second, there is a maximum density, ρ_{max}, which reflects the confluent density at which MEC stop growing, as observed both *in vivo* and *in vitro*. We postulate that if a sprout density drops to the minimum, then the sprout cannot further elongate (therefore $\tilde{v}$ = 0) and new buds cannot start growing off of it. Thus, if the proliferation rate and cell redistribution rate are not great enough in comparison to the migration rate, the vessel will not be able to elongate, because additional cells will not be available to maintain a contiguous vessel. If ρ reaches ρ_{max}, then the cells stop proliferating. For the parent vessel, $\rho = \rho_{max}$ at the time of initial budding (t=0). Branches bud off existing sprouts and loops with a probability density, p_b, the probability per unit time per unit length of vessel. p_b is uniform over all vessels and for all positions and times. No branches can start growing off a sprout or loop whose density is ρ_{min}. Anastomosis occurs whenever a sprout tip runs into another sprout (anywhere along the length) or a loop. When this happens, the tip "dies" ($\tilde{v}$ = 0) and only ρ is continually calculated for purposes of cell redistribution calculations to sprouts which are growing off of this sprout.

The equations used to model this situation are as follows: A stochastic ordinary differential equation (see Box 8, this volume) was used to describe the velocity, $\tilde{v}_i$, of the tip cell:

(1) $$d\tilde{v}_i(t) = - \beta\, \tilde{v}_i(t)\, dt + d\tilde{W}_i(t) + \tilde{\psi}\, dt$$

W is the Wiener process, a white noise process, β is the magnitude of resistance to random fluctuations caused by W, ψ is a drift function, and t is time. The subscript i on the dependent variables indicates the ith sprout. A stochastic description of the motility was desired because of the apparent random walk quality of cell migration. Equation 1 is a modification of the Ornstein-Uhlenbeck process (Uhlenbeck and Ornstein, 1930) as interpreted by Doob (1942). The Ornstein-Uhlenbeck process is the simplest type of continuous, autocorrelated, random motion process. The velocity process described by Equation 1 can be thought of as the sum of three separate components: The first term, $-\beta \tilde{v}_i(t)$, tends to oppose velocity change, while the second provides random fluctuations. W, the white noise, has the property that W(t)-W(s) is Gaussian distributed with a mean of 0 and a spectrum, or variance, of $\alpha|t|$, α a constant. Thus, the velocity process for random motility ($\tilde{\Psi}$= 0) is characterized by two parameters, α and β. α and β represent the magnitude of the random fluctuations and the magnitude of the resistance to the fluctuations, respectively. The function $\tilde{\Psi}$ in the third term is a drift function which describes the directional bias the cell displays in an attractant gradient. We have assumed that the cell effectively responds to a spatial gradient of bound receptors, ∇N_b, across its surface (Zigmond, 1977), although we make no statement about its actual underlying mechanism of perception of a chemical gradient. ∇N_b will be dependent on the attractant gradient, ∇a, and using the chain rule ∇N_b is equivalent to $\nabla a \, dN_b/da$. Assuming that the receptor binding occurs linearly with attractant concentration in the concentrations of interest, we have absorbed dN_b/da into the proportionality constant, κ. We define κ as the chemotactic sensitivity. We also assumed that the effect of a gradient increases as the direction in which the cell is moving is farther down the gradient (ϕ increasing, where ϕ is the angle between the direction in which the cell is currently migrating and the direction of the attractant source; see Figure 1). The simplest way to incorporate this was to let $|\tilde{\Psi}|$ be proportional to $\sin|\phi/2|$, . Thus, the form given to $\tilde{\Psi}$ is

$$(2) \qquad \tilde{\psi} = \kappa \, \nabla a \, \sin \left| \frac{\phi_i}{2} \right|$$

With $\tilde{\Psi}$=0, Equation 1 is a form of the Langevin equation. Dunn and Brown (1987) originally suggested the use of the Ornstein-Uhlenbeck process for analyzing cell random motility, and used a discretized version to describe motility of chick heart fibroblasts, showing that the description is reasonable for those cells.

The position vector, $\tilde{x}_i(t)$, which denotes the trajectory of the sprout, is calculated by integrating $\tilde{v}_i$:

$$(3) \qquad \frac{d\tilde{x}_i(t)}{dt} = \tilde{v}_i(t)$$

The rate of change of cell density, ρ_i, is given by Equation 4:

$$(4) \qquad \frac{d\rho_i(t)}{dt} = \left(k_g' - \frac{s_i(t)}{L_i(t)} \right) \rho_i(t) + k_b \left(\rho_p(t) - \rho_i(t) \right) - \sum_{j=1}^{n_i} k_b \left(\rho_i(t) - \rho_j(t) \right)$$

where
$$k_g' = k_g \left(1 - \frac{\rho_i(t) - \rho_{min}}{\rho_{max} - \rho_{min}} \right), \qquad s_i(t) = \| \tilde{v}_i(t) \|, \qquad \frac{dL_i(t)}{dt} = s_i(t)$$

k_g is the proliferation rate constant, k_b is the redistribution rate constant, s_i is the instantaneous cell speed, and L_i is the instantaneous sprout length. The rate of change of ρ_i depends on the proliferation rate of the cells in the sprout (first term on right hand side), the rate at which the sprout is elongating (second term), and the redistribution of cells from the parent vessel to sprout i (third term) and from sprout i to sprouts j growing off of sprout i (fourth term). Redistribution of cells by movement from sprout to sprout is necessary to account for the data of Sholley *et al.* (1984). They demonstrated that angiogenesis can occur in the absence of MEC proliferation by the redistribution of existing cells to new sprouts. The form of the cell proliferation rate (first term) was chosen to give a decreasing proliferation rate as cell density reaches the maximum, much as is seen in cell culture.

For an attractant gradient, we have used one plane of a three-dimensional spherical steady state gradient created by a sphere at concentration a_0 with radius R (see Figure 1). The attractant concentration field, $a(\tilde{x})$, is

$$(5) \qquad a(\tilde{x}) = \frac{a_0 R}{\| \tilde{x} - \tilde{x}_a \|}$$

$\tilde{x}_a$ is the position of the attractant source. The concentration gradient for this field is proportional to the inverse of distance from the center squared. This gradient was chosen because it reflects a steady state, and because it might be a reasonable approximation of small, round tumor or other angiogenic condition.

For the random motility velocity process ($\tilde{\Psi} = 0$), it can be shown (Doob, 1942) that the expected square speed, which we will denote $<s^2>$, and expected square displacement, $<D^2>$, in two dimensions are given by

$$(6) \qquad <D^2> = E\{[\tilde{x}(t)-\tilde{x}(0)]^2\} = \frac{2\alpha}{\beta^3}(\beta t - 1 + e^{-\beta t})$$

$$(7) \qquad <s^2> = E\{\tilde{v}(t)^2\} = \frac{\alpha}{\beta}$$

From these we can define two new parameters in terms of α and β which are more intuitive in thinking of cell paths. We define S, the speed, as the root mean square speed, and P_v, the persistence time in velocity, as the inverse of β, the magnitude of resistance to fluctuations:

$$(8) \qquad S = \sqrt{\frac{\alpha}{\beta}} \qquad\qquad P_v = \frac{1}{\beta}$$

These definitions can be compared to similar definitions in Rivero *et al.* (1990) and Othmer *et al.* (1988). These definitions transform the equation for expected square displacement in two dimensions into the following form:

$$(9) \qquad <D^2> = 2S^2 P_v^2 \left(\frac{t}{P_v} - 1 + e^{-t/P_v} \right)$$

We have measured values of S and P_v in terms of Equation 9 as described below.

To simulate angiogenesis, the equations are solved numerically in sequence for each sprout at each time step. After each step, the proximity of the tip of each sprout to the vessels around it is measured to determine whether anastomosis has taken place. If it has, then a loop has formed and the velocity of this tip is set to zero for the remainder of the simulation. The probability of a new sprout growing off a bud at each distance step is calculated, and new sprouts are generated where indicated. Thus, the number of sprouts growing in the network is continually changing with time due to tip birth (branching) and tip death (anastomosis).

The values assigned to the parameters are summarized in Table 1. S and P_v were measured experimentally and κ was estimated by simulations, as explained below.

Table 1. Summary of parameter values for angiogenesis simulations.

Parameter	Symbol	Value(s)	Comment/Reference
Speed	S	40 μm/h	Measured as explained in Section 3
Persistence time	P_v	3 h	Measured as explained in Section 3
Chemotactic sensitivity * a_0	κa_0	1600 μm²/h²	Estimated as explained in Section 4
Growth rate constant	k_g	0.02 h⁻¹	Estimated as less than the normal growth rate (Sholley *et al.*, 1984), since all cells in sprout are allowed to proliferate
Redistribution rate constant	k_b	S p_d	Expected rate cells could move from sprout to sprout
Probability of initial budding	p_d	0.1 μm⁻¹	Sholley *et al.* (1984), Zawicki *et al.* (1981)
Probability of branching	p_b	0.0001 μm⁻¹ h⁻¹	Sholley *et al.* (1984)
$\rho_{min}/(\rho_{max}-\rho_{min})$		0.1	Simulations with no cell proliferation, based on Sholley *et al.*(1984)

3. Measurement of Speed and Persistence Time

Human MEC were isolated from subcutaneous fat and cultured as described (Stokes *et al.*, 1990). 35 mm polystyrene petri dishes were treated with a 1% gelatin solution in 0.9%

NaCl overnight, and washed with Medium 199 (M199). MEC were plated in a petri dish at a density of approximately 400 cells/cm^2. The cells were allowed to settle and attach to the dish for at least two hours prior to videotaping. The medium in a petri dish was covered with 2 ml of light mineral oil to provide protection of the medium and MEC from environmental contaminants. The petri dish was placed in a Leiden chamber, a type of microscope stage incubator with a temperature controller, and the temperature maintained at 37 °C.

The chamber was placed on an inverted microscope and the cells were observed with phase contrast optics with a 4x objective and 1x TV projection lens. A video camera was attached to the microscope, leading to a time-lapse video recorder. A single field was taped for up to 72 h using time-lapse at 1/240 real time. At the density the cells were plated, approximately 15 to 20 cells per field were present. Cell displacement with time was measured with a computerized image analysis system, the Imaging Technologies, Inc., Series 151 system using software subroutines provided by Mnemonics, Inc. (Mt. Laurel, New Jersey).

S and P_v are obtained by using nonlinear regression to best match the predictions of Equation 9 with the experimental data. First, $<D^2>$ was calculated as a function of time increment, t, for each cell track. For a given time increment length t, $<D^2>$ is the sum of n displacements which occurred sequentially during time steps of length t, divided by n. Nonlinear regression was then used to find the best values of S and P_v for each cell trajectory's ($<D^2>$,t) data. We found that a function of the form of Equation 9 can represent the data very well for small to intermediate t (compared to the full duration of the cell path). However, the data can become unpredictable when $<D^2>$ is the average of only a few points (t approaching the full duration of the cell path). This is to be expected, because at longer time increments there are fewer and fewer displacement increments to average. Hence, values of S and P_v reported are calculated from the experimental data $<D^2>$ vs. t for t up to duration/2.

Migration was measured under control conditions, consisting of M199 buffered with 10 mM HEPES and containing 0.1 percent bovine serum albumin. For 18 cells, the mean value of S obtained was 22 μm/h, and the mean value of P_v was 6 h. Migration was also measured under stimulatory conditions, consisting of the growth medium for MEC. This medium contains 10 percent fetal calf serum, endothelial cell growth factor, and 5 μg/ml heparin in HEPES-buffered M199. Under these conditions for 8 cells, the mean speed doubled to 42 μm/h and the persistence time decreased to 2.9 h. We can analyze the net effect of these changes on random motility through their relationship to the random motility coefficient, μ. μ is defined in a previous mathematical model of cell migration (Alt, 1980; Lauffenburger, 1983). It is a measure of the dispersion of a population of cells, analogous to the diffusion coefficient for the diffusion of a molecular species. For migration in two dimensions, Alt (1980) has shown that μ = speed^2 x persistence time / 2. Thus, μ roughly doubles under stimulating conditions.

4. Estimation of Values for the Chemotactic Sensitivity

We desired to define values of κ which were physiologically reasonable for use in the angiogenesis simulations. This was accomplished by first simulating cell paths with Equations 1 to 3 for various values of κ. Then statistics for these paths were compared to the same for previous experimental results in which MEC migrated in gradients of acidic fibroblast growth factor (Stokes *et al.*, 1990). The comparisons were made using the chemotactic index, CI. CI is essentially the net distance travelled towards a source or target divided by the total distance travelled. In terms of velocities, CI = $|\tilde{v}_c|/S$. $\tilde{v}_c$, the chemotactic velocity, is the velocity directed up a gradient in addition to that expected from random motility alone. S is an average cell speed. $\tilde{v}_c$ can be defined in terms of the chemotaxis coefficient, χ, by $\tilde{v}_c = \chi \nabla a$. χ is a phenomenological parameter which describes the drift of a cell population in a gradient due to directional bias (Alt, 1980; Lauffenburger, 1983). We have measured values of χ for migration of MEC in gradients of aFGF (Stokes *et al.*, 1990), which we can now use to estimate reasonable values of κ.

In our earlier experiments (Stokes *et al.*, 1990), the maximum chemotactic response was measured when in concentrations of aFGF around 10^{-10} M in which $\nabla a = 3.5 \times 10^{-15}$ M/μm, giving $\chi = 2600$ cm^2/s-M. Using the speed of the stimulated migration, S ≈ 40 μm/h, which would be expected in the presence of an angiogenic stimulus, we can calculate that CI was 0.08 in those experiments. The gradient is substantially larger in our simulations, $\nabla a \approx 2 \times 10^{-14}$ M/μm. Using $\chi = 2600$ cm^2/s-M, and this ∇a, and again assuming S = 40 μm/h, CI should be about 0.5 in the simulations. Using this criterion, single cell simulations for various values of the ratio $\kappa|\nabla a|/S^2$ (obtained in making the equations dimensionless) predict that values of this ratio up to and around 1.5 are physically realistic to give CI ≤ 0.5. With S=40 μm/h, this value corresponds to values of κa_0 of about 2400 μm^2/h^2. Since we have not attempted to show analytical relationships between κ and χ or CI, however, it is not necessary to attempt to elucidate exact values of κ. While we have not proven that the mechanisms underlying χ and κ are the same, we can be assured that the net directed movement predicted by both in the same gradient are the same.

5. Results of Angiogenesis Simulations

At least a dozen simulations were done for each set of parameter values. Only one representative simulation is shown here for each set of parameters illustrated. The angiogenesis simulations we show are dimensional with units on the axes in millimeters. A simulated time period of 4 days was chosen because it is long enough to reveal resulting morphologies, it is significantly longer than cell persistence time, and the budding probabilities over this time period are finite. Longer times were not chosen because several investigators have reported that vessel differentiation and regression becomes more significant as growth continues (Clark and Clark, 1939; Uhlenbeck and Ornstein, 1930). The parent vessel is the horizontal line in Figure 2 at y=0. The attractant source for all simulations is placed at x = 1.0 mm, y = 2.0 mm. The initial vessels were generated between x = 0.0 and 2.0 mm, checking for the presence of a sprout every 0.01 mm.

Figure 2a contains a sample simulation for the base case: only random migration, S =40 μm/h, P_v = 3 h, κa_0=0 and all other parameter values as shown in Table 1. These S and P_v correspond to the values measured for stimulated migration above. The initial velocity (direction) of the sprouts is random. As these simulations reveal, purely random motility with random initial directionality does not result in networks similar to those observed *in vivo* (Eddy and Casarett, 1973; Zawicki *et al.*, 1981). Most significant to this conclusion is that the directionality of vessel growth is haphazard, not resembling the directed vessel growth observed *in vivo*. In simulations not shown, we found this result even when the vessels all were forced to begin growing in the general direction of attractant source and the speed and persistence time were increased (Stokes, 1989). Because of this disagreement, it is not useful to make any quantitative comparisons of quantities such as vessel lengths and network expansion rates.

Figure 2b reveals the result when a moderate level of chemotaxis is incorporated in the simulations. With S = 40 μm/h and P_v = 3 h, and the initial directions the sprouts begin are

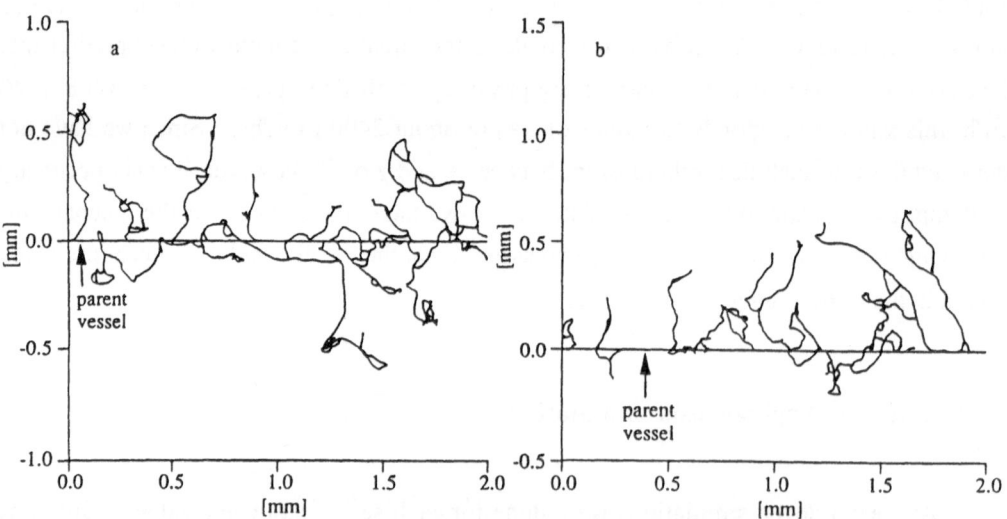

<u>Figure 2.</u> Angiogenesis simulation with a) random motility only (κa_0 = 0), and b) moderate chemotaxis (κa_0 = 1600 μm²/h²). See Table 1 and text for other parameter values.

random, κa_0 was given a value of 1600 $\mu m^2/h^2$. This level corresponds to CI of about 0.4. In Figure 2b, and other simulations not shown, most of the growth is in the general direction of the attractant source, similar to angiogenesis *in vivo*. In this case, the average network expansion rate for the simulation was 0.16 mm/day, and the average for the 12 simulations done for this set of parameter values is 0.17 mm/day. This is calculated as the maximum vessel network coordinate divided by the duration of the simulation, 4 days. Physiological values of this quantity are reportedly in the range of 0.09 to 0.3 mm/d (Clark and Clark, 1939; Zawicki *et al.*, 1981) so the simulations are in agreement. In addition, the average vessel lengths for the vessels between branches in the simulation is 0.18 mm, and 0.15 mm is the average for the 12 simulations. These are also similar to *in vivo* networks, where most vessels are between 0.1 and 0.3 mm long (Zawicki *et al.*, 1981). Thus, with experimentally measured or extimated values of S, P_v, and κa_0, networks can be simulated which appear to be physiological both qualitatively and quantitatively. As we showed earlier, we can expect MEC to be able to exhibit chemotactic sensitivities for values of κa_0 of about 2400 $\mu m^2/h^2$. Thus, these networks are based on reasonable levels of chemotaxis for MEC.

6.   Discussion

A probabilistic model of angiogenesis has been developed to investigate the role of MEC chemotaxis in microvessel network formation rate and morphology. The model predicts that a moderate level of chemotaxis is necessary to account for the directed growth of networks; random motility alone cannot give directed growth. Experimentally measured values for speed and persistence time led to quantitative agreement in rates of network expansion and vessel lengths when the simulations were compared to *in vivo* observations. The level of chemotaxis necessary for formation of reasonable networks was similar to that level measured in our earlier experiments (Stokes *et al.*, 1990). The results of this model support the hypothesis that MEC motility and chemotaxis are critical in determining the rate of growth and morphology of microvessel networks.

Acknowledgements

This work has been funded by NIH grant GM-41476 to D.A.L. and S.K.W. and a fellowship from the American Association of University Women to C.L.S.

REFERENCES

Alt, W. (1980) Biased random walk models for chemotaxis and related diffusion approximations. J. Math. Biology 9:147-177.

Clark, E. R. and E. L. Clark (1939) Microscopic observations on the growth of blood capillaries in the living mammal. Am. J. Anat. 64:251-301.

Doob, J. L. (1942) The Brownian movement and stochastic equations. Ann. Math. 43:351-369.

Dunn, G. A. and A. F. Brown (1987) A unified approach to analyzing cell motility. J. Cell Sci. Suppl. 8:81-102.

Eddy, H. A. and G. W. Casarett (1973) Development of the vascular system in the hamster malignant neurilemmoma. Microvasc. Res. 6:63-82.

Folkman, J. and M. Klagsbrun (1987) Angiogenic factors. Science 235:442-447.

Lauffenburger, D. A. (1983) Measurement of phenomenological parameters for leukocyte motility and chemotaxis. Agents Actions Suppl. 12:34-53.

Othmer, H. G., S. R. Dunbar and W. Alt (1988) Models of dispersal in biological systems. J. Math. Biol. 26:263-298.

Rivero, M. A., D. A. Lauffenburger, R. T. Tranquillo and H. M. Buettner (1990) Transport models for chemotactic cell populations based on individual cell behavior. Chem. Eng. Sci. (accepted for publication).

Sholley, M. M., G. P. Ferguson, H. R. Seibel, J. L. Montour, and J. D. Wilson (1984) Mechanisms of neovascularization: Vascular sprouting can occur without proliferation of endothelial cells. Lab. Invest. 51:624-634.

Stokes, C. L. (1989) Quantitative studies of endothelial cell motility and chemotaxis in angiogenesis. Ph. D. Thesis, Department of Chemical Engineering, University of Pennsylvania.

Stokes, C. L., M. A. Rupnick, S. K. Williams, and D. A. Lauffenburger (1990) Chemotaxis of human microvessel endothelial cells in response to acidic fibroblast growth factor. (Submitted to Lab. Invest.).

Uhlenbeck, G. E. and L. S. Ornstein (1930) On the theory of Brownian motion. Phys. Rev. 36:823-841.

Zawicki, D. F., R. K. Jain, G. W. Schmid-Schoenbein and S. Chien (1981) Dynamics of neovascularization in normal tissue. Microvasc. Res. 21:27-47.

Zigmond, S. H. (1977) Ability of polymorphonuclear leukocytes to orient in gradients of chemotactic factors. J. Cell Biol. 75:606-616.

A MODEL FOR TRAIL FOLLOWING IN ANTS:
INDIVIDUAL AND COLLECTIVE BEHAVIOUR

V. Calenbuhr   J.-L. Deneubourg
Unit of Theoretical Behavioural Ecology, C.P. 231,
Université Libre de Bruxelles, B-1050 Bruxelles

Abstract

A phenomenological model for osmotropotaxis is presented based on the physico-chemical properties of a scent trail and on a perception-reaction function. The model reproduces qualitatively experimental results and is to be used to test the influence of the physico-chemical properties of scent trails on ant behaviour. Moreover, we investigate how osmotropotaxis and interaction between individuals can organize trail traffic as found in army ant and termite species.

1. Introduction

A large part of a social insect colony's organisation is based upon chemical communication between individuals (Wilson, 1971; Eder & Rembold, 1987). Examples range from recognition of nestmates to the phenomenon of mass-communication in foraging. Most of these phenomena are related to interindividual attraction or to trail following, as the following examples show. In ants, the emission of a chemical substance (pheromone) is used, for example, to alert nearby co-workers in case of alarm. A trail pheromone, laid by an ant homing from a successful foraging excursion, is able to induce recruitment of ants in the nest. On higher levels of organisation, i.e. phenomena in which not only one or a few individuals are involved but a large number of them, complex patterns in the temporal and the spatial domain can often be observed. Army ant and termite species following chemical trails form large foraging columns with lanes in which all individuals walk in the same direction (Topoff, 1980; Jander & Daumer, 1974; Schneirla, 1940), see Fig.(1a). When army ants happen to meet an obstacle it sometimes happens that they move around it and end up forming circles that revolve endlessly around the obstacle (Schneirla, 1940), see Fig.(1b). *Iridomyrmex humilis* workers are able to use trail information to make decisions when confronted with the problem of choosing the shorter of two branches at a trail bifurcation (Goss et al., 1989) or in constructing the shortest network connecting different nests (see Aron et. al., this volume, section III). Clearly, the larger the number of individuals involved in a certain phenomenon the more complex this phenomenon is going to appear and the more orderly compared to the erratic behaviour of a single ant. Large scale behaviour features phenomena that are not found at the individual level and that are far beyond the capabilities of one single ant. One is thus confronted with the question, "what are the rules at the individual level that lead to these complex patterns at higher levels of organisation"? We present a simple model of individual behaviour that seems to be promising in bridging the gap between these two extremes.

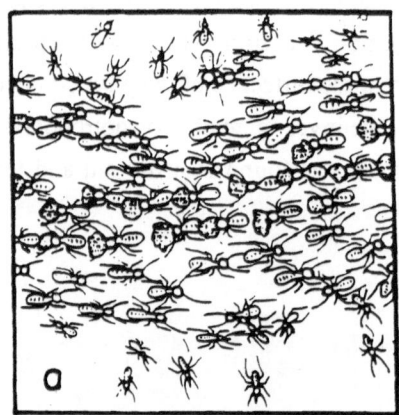 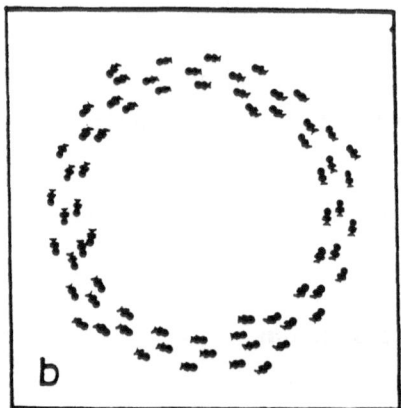

Figure 1a. Organization of a foraging column of *Lacessititermes lacessitus* (Redrawn from Jander & Daumer, 1974).
Figure 1b. Abstraction of a revolving army ants pattern (From Deneubourg & Goss, 1989, in press).

2. Formulation of the problem

Chemical trails are laid by individuals involved, for example, in foraging. The trails contain different types of information. A pheromone perceived in the nest is able to induce recruitment. A somewhat similar situation is the case of alarm where a large number of individuals are attracted to a site of an emergency. Outside the nest the trail acts as a means of orientation to find a food source or to return to the nest. Since the trail is often reinforced when homing from a successful foraging trip, it also contains information about the quality of the food source (Hangartner, 1969; Wilson, 1971).

   Ants following scent trails use two mechanisms for orientation: osmotropotaxis and klinotaxis (Hangartner, 1967; Leuthold, 1975). Osmotropotactic orienting ants determine the concentration of a pheromone with the tip of each antenna (and so measure the concentration gradient) and turn to the higher concentration (Kühn, 1919; Fraenkel & Gunn, 1940; Bell & Tobin for a recent review of these concepts, 1982). It is not known, however, what the detailed motion actually looks like: whether they scan the trail smoothly or only after certain periods of sampling; whether they change the direction always by the same amount or whether they can adjust the amount of turning. Klinotaxis (see Vicker, this volume, section II/3) on the other hand (the mechanism of a dog tracking prey) also plays a role. In klinotactic orientation the animal makes several concentration measurements sequentially. Since only single measurements are taken (instead of two for the case of osmotropotaxis) the animal does not determine the concentration gradient at one point in space, but concentration differences in time and space. Therefore klinotaxis gives the information whether the animal has to continue or to back up in order to follow a gradient. It gives no directional information, however, in the sense that it provides no means of telling in which direction one has to turn in order follow a gradient. A thorough discussion of this subtle point can be found in Bell & Tobin (1982) and the references cited therein. From that it follows that klinotaxis is not sufficient for trail orientation, as long as other senses like vision for example are not involved. Preliminary results of computer simulations show that this is indeed the case. This is the reason why we concentrate on osmotropotaxis rather than klinotaxis. We will present a simple model for osmotropotactic behaviour in response to a chemical stimulus at the individual level and investigate its properties in trail following.

Considering many trail following individuals one may conceive of additional rules that have to be followed by the insects. However, if trail following is still to be the underlying mechanism then the basic rules must also remain valid even if modified. In order to test whether it is likely to induce the emergence of a collective pattern on a large scale simply by having many individuals acting as trail followers our model will be tested on an ensemble of identical individuals.

Since the number of interactions increases with system size, e.g. the number of individuals, the problem becomes increasingly complex the more individuals are acting together. Therefore the model, i.e. the rules that determine the behaviour in response to a stimulus, is on a phenomenological level and we do not take into account physiological or neurological processes taking place at the level of the individual. Thus our insects are considered as black boxes that map an input condition (chemical stimulus) onto an output (behavioural response). Both aspects, the physical chemistry of the pheromone and the mapping function will be discussed.

3. The model

3.1. The ants' equations

Suppose the ants perceive a pheromone concentration at a certain moment with the tips of their antennae. The ant will then change direction and move, after which a new measurement is performed. Assuming this simple algorithmic behaviour, the ants' movements are described by the following equations of motion:

(1a,b,c) $\Delta x = (v \cos \theta) \Delta t$; $\Delta y = (v \sin \theta) \Delta t$; $\Delta \theta = F(C_l, C_r) \Delta t$

x and y denote the x- and y-coordinates of an ant's head. $F(C_l, C_r)$ is the function that relates a concentration measurement to the change of direction and takes into account the different chemical environment at the tip of the left and right antennae (C_l designates the concentration perceived at the left antenna tip and C_r the one at the right antenna). θ defines the direction of the body axis of the insect with respect to the trail-normal, see Fig.(2) (the trail corresponds thus to the y-axis). v is the speed of the animal, which we will assume to be constant.

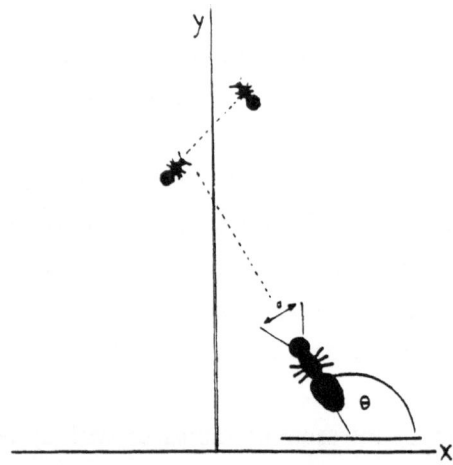

Figure2. Geometry of the model and movement; d = distance between tips of antennae.

It is not known whether the ants make concentration measurements at fixed intervals, use the average of many measurements or even perform continuous measurements. Taking the limit of small Δt one obtains:

(2a,b,c) $\dot{x} = v \cos \theta$; $\dot{y} = v \sin \theta$; $\dot{\theta} = F(C_l, C_r)$

Equations (2) describe continuous measurements and reactions. To take into account some intermediate behaviour between these extremes, C_l and C_r can be considered as some functional over a sampling time:

(3) $C_l = A(t_2 - t_1) \int_{t_1}^{t_2} w(\tau) \, c_l(x, \tau) \, d\tau$

where $A(t_2 - t_1)$ is some averaging factor, $w(\tau)$ is a weighting factor and c_l is the actual concentration at point x and time t (in contrast to C_l, which designates the concentration used as input signal). A similar expression holds for C_r. In this case $F(C_l, C_r)$ becomes an operator. It would also be possible to introduce several time-delays due to processing of information. We shall yet not go into these problems here (see Tranquillo, this section). As mentioned in the introduction we are interested in a simple phenomenological description of the phenomenon of trail following. The introduction of time delays and sampling times allows, however, for extensions of the model and for comparisons with experiment if wanted. In the following we will mostly use the discrete time version of the model in which measurements are performed after each time interval, i.e., the concentration measured last will be used as an input signal $(C_l = c_l)$.

The mapping function $F(C_l, C_r)$ must assure that the ants can stay on a trail and follow it sinusoidally, as observed experimentally (Hangartner, 1967). We assume that there is some maximal turning rate $\dot{\theta}_{max}$ and that the actual correction of direction is somehow related to the concentration difference ΔC between the tips of the two antennae. Further, since the rate of change of ΔC decreases with increasing distance from the trail, to induce a sinusoidal walk the turning rate must be greater off the trail than close to the trail in order to "swing" the insect's body around. Thus $F(C_l, C_r)$ will be of the form:

$$(4) \qquad F(C_l, C_r) = \frac{\Delta C}{|\Delta C|} \; \frac{\dot{\theta}_{max} \, |\Delta C|^n}{|\Delta C|^n + Ref^n}$$

Ref is a positive reference parameter. Since we arrived at (4) by rather heuristic arguments some remarks are noteworthy. Firstly $F(C_l, C_r)$ is a nonlinear function. This is in accordance with the findings of Poggio & Reichardt (1976), who state that for orientational problems the mapping functions are required to be nonlinear functions of the input arguments. Secondly, the shape of $F(C_l, C_r)$ as a function of the stimulus shows a behaviour that is typical for excitation/saturation phenomena. Curves like this are generally found in connection with cooperative systems. One certainly can assume that the processing of input information will be performed in a cooperative manner. Seen like that, the exponent n in (4) tunes the sharpness of the output curve due to internal dispositions. Further, computer simulations revealed that a response with large n ($>= 5$) or a response linear in ΔC (e.g., without saturation) does not lead to effective trail following (see discussion below). See also Tranquillo (this section) and Gruler (this section) for their discussion of stimulus-reponse relationships.

Thus the behaviour of the animals can be summarized as follows. At each unit time step the pheromone concentration is perceived with the tip of each antenna. This concentration difference is then used to change direction (towards the higher concentration) using (1c). Finally the animal advances one unit steplength (1a,b) and the cycle begins anew.

3.2. The physical chemistry of the pheromone - The case of trail following

Trail following experiments are usually performed on circular trails (see e.g. Van Vorhis Key & Baker, 1982; Topoff et al., 1972). A small quantity of pheromone, frequently in hexane solution, is applied to cardboard or filter paper in the form of circles with diameters of about 10 cm. The solution is readily adsorbed and the pheromone as well as the solvent evaporate and diffuse in the course of time. Compared to these artificial trails, natural trails have in common that they are generally also laid on adsorbing ground, for example soil and sand. They differ, however, in that they are more or less linear, are frequently reinforced and are broad rather than thin and linelike.

Although there seem to be cases where the pheromone, be it for alarm or for trails, consists of only one active substance (Van Vorhis Key & Baker, 1982), it is generally a mixture of different substances (Morgan, 1984; Eder & Rembold, 1987). In the following we will use two assumptions. First we assume that we have only one active substance. Second, we assume that the concentration profile is constant in time. Our last assumption was suggested by an experimental observation of Van Vorhis Key & Baker (1982), who noted that the emission rate of pheromone is constant over a relatively long time compared to the time of a trail following experiment (which lasts a couple of minutes). Our approximation will show that the concentration appears almost constant for the ants, provided that certain conditions are fulfilled.

Having in mind possible future applications such as the formation of foraging columns we will approximate the case for a linear trail. The solution of the diffusion equation for an instantaneous infinitely long line source in a semi-infinite space is given by:

$$(5) \qquad\qquad C(r,t) = \frac{Q}{2\pi Dt} \exp\left\{ -r^2 \middle/ 4Dt \right\}$$

Q is the amount of pheromone initially deposited on the trail, D is the diffusion coefficient of the pheromone, t is the time of diffusion and r measures the radial distance from the center of the trail.

To obtain the concentration off the trail due to a continuously emitting line source one has to sum over the emission time t'. From (5) one thus obtains:

$$(6) \qquad C(r,t) = \int_0^t \frac{M(t')}{2\pi D(t-t')} \exp\left\{ -r^2 \Big/ 4D(t-t') \right\} dt'$$

M(t') denotes the release rate of the pheromone. For not too long a time the release rate of pheromone can be expected to be constant and one obtains after two variable transformations an expression known as the exponential integral function,

$$(7) \qquad C(r,t) = \frac{M}{2\pi D} \int_u^\infty \frac{\exp\{-v^2\}}{v} dv \ , \qquad u = \frac{r}{\sqrt{4Dt}}$$

whose solution is (Abramowitz & Stegun, 1970):

$$(8) \qquad C(r,t) = \frac{M}{2\pi D} \left[-\gamma - \frac{1}{2}\ln u - \sum_{n=1}^\infty \frac{(-1)^n u^{n/2}}{n \, n!} \right]$$

where γ is Euler's constant.

Inserting suitable parameters it is easily verified that eqn.(8) predicts changes of concentration of about 30% for the duration of an experiment (a few minutes) and close (up to ~0.25 cm) to the center of the trail. This seems to be rather high at first sight. Considering that the behaviour of ants in experiments only changes markedly if one changes the amount of deposited pheromone over some orders of magnitude (Pasteels et. al., 1986) we can consider the concentration profile as constant near the trail and on the time scale of the duration of an experiment, which justifies the use of (5) instead of (8) which is actually more precise, but is also more time consuming in computer simulations. More important, however, computer simulations show that the behaviour does not change qualitatively when using a dynamic concentration profile.

3.3. Discussion of the phase-portrait of the equations of motion

The concentration profile can be considered constant in time near the center of trail. This renders the set of equations of motion (1, 4) autonomous and the product Dt which appears in (5) in the expression for the concentration can be regarded as a parameter that shapes the concentration profile. In this case, an analysis of the trajectories of the equations of motion in phase-space will increase our understanding of their behaviour.

We consider (1a) and (4) only for the case where the flexibility exponent n equals one. Thus we have:

$$(9) \qquad \dot{x} = v \cos \theta \quad ; \quad \dot{\theta} = \frac{\Delta C}{|\Delta C|} \frac{\theta_{max}}{\Delta C + \text{Ref}} \Delta C$$

with

$$(10) \qquad \Delta C = B \left[\exp\left\{ -\frac{(x - (d/2 \sin \theta)^2}{4Dt} \right\} - \exp\left\{ -\frac{(x + (d/2 \sin \theta)^2}{4Dt} \right\} \right] \quad , \qquad B = \frac{Q}{2\pi Dt}$$

Considering only the first two quadrants the system (9) has only one stationary point in the phase plane (x, θ), namely $(0, \pi/2)$. To consider the behaviour of perturbations δx and $\delta \theta$ around this stationary point we develop (9) in a Taylor series. Retaining only linear terms we arrive at:

$$(11) \qquad \dot{\delta x} = -\delta \theta \, v \quad ; \quad \dot{\delta \theta} = \delta x \, R$$

with

$$(12) \qquad R = \dot{\theta}_{max} B \frac{1}{\text{Ref}} \frac{d}{2Dt} \exp\left\{ \frac{-(d/2)^2}{4Dt} \right\} .$$

The eigenvalues λ_1 and λ_2 of the solution of the characteristic equation

$$(13) \qquad \text{Det} (S - \lambda I) = 0 \quad ,$$

where S is the matrix of the first order partial derivatives of the Taylor-expansion

$$(14) \qquad S = \begin{pmatrix} 0 & -v \\ R & 0 \end{pmatrix} \quad ,$$

and I is the unit matrix, are complex conjugate without real part, namely:

$$(15a,b) \qquad \lambda_1 = i \sqrt{Rv} \quad ; \quad \lambda_2 = - i \sqrt{Rv} \quad .$$

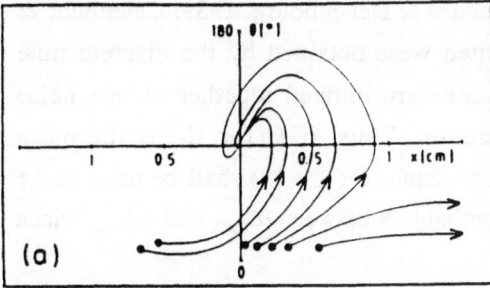 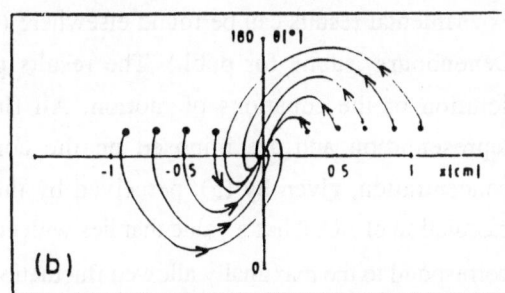

Figure 3. Sketch of the trajectories in phase space; Concentration profile fixed after 10s diffusion; $\Delta t = 0.01s$; $D = 0.01cm^2/s$; $v = 0.15cm/s$; $d = 0.5cm$; $Q = 1.2 \times 10^{-11}g/cm$; Ref$= 5.0 \times 10^{-12}g/cm^3$; $\dot{\theta}_{max} = 30°/s$; $n = 1$;
Initial conditions: Starting angles : a: $10°$; b: $110°$.

Thus, considering only linear terms these results predict a center, e.g. neutral stability. Computer simulations, however, suggest that the system displays the behaviour of a stable focus (in a few cases a stable node). A center can always turn into a focus or a node in the presence of non-linear higher order terms, as seems to be the case here. Taking into account higher derivatives does not lead to definite analytical conclusions about the phase portrait. This is mainly due to the fact that most of the higher derivatives in the Taylor-expansion disappear. No comparisons can be made between this and other systems, whose behaviour is known. Therefore the only tool to rely on is a sketch of the calculated phase portrait, which is depicted in Fig.(3), and which shows that the system is a stable focus. This indicates that an individual whose motion is governed by (9) will always tend to the center of the trail. Any perturbation (as long as it is not too great) will be damped out. This result is important because so far we were dealing with rather idealized conditions. The concentrations used were given by the solution of the diffusion equation, which is certainly an oversimplification. Although we don't expect that the concentration profile will be changed considerably by the motion of the ants (M. Mareschal, pers. comm.) there will be certainly distortions of the concentration profile in nature due to wind and turbulences besides being subject to fluctuations in density. In general concentration fluctuations are δ-correlated. This is true for homogeneous concentration profiles as well as for concentration profiles that are under the influence of a flux (F. Baras, pers. communication; Calenbuhr & Deneubourg, in prep.). In the next chapter we disuss the results of computer simulations using (1) and (4) under different conditions.

4. Results obtained by computer simulations

Since the model contains a considerable number of parameters, a great variety in its behaviour can be expected. Here we will present only a few results. More details and comparisons with experimental results can be found elsewhere (Calenbuhr & Deneubourg, 1989; Calenbuhr & Deneubourg, subm. for publ.). The results presented were obtained by the discrete time solution of the equations of motion. All fluctuations are lumped together in one noise representation and are imposed on the concentration. Thus, if C_l (or C_r) is the mean concentration, given by (5), perceived by the left antenna, $C_l(1 + \varepsilon)$ shall be used in the calculation of ΔC. ε has a value that lies with equal probability between $-\varepsilon_{max}$ and $+\varepsilon_{max}$ which correspond to the maximally allowed fluctuations.

In trail following experiments one is mainly concerned with two questions. What is the probability of losing the trail and what is the influence of the strength of the scent field, e.g. the amount of pheromone deposited? These two experiments are easily conducted, whereas other factors, like the speed, the distance between the antennae and so on are difficult to control, since they are linked to the size of the animals and cannot be measured independently.

In experiment it is found that the fraction of ants still on the trail decreases exponentially as a function of the distance traveled on the trail. For the mean distance travelled by a certain number of ants as a function of the amount of pheromone deposited, a peaked curve is obtained (Evershed et al., 1981; Pasteels et al., 1986; Detrain, 1988). Both results were also obtained by computer simulations of the model, see Figs.(4, 5). One remark should be added concerning the peak height of the curve in Fig.(5). The trail following capacities of the individuals are not always independent of the starting angle (as in experiment). The mean distance at the left side of the maximum is determined by the distance of those individuals that rest rather long on the trail and those whose starting conditions are so unfavourable that they leave the trail after a few steps, if they are able to respond at all. The more the individuals become independent of the starting conditions the longer the distance they contribute to the mean. That means that the height of the peaks in Fig.(5) is determined by the run-time of the computer simulation (since these individuals did not always leave the trail) and is thus underestimated. This does not influence, however, the validity of the qualitative content and the validity of the results. Moreover, here we are treating only linear trails since we have certain applications of the model in mind. For the circular trails the single humped curve is also obtained. But in this case its height is independent of computer run-time because all individuals leave the trail much earlier. The question arises whether there are other mechanisms that are responsible for leaving the trail or whether the trail following capacities are really that much better for linear trails. Recent experiments (Detrain, in prep.) indicate that this might indeed be the case. As predicted by the model, mean distances almost one order of magnitude longer for linear trails have been obtained in the experiments conducted recently by C. Detrain. Further experiments are needed and are being prepared.

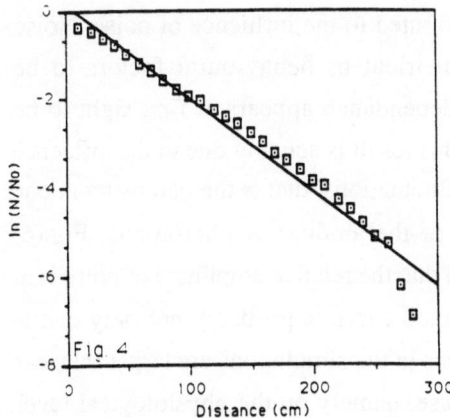 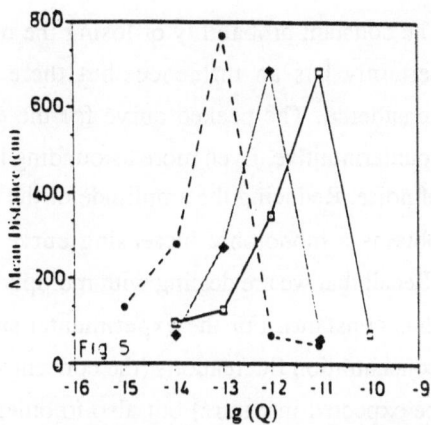

Figure 4. Natural logarithm of fraction of ants still following the path as a function of distance; linear trail; Plot-fitting: $\ln(N/No) = 0.197 - 0.021 \cdot$ distance; Concentration profile fixed after 10s diffusion; $\Delta t = 1.0s$; $D = 0.01 cm^2/s$; $v = 0.15 cm/s$; $d = 0.5 cm$; $Q = 1.2 \times 10^{-11} g/cm$; Ref $= 5.0 \times 10^{-12} g/cm^3$; $\dot{\theta}_{max} = 30°/s$; ε_{max}, $= 0.85$; $n = 1$; number of simulations $= 1000$; starting angle $= 20°$.

Figure 5. Mean trail distance followed as a function of the amount of pheromone deposited; linear trail;
Parameters: idem fig.(4), except:
Ref values: $\bullet = 5 \times 10^{-13}$ (g/cm^3); $\blacklozenge = 5 \times 10^{-12}$ (g/cm^3); $\square = 5 \times 10^{-11}$ (g/cm^3); number of simulations $= 100$;
equiprobable distribution of starting angles; Q is in units of (g/cm).

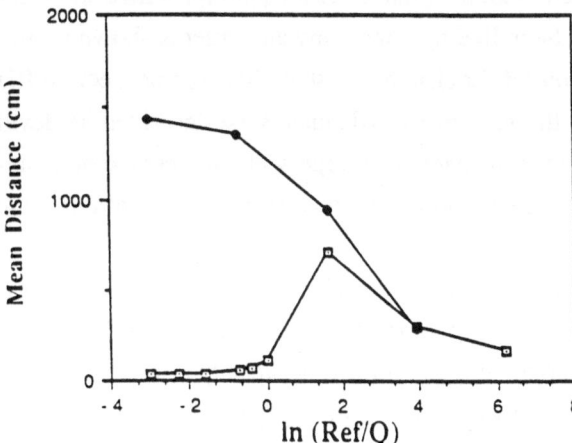

Figure 6. Mean distance followed on the trail as a function of the amount of pheromone deposited for different values of the noise amplitude ε; linear trail; Parameters: idem fig.(4), except: number of simulations $= 100$; equiprobable distribution of starting angles; Noise amplitudes: $\blacklozenge$: $\varepsilon = 0.25$; $\square$: $\varepsilon = 0.85$; Ref/Q is in units of cm^{-2}.

The constant probability of losing the trail is often attributed to the influence of noise. Noise certainly has an influence, but there are also geometrical or behavioural factors to be considered. The peaked curve for the concentration dependance appears at first sight to be counterintuitive. Even more astounding is the fact that this result is actually due to the influence of noise. Reducing the amplitude of the concentration fluctuations, that is the parameter ϵ, one obtains a monotonic increasing curve as a function of the amount of pheromone, Fig.(6) (Recall that we are dealing with multiplicative noise and that the relative amplitude of noise is as thus constant). For the experimental situation the peaked curve is probably not only due to concentration fluctuations (the concentration fluctuations in the simulations are larger than can be expected in nature) but also to other sources of noise, namely on the physiological level. When fluctuations were imposed on the parameter Ref rather than ΔC qualtitatively the same result was obtained. Namely, the peaked curve turned into a monotonic increasing one.

5. Application of the model to an ensemble of individuals

As was pointed out in the introduction we are interested in finding out what the individual rules are supposed to be to obtain collective behavioural patterns. We will have a closer look at those patterns already mentioned, i. e. those of the army ants or termites (see Figs.(1a, b)) and the network problems of *Iridomyrmex humilis* (cf. Aron et al., this volume, section III). These patterns are of two different kinds. First, there are patterns where short range order is repeated over the whole pattern. Seen like that, the army ant patterns shown seem to be similar to e.g., collective cell orientation (cf. Edelstein - Keshet, this volume, section III). On the other hand there are cases, where the pattern can adequately be described as developing a preference (choice) for certain regions in space on a large scale: the choice of a shorter branch at a trail bifurcation (Goss et al., 1989) and the shortest network problem of *Iridomyrmex humilis* as described by Aron et al. (idem).

The common feature shared by all these patterns is that they are based on trail following. In the network problems of *Iridomyrmex humilis* questions of individual perception/stimulus-response and trail following capacity bring at least two different points into play. These are the probability of choosing one of the two branches at a trail bifurcation and the probability of losing the trail. The second is eliminated using bridges, which reduces the complexity of individual behaviour. In the case of Aron et al. a phenomenological description is used. On the other hand one can expect to obtain the same results for this experiment by numerical integration of the equations of motions of the model for individual trail following presented here. This is a good test for the validity of models that are located on a more phenomenological level, but this is also quite time consuming. Even if a quantitative agreement between the mechanistic and phenomenological approach cannot be expected, and is in general actually not obtained (as is for example the case for the binary choice function), the results must coincide qualitatively. Thus, inconsistencies of either viewpoint can be covered up. Already in the case of qualitative agreement an important step in understanding a complex self organizing system has been taken.

The goal of this section is to investigate how osmotropotactic trail following and interaction between individuals can organize trail traffic. This order is characterized on a short length-scale (perpendicular to the trail axis) and is repeated along the axis of the trail. Further, for the time being, the mechanical model seems to be the only description of the phenomenon at hand.

If trail following is also at work in army ants and termites exploring a food source, then it can be assumed that the model may be regarded as a kind of starting point for more complex interactions and behaviour. There are at least two basic approaches to the problem, which will be outlined in the following.

The first approach assumes that the insects follow the trail but that inter-individual collisions between animals moving in opposite directions are inevitable. One might suspect that the system will finally be lead to an ordered state (multi-lane traffic) due to the interplay of the attraction to the trail on the one hand and the repulsion between individuals on the other hand. Yet, if the traffic density is high enough, there will be so many collisions that the system will hardly display any order.

The second approach takes into account that army ants and termites lay trail pheromone continuously ("they trail"). The chemical signal perceived by one individual will be thus strongly determined by the individual moving ahead. In this case one would have "chains" of insects following each other. In this case the system displays already some order (the chains of insects). However, for the sake of a realistic model, collisions have still to be taken into account.

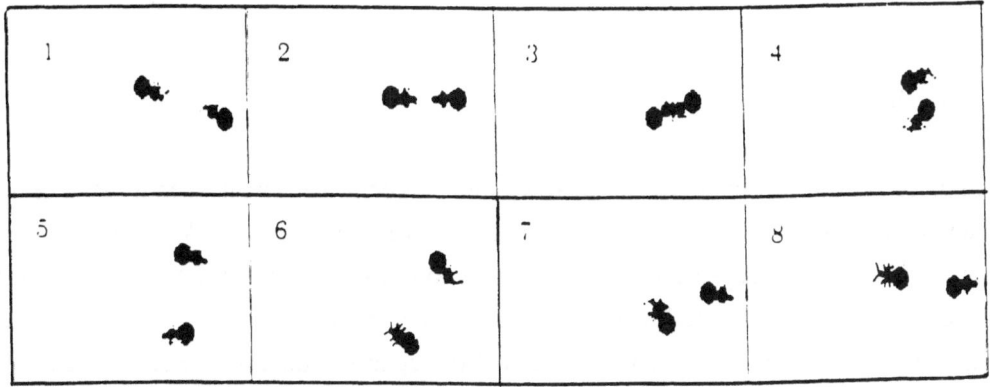

Figure 7. Dodge-manoeuvre of two ants.

In the following we will discuss results obtained using the first hypothesis, in which collisions play a key role. E.g.the individuals react to a chemical stimulus according to eqns. (1) and (4), they do not trail and if they collide they try to resolve the problem by turning to that direction that lies opposite to the direction of contact. E.g., if ant 1 touches ant 2 with, for example, its right antenna, then ant 1 will turn to the left. Ant 2 on the other hand, which feels the touch at its left legs will turn to the right, see Fig.(7). The individuals thus respond either to the chemical stimulus or to a collision. In order to get started we assume that the ants have no information concerning the geometry of a collision (this might be justified, because they are often blind; but see also below!). In that case it is sound to assume that the correction angle to disentangle the collided ants is constant.

Computer simulations using these two rules show the following results:

a) a swarm of ants moving all in the same direction remains a relatively long time together (typically 15 - 20cm) before it spreads. The larger the number of starting ants the smaller is the distance over which a synchronized behaviour can be observed.

b) Suppose the trail center is saturated with ants heading in one direction. Each time an ant reaches the nest (for example), it reverses its direction and walks to the source. The ant that has reversed its direction has been observed to move along either side of the central line of ants. However, this structure (three lanes) is unstable and only observed on a length-scale of a few centimeters. Under these conditions the build-up of two parallel lines of ants in either direction has not yet been observed.

For a large number of parameter combinations tested we were not able to obtain a structured traffic flow. This might indicate that the first hypothesis is not a good candidate to explain trail structuration. On the other hand a few simulations using the second hypothesis have been conducted and give very encouraging results. These preliminary results could suggest that this second mechanism could be a key mechanism in the organisation of trail traffic. However, this is future work.

Discussion

Compared to the actual biological problem an animal is confronted with when it has to deal with orientational problems the model is certainly an oversimplification. Anyway, despite this limitation, or maybe because of it, it allows us to draw some general conclusions. Firstly, concerning individual trail following, the results of the model suggest that the chosen function that relates a chemical stimulus to a behavioural response should not be considered as THE function but only as a representative of a class of functions with similar properties. For example, the introduction of saturation at the level of the antennae does not change any results as long as the flexibility of the response is not diminished considerably. Increasing, however, the stiffness of the response (larger value of the parameter n) reduces the trail following capacities of the individuals. In the limit of large n eqn.(4) becomes a step function, which allows the individual to respond only when the stimulus reaches a certain value, after which almost always the same change of direction is performed. We attribute the parameters n and Ref to a kind of internal disposition.

Computer simulations show that it is possible to find parameters that allow the individual to perform trail following also when using threshold values and a constant turning rate. Comparison of the results obtained for flexible and rigid responses show, however, that a flexible response yields a much better trail fidelity and, even more important, the animals can stay on the trail for a large number of parameter combinations (e.g. change of concentration, speed and so on).

What has been said concerning the simplicity of the model for trail following is even more true for the case of collective behaviour. At first sight, the problem to obtain parallel lanes and structuration seems to be very much related to problems like the alignment of liquid crystals, or the formation of oriented cell agregations (see L. Edelstein-Keshet, this volume, section III) This, however, is not the case. While liquid crystals and cell aggregates can be modelled by means of a continuous description, our ants cannot be treated this way. Ants have a preferred direction of motion, they have a complicated shape, their directions fluctuate (which is required for the model to function) and they are not describable as continuum since it deprives them of their liberty to behave and to respond.

The simplest type of interaction that is conceivable does not yet lead to the desired structures, but it already displays some tendencies. We have shown, however, that it is possible for example to obtain a situation where all individuals walk in circles revolving in one direction without introducing new rules of behaviour. In this case (the second approach, see above) the ants detect the pheromone emitted by nestmates and react as discussed if they collide. It appears to be rewarding to make the animals be capable to respond more smoothly to a collision. This could be introduced by contacts, perceived by the antennae.

It is as yet still too early to say which behavioural rules and which physico-chemical requirements are necessary to obtain collective pattern formation for the case of army ants and termites. The model shows, however, that the same basic rule might be at work for different phenomena in different species. It is not only good for individual trail following but it also displays some features of traffic organisation similar to those found in such complex phenomena like the foraging columns in army ants and termites. These might be achieved by tuning parameters to the posed problem and by simple physical constraints (for example there cannot be more than one individual at one place).

Even if there are some details to be added, the individuals are still rather simple objects. The presented model shares features of an emerging point of view in the theories of complex and self-organizing systems which challenges the classical biological view that complex phenomena, displayed by systems made up of many individuals, are due to individual complexity.

Acknowledgements:

We would like to thank W. Alt and the organizing committee for inviting us to the conference. We would like to express our gratitude especially to J. M. Pasteels without whose discussions this paper wouldn't exist. We also thank R. Beckers and J. W. Turner for many fruitful discussions. A special thanks to S. Goss for helpful considerations concerning the computer work and for proof reading the English manuscript of the paper. Last but not least we are indebted to P. Kinet for technical help in preparing some of the figures.

V. Calenbuhr gratefully acknowledges a grant from the Communauté Française de Belgique and the DAAD (Deutscher Akademischer Austauschdienst). This work is supported in part by the Belgian Program On Interuniversity Attraction Poles.

REFERENCES:

Abramowitz M. and Stegun I. A. (1970) Handbook of Mathematical Functions. New York: Dover Publications.

Bell W. J. and Tobin T. R. (1982.) Chemo-orientation. Biol. Rev. 57: 219-260.

Calenbuhr V. and Deneubourg J.-L. (1989) Modélisation du comportement du suivi de la piste chez les fourmis. Actes coll. Insectes Sociaux 5: pp. 207-214.

Deneubourg J.-L. and Goss S. (1989) Collective patterns and decision making. EEE in press.

Detrain C., Pasteels J. M., Deneubourg J.-L. (1988) Polyéthisme dans le tracé et le suivi de la piste chez *Pheidole Pallidula* (Formicidae). Actes coll. Insectes Sociaux 4: pp. 87-94.

Eder J. and Rembold H., (Edits.) (1987) Chemistry and biology of social insects. München: Verlag J. Peperny.

Evershed, R. P., Morgan, E. D., Cammaerts, M. C. (1981) 3-ethyl-2,5-dimethylpyrazine, the trail pheromone from the venom gland of eigtht species of *Myrmica* ants. Insect Biochem. 12: pp.383-91.

Fraenkel G. S.and Gunn D. L. (1940) The orientation of animals. Oxford University Press. Revised edition: (1961) New York: Dover Publications.

Goss S., Aron S., Deneubourg J.-L., Pasteels J. M. (1989) Self-organized short-cuts in the argentine ant. Naturwissenschaften 76, 579-581.

Jander R. and Daumer K. (1974) Guide- line and gravity orientation of blind termites foraging in the open (Termitidae : *Macrotermes, Hospitalitermes*). Ins. Soc. 21, no 1: pp. 45-69.

Kühn A. (1919) Die Orientierung der Tiere im Raum. Jena:Gustav Fischer Verlag.

Hangartner W. (1967) Spezifiät und Inaktivierung des Spurphheromons von *Lasius fuliginosus Latr.* und Orientierung der Arbeiterinnen im Duftfeld. Z. vergl. Physiol. 57: 103-136.

Hangartner W. (1969) Structure and variability of the individual odour trail in *Solenopsis geminata* Fabr. (Hymenoptera, Formicidae). Z. vergl. Physiol. 62: 111-120.

Leuthold R. H. (1975) Orientation mediated by pheromones in social insects, pp. 197-211. In: C. Noirot, P.E. Howse, J. Le Masne (Edits.) Pheromones and defensive secretions in social insects. Dijon: University of Dijon Press.

Morgan E. D. (1984) Chemical words and phrases in the language of pheromones for foraging and recruitment in insect communication, pp. 173-183. In: T. Lewis (Edit.) Insect communication. London: Academic Press.

Pasteels J. M., Deneubourg J.-L., Verhaeghe J.-C, Boevé J.-L., Quinet, Y. (1986) Orientation along terrestrial trails by ants. In: T. L. Payne, M. C. Birch, C. E. J. Kennedy (Edits.) Mechanisms in insect olfaction. Oxford: Clarendon Press.

Poggio T.and Reichardt W. (1976) Visual control of orientation behaviour in the fly. Part II: Towards the underlying neural interactions. Quart. Rev. of Biophys. 9, no. 3: 377-438.

Schneirla T. C. (1940) Further studies on the army-ant behaviour pattern. Mass organization in the swarm-raiders. J. Comp. Psychol. 29: 401-461.

Topoff H., Lawson K., Richards P. (1972) Trail following and its development in the neotropical army ant genus *Eciton* (Hymenoptera: Formicidae: Dorylinae). Psyche 79, no. 4, pp. 357-364.

Topoff H. (1980) Le comportement social des fourmis légionnaires. In: Les Societés animales. Edited by: Pour la Science S.A.R.L., Paris.

Van Vorhis Key S. E. and Baker T. C. (1982) Trail-following responses of the Argentine Ant, *Iridomyrmix humilis* (Mayr), to a synthetic trail pheromone component and analogs. J. Chem. Ecol. 8, no. 1: 3-14.

Wilson E. O. (1971) The social insects. Cambridge, Ma.: The Belknap Press of Harvard University Press.

Conclusion

During the past live sciences have flourished with the growth of knowledge of the chemistry of living matter. Increasingly, however, live sciences are dependent on the mathematical sciences for a deeper and more quantitative understanding of important biological and medical problems. This tendency is evident and expressed by the articles of this section, of this book.

A detailed mathematical analysis makes sens for reliable experimental data. Thus it is not surprizing that the most experiments are highly sophisticated: Tracking the object of interest, picture analysis. Now the scientists studying biological objects with such sophisticated machines are then always faced with the important decision of how to report the data collected. Thousands of data points describing the object of interest accumulate in notebooks, discs, etc and often only the average values of these data are actually reported in the literature. Here it is shown that important information can be extracted from these data if they are all presented and analyzed according to well established methods in physical and mathematical sciences. This mathematical manipulations reveal important information that is not immediately evident. On the other hand if a law for a simple situation is revealed then this law can be used to make predictions for more complex situations.

How we shall proceed to find general laws for our objects of observation is shown in this section: A bright idea is the starting position. For example for magnetic bacteria it was a fundamental physical law – the torque balance equation. By means of averaged trajctories **Petermann et al.** were able to show that the applied magnetic torque equals the viscous torque. Other examples are shown when active cellular processes are involved. The cells, the microorganisms, the organisms, etc., collect information from their environment and guid their locomotory machinery. Thus the information theory can be used as a framework for understanding directional responses. **Gruler** as well as **Tranquillo** derived a rate equation for the direction of migration by assuming the cell acts like an automatic controller. The solutions of the stochastic differential equation are verified by experimental results. Not only averaged values like the dose-response curve are predicted but also the correct shape of the angle distribution function or of the angle autocorrelation function. The results of **Crenshaw, Stokes et al.** as well as **Calenbuhr and Deneubourg** can also be seen in this framework.

IDENTIFYING TAXIS AND KINESIS

(Coordinator: Michael G. Vicker)

Introduction

The contributions to this section address mainly two topics: 1) the problems gathering around the definitions, concepts and measurements of cellular taxis and kinesis and 2) the mathematical and behavioural analysis of accumulation under different types of directed signals in neutrophil leukocytes and the morphogenetic amoeba *Dictyostelium discoideum*. The behaviour of these classical model systems of eukaryotic cell chemotaxis has clinical and developmental significance.

How the study of directional locomotion in cells began and developed within the last 100 years has significantly depended on the type of phenomena and organism under investigation. Unfortunately, this fostered a tendency to restrict the conceptual scope of investigation to essentially two catagories of cell behaviour: taxis and kinesis, which have became elegantly but rather rigorously defined (Kühn 1919, Fraenkel and Gunn 1961). However, suspiciously odd differences of opinion arose between observers of crawling and swimming cells, e.g., whether taxis was a reaction (Keller *et al.* 1977) or simply an effect (Diehn *et al.* 1977).

This restriction also heralded two, complementary problems. First, taxis became the choice behaviour of study in the neutrophil leukocyte despite over 30 years of warnings that taxis might not play such a dominant role in inflamatory reactions in vivo (e.g. Harris 1954, Wilkinson 1982). Yet interest generally tended to focus on "taxis", perhaps because it alone was thought capable of producing orientation and accumulation, perhaps because the term was simply a convenient euphemism for behaviour showing any element or combination of orientation, attraction or accumulation generally. Second, the chemotactic

stimulus itself became universally, yet inexplicably, identified with an attractant spatial gradient: the "chemotactic
gradient". Thus, students of biological locomotion found themselves in possession of a stringent catechism of stimulus and
cell reaction long before experiments were conceived in order
to answer those very questions. These missunderstandings have
led to a growing discrepancy between the precision of definitions of cell behaviour and the reliability of the experiments
themselves.

One way out of this biological and clinical monotony appears
to have come, somewhat surprisingly, from mathematical analyses, modelling and simulation of locomotory behaviour. Thus, a
systematic, global framework including the *possible* facets of
behaviour has begun to acquire recognition, at least in the
mind of the theoretician. These investigations are of a notor-
ious species, because of the abstractness of the research and,
as in this case, because of the irritating fact that some of
the predicted behaviour has yet to be experimentally demonstrated. Nevertheless, while not every model can claim to represent
a real process or phenomenon, many models have helped both to
identitify possible factors needing attention in locomotion
research and to gather them into a practicable, analytical
system.

1. Taxis and Kinesis in Cell Accumulation

There are currently three different views about the mechanism of chemotaxis employed by crawling cells. 1) Zigmond and
her coworkers (cf. Zigmond 1978) have proposed that neutrophils
read spatial gradients of attractants by detecting concentration differences across the length of the cell: the so-called
"spatial mechanism". 2) Gerisch *et al.* (1975) put forward an
alternative gradient-reading mechanism in which cells of *D.
discoideum* project trial pseudopods in various directions: the
"temporal" mechanism. An advancing pseudopod detects either an
increasing or decreasing attractant concentration and thereby
becomes dominant or is withdrawn, respectively. 3) Vicker (cf.
this section) rejected gradient-reading in both cell types and
suggested that taxis was induced by directed temporal (e.g.,
pulse) signals, but that accumulation may also occur in spatial

gradients due to orthokinesis and to klinokinesis. These diver-
gent views have necessitated more explicit tests, analyses and
definitions of cell behaviour.

The first paper in this section by **Tranquillo and Lauffenburger** is a major contribution to cell behaviour analysis aimed
at rendering neutrophil leukocyte behaviour into mathematical
language. Their paper develops one of the most elaborate and
precise models currently available of cell, or rather population, reactions in a signal field. Here the authors discuss
derivation of the chemotaxis flux expression and apply their
results experimentally. The goal of their efforts is a rigorous
and accurate description of neutrophil responses to inflamatory
situations based on events beginning with the stochastics of
cell surface receptor-attractant interactions and proceeding to
take into account some of the most important behavioural elements which might affect directed locomotion. Their analysis
employs real cell locomotion parameters to provide and test
explicit models of population behaviour in clinically applicable "under-agarose" assays. At the highly complex level of
cell and intercellular signalling and cooperative cell interactions, which co-determine cell behaviour in this system, this
mathematical analysis has played an important role in planning
experiments and deciphering their results.

In many ways some of the most important elements of patho-
genic processes are not unlike those of morphogenic events. The
second paper by **Vicker** concentrates on the behaviour of *Dictyostelium* amoebae, one of the most intriguing models of morphogenesis and taxis. The questions here focus on the signal field
requirements for taxis and kinesis. Essentially, the spatio-
temporal patterns of the stimulus distribution in a developing
field are calculated and simulated and used to predict what
forms of behaviour might be expected if cells could read one of
these particular types of stimuli. These predictions are then
compared the actual cell responses to developing and mature
taxin spatial gradients, i.e. under temporal and spatial signal
fields. Especial care is taken in regard to the quality of the
signal field. Of course, one critical point in studies of this
sort is just how good do the experimentally generated signal
fields match the theoretically desired ones.

2. A Reappraisal of the Concepts of Taxis and Kinesis

Some of the problems in the present concept of taxis and kinesis are attacked in the third paper by **Doucet and Dunn**, who approach the controversy from the standpoint of dynamical sys- tems theory. However, the reader will also soon discover that their presentation includes a rich series of critical and wide- ranging views concerning many of the central issues of the measurement and description of biological locomotion and of the possible types of information in signal fields.

This section includes a glossary of terms by **Tranquillo and Alt** in which it is hoped to bring the definition of known and anticipated forms of behaviour to a point. Although the task of assembling these terms fell to them, the impetus for this contribution developed out of exchanges between all the members of this section and other conference participants besides, who became enticed by the theme during their less structured, more fluid "free-time". One of the main aims of the glossary was to provide a basis for continuing discussion, intended to encour- age new concepts and experiments. These are probably of more value in themselves than any realizable list of "accepted" definitions, despite the obvious merits of standardized terms. Indeed, it is the idea of the evolution of these definitions which ought to be accepted. Each of the contributions included in this section are intended as part of this process.

DEFINITION AND MEASUREMENT OF CELL MIGRATION COEFFICIENTS

Robert T. Tranquillo and Douglas A. Lauffenburger[*]
Departments of Chemical Engineering
University of Minnesota, Minneapolis MN 55455, USA
[*]University of Illinois, Urbana IL 61801, USA

Abstract

We demonstrate the measurement of the cell migration coefficients defined in a constitutive cell flux expression first proposed by Keller and Segel: the random motility coefficient, μ, and the chemotaxis coefficient, χ. Discussion begins with more recent derivations leading to refined flux expressions defining μ and χ. Cell density profiles obtained for leukocytes in the linear under-agarose assay are then examined for consistency with theoretical predictions based on the flux expression derived by Alt, yielding values of μ and χ which can be interpreted in terms of fundamental cell movement parameters using the more recent theory. Finally, experimental complications inherent in current population assays are discussed and a novel migration assay which is designed to circumvent the complications is proposed and analyzed.

1. Introduction

Clinical and scientific investigations of leukocyte chemotaxis would be greatly aided by an ability to objectively characterize the intrinsic chemokinetic and chemotactic properties of cell populations responding to a given chemoattractant (CA) from simple population assays (Zigmond and Lauffenburger, 1986). They utilize uniform CA concentrations to assay chemokinesis and CA concentration gradients to assay chemotaxis (see Box 14, this section, for definitions of cell migration terms). Although it is not the most widely used, the under-agarose assay (Fig. 1) has the advantage of allowing direct observation of individual cell movement for correlation with the population response. The CA concentration and concentration gradient in the under-agarose assay vary greatly in time and space (Lauffenburger and Zigmond, 1981). It is known that cell speed depends on the CA concentration (Maher et al, 1984) and directional orientation bias depends on the CA concentration gradient (Zigmond, 1977). It follows that the typical chemotaxis assay observation, a single measurement of the migrated cells after some incubation period, reflects a cumulative response determined by the transient CA concentration field during the assay. The traditional scoring quantities stated in Fig. 1, therefore, depend heavily on factors that determine this field, such as the well separation distance, the incubation time, and the CA diffusion coefficient. Also unsatisfactory, the dependence of these quantities on the CA well concentration is consequently ambiguous.

In contrast, the random motility coefficient, μ, and the chemotaxis coefficient, χ, which were first defined in the constitutive cell flux expression proposed by Keller and Segel (1971) are objective, not depending on such irrelevant aspects of the assay protocol, but, in principle, reflecting the intrinsic chemosensory movement of the cells. Further, they offer the possibility of a precise quantitative characterization of the relative contributions of chemokinesis and chemotaxis to migration of a cell population.

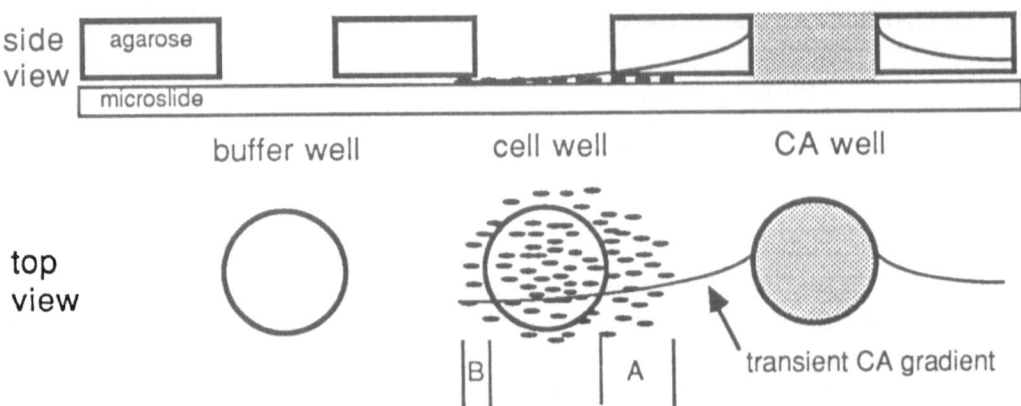

Figure 1. The conventional under-agarose assay. A cell suspension is placed into a cylindrical well located between two wells, into which are added either the CA solution or a buffer solution. During the incubation period, the cells migrate on the substratum under the agarose in response to the transient CA concentration gradient created by CA diffusion through the agarose. Most of the traditional scoring quantities are based on the greatest distance migrated towards the attractant well, A, and "control" well, B, during the incubation period: the leading front distance, A; the chemotactic ratio, A/B; the chemotactic difference: A-B. The use of the "control" response, B, in the latter two quantities reflects a recognition that the final pattern of migrated cells reflects chemokinesis as well as chemotaxis.

2.   Constitutive Cell Flux Expressions

The seminal Keller-Segel expression for the mean cell flux, J_c, of a chemotactic cell population migrating in a steady spatial gradient of CA is (Keller and Segel, 1971):

$$(1) \qquad J_c = -\mu \frac{\partial c}{\partial x} + \chi c \frac{\partial a}{\partial x}$$

where c is the cell density and a is the CA concentration. Their expression derives from an analysis of a model cell which moves in one dimension with constant step size and changes direction with a frequency dependent on the local CA concentration ($\partial a/\partial x$ defines the local magnitude of the CA spatial gradient). Although the expression does not faithfully represent some aspects of known cell behavior nor are the definitions of μ and χ related to mechanistic properties of individual cells (Lauffenburger, 1985), their result does illustrate the phenomenology: The first term in which μ appears has the same functional form as molecular Fickian diffusion, as expected since random movement exhibited in the absence of a CA gradient is an exact analogy on a suitably long time scale. The second term containing χ can be rewritten as the product $V_d c$, where V_d is a drift velocity, so that the chemotactic component of a total cell flux is analogous to the convective component of a total molecular flux.

Cell flux expressions based on probabilistic models which incorporate various details of cell behavior, such as finite cell speed, directional persistence in movement, and bias in turning or directional orientation in a CA gradient, were subsequently developed by Patlak (1953),

Nossal (1980), and Alt (1980). The model formulated by Alt, applicable for movement in n dimensions, derives from a diffusion approximation for J_c based on the following underlying assumptions of stochastic cell movement:

A1) a cell exhibits piecewise linear movement, with mean speed $s(t,\underline{x})$ independent of the direction of movement, θ
A2) a cell moving in direction θ turns with transition probability per unit time $\beta(t,\underline{x},\theta)$
A3) a $\theta \to \eta$ direction change is governed by the conditional probability distribution $\kappa(\eta;t,\underline{x},\theta)$

Under the "smallness" conditions that the mean run time is much less than the observation time, the mean run length is much less than the characteristic system dimension, and the CA gradient is significantly shallow in time and space so that parameters defined in the receptor-mediated motility and CA gradient sensing mechanisms are linearly dependent on gradients in bound receptors, Alt derives the following expressions:

$$(2) \qquad J_c = -\frac{\mu}{s}\nabla(sc) + \chi c\nabla a \quad \text{where } \mu = \frac{s^2}{n(1-\psi)\beta_0} \quad \text{and} \quad \chi = \frac{s^2}{n}\left(\beta_0^{-2}\frac{d\beta_0}{da} + \alpha_1 + \frac{\phi}{s}\right)$$

n is the number of dimensions in which cell movement occurs, ψ is an index of directional persistence given by the mean cosine of the turn angle distribution, $\beta_0 = \lambda/v$ is the mean turning frequency which is generally a function of the experienced CA concentration (λ is the first-order rate constant for degradation of a proposed intracellular motility regulator and v is a parameter which describes the Poisson distribution to which the regulator concentration is reset upon falling to a lower threshold value α whereupon translocation transiently ceases), $\alpha_1 = -1/\lambda \cdot d\alpha/da$ describes a characteristic increase in run time with increasing experienced CA concentration, and ϕ characterizes the tendency for turns to be directed up a CA spatial gradient (ϕ is a complicated function of the parameters defined in the receptor-mediated motility mechanisms assumed by Alt).

For leukocytes, early experimental data for turn angle distributions over arbitrary time intervals suggest that ψ and β_0 do not depend on CA concentration (i.e. klinokinesis can be neglected). Along with the assumption that $\alpha_1 = 0$ and defining a directional persistence time $P = [(1-\psi)\beta_0]^{-1}$, eq. (2) can be rewritten as:

$$(3) \qquad J_c = -\mu\nabla c + \left[-\frac{1}{2}\frac{d\mu}{da} + \chi\right]c\nabla a \quad \text{where } \mu = \frac{Ps(a)^2}{n} \quad \text{and} \quad \chi = \frac{\phi(a)s(a)}{n}$$

This is identical to the Keller-Segel expression except for the term involving $d\mu/da$, which represents a cell flux that can occur in a CA spatial gradient solely from cell speed variations (i.e. orthokinesis). Thus, there is an account for the simultaneous contributions of chemokinesis (more precisely, orthokinesis) and chemotaxis, to population drift in a CA spatial gradient, an important and necessary distinction that the Keller-Segel expression does not make. Recently, investigators have begun to characterize the turning behavior of leukocytes in uniform CA concentrations in terms of P, which will allow a critical assessment of the neglect of klinokinesis that was justified from less objective turn angle distribution data (e.g. Farrell *et al*, 1990). A heuristic derivation of eq. (3) can be found in Rivero *et al* (1989).

Interestingly, for the case of bacteria where experimental data indicate that s and β_0 do not depend on CA concentration (i.e. no orthokinesis) and $\phi = 0$ (i.e. no turning bias), then $\chi = \alpha_1 s^2/n$, consistent with the recognition that bacterial "chemotaxis" is really a kinesis, specifically, an adaptive klinokinesis: biased movement in a CA spatial gradient results from an alteration of motility in response to an experienced CA gradient, the bacterium exhibiting a longer run time when experiencing an increasing CA concentration (see Doucet and Dunn and Box 14, this section).

3.   Experimental Validation

The validation of J_c proceeds by incorporating it into a mathematical model comprised of conservation equations for leukocytes and CA in the under-agarose migration assay. The general procedure is to first characterize $\mu(a)$ in a series of random motility assays by incorporating the CA at some known uniform concentration a^* directly into the agarose in lieu of using a CA well to establish a CA gradient. Otherwise, the assay is conducted in the same way. Analysis of the spatial profile of migrated cells then directly yields $\mu(a^*)$. Once the function $\mu(a)$ has been characterized, the theoretical chemokinesis contribution to the migration of the cell population in a chemotaxis assay is fully specified allowing quantitative determination of $\chi(a)$ if qualitative agreement exists. This procedure has been successfully applied to the case study of human neutrophils migrating in response to the chemotactic peptide summarized below (Tranquillo et al, 1988), and more recently to human microvessel endothelial cells (Stokes et al, 1990) and rat alveolar macrophages (Farrell et al, 1990).

In the absence of a CA gradient for a random motility assay which uses simplifying linear geometry (parallel linear vs radial wells), J_c simplifies considerably yielding the diffusion equation for $c(x,t)$ (cell death is negligible during the assay). Denoting $x = 0$ as the edge of the cell well, the assay is designed to be consistent with the conditions $c(x,0) = 0$, $c(0,t) = c_0$ (see Discussion), and $c(\infty,t) \rightarrow 0$. The slope from a plot of $erfc^{-1}(c/c_0)$ versus x at a fixed t allows estimation of $\mu(a^*)$. Generally $r > 0.99$ was found for over the range of a allowing the function $\mu(a)$ to be characterized as seen in Fig. 2.

For a chemotaxis assay, the CA gradient is predicted from the solution $a(x,t)$ to the diffusion equation with conditions $a(x',0) = 0$, $a(0,t) = a_0/2$, and $a(\infty,t) \rightarrow 0$, as justified by

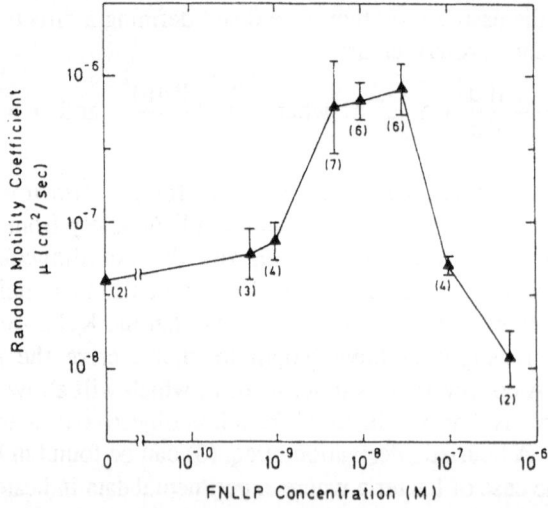

Figure 2. Dependence of the random motility coefficient for human neutrophils on the concentration of chemotactic peptide (from Tranquillo et al, 1988, Fig. 5).

proper assay design, where x' = 0 denotes the edge of the CA well. This assumes negligible alteration of the CA gradient by the cells (see Discussion). In order to proceed, it is necessary to specify the functional form of $\chi(a)$, since the CA concentration varies greatly in the under-agarose assay as noted above and, consequently, $\chi(a)$ cannot generally be approximated as constant. Using the definition of μ in eq. (3), note that with $\mu(a)$ characterized and P independently measured (a constant in the assumed absence of klinokinesis), then s(a) is characterized. From the definition of χ in eq. (3), the specification of $\chi(a)$ thus requires only specification of $\phi(a)$. However, $\phi(a)$ is a lumped parameter, involving three additional parameters associated with the receptor-mediated motility mechanisms assumed by Alt. In order to circumvent the additional complication of proposing or measuring the CA-dependence of each of these, we resort to the phenomenological description of chemotaxis presented in §2.

The drift velocity, V_d, of cells migrating in a one-space dimension CA gradient can be expressed as:

(4) $V_d = (2f(a) - 1)s(a)$

where $f(a)$ is the fraction of cells moving towards higher (versus lower) CA concentration ($f = 0.5$ corresponds to equal movement up- and down-gradient, i.e. random movement, and $f = 1$ corresponds to movement up-gradient only). Zigmond (1977) has determined $f(a)$ for rabbit neutrophils migrating in a quasi-steady linear gradient of chemotactic peptide. It can be empirically described by a saturating function of the difference in the number of bound CA receptors across the cell dimension (Zigmond, 1981; Rivero et al, 1989):

(5) $f = \dfrac{1}{2}\left[1 + \chi_0 N_T(a)\dfrac{dp}{dx}\left(1 + \chi_0 N_T(a)\dfrac{dp}{dx}\right)^{-1} \right]$ where $p(a) = \dfrac{a}{K_d + a}$

N_T is total number of chemotactic peptide receptors on the cell surface, p is the mean fraction of receptors bound with CA, and K_d is the equilibrium dissociation constant for the receptor-CA complex. χ_0 is a constant termed the chemotactic sensitivity which reflects the extent of directional orientation bias in a CA spatial gradient; specifically, χ_0 equals the reciprocal of the difference in bound receptors per spatial distance for $f = 0.75$. Note, however, that eq. (5) does not have a direct mechanistic interpretation since receptors are not uniformly distributed on a migrating neutrophil, as implicitly assumed. Substituting the small CA spatial gradient linearization of eq. (5) into eq. (4) and recalling from §2 the phenomenological equivalence of V_d and $\chi\partial a/\partial x$ in eq. (3), the working expressions become:

(6) $J_c = -\mu\dfrac{\partial c}{\partial x} + \left[-\dfrac{1}{2}\dfrac{d\mu}{da} + \chi\right]c\dfrac{\partial a}{\partial x}$ where $\mu = \dfrac{Ps(a)^2}{2}$, $\chi = \chi_0\dfrac{N_T(a)K_d}{(K_d + a)^2}\sqrt{\dfrac{2\mu(a)}{P}}$

J_c is thus completely specified except for χ_0 since $\mu(a)$ is characterized, a is predicted, and both K_d and $N_T(a)$ are known from independent CA-receptor binding assays. Predictions of cell density profiles were obtained using a finite difference approximation method (Crank-Nicholson) to solve the cell conservation equation with c(x,t) subject to the same conditions as for the random motility assay. Experimental and predicted profiles for four different CA well concentrations a_0 are plotted in Fig. 3 for $\chi_0 = 4 \times 10^{-6}$ cm. A single value for χ_0 yields a satisfactory account both qualitatively and quantitatively over this range of a_0 supporting the validity of the above forms for J_c, μ, and χ for the case study of human neutrophils and chemotactic peptide. It has been shown that the values of μ and χ determined in this manner compare well with those based on measurements of P, s, and f from individual leukocyte movement and using the relationships in eqs. (4)- (6) (Farrell *et al*, 1990).

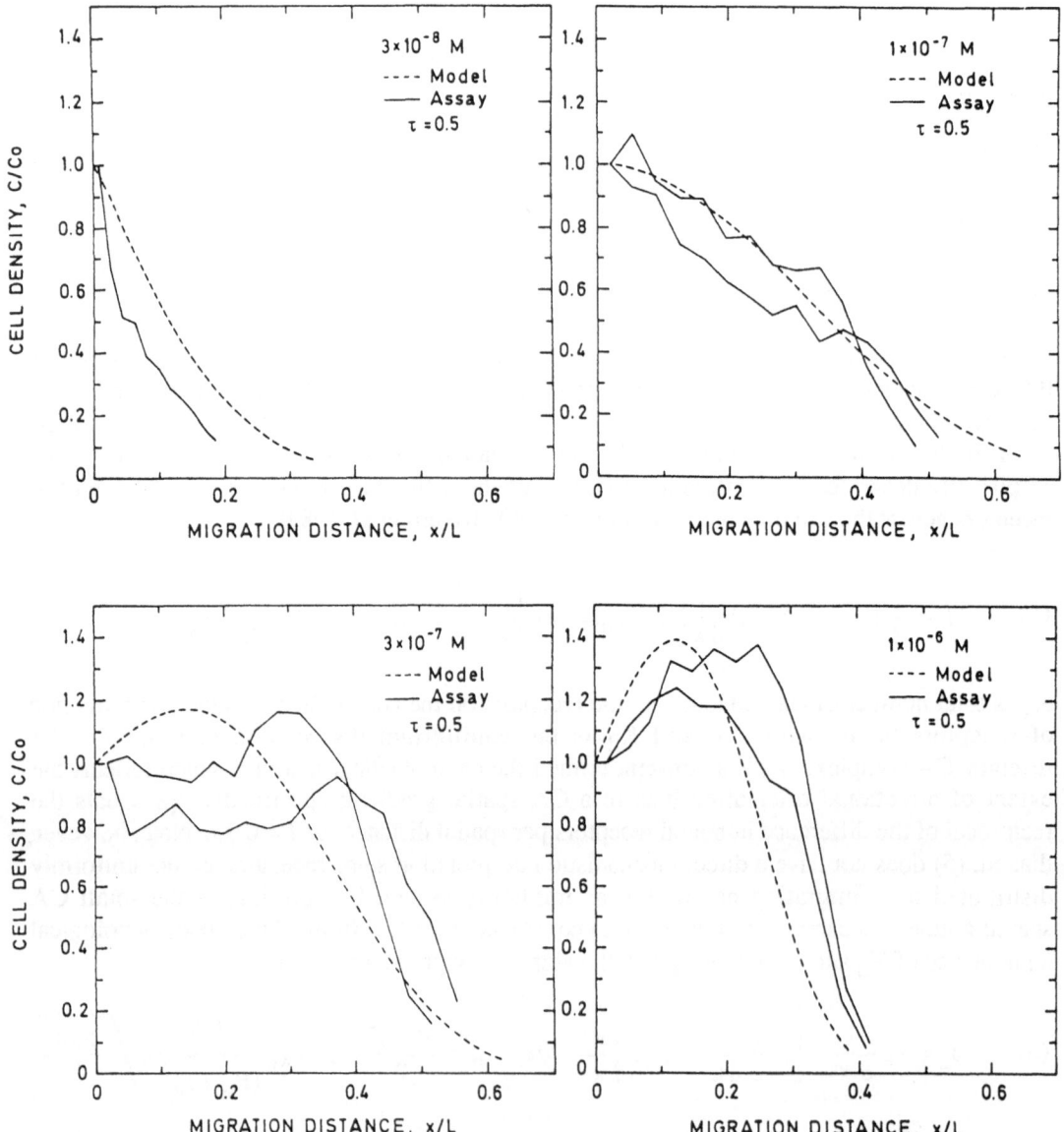

Figure 3. Comparison of experimental and theoretical cell density profiles for chemotaxis. The experimental profiles from the assay are representative results from independent trials. The same dimensionless incubation time $\tau = tD/L^2 = 0.5$ applies to all profiles, where t is the incubation time, D is the CA diffusion coefficient, L is the well separation distance (from Tranquillo *et al*, 1988, Fig. 6).

4.   Discussion

With estimates for μ and χ it is possible to design the assay (i.e. choose τ for a given a_0) to maximize the chemotaxis component of J_c. Interestingly, the chemokinetic (i.e. orthokinetic) component of the flux, associated with the second term of J_c in eq. (6), is seen in Fig. 4 to contribute significantly in theory to the cell density profile for this case study, although this may be unresolvable in practice given typical experimental variability.

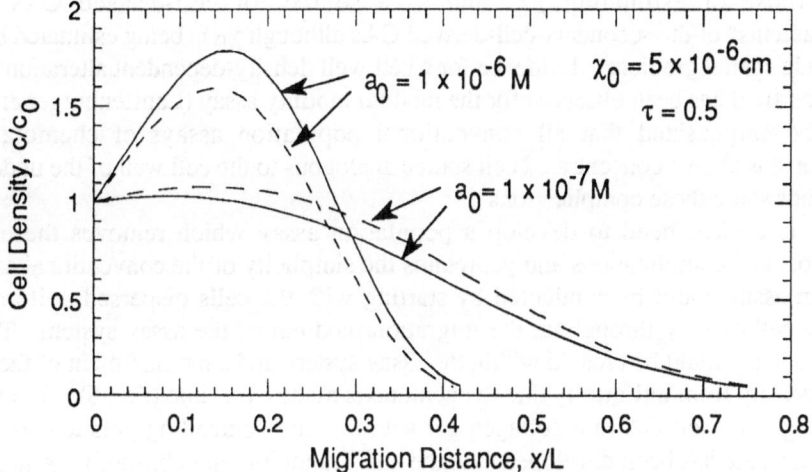

Figure 4. Contribution of chemokinesis to the theoretical cell density profile. Predicted profiles including (solid lines) and excluding (dashed lines) the flux term identified with chemokinesis (i.e. orthokinesis) are compared at two different values of CA well concentrations a_0. Inclusion can enhance (1×10^{-6} M) or diminish (1×10^{-7} M) the profile depending on a_0, because the sign of this term changes (yielding a flux up the CA spatial gradient) wherever $a > a^*$, where $\mu(a^*) = \mu_{max}$ (see Fig. 2). For fixed τ, $a > a^*$ at all x for a greater proportion of time as a_0 increases.

It is appropriate to address possible reasons for the qualitative discrepancies evident in Fig. 3. There are several theoretical possibilities. Klinokinesis, which has been omitted in applying the general cell flux expression to leukocytes, might be contributing significantly to their migration in a CA spatial gradient. Also, the CA spatial gradient $\partial a/\partial x$ is sufficiently steep at early times that based on the known dependence of orientation bias f on $\partial a/\partial x$ for this case study, the assumed linear dependence is not valid then (cf. Farrell et al, 1990). Moreover, the diffusion approximation, consistent with small directional bias, is not valid when $f \to 1$, which occurs with these steep gradients for optimal a_0. One biological explanation might be that the cell population is very heterogeneous with respect to cell movement and turning properties. This would necessitate modeling subpopulations each with its own values of μ and χ and summing the predicted profiles. Another biological explanation is that the directed turning of cells in the assay may be influenced by the temporal component of the CA gradient, a phenomenon which continues to be elucidated (Lauffenburger et al, 1987).

While these inherent limitations in the theory and complexity of the biology might completely explain the discrepancies, it may be that they are mainly due to one or more complications related to the necessity of maintaining a cell monolayer at the cell well edge so as to justify the imposed boundary condition $c(0,t) = c_0$. Cell counting is not accurate near $x = 0$ because of the high cell densities which occur there consequently. More significantly, there is no account for cell-cell collisions in deriving J_c, and collisions occur with high frequency near $x = 0$ again because of the high cell densities. Also, the cells are likely potentiating the CA concentration field from that predicted on the basis of simple diffusion in a number of ways. Neutrophils both degrade the chemotactic peptide and release cell-derived CAs after peptide stimulation. Thus, the cells concentrated in the cell well may be simultaneously acting as a sink for the diffusing peptide (steepening the peptide spatial gradient beyond that predicted for the purpose of estimating χ_0) and as a source for cell-derived CAs (creating "reverse-gradients" of the secondary cell-derived CAs although χ_0 is being estimated based only on the peptide spatial gradient). Evidence for a cell well density-dependent alteration of the CA concentration field has been observed for the random motility assay (Lauffenburger *et al*, 1988). It should be emphasized that all conventional population assays of chemotaxis use a configuration based on a concentrated cell source analogous to the cell well of the under-agarose assay and thus share these complications.

There is a clear need to develop a population assay which removes these high cell density-associated complications and yet retains the simplicity of the conventional assays. For example, an assay could be conducted by starting with the cells dispersed uniformly at the desired low cell density throughout the migration medium of the assay system. Then a CA diffusion gradient would be created within the assay system and a measurement of the departure of the cell density from uniformity due to chemotaxis made over time (Fig. 5). An assay based on dispersing cells and CA in a collagen gel solution and controlling gelation so as to yield such an initial state has been developed (Moghe and Tranquillo, unpublished). A mathematical model for it based on the Keller-Segel flux expression has been analyzed (see Box 13, this section).

a.   b.

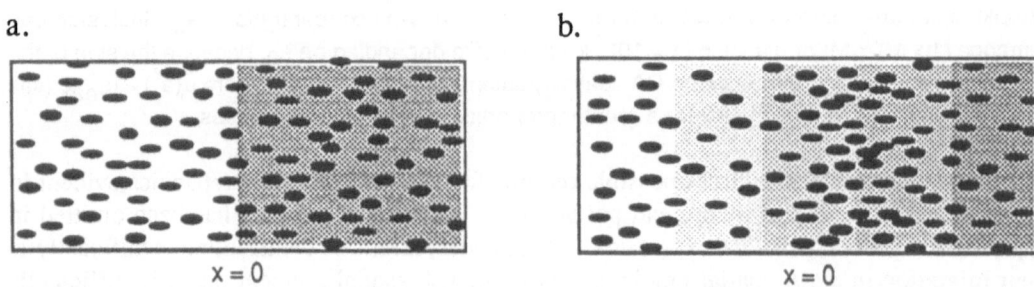

$x = 0$ $x = 0$

<u>Figure 5.</u> Novel dispersed-cell chemotaxis assay. Filled ellipses represent cells. Filled areas indicate the CA concentration. (a) the initial state of the assay. (b) a representative state after incubation showing chemotactic cell accumulation.

5. Conclusions

A firm theoretical foundation has evolved for the phenomenological description of leukocyte chemotaxis since Keller and Segel first proposed a constitutive cell flux expression. The random motility and chemotaxis coefficients have precise definitions in terms of meaningful parameters of individual cell movement in refined variants of this expression.

Although these expressions have been gainfully employed in the analysis of conventional population chemotaxis assays, a new assay configuration more amenable to treatment with the theory is warranted, such as that proposed herein.

Acknowledgements

The authors thank Dr. Sally Zigmond and Dr. Bob Nelson for providing resources and assistance in the experimental work described herein, and Dr. Wolfgang Alt for many helpful comments on the manuscript.

BOX 13

APPROXIMATE CELL DENSITY PROFILE IN A NOVEL CHEMOTAXIS ASSAY: PERTURBATION SOLUTION OF DIFFERENTIAL EQUATIONS

Robert T. Tranquillo

Here an analysis of the mathematical model for the novel assay proposed in Fig. 5 is given. It is based on the Keller-Segel flux expression, eq. (1), (equivalently, the expression derived by Alt, eq. (3), excluding the chemokinesis term) assuming μ and χ are constant. The validity of this assumption for the limiting case of "weak" chemotaxis described below will require complementary numerical studies using known functional forms $\mu(a)$ and $\chi(a)$.

The initial conditions consistent with Fig. 5 are:

$$
\begin{aligned}
(1) \qquad & c = c_0 \qquad && x \in (-\infty, +\infty) \\
& a = a_0 \qquad && x > 0 \\
& a = 0 \qquad && x < 0
\end{aligned}
$$

These are simply a statement that the cells are initially present uniformly everywhere in the geland that CA is initially present uniformly in one-half of the gel. Assuming the physical walls of the assay chamber are sufficiently far from $x = 0$ so that "end effects" are insignificant, during the assay then the appropriate boundary conditions are:

$$
\begin{aligned}
(2) \qquad & c \to c_0 \qquad && x \to \pm\infty \\
& a \to a_0 \qquad && x \to +\infty \\
& a \to 0 \qquad && x \to -\infty
\end{aligned}
$$

These are simply a statement that nothing has changed from the initial conditions far away from the $x = 0$ boundary because there is no CA gradient there (the CA gradient is always greatest at $x = 0$, where the CA concentration remains constant at $a_0/2$, and decreases to zero symmetrically on either side).

Assuming the CA dynamics are governed solely by diffusion, consistent with the low cell density configuration of the novel assay, the diffusion equation solved subject to the above conditions on $a(x,t)$ yields:

BOX 13

(3) $a(\varphi) = 1 + \mathrm{erf}(\varphi)$

where

$\quad\quad a \;=\; a/(a_0/2)$, a dimensionless CA concentration
$\quad\quad \varphi \;=\; x/(4Dt)^{1/2}$, a combined independent variable

Substituting J_c of eq. (3) into the general cell conservation equation yields (with μ and χ constant):

(4) $0 = \Delta \dfrac{d^2 c}{d\varphi^2} + \left(\varphi - \dfrac{\beta}{2} e^{-\varphi^2}\right) \dfrac{dc}{d\varphi} + \beta\varphi e^{-\varphi^2} c$ where $c(\pm\infty) = 1$

where

$\quad\quad c \;=\; c/c_0$, a dimensionless cell density
$\quad\quad \Delta \;=\; \mu/(2D)$, cell random motility relative to CA diffusion parameter
$\quad\quad \beta \;=\; \chi a_0/(D\pi^{1/2})$, cell chemotaxis relative to CA diffusion parameter

An approximate solution to eq. (4) can be obtained by defining $\varepsilon = \beta/\Delta$, reflecting the relative magnitude of chemotaxis, and solving in the limit of $\varepsilon \to 0$ with regular perturbation theory. This limit is seen to be the case where chemotaxis is a small contribution to the cell flux, e.g. when a_0 is suboptimal. The regular perturbation solution proceeds by assuming an asymptotic series for the solution $c(\varphi)$:

(5) $c(\varphi) = \displaystyle\sum_{i=0}^{\infty} c_i(\varphi)\varepsilon^i$

where it has been assumed that the gauge functions multiplying the $\{c_i(\varphi)\}$ are integer powers of ε, though this is not always necessarily the case (Nayfeh, 1981). Grouping coefficients of terms $O(\varepsilon^i)$ after substituting eq. (5) into eq. (4) yields the following two lowest-order equations:

(6) $O(\varepsilon^0)$: $0 = \dfrac{d^2 c_0}{d\varphi^2} + \dfrac{1}{\Delta}\varphi\dfrac{d c_0}{d\varphi}$ where $c_0(\pm\infty) = 1$

(7) $O(\varepsilon^1)$: $0 = \dfrac{d^2 c_1}{d\varphi^2} + \dfrac{1}{\Delta}\varphi\dfrac{d c_1}{d\varphi} - \dfrac{1}{2}e^{-\varphi^2}\dfrac{d c_0}{d\varphi} + \varphi e^{-\varphi^2} c_0$ where $c_1(\pm\infty) = 0$

Eq. (6) yields $c_0(\varphi) = 1$, the first term of eq. (8), and eq. (7) yields $c_1(\varphi)$ given by the bracketed term of eq. (8).

(8) $c(\varphi) = 1 + \dfrac{\beta}{\Delta}\left[\dfrac{\sqrt{\pi}}{\dfrac{2}{\Delta} - 4}\left(\mathrm{erf}\left(\dfrac{\varphi}{\sqrt{2\Delta}}\right) - \mathrm{erf}(\varphi)\right)\right] + O(\varepsilon^2)$

A predicted cell density profile based on eq. (8) is presented in Fig. 1. It is of interest to note that μ and χ can be obtained in a single experiment from this result: μ is found from

BOX 13

measuring the value φ^* associated with c_{max} and determining from eq. (8) the value of Δ which satisfies the extrema condition $dc/d\varphi = 0$. χ is then calculated from the value of β/Δ required to match c_{max} (the same procedure applies using c_{min} and $-\varphi^*$, by symmetry). Thus, μ and χ can be determined from only the location and amplitude of the maximum (or minimum) of cell density. Similar perturbation analyses and comparison to numerical solutions are available for the linear under-agarose assay (Rothman and Lauffenburger, 1983).

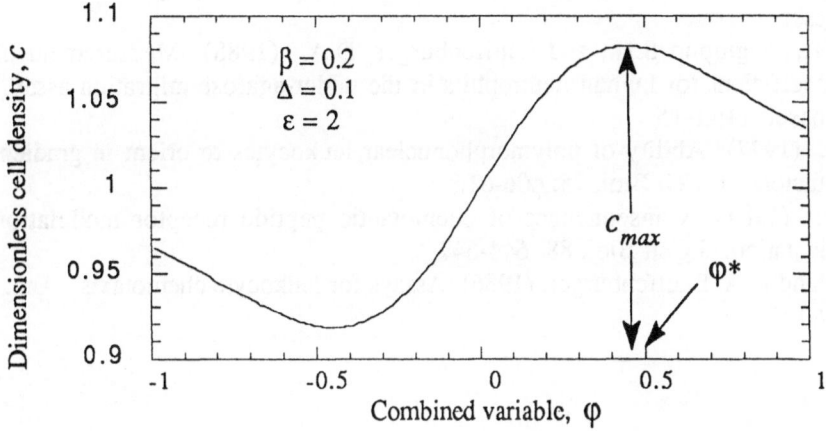

Figure 1. Prediction of cell density profile in the "weak" chemotaxis limit for the dispersed cell assay using the regular perturbation theory solution, eq. (8).

REFERENCES

Alt, W. (1980) Biased random walk models for chemotaxis and related diffusion approximations. J. Math. Biol. 9: 147-177

Farrell, B.E., Daniele, R.P., and Lauffenburger, D.A. (1990) Quantitative relationships between single-cell and cell-population model parameters for chemosensory migration responses of alveolar macrophages to C5a. Cell Motility Cytoskel. (in press)

Keller, E.F. and Segel, L.A. (1971) Model for chemotaxis. J. Theor. Biol. 30: 225-234

Lauffenburger, D.A. (1985) Chemotaxis: Analysis for quantitative studies. Biotech. Prog. 1: 151-160

Lauffenburger, D.A. and Zigmond, S.H. (1981) Chemotactic factor concentration gradients in chemotaxis assay systems. J. Immunol. Methods. 40: 45-60

Lauffenburger, D.A., Farrell B.E., Tranquillo, R.T., Kistler, A., and Zigmond, S.H. (1987) Gradient perception by neutrophil leukocytes, continued. J. Cell. Sci. 88: 415-416

Lauffenburger, D.A., Tranquillo, R.T., and Zigmond, S.H. (1988) Concentration gradients of chemotactic factor in chemotaxis assays. Meth. Enzymol. 162: 85-101

Maher, J., Martell, J.V., Brantley, B.A., Cox, E.B., Neidel, J.E., and Rosse, W.F. (1984) The response of human neutrophils to a chemotactic tripeptide studied by microcinematography. Blood 64: 221-228

Nayfeh, A. H. (1981) Introduction to perturbation techniques, Wiley, New York

Nossal, R. (1980) Mathematical theories of topotaxis. Lect. Notes Biomath. 38: 410-439

Patlack, C. (1953) Random walk with persistence and external bias. Bull. Math. Biophys. 15: 311-338

Rivero, M.A., Tranquillo, R.T., Buettner, H.M., and Lauffenburger, D.A. (1989) Transport models for chemotactic cell populations based on individual cell behavior. Chem. Eng. Sci. 44: 2881-2879

Rothman, C. and Lauffenburger, D.A. (1983) Analysis of the linear under-agarose leukocyte chemotaxis assay. Ann. Biomed. Eng. 11: 451-477

Stokes, C.L., Rupnick, M.A., Williams, S.K., and Lauffenburger, D.A. (1990) Chemotaxis of human microvessel endothelial cells in response to acidic fibroblast growth factor. Lab. Invest. (in press)

Tranquillo, R.T., Zigmond S.H., and Lauffenburger, D.A. (1988) Measurement of the chemotaxis coefficient for human neutrophils in the under-agarose migration assay. Cell Motility Cytoskel. 11: 1-15

Zigmond, S.H. (1977) Ability of polymorphonuclear leukocytes to orient in gradients of chemotactic factors. J. Cell Biol. 75: 606-616

Zigmond, S.H. (1981) Consequences of chemotactic peptide receptor modulation for leukocyte orientation. J.Cell Biol. 88: 644-647

Zigmond, S.H. and D.A. Lauffenburger. (1986) Assays for leukocyte chemotaxis. Ann. Rev. Med. 37: 149-155

SIGNALS FOR CHEMOTAXIS AND CHEMOKINESIS IN CELLS OF *DICTYOSTELIUM DISCOIDEUM*

Michael G. Vicker

Department of Biology, University of Bremen,
2800 Bremen, Federal Republic of Germany

Abstract

Tactic behaviour and its external signal requirements were examined in two ways. 1) The development of a stimulus spatial gradient was simulated in order to predict how cells might behave if able to read either the spatial concentration or relative spatial gradients. 2) Cells of *Dictyostelium discoideum* were exposed to two types of cyclic AMP signal fields while migrating in a micropore filter: a directed pulse lasting 40 s or a stable spatial gradient (which developed from an initially isotropic concentration after removal of the cAMP from one side of the filter). The cell population shifted up-field after a pulse, but down-field in a spatial gradient with no pulse. The optimum for cell motility is 30-50 nM cAMP in a gradient, but ≤ 1 nM as a pulse. The results demonstrate that taxis requires a directed, temporal stimulus and is not induced by cAMP spatial gradients.

1. Introduction

Directional locomotion of relatively slow, crawling cells, like neutrophil leukocytes or the morphogenic myxamoeba *Dictyostelium discoideum* has been classified in terms of reactions to a stimulus that are defined as orthokinesis, klinokinesis and taxis (Keller *et al.* 1977). However, in both of these classical models of tactic behaviour, either type of reaction may induce cells to undergo accumulation, which may be attributable, respectively, to directed turning due to taxis and to biased-random locomotion arising in a spatially anisotropic stimulus field during either kinesis. What characterizes directional reactions and what are the differences between their stimuli?

Several problems inhibit the identification of cell behaviour with its stimulus and these are exacerbated by old tendancies to regard taxis synonymously with spatial concentration gradient stimuli, i.e. "chemotactic gradients". Investigators merely aim at "how do cells read gradients?", although gradient development always includes temporal components and is unpredictable in low viscosity media. Predictable stimulus propagation (e.g., by the laws of diffusion) requires a medium capable of suppressing convection and turbulence such as a gel or filter labyrinth (Vicker 1981), which are rarely used. Difficulties also arise in regard to field boundary conditions and gradient maintainance, e.g. gradients are usually unstable and tend to weaken considerably during the experiment (Tranquillo and Lauffenburger, this section). Finally, global phenomena, including differences in behaviour occurring at individual cell and population levels, usually pass unrecognized.

2. Materials and Methods

Amoebae of *D. discoideum* NC-4(H) were grown on nutrient agar plates at 22°C with *E. coli* as food (Vicker *et al.* 1984). After 48 h the cells were washed in water twice by 5 min centrifugation at 250 g at 2°C, then either plated on non-nutrient agar for 16 h at 6°C or for 6 h at 22°C to allow development to the early aggregative stage. The cells were then washed once, resuspended in water and kept on ice less than 30 min before use.

In some experiments cells were seeded onto one surface of a nitrocellulose filter of 5 μm pore size nominally 150 μm-thick (Sartorius, Göttingen), which they readily penetrate. The filter was placed in a chamber (Fig. 1), which was then laid horizontally, and $1.5 \cdot 10^6$ cells in 0.5 ml developmental buffer (DB, Fontana and Devreotes 1984) was added to the upper compartment. Excess fluid was drained through the filter by gentle suction and the cells were allowed to adhere for 5 min at 22°C before the filter chamber was righted for the experiment.

In other experiments, 5 min after adding cells to one filter surface, the process was repeated by inverting the chamber and

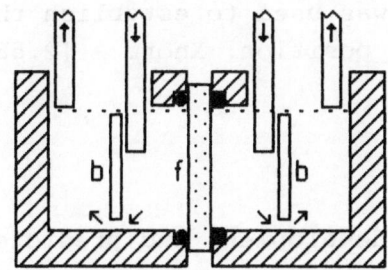

Figure 1. A cross section of the micropore-filter chamber made of two lucite blocks bolted together. Each compartment contains 0.7 ml of DB, and is separated by a 13 μm ⌀ micropore filter (f) pressed between a rubber O-ring and a flat neoprene gasket. To maintain spatial gradients, media was injected into each compartment by peristaltic pump at 0.5 ml/min and removed at 2 ml/min from a single external pool connecting all compartments, thus assuring a uniform fluid level. Chambers could be connected in parallel by glass tubes (2 mm internal ⌀). The input flow was guided across the filter by a glass barrier (b) to maximize gradient quality. The reproducibility of spatial gradient development was analysed by constructing gradients using phenol red. These gradients stabilized in <10 min to a steepness of at least 100:1 (data not shown).

adding an equal number of cells to the other. The population was allowed to evenly permeate the filter by incubation for 30 min in DB and then 30 min in 30 nM cyclic AMP (cAMP, Boehringer, Mannheim) in DB at 22°C. The solutions were infused into both compartments, which reduced the tendancy of cells to concentrate at the filter surfaces (Vicker et al. 1984). The presence of 2 mM caffeine (Sigma, Munich) and 2 mM dithioerythritol (DTE, Boehringer) suppressed cAMP synthesis and phosphodiesterase activity, respectively.

The filters were fixed in formaldehyde, stained in Giemsa, cleared in Permount (Fisher, Philadelphia, USA) and analysed by measuring the cell number in 5 fields in optical sections through the filter using a 40x objective with a numerical aperature of 0.75 (Vicker et al. 1984). With care, a 15 μm section prevented a cell from being counted more than once. The

Wilcoxon rank test was used to establish the significance of shifts in the median position: Xnorm ≥ |2.56| at the 1% level.

3. Results

3.1 Simulated field development and predicted cell behaviour

Ideally, the development of a concentration gradient across a narrow field between a constant source of cAMP and a constant sink proceeds by molecular diffusion until a linear gradient is established (Fig. 2A). The exponentially increasing [cAMP] distribution confronts a cell with several putative stimuli including $\partial C/\partial t$, the temporal gradient, $\partial C/\partial x$, the spatial concentration gradient, and $(\partial C/\partial x)/C$, the relative spatial gradient.

The spatial concentration gradient (Fig. 2B) is often taken to be the tactic stimulus by reasoning that a cell perceives it by comparing concentrations either between two spaced receptors (Zigmond 1978) or with one receptor as it transverses the field (Gerisch et al. 1975). During gradient development its steepness at first increases up-field and then decreases until equilibrium when it is a constant everywhere. Down-field, the value essentially only increases. Thus, the field is characterized by two regions in which the stimulus differs spatially and temporally. If cells respond exclusively to the especially steep stimulus occuring only during development, then taxis will be stronger or perhaps only observed up-field. But the response may be short-lived, because the stimulus is very brief. If cells respond to a stimulus as weak as the linear gradient, one might expect the same amount of taxis throughout the field with the caveat that a suboptimal [cAMP], e.g. down-field from an optimal stimulus, would have little or no effect. This stimulus will persist for the life of the gradient.

The relative spatial gradient (Fig. 2C) appears a more realistic stimulus, at least after a short time, because of cellular adaptation and desensitization to tactic stimuli. A hyperbolically increasing stimulus sweeps down-field early in development. However, within a few seconds a region comparably steep to the early profiles of the relative gradient appears

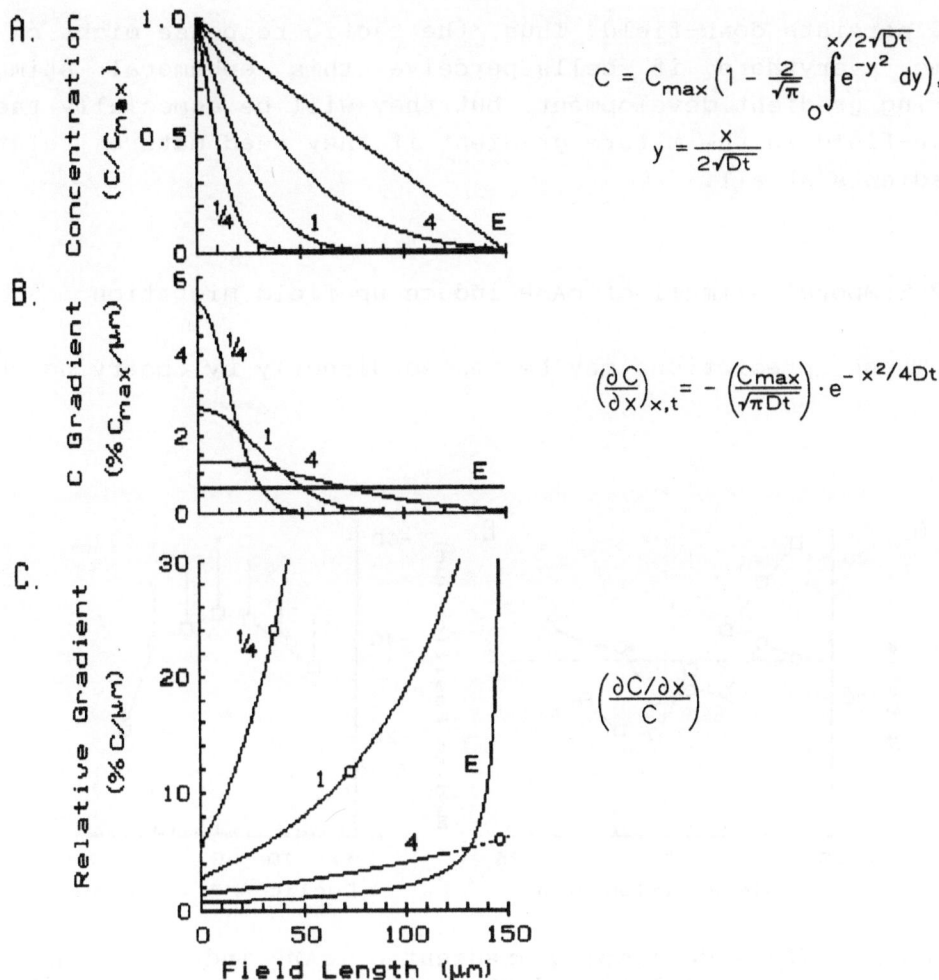

$$C = C_{max}\left(1 - \frac{2}{\sqrt{\pi}} \int_0^{x/2\sqrt{Dt}} e^{-y^2}\, dy\right),$$

$$y = \frac{x}{2\sqrt{Dt}}$$

$$\left(\frac{\partial C}{\partial x}\right)_{x,t} = -\left(\frac{C_{max}}{\sqrt{\pi Dt}}\right)\cdot e^{-x^2/4Dt}$$

$$\left(\frac{\partial C/\partial x}{C}\right)$$

Figure 2, <u>(A)</u> The cAMP distribution at 1000 points was generated numerically at ¼ s, 1 s, 4 s and E (equilibrium) using the expression of Brdička (1963) to the figure's right. C is the cAMP concentration [cAMP], $D=4.44\cdot 10^{-6}\,cm^2\,s^{-1}$ is the cAMP diffusion constant (Dworkin and Keller 1976), e is the base of the natural log., t is in seconds, x is the field position, $C_{min}=0$ and C_{max} is a constant. The "sink" affects the development even before 4 s. <u>(B)</u> The spatial concentration gradient (Brdička 1963). <u>(C)</u> The relative spatial gradient. The flat tail of the 4 s curve is spurious, since it becomes transformed into the E curve. The circle in each line denotes $10^{-2}\cdot C_{max}$ (off scale on E) to illustrate that high steepnesses at low [cAMP], although calculable, may have no physiological significance. The sign of the slopes has been rectified. From Vicker (1989).

and persists down-field. Thus, the tactic response might be the
same everywhere if cells perceive this ephemeral stimulus
during gradient development, but they will be especially tactic
down-field in the mature gradient if they read mature relative
gradients at all.

3.2 Temporal stimuli of cAMP induce up-field migration

These predictions may be tested directly by observing cell

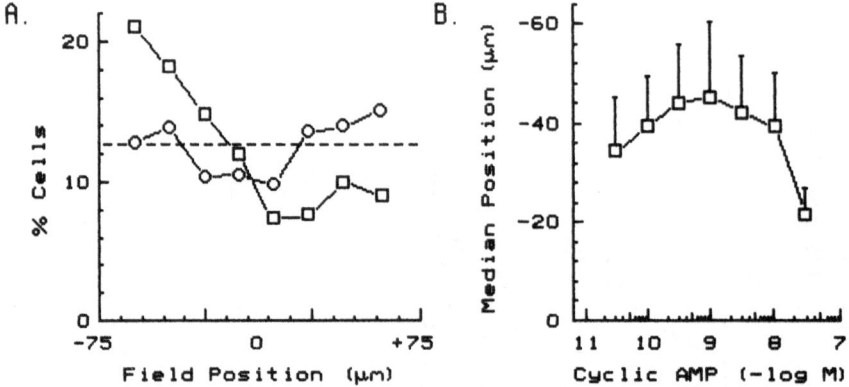

Figure 3. (A) A developing gradient of cAMP induces an up-field
shift in the population distribution. Cells were initially
evenly distributed in the filter. Each chamber was then filled
with 2 mM caffeine and 2 mM DTE in DB and cAMP was rapidly
mixed in the compartment at -75 µm. After 40 s the chambers
were drained, washed 3x in DB and incubated in DB for a further
20 min before fixation. The cAMP pulses were 1 pM cAMP
(circles: control with 'edge effect') or 1 nM (squares). The
dashed line marks the nominal, even distribution. (B) The cAMP
concentration-dependence of the up-field shift in populations
of *Dictyostelium*. Cells were added to the upper surface of a
horizontally placed filter, which was then soaked 5 min at 21°C
in 2 mM caffeine and 2 mM DTE in DB. 0.2 ml cAMP with caffeine
and DTE was added to the lower compartment without wetting the
filter. The chamber was inverted and the cAMP solution was
gently aspirated through the filter within 1 s, thus, exposing
each cell to the same [cAMP] simultaneously in all filters. The
cells were freed of cAMP after 40 s by 3 washes in DB and then
incubated 14 min in DB. The vertical bars represent the Xnorm.

behaviour in the narrow, bounded confine of a penetrable field. A cell population was exposed to a pulse of cAMP while initially evenly distributed across the filter field, then washed and incubated a few minutes to allow migration. The results indicate that the whole population shifts up-field in response to the stimulus (Fig. 3A), i.e. the response was stronger neither up-field nor down-field.

The experiment was repeated with two changes: the initial cell position was the down-field surface of the filter and the signal was applied by aspirating cAMP across the filter from the other side in order to confront every cell simultaneously with a cAMP signal of known concentration, direction and duration. Fig. 3B demonstrates a concentration optimum for cAMP at ≤ 1 nM. The directed reaction to a pulse is known to be greatest within the first few minutes, but subsequently the accumulated cells are relatively slow moving and dispersion lags. Thus, the cells may be found still accumulated after only one signal.

3.3 Spatial gradients induce down-field accumulation

Admittedly, a pulse is not a purely temporal stimulus, since a spatial gradient develops and persists for almost 40 s before the cAMP is washed away. In order to quantify the contribution of the spatial gradient, a gradient was developed across a filter that was evenly permeated with cells. At first, cAMP was infused on both sides of the filter for 30 min. By this time any cellular adaptive reactions, which might have been stimulated, would have ceased. The down-field compartment was then infused with DB alone, thereby generating a spatial gradient as [cAMP] decreased across the whole field. Whether this negative but directional change in the cAMP concentration might also act as a stimulus and effect directed cell movement seems highly unlikely based on the known adaptive behaviour of *Dictyostelium*. Within 45 min of gradient development significant migration of the population had occurred down-field, as indicated by the shift in the mean and median cell positions and the cell distribution, which remained stable for at least 2 h (Fig. 4, Table 1).

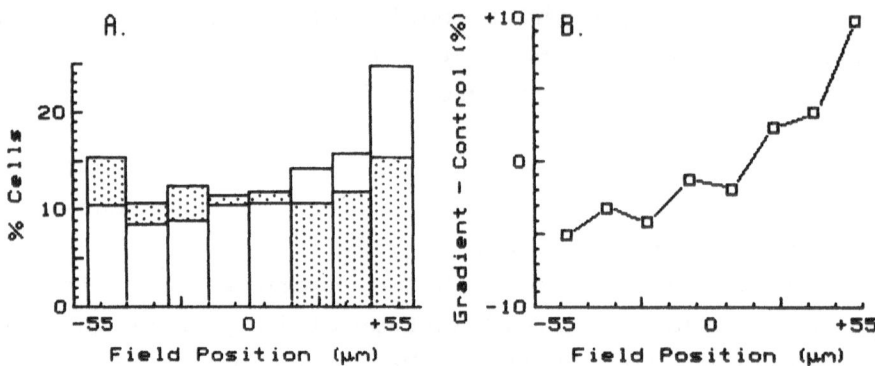

Figure 4. <u>(A)</u> A cAMP spatial gradient induces the down-field migration and accumulation of cells. A spatial gradient signal was generated across an evenly distributed cell population, which had been incubated in a filter infused with 30 nM cAMP on both sides for 30 min. Thereafter, the up-field side (at −75 μm) of the 'gradient' filter was infused with cAMP and the other side with DB for 45 min before fixation (open bars). Both sides of the 'control' filter were infused with cAMP (stipled bars). <u>(B)</u> The same data as in (A) are presented as the difference in the cell numbers between the positive gradient and the isotropic control at each position along the field.

Table 1. The behaviour of amoebae populations in isotropic concentrations and spatial gradients of cAMP while migrating within a bounded filter field.

[cAMP]	Exp.	Cells	Mean (μm)	Median (μm)	Xnorm
Isotropic	a_1	481	−1.21	−1.56	−0.57
	a_2	345	1.45	1.30	0.59
	b	969	4.21	4.48	2.56
	c_1	2246	−4.33	−6.81	−1.48
	c_2	1740	−0.51	−1.83	−0.41
Gradient	a_1	444	13.1	18.2	6.38
	a_2	239	10.0	16.4	3.60
	a_3 *	435	13.8	19.1	6.49
	b	1981	28.3	33.0	23.8
	c_1	2282	8.16	15.8	3.17
	c_2	1687	12.8	23.2	5.26

Evenly distributed cell populations were treated with 30 nM cAMP as an isotropic field or as a spatial gradient, generated as in Fig. 4, for 2 h or *45 min. The resulting mean and median positions of the populations are shown. Positions up-field from the middle of the filter are indicated by a negative sign.

4. Discussion

Neither the migration pattern of *Dictyostelium* in a developing cAMP gradient (Fig. 3A) nor that in a mature gradient (Fig. 4, Table 1) are consistent with views that these cells can read spatial gradients. This reading failure includes even the steepest regions in developing or equilibrium stimulus fields as either the spatial concentration gradient or the relative spatial gradient. None of the predictions obtained from the numerical simulations of gradient reading (Fig. 2) proves relevant to real cell behaviour. The down-field migration of a population (the resulting cell distribution and the change in its mean and median cell positions compared to the initial even distribution), which follows exposure to a spatial gradient lacking a positive temporal component, is probably due to orthokinesis (Vicker *et al.* 1984, Vicker 1989). This induces accumulation where the cells move relatively slower (Wilkinson *et al.* 1984). Klinokinesis, although demonstrable in a spreading population of *Dictyostelium*, does not generate cell accumulation in an evenly occupied, bounded field (at least in the range of cAMP concentrations tested so far) presumably because, unlike the case of motile bacteria, the locomotory reactions of *Dictyostelium* are fully adapted in stable, spatial gradients (Vicker *et al.* 1984).

The results reported here provide strong evidence that a spatial gradient stimulus induces kinesis but not taxis. Similar results have been obtained with neutrophil leukocytes (Vicker *et al.* 1986). The up-field migration following a directed pulse is probably taxis, since time-lapse films (Gerisch *et al.* 1975, Swanson and Taylor 1982) clearly indicate that the response of *D. discoideum* to a directed pulse or impulse of cAMP is a directed turn, i.e. a prompt, directed projection of a pseudopod. There appears no occasion of "testing" of the gradient by trial pseudopods, which strongly argues against the temporal gradient-reading hypothesis put forward by Gerisch *et al.* (1975).

The effects of temporal and concentration stimuli may also be distinguished by their [cAMP]-dependence, which has not been remarked previously. The optima for kinesis (both speed and persistence) in a spatial gradient or isotropic concentration is 30-50 nM cAMP (Vicker *et al.* 1984, and unpublished), but

that for taxis after a pulse of cAMP is $\leq$ 1 nM, a value similar
to that found by Nanjundiah (personal communication). This
difference probably rests on the adaptive status of the cells
to cAMP and, therefore, is evidence for different types of
response to gradient and positive temporal stimuli. Present
concepts of dose-response relationships may, thus, be inadequate in regard to reactions to temporal signals. Siegert and
Weijer (1989) observed that the amplitude of the tactic
response of D. discoideum is the same at natural and at very
low [cAMP]-pulse amplitudes, and they suggested that the
response is, therefore, proportional to the relative change in
signal strength. These observations may support the view
presented here that the stimulus for taxis is a temporal rather
than spatial one.

These results have critical implications for current models
of chemotaxis and morphogenesis. Although a Dictyostelium population in a bounded field reacts to cAMP in a gradient in that
it accumulates where cells move slower, this expresses a
population-level effect derived from a biased-random locomotory
reaction by individual cells to the locally different [cAMP],
not to the gradient per se. The individual cell reacts,
essentially, only to the [cAMP] it encounters, as in an
isotropic field. If the population is spreading from one point
in a spatial gradient field it demonstrates a brief accumulation response since, unlike the case in bacteria, the cells are
adapted to the stimulus (Vicker et al. 1984). Cell orientation
will also occur temporarily under this form of klinokinesis of
stimulus-adapted cells, but not during orthokinesis (contra
Keller et al. 1977). The reaction to a brief, directed pulse of
cAMP by de-adapted cells is different from that to a spatial
gradient stimulus in two ways. 1) Continuous cAMP stimulation
is accompanied by desensitization and adaptation of important
locomotory responses. Second, a directed pulse affects each
cell at one end before it passes distally. The specificity of
temporal stimuli in taxis probably rests on these points.
Therefore, it seems worth suggesting that non-linear cellular
adaptive reactions, feedback and/or oscillatory systems determine signal perception: e.g., by "resetting" some self-organ-
izational feature of cytoskeletal dynamics, critically responsible for local pseudopod projection.

REFERENCES

Brdička, R. (1963) Grundlagen der Physikalische Chemie. Deutscher Verlag der Wissenschaft, Berlin.

Dworkin, M. and Keller, K.H. (1976) Solubility and diffusion coefficient of adenosine 3':5'-monophosphate. J. Biol. Chem. 252, 864-865.

Fontana, D.R. and Devreotes, P.N. (1984) cAMP-stimulated adenylate cyclase activation in *Dictyostelium discoideum* is inhibited by agents acting at the cell surface. Devel. Biol. 106, 76-82.

Gerisch, G., Hülser, D., Malchow, D. and Wick, U. (1975) Cell communication by periodic cyclic-AMP pulses. Phil. Trans. R. Soc. Lond. B 272, 181-192.

Keller, H.U., Wilkinson, P.C., Abercrombie, M., Becker, E.L., Hirsch, J.G., Miller, M.E., Ramsey, W.S., Zigmond, S.H. (1977) A proposal for the definition of terms related to locomotion of leukocytes and other cells. Clin. exp. Immunol. 27, 377-380.

Siegert, F. and Weijer, C. (1989) Digital processing of optical density wave propagation in *Dictyostelium discoideum* and analysis of the effects of caffeine and ammonia. J. Cell Sci. 93, 325-335.

Swanson, J.A. and Taylor, D.L. (1982) Local and spatially coordinated movements in *Dictyostelium discoideum* amoebae during chemotaxis. Cell 28, 225-232.

Vicker, M.G. (1981) Ideal and non-ideal concentration gradient propagation in chemotaxis studies. Exp. Cell Res. 136, 91-100.

Vicker, M.G. (1989) Gradient and temporal signal perception in chemotaxis. J. Cell Sci. 92, 1-4.

Vicker, M.G., Schill, W. and Drescher, K. (1984) Chemoattraction and chemotaxis in *Dictyostelium discoideum*: myxamoeba cannot read spatial gradients of cyclic adenosine monophosphate. J. Cell Biol. 98, 2204-2214.

Vicker, M.G., Lackie, J.M. and Schill, W. (1986) Neutrophil leukocyte chemotaxis is not induced by a spatial gradient of chemoattractant. J. Cell Sci. 84, 263-280.

Wilkinson, P.C., Lackie, J.M., Forrester, J.V. and Dunn, G.A. (1986) Chemotactic accumulation of human neutrophils on immune-complex-coated substrata: analysis at a boundary. J. Cell Biol. 99, 1761-1768.

Zigmond, S.H. (1978) Chemotaxis by polymorphonuclear leukocytes. J. Cell Biol. 77, 269-287.

DISTINCTION BETWEEN KINESIS AND TAXIS
IN TERMS OF SYSTEM THEORY

Paul G. Doucet   and   Graham A. Dunn

Faculteit Biologie   MRC Cell Biophysics Unit
Vrije Universiteit   King's College London
De Boelelaan 1087   26-29 Drury Lane
AMSTERDAM, Netherlands   LONDON WC2B 5RL, UK

Abstract

We describe how a distinction between kinesis and taxis might be
made in terms of dynamical systems theory. The essential feature
is that the stimulus is treated as an independent input variable.
This necessitates that the whole input field is regarded as the
stimulus. In our view, whether an organism is kinetic or tactic
for a particular stimulus depends on what aspects of the input
information can be transferred to the response: a knowledge of
the mechanism by which the information is transferred, or of the
exact form in which the information is used in the response, is
not required to make the distinction, although the final aim is
usually to understand the mechanism.

1. Introduction

Advancement in any scientific field depends, to a large extent,
on the progressive refinement of concepts. Biological motion is
so varied and complex that it sometimes seems a hopeless task to
develop concepts that are both precise and realistic. What is
needed are concepts that keep within the framework of accepted
(or, at least, acceptable) biological usage but are robust enough
to have an obvious and unambiguous meaning when applied to a wide
range of observable phenomena. Here we suggest ways in which a
robust basis might be developed for the concepts of kinesis and
taxis. It is hoped that any clarification that we thus accomplish
will be useful for the practical scientist and will not merely
constitute a new terminology.

Our secondary aim is to accentuate sources of ambiguity and
obscurity in the current usage of these terms. We will have
succeeded in this if we can convince people of these
difficulties and thereby encourage them to explain what they mean
by the terms they use. The hope is that a new dynamic and
adaptive terminology might then develop naturally.

We will first examine some of the range of accepted usage of the
terms. They had already been in use for some years when Fraenkel
and Gunn (1940) incorporated them into their now classic and
comprehensive survey of the orientation of animals. They define
the terms as:
'Undirected locomotory reactions, in which the speed of movement
or the frequency of turning depend on the intensity of
stimulation, we call <u>kineses</u>.'
'The term <u>taxis</u> is used today for directed orientation reactions.
Thus, positive and negative photo-taxis mean respectively
movement straight towards or straight away from the light. We use
the word only for reactions in which the movement is straight
towards or away from the source of stimulation.'

Keller et al. (1977) applied the terms to chemical effects on
cell locomotion:
'<u>Chemokinesis</u>. A reaction by which the speed or frequency of
locomotion of cells and/or the frequency and magnitude of turning
(change of direction) of cells or organisms moving at random is
determined by substances in the environment. Chemokinesis is said
to be positive if displacement of cells moving at random is
increased and negative if displacement is decreased. Two forms
of kinesis have been distinguished: <u>ortho-kinesis</u>, a reaction by
which the speed or frequency of locomotion is determined by the
intensity of the stimulus, <u>klino-kinesis</u>, a reaction by which the
frequency or amount of turning per unit time is determined by the
intensity of the stimulus.'
'<u>Chemotaxis</u>. A reaction by which the direction of locomotion of
cells or organisms is determined by substances in their
environment. If the direction is towards the stimulating
substance, chemotaxis is said to be positive, if away from the

stimulating substance, the reaction is negative. If the direction
of movement is not definitely towards or away from the substance
in question, chemotaxis is indifferent or absent. Positive
chemotaxis can result in attraction towards the stimulating agent
or in retaining the cells in high concentrations of the active
substances by avoidance of low concentrations.'

2. The problems

One source of confusion in using these definitions of kinetic and
tactic behaviour is that they describe the behaviours at several
different levels of causality. Thus Fraenkel and Gunn distinguish
between the intensity of a stimulus in a kinesis and the
direction of stimulation (in relation to the direction of motion)
in a taxis. This distinction is at the level of the information
content of the stimulus. Both groups of authors describe the
behaviour at the level of the short term responses. Keller et al.
imply that the long term responses (i.e. the consequences of the
short term responses) must also be incorporated in the
definition: the long term displacement in the case of a kinesis
and the attraction to or repulsion from a stimulating agent in
the case of a taxis.

The sort of questions commonly generated by this source of
confusion are: Is 'kinesis' the name of a stimulus; a property
of an organism; a reaction or response; or some relation or set
of relations between these? If it is the name of a stimulus then
is the stimulus a static set of environmental properties or
conditions; is it a change in these or in one specified variable;
or is it a difference or comparison between two sets of
properties? If it is the name of a property of the organism then
does it refer to receptors; to mechanism or algorithm; or to
observed and/or theoretical relations between stimulus and
response? If it is the name of a response then does it refer to
a whole pattern of translocative behaviour; to a change in some
property of a pattern; to a comparison of two patterns; or to a
single displacement or turn?

Another area of difficulty in using the current definitions
arises at a more practical than philosophical level: how can we
precisely define and measure mathematical quantities such as
frequency of turning, speed of travel and even position in space?
Examples and possible methodological approaches to reducing these
difficulties have been discussed in the Workshop (see e.g. Dunn
and Brown, section I/1 and Box 4, section II/1).

3. A Dynamical Systems Theory Approach

We will now examine the system of stimulus-organism-response to
see if any robust concepts lurk within the current usage of the
terms kinesis and taxis. We will also examine the implications
of any concepts for what they tell us about the mechanisms of
behaviour.

Stimulus:
What we are basically interested in is how an organism responds
to an external stimulus. This suggests viewing the organism as
a <u>transducer</u>, and a description in terms of dynamical systems.
Let us examine this approach. The current response value may
depend not only on the current stimulus value but also on earlier
stimulus values experienced by the organism (note that we are
speaking here of a specific external factor and ignoring other
external factors and all internal ones like age). In the study
of locomotory behaviour, this '<u>memory</u>' effect is the basis for
a phenomenon usually called <u>adaptation</u>. Dynamical systems theory
formulates such phenomena with the aid of <u>state variables</u> which
record the stimulus (or input) history insofar as this history
influences the response (or output). Often, state and output
variables coincide and one relationship remains which expresses
how the state's rate of change depends on the input and the
state. State and input variables can never coincide: input
variables are required to change independently (that is, they are
not allowed to be influenced by any of the other system
variables). In the parlance of <u>systems theory</u>, a system without
a state variable (where output is simply a function of input) is

called a <u>static</u> system (note that this does not imply that nothing changes).

The environment in which the organism finds itself contains not just a single stimulus but an entire <u>stimulus field</u>: $g = g(t,\mathbf{x})$ a function of possibly one temporal and two or three spatial coordinates. The current stimulus value is codetermined by the stimulus field and by the state variables (position, velocity etc.) and so this cannot be treated as an input variable; the whole field must be the input variable. What we want to know is how the organism reacts to this field: the field is the cause and the translocative behaviour is the effect.

This leads to the following scheme:

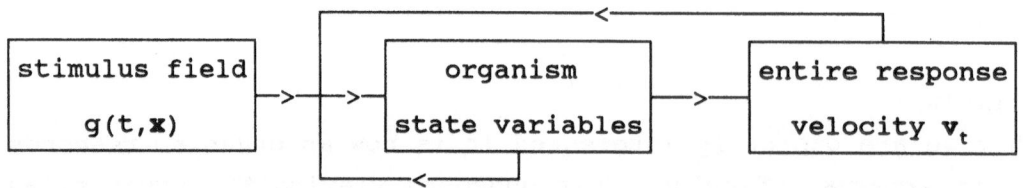

where the resulting velocity vector is a functional
$$\mathbf{v}_t = F\{g(\bar{t},\bar{\mathbf{x}}): \bar{t}{\le}t, \ \bar{\mathbf{x}} \ \text{near} \ \mathbf{x}_{\bar{t}}\}$$
of the stimulus field along the motion path $\mathbf{x}_{\bar{t}}$ for past time $\bar{t}{\le}t$.

This scheme has a number of consequences. Let us suppose that we are dealing with the simple situation of a concentration distribution that does not change with time. The stimulus field is itself a scalar field but it **defines** a vector field such as the field constructed by assigning to each point in space the local concentration gradient $\nabla_{\mathbf{x}}g(t,\mathbf{x})$ (it also defines other vector fields, tensor fields and much else). It is at this point that the organism's response faculties come in. It may be able to perceive the local gradient (its magnitude and/or direction) by comparing measurements of multiple sensors, or perhaps by other means, and act upon it. This means that it is able to distill a **directional component from the stimulus field** (and act upon it). This is what we would call a <u>taxis</u> regardless of the action taken, provided that some action is taken. If it is not able to do so, it may still be able to detect and act upon the **scalar stimulus values** it encounters and we would call this a <u>kinesis</u>. It is also possible that the organism could perceive only the magnitude of the gradient (which does not contain directional information), or some other property of the stimulus field, and a general guideline is that a taxis refers only to those cases in which at least some directional information is acted upon.

There are a few crucial points in this view of the problem:

1) We are assuming that it is not so much the response to an instantaneous single stimulus that is interesting, as the entire response to an entire stimulus field.

2) We allow the stimulus as experienced by the organism to change with time but we do not allow the organism to (co)determine the way it changes, insisting on independent stimulus changes and avoiding the complication of self-generated stimuli. However, perception of a stimulus and storage of information in internal state variables might well depend on the organism's moving state, see below.

In this view, kinesis and taxis (and perhaps other terms) are names of transducer properties. They tell us something about the whole-animal mechanism and not (at least not directly) about the stimulus, or the response pattern, or the structure of the receptors. Grammatically speaking, adjectives would perhaps be more suitable than nouns. A statement like 'this is a tactic mechanism or property of the organism' is clearer than 'this (?) is a taxis'.

Mechanism:
It looks as if an organism without an adaptation-like memory can thus be described by a static system. Consider, however, the possibility (which may be quite common) that an organism may be sensitive to the rate of change of a stimulus as it experiences

it. Mathematically, this rate of change can be defined for an
instant in time, and so no memory is needed, but a real receptor
system may require a short-term memory to detect it. What the
organism perceives is the total derivative of the stimulus along
the path, $dg/dt = \partial_t g + \nabla_x g \cdot v_t$. This depends on three ingredients:
the <u>temporal rate of change</u> of the stimulus field; the <u>spatial
gradient</u> (its magnitude and direction) of the stimulus; and the
<u>current velocity</u> (speed and direction) of the organism. The first
two terms are input but the third is not; since it does not
change independently from the motion, it must be considered a
state variable and the system is therefore not static. In a
uniform stimulus field that changes with time, the second term
is zero and so this organism would not perceive any directional
information and would be capable only of showing a kinesis. But
in a non-uniform, stationary, stimulus field, the first term is
zero and the time derivative along the organism's path
(co)depends on the local slope (magnitude and direction) of the
stimulus field. Hence the mechanism qualifies as tactic even
though the animal may have no way of 'knowing' the local spatial
gradient $\nabla_x g$ (but it could know its own speed and, in some
circumstances, it may know the magnitude of $\nabla_x g$ - marine
organisms might know the magnitude of the pressure gradient - and
so it could 'deduce' its direction of travel in relation to the
external field). This single mechanism that can operate in either
a kinetic or tactic manner, depending on the stimulus field,
raises the question of whether the mechanism should be called
tactic regardless of the stimulus field or only when the stimulus
field contains the necessary directional information. The first
is preferable because the kinesis/taxis distinction then refers
only to transducer properties without taking account of whether
the properties are realised in a particular situation.

A point we have deliberately left open is the precise way in
which an organism manages to extract <u>directional information</u> from
the stimulus field. The reason is that we set out to examine
whole-organism behaviour, ignoring - if possible - all
physiological details. Is it indeed possible to ignore details
of the mechanism? If a single stimulus is itself directional

(collimated light, gravity, magnetism etc.) its detection is a matter of having the proper equipment; if it is not (chemical concentration, hydrostatic pressure etc.) for a tactic response an organism needs special ways of extracting this information. One way to accomplish this is to compare the signals from multiple receptors (taxis based on spatial signal differences) and another is to wave a single receptor around and to compare the successive signals (taxis based on temporal signal differences). A third way is simply to move and compare successive signals and this is traditionally known as a kinesis with memory. But consider a hypothetical snake following a smell, waving its head from side to side. In the traditional view, one could justify two ways of description.

1) the head traces a zigzag path, comparing successive scalar observations - kinesis with memory?

2) the head movements are considered details superimposed on the snake's path - taxis based on temporal signal differences?

This example illustrates that the distinction between some types of taxis and kinesis depends, in the traditional view, on arguments of temporal and spatial scale: on how we define the reference point for recording the animal's position. In our view, the manner of extracting the directional information is important, of course, but irrelevant to this issue. We prefer to regard all mechanisms that extract directional information from the stimulus field at the global level as tactic. In this way, the distinction between kinesis and taxis does not become complicated by rather arbitrary choices of scale or by hypotheses about the geometry and kinetics of possible sensing mechanisms.

Response:

Besides details of mechanism, a second point that we have neglected is the details of the response. In our view, whether an organism is kinetic or tactic for a particular stimulus depends on what aspects of the input information can be transferred to the response. It seems to us that knowledge of the mechanism by which the information is transferred, or of the exact form in which the information is used in the response, is not required to make the distinction.

But what does it mean to say that information has been transferred to the response? At an operational level, what are we expecting to observe in $\mathbf{x}_t$ in order to know what kind of information the organism has used? It is clear that, if the organism can act upon scalar information, then certain transition probabilities in the response will depend on <u>positional information</u>: the pattern of behaviour will be systematically <u>non-uniform</u>. Furthermore, if the organism can act upon <u>directional information</u>, then certain transition probabilities will depend on the organism's direction of travel in relation to the local environmental direction: the pattern of behaviour will be systematically <u>anisotropic</u>.

Unfortunately, non-uniformity and anisotropy of the behaviour pattern cannot generally be used as identifying characteristics of a kinesis and a taxis respectively. Organisms sensitive only to the scalar value of the stimulus may nevertheless show a transient, anisotropic, uni-directional behaviour pattern, sometimes called <u>kinetic drift</u>, as they approach a steady-state distribution. Similarly, organisms sensitive only to direction may show a non-uniform behaviour pattern if directions converge and diverge in the stimulus field. In a stimulus field consisting of a uniform gradient, however, non-uniformity of behaviour implies a kinetic mechanism but it can still be very difficult to determine whether a weak tactic mechanism might also be present (see Tranquillo, this section).

In spite of the difficulties, the underlying concept is very simple if we base the distinction between kinesis and taxis on how an input can be handled rather than purely on a classification of the output. Guided by this simple concept, it is possible to analyse the kinetic mechanism experimentally in uniform, stationary scalar fields and hence to predict, using either theory or modelling, the pattern of purely kinetic behaviour in non-uniform stationary environments. Any discrepancy between observation and prediction would then indicate that a tactic mechanism is present. It may also be possible, though more difficult, to observe the response to the rate of change of uniform scalar fields and hence to predict any temporal tactic behaviour in the non-uniform stationary environment. Any remaining discrepancy would then indicate that a spatial tactic mechanism might also be present.

Finally, it should be mentioned that there is a danger in referring to a behaviour pattern as a kinetic response or even as a putative kinetic response in the absence of any information about the mechanism or the stimulus field. A so-called random pattern of motility, for example, can be generated if the organism moves in the direction of a randomly changing directional variable. On the other hand, an internally stored positional 'search plan' may give rise to an apparent kinesis even though external positional information is absent (see Hoffman, Section II/2).

4. Towards a Classification of Behavioural Mechanisms

We suggest that the kinetic/tactic distinction refers to the basic subdivision between behavioural mechanisms using positional-only and directional-only information. Combinations are possible: cellular chemotaxis is commonly dependent on both the magnitude of the gradient and the chemical concentration. In this context taxis is used as a concept dominating kinesis and the term pure taxis is needed to distinguish directional-only mechanisms. The alternative is to refer to mixtures explicitly as mixtures and reserve the term taxis for directional-only mechanisms.

Perhaps we should emphasise here that we do not consider this classification exhaustive, even as the lowest stratum of a much larger system. For example, there is also a class of directional properties of the stimulus field that do not have a unique direction. At the simplest level these are known as second order tensor fields. Examples of tensor fields are the curvature of the substratum and the mechanical strain in a fibrillar gel. Such fields commonly give rise to bi-directional responses, traditionally known as contact guidances in cell biology.

Also at this lowest stratum of the classification we have to deal with mechanisms for responding to boundaries or singularities in the stimulus field: lines or surfaces at which the field changes abruptly. It may be possible to extend the kinesis/taxis/guidance scheme to deal with these singularities. For example, a boundary separating regions of different level contains positional information:- one side of the boundary is different from the other - and the boundary also specifies a unique direction at each point along it: the direction normal to the boundary. Mechanisms for responding to the first and second types of information could be classed as kinetic and tactic respectively. Another sort of boundary is tensor like e.g. a sudden change in inclination of the substratum: both sides of the boundary are similar but the boundary specifies a unique axis at each point along it and a mechanism for responding to this could be classed as a guidance. Multiple parallel boundaries of either type could be treated individually in this manner or treated together as a guidance field. Again this is a question of spatial scale and an array of parallel grooves is usually treated as a guidance field in cell biology if their separation is small in comparison with cellular dimensions.

An essential way in which our proposed classification differs from many others is that we do not consider the details of the response to be relevant for distinguishing kinesis from taxis except in so far as they can be useful diagnostics for deciding whether the organism responds to directional information. Details of the response are only required in the higher strata where we may wish to distinguish positive kinesis from negative kinesis or ortho-kinesis from klino-kinesis. Similarly, we do not require that a tactic response consists of a uni-directional pattern of translocation. If an organism persistently moves at right angles to an environmental direction then we would call this a tactic mechanism even though the resultant behaviour pattern is bi-directional in two dimensions (this has been called a diataxis by some authorities).

We should finally mention that some authorities take a very different view of the kinesis/taxis distinction. For example, Lapidus and Levandowsky (1981) describe a particular response in which '...individual cells swim in a random walk, without orienting themselves toward the light; but when, by chance, they do swim toward the light, their mean free path is increased, either by increased swimming speed or inhibition of turning, or both. ...' and they add that this '...type of behaviour is indeed a kinesis by our definition, since the organisms do not actively aim toward the light.' From the point of view of Fraenkel and Gunn, this response is to the direction of the stimulus and would not be considered a kinesis since they state explicitly in their introductory remarks that the kinesis response depends on the intensity of stimulation and add that this is established usage. We concur with this latter view and consider that the distinction made by Lapidus and Levandowsky should be incorporated at some higher stratum in the classification of tactic mechanisms.

REFERENCES

Fraenkel G.S. & Gunn D.L. (1940) THE ORIENTATION OF ANIMALS: KINESES, TAXES AND COMPASS REACTIONS. Oxford University Press (Dover edition 1961)

Keller, H.U., Wilkinson, P.C., Abercrombie, M., Becker, E,L., Hirsch, J.G., Miller, M.E., Ramsey, W. Scott & Zigmond, Sally H. (1977) A proposal for the definition of terms related to the locomotion of leucocytes and other cells. Clin. exp. Immunol., 27: 377-380.

Lapidus, I.R. & Levandowsky, M. (1981) Mathematical models of behavioural responses to sensory stimuli by protozoa. In Levandowsky, M. & Hunter, S.H. (eds) Biochemistry and Physiology of Protozoa, 2nd ed. Academic Press, Inc., Orlando, 4: 235-260.

BOX 14

GLOSSARY OF TERMS CONCERNING ORIENTED MOVEMENT

Robert T. Tranquillo, Wolfgang Alt

Following the spirit of the contribution by Doucet and Dunn (this section) the following glossary is not meant to propose a fixed list of accepted definitions, but rather to provide a "formulated" basis for a continuing discussion of concepts, experiments, methods of analysis and mathematical models for various kinds of oriented movements. The descriptive or symbolic terms to be used should, on one hand, possess a far–reaching generality and, on the other hand, serve as a basis for more and more detailed distinctions.

From the previous contributions and discussions during the Workshop and within this chapter, we have extracted three cases which might serve as guiding examples for verifying the utility of this glossary: (1) the bacterial swimming response in temporal and/or spatial chemoattractant gradients, as mentioned by Tranquillo and Lauffenburger (this section) , (2) the return–orientation and search behavior of desert isopods, as described by Hoffmann (section II/1) and (3) the orientation of spermatozoa by helical motion (Crenshaw, section II/2).   In particular, the first example illustrates that *temporal(ly differential) kinesis* can yield *taxis*, as we define it, *without directed turning*. The second example includes the possibility that locomotion is controled by internally stored (*positional or directional*) information with or without "direct" reference to external stimuli. Therefore, although the organisational scheme below as well as the glossary itself take a *stimulus field* as the primary "input" for regulating individual movement, the description of observed "outputs" relies on three distinct conceptual classes, namely internal signal kinetics (S), induced individual motion responses (I) and subsequent (spatial) population or probability density distributions (P), see also Table 1 below. We have tried to classify and explain each of the selected terms within (at least) one of these classes.

We have avoided prefexes such as *chemo–* or *photo–*, because specific *stimuli* shall not be discussed here. All words used in the definitions that appear elsewhere in this glossary are *italicized* for cross referencing. We suggest the reader to begin with the definitions proposed for *kinesis* and *taxis*.

Table 1:   **BOX 14**

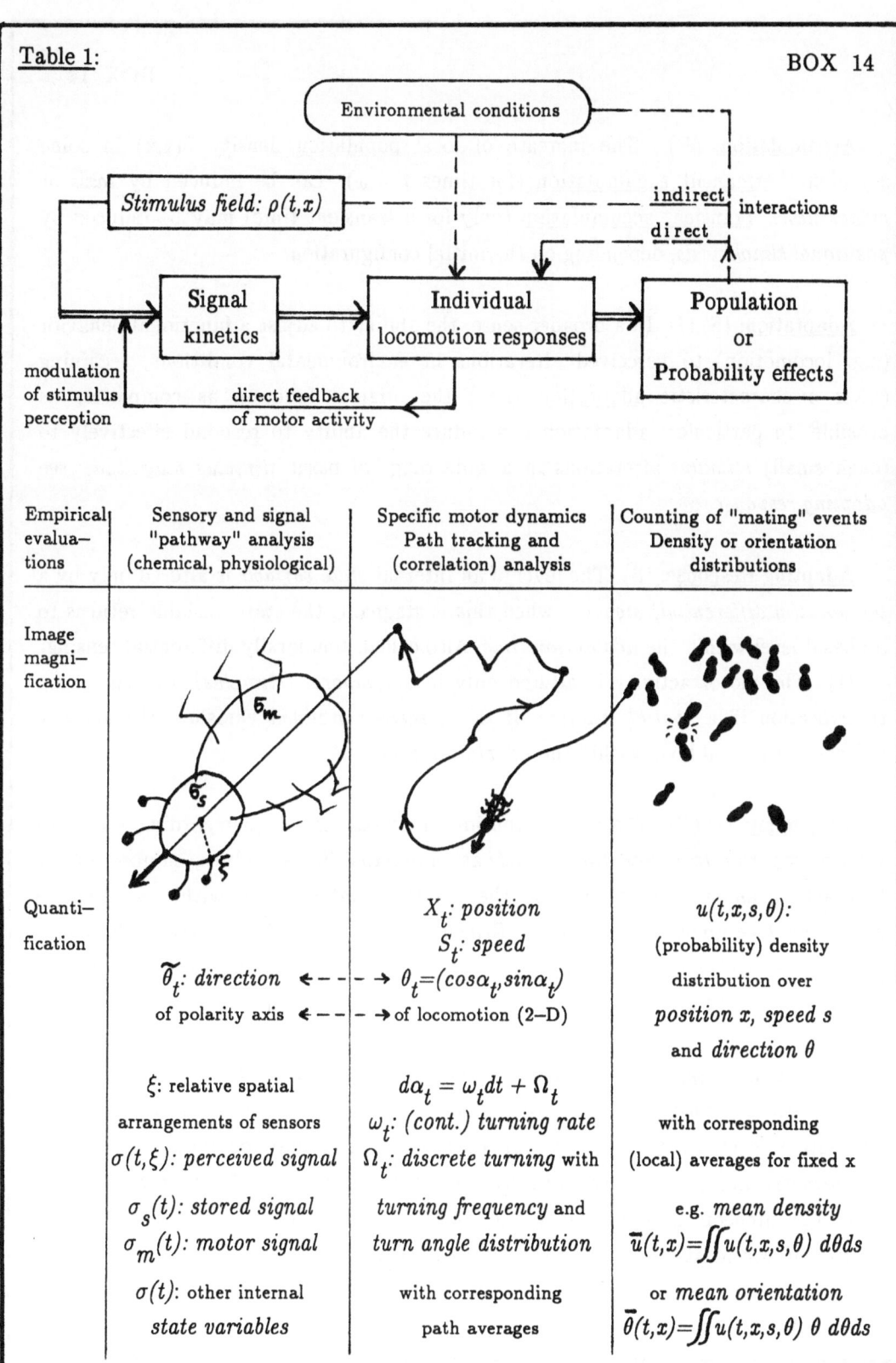

Empirical evaluations	Sensory and signal "pathway" analysis (chemical, physiological)	Specific motor dynamics Path tracking and (correlation) analysis	Counting of "mating" events Density or orientation distributions
Image magnification			
Quantification		X_t: position S_t: speed	$u(t,x,s,\theta)$:
	$\widetilde{\theta}_t$: direction of polarity axis	$\theta_t=(cos\alpha_t,sin\alpha_t)$ of locomotion (2–D)	(probability) density distribution over position x, speed s and direction θ
	ξ: relative spatial arrangements of sensors $\sigma(t,\xi)$: perceived signal $\sigma_s(t)$: stored signal $\sigma_m(t)$: motor signal $\sigma(t)$: other internal state variables	$d\alpha_t = \omega_t dt + \Omega_t$ ω_t: (cont.) turning rate Ω_t: discrete turning with turning frequency and turn angle distribution with corresponding path averages	with corresponding (local) averages for fixed x e.g. mean density $\bar{u}(t,x)=\iint u(t,x,s,\theta)\, d\theta ds$ or mean orientation $\bar{\theta}(t,x)=\iint u(t,x,s,\theta)\, \theta\, d\theta ds$

BOX 14

Accumulation: (P)   The increase of local population density $\bar{u}(t,x)$ in some x–region. Permanent accumulation (for times $t \to \infty$) can be induced by *taxis* or *orthokinesis*. Transient accumulation (only for a transient time) may be induced by *positional klinokinesis*, depending on the initial configuration.

Adaptation: (S),(I) In a broader sense, the ability to adjust a functional behavior (e.g. locomotion) to perceived alterations in environmental conditions, regaining (after a characteristic adaptation time) the original function as completely as possible. In particular, adaptation can induce the ability to respond effectively to (even small) *stimulus* alterations in a wide range of basal *stimulus magnitude*: see *adapting response*.

Adapting Response: (S) The level of an internal *state variable* is altered only by a *temporal(ly differential) signal* ; when this is stagnant, the state variable returns to its basal level within the *adaptation time*. Also called: temporally differential sensing.

(I) The motor activity is altered only in a *temporal (stimulus) gradient* or by translocation in a *spatial gradient*. If the *perceived gradient* vanishes, the activity returns to its basal level within the *adaptation time*.

Alignment: (I),(P) Biased distribution of *directions* $\pm\theta$ (disregarding the sign) with respect to a *tensorial stimulus field* as an external line structure, induced e.g. by (contact) *guidance*. Could also be the result of interactions within a (swarm) population (see Edelstein–Keshet and Ermentrout, Part III). The degree of alignment can be (inversely) quantified by the variance of the $\pm\theta$ – distribution. See also *(directional) orientation*.

Attraction (Repulsion): (I) The tendency of individuals to move towards (away from) a *stimulus* source, or up (down) a *spatial stimulus gradient* ; Can result, on the population level, in an *accumulation* or a *drift* and may be induced by *taxis*, *orthokinesis* and (transiently) also by *positional klinokinesis*. An attraction index (e.g. McCutcheon index) relates the (polar or orthogonal) component of the *positional* translocation vector to the total path length.

BOX 14

Dispersal: (P) Spreading of the *positional* distribution of an initially bounded population, also meaningful for a movement with *drift* ; Can generally be quantified by the variance (or covariance matrix) around the mean location of the population.

(I) The individual dispersal from a starting position X is usually quantified by the <u>mean squared displacement</u> $\mathscr{D}(t)=\langle\,|X_t-X|^{\,2}\rangle$ (see e.g. Alt or Klafter et al., section II/1). In case of individual *random locomotion* its linear asymptote (for t → ∞) is related to the *random motility* coefficient (slope = 2 · product of *speed* and *asymptotic persistence length*) and to the (mean) *persistence time* (= intersection of linear extrapolation with abscissa).

Drift: (P) A local mean displacement (shift in the *positional* distribution of a population) per unit time, quantified by the mean *velocity*. Often associated with individual *attraction* or *repulsion*.

Guidance: (I) The dependence of individual movement *direction* on a structured external field (as a line field or a fibrous matrix) which may be anisotropic, but without a *directional orientation*. Further distinctions concerning the θ_t- dependent response to such a *tensorial* (versus a *directional*) *stimulus field* may be carried over from analogous specifications of *taxis*, with θ_t replaced by $\pm\theta_t$.

Kinesis: (I) The dependence of individual movement on a *scalar stimulus* $\rho(t,X_t)$ or other information about *time t* and *position* X_t along its path. A *kinesis* can be characterized as

a) **Positional** — an (almost) instantaneous response induced by a purely *positional signal* (Kinesis–A: the organism stays always adapted) ;

b) **Temporal(ly differential)** — an *adapting response* induced by a *temporal(ly differential) signal* (Kinesis–S: the organism is stimulated and starts to adapt).

An increasing *positional signal* or a positive *temporal(ly differential) signal*, resp., can induce a <u>positive/negative</u>

1) **Orthokinesis** — by in/decreasing the (mean) *speed* ;

2) **Klinokinesis** — by de/increasing the (continuous) *turning rate* or the (discrete) *turning frequency* and/or the absolute magnitude of *turn angles*.

BOX 14

Orientation: (I) In a broad sense, the ability of an organism to adjust its movements (*direction, speed* or even the *positional* distance) in relation to its environment (*stimulus*, markers, distribution of food, social group). In a narrower sense, according to the original meaning, *directional orientation*.

(Directional) Orientation: (I),(P) Biased *directional* distribution of θ with respect to a *directional stimulus*, as a *spatial gradient*. The <u>degree of orientation</u> can be quantified by the modulus of the *mean orientation* vector and (inversely) by the variance of the θ – distribution. Directional orientation may be induced by *taxis*, and also by any *kinesis* during *(transient) accumulation*. See also *alignment*.

Path Integration: (S),(I) Ability of a moving individual to accumulate *positional* or *directional* components of information along its path, and to use the corresponding *stored signal* for *orientation* or for controling its *search intensity* (see Hoffmann, section II/1).

Persistence (of Direction): (I) The tendency of an individual (exhibiting *random locomotion*) to maintain its *direction* for a characteristic (mean) *persistence time*; Can be quantified by the *directional* <u>autocorrelation function</u> $G(\tau)=\langle \theta_t \cdot \theta_{t+\tau} \rangle$, i.e. the mean cosine of angular changes after a time τ (see Alt, section II/1).

(Directional) Persistence Index: (I) The mean cosine of the *turn angle* distribution quantifying the *directional persistence* after a discrete turning event.

(Asymptotic Directional) Persistence Length: (I) The final mean displacement of a *randomly locomoting* individual projected to its initial *direction* vector θ_0 (see Alt, section II/1).

(Directional) Persistence Time: (I) A characteristic time during which a *randomly locomoting* individual loses its *directional* "memory". Can be quantified as the (mean or asymptotic) decay time of the *directional persistence*, i.e of the directional autocorrelation function $G(\tau)$; Can also be calculated from the asymptote of the mean squared displacement function $\mathscr{D}(t)$, see *dispersal*.

BOX 14

Persistence of Speed: (I) In analogy to *(directional) persistence*, the tendency to maintain the individual locomotion *speed* S_t ; Can be quantified by the *speed* autocorrelation function.

Phobic Response: (I) Locomotion response to a suddenly *perceived stimulus gradient* which, in the case of a *spatial* stimulus gradient, leads to an avoidance of the region just entered. Induced by (and sometimes synonymously used for) negative *temporal(ly differential) klinokinesis* with reversed *turning* (i.e. with negative *persistence index*) or negative *taxis without directed turning*.

Random Locomotion: (I) Individual stochastic movement with constant *speed* distribution, *turning frequency* and with unbiased *turn angle* or *turning rate* distribution ; Can be quantified on population level via the *random motility* (or diffusion) coefficient (see Tranquillo and Lauffenburger, this section II/3). Leads to *dispersal* and an asymptotically homogeneous *search intensity*.

Search Intensity (in a region: area or volume): (I) The fraction of "encountered" region along the locomotion path, related to the total region "encircled" by the path up to time t. For *random locomotion* it can be (inversely) quantified by the *(asymptotic directional) persistence length*, (see Alt, section I/1).

Signal: (S) Carrier of information, which is perceived from a *stimulus field* and possibly *stored* by the individual, and which governs the individual (locomotion) response ; Can be <u>external</u> (as bound receptor ligands) or <u>internal</u> (as second messangers or hormones) . Depending on geometry and kinetics of *stimulus* perception, signal transduction and motor response, various types of *signals* carrying various <u>components of information</u> may be present at time t:

a) **Spatial** — containing evaluations $\sigma(t,x)$ about the *stimulus* $\rho(t, X_t + \xi)$ at   (at least) two distinct locations ξ around the individual, the vectors ξ being related to direction (of polarity axis) θ_t ; Can induce a *directional signal*.

b) **Temporal(ly differential)** — containing evaluations $\sigma(\tau)$ about the *stimulus* $\rho(\tau, X_\tau)$ at (at least) two distinct times $\tau \leq t$; Can permit an (approximate) evaluation of the *perceived stimulus gradient*.

BOX 14

c) **Positional** — containing evaluations $\sigma(t)$ about the individual *position* X_t or some momentary properties $\rho(t,x)$ there, e.g. the *stimulus magnitude* or a food location probability.

d) **Directional** — containing evaluations $\sigma(t)$ about the individual *direction* θ_t or its relation to a *directional stimulus* field, e.g. a *spatial stimulus gradient* at *position* X_t.

e) **Tensorial** — analogous to *directional.*

Stimulus: (S) A physico–chemical quantity (also called *external signal*) acting on an individual and inducing a stimulation (by signal perception). A (temporally and/or spatially changing) <u>stimulus magnitude</u> $\rho(t,x)$ defines a <u>stimulus field</u> (also called an *external signal field*) whose various components (of information) may be perceived as (*internal*) *signals*. It can be

a) **Scalar** — as temperature or chemical concentrations

b) **Phasic** — as polarized light

c) **Directional (Vectorial)** — as light rays or an electromagnetic field

d) **Tensorial** — as curvature or tension of a substratum, or structural matrices, or (bi–directional) guiding lines.

(Stimulus) Gradient: (S) Differential (or approximating difference) values of a *(scalar) stimulus field* $\rho(t,x)$ with respect to a fixed reference frame. It can be

a) **Spatial** — the vectorial quantity $\nabla_x \rho(t,x)$

b) **Temporal** — the scalar quantity $\partial_t \rho(t,x)$

   or with respect to the individual moving coordinate X_t:

c) **Perceived (along the motion path)** — the scalar quantity $D_t \rho(t,X_t) := \partial_t \rho(t,X_t) + V_t \cdot \nabla_x \rho(t,X_t)$ depending on the individual <u>velocity</u> vector $V_t = S_t \cdot \theta_t$.

Taxis: (I) The dependence of individual movement on a *directional stimulus* or *signal* related to the movement *direction* θ_t. Can be induced by "direct" (e.g. *spatial*) or "indirect" (e.g. *temporal*) determination of the movement *direction* θ_t, or some other evaluations about θ_t, in relation to an external *directional* field (e.g. a *stimulus gradient*) or an internal(ly stored) *directional* information.

BOX 14

In principle, this definition would include the possibility of

1) **Orthotaxis** — the dependence of (mean) *speed* on *direction* θ_t

2) **Klinotaxis** — the dependence of *turning* behavior on the current *direction* θ_t

    and, as a particular case of (almost instantaneous) *turning* response:

3) **Topotaxis** — the direct adjustment of movement *direction* θ_t towards a given location.

The use of the term *taxis* mostly extends to cases (2) and (3) only. However, also in these cases, the following distinction has to be made:

a) **Taxis without _directed turning_** — with a turning (or speeding) response as in *klinokinesis* (or *orthokinesis*), but depending on locomotion *direction* θ_t. The *directional signal* inducing the response might be *spatial* or *temporal* (examples are bacteria changing their *turning frequency* or microorganisms changing pitch and radius of their helical path in a *perceived stimulus gradient* — both special cases of *temporal klinokinesis*)

b) **Taxis with _directed turning_** — directional bias of the *turn angle* distribution or of the (continuous) *turning rate* ; Can be induced only by a *directional signal*, but also by others than a *spatial signal* (an example is amoeboid locomotion with membrane protrusions in different directions, which probably are related to a combined *spatial–temporal signal*)

Often the term *taxis* is restricted to case (b), obviously because cases (a) different from *temporal klinokinesis* have not been noticed so far. In many situations the *temporal* component in a *directional signal* cannot be neglected (as in amoeboid motion, see above).

GENERAL DISCUSSION

1. Concepts of Taxis and Kinesis

The glossary of terms by **Tranquillo and Alt** is aimed at
providing a coherent set of biomathematical definitions of the
most common terms relevant to directed locomotion. In doing so
the authors have provided support to future efforts in the
interpretation of cell and organismal behaviour. The contribution by **Doucet and Dunn** is an attempt to lay theoretical foundations for the characterization of taxis and kinesis. Their
methods and results differ from those employed by preceding
authors, because by applying theory to the problem they enjoy
the unique advantage of being able to deal with the global
aspects of signals and directed behaviour in an objective and
systematic way. Compare their perspective with the previous and
widely held assumptions that the signal for directed locomotion
is a stimulant amplitude gradient and that the outcome of kinesis is limited to random distribution. In the traditional view
of the problem, questions about the roles of different signals
and behaviour strategies are placed beyond the practical
horizon, and the possibility that accumulatory movements may
have a number of causes besides directed turning has, therefore, only been sporadically discussed and its significance has
not been fully grasped.

1.1. Differences in Current Terminology and Usage

The definitions of directed locomotory reactions still pro-
voke differences of opinion which, far from being inconsequen-
tial semantic excercises, evidently signal fundamental difficulties within the concepts of taxis and kinesis. Thus, among
the most authoritative reviewers of the subject, taxis is
viewed variously as a reaction (Fraenkel and Gunn 1961, Keller
et al. 1977) or an effect (Diehn *et al*. 1977) and kinesis is an
undirected reaction (Fraenkel and Gunn 1961, Diehn *et al*. 1977,
Keller *et al*. 1977) or it may lead to accumulation, but not

orientation (*op. cit.*). It is differentiated from random locomotion (Keller *et al.* 1977). A kinetic response may be associated with adaptation to the stimulus (Fraenkel and Gunn 1961,
Keller *et al.* 1977) or not (Diehn *et al.* 1977) and induced by a
stimulus gradient (Fraenkel and Gunn 1961) or the stimulus
magnitude (Diehn *et al.* 1977, Keller *et al.* 1977). Taxis is
universally attributed to gradient reading (cf. Fraenkel and
Gunn 1961); and accumulation, orientation and attraction in
leukocytes and *Dictyostelium* are virtually always attributed to
taxis (cf. Zigmond 1978, Fisher *et al.* 1990). Despite definitions enjoining us to accept taxis as a directed reaction,
accumulation is applied by many if not most workers, e.g. those
dealing with bacteria, in the sense of taxis, intrinsically
(cf. Koshland 1977).

Unfortunately, some of these views are in need of modification. Undirected individual cell reactions such as kinesis may
well lead to a directed effect of the whole population.
Undirected, random locomotory reactions may lead to transient
or lasting accumulation under klinokinesis depending on whether
the cells become adapted to the stimulus or not. During
klinokinesis cells will demonstrate accumulation and orientation which is unrelated to that induced by taxis. Accumulation
generated by orthokinesis will not be accompanied by cell
orientation.

1.2. Synthesized, Alternative Definitions: an Exercise

As an illustration of how current terminology is virtually
ordained to present biologists and theoreticians with one or
another dilemma, I shall list here three possible alternative
definitions of taxis in different senses, each based on current
definitions and usages. On occasion, some workers tend to blur
the distinctions between them, which only enhances the confusion.

The first alternative retains the notion that taxis is a
directed individual locomotory reaction. Thus, taxis refers
specifically to the *directed turning* (directed step) element of
the reaction to a directed signal field, although enhanced
motility and accumulation may follow.

In the second alternative taxis is meant only as a classifi-

cation of spatial pattern: an effect. Unlike the first definition, it does not refer to mechanism and means nothing more
than the non-random distribution or accumulation generated in a
motile population. Taxis would, therefore, mean *accumulation*
and the locomotion, oriented or not, leading to it in a
population, whether induced by klino- or orthokinesis or by
directed turning. This definition is incompatible with the
views of Fraenkel and Gunn (1961) who considered taxis to be a
reaction, characterized by the "regimented" behaviour of
individuals in the population. This does not occur during
ortho- or klinokinetic responses in a spatial gradient signal
field, where the directional response is above the level of the
individual.

A third alternative is possible by forming a hybrid definition. Thus, all accumulative responses would be considered
tactic ones, but it would be necessary to make a distinction
between those exhibiting directed turning from those arising
from random locomotion (e.g., kinesis in a spatial gradient
field). For the sake of the argument, the former may be termed
"telo-taxis" (Fraenkel and Gunn 1961) to indicate that the
individuals themselves react to the directional information in
the stimulus field. The latter may be termed the *epitaxes*,
i.e., klino- and orthotaxis to indicate stochastic reactions to
the local, directionless concentration or amplitude signal
information in the stimulus field, which result in accumulation
through a mechanism acting at the level of the population. As
with the second alternative above, the differences between the
mechanisms of accumulation and of turning are decisive and
clearly distinguished.

1.3. Definitions as Tools and Fallacies: Conclusions

Unlike the argument of **Doucet and Dunn**, the point of these
definitions is based less on information content of the signal
and rather more on (relatively) well characterized aspects of
the behaviour of the individual. If there are fundamental
differences between these ways of tackling definition of directional behaviour, they may lie in a) the relegation by Doucet
and Dunn of the role of signals generated by the activity of
the organism itself (e.g., Lauffenburger 1982), which unfortun-

ately enough appears to fall between system-theoretical chairs, and b) and the conspicuous position which I have assigned to directional behaviour, e.g. directed turning. However, primacy in consideration ought to be reserved both for the signal, because it imposes a limit on the potential range of the response, and for the organismal response (at both the level of the individual and the population), because that is the chief object of interest, after all.

The differences of opinion mentioned here should alert Readers that they should not expect to find a frictionless consensus on behavioural terminology crystallizing before their eyes, but at least the radians of disaggreement have narrowed to the same (new) sector of the critical circle. It is still too early to tell whether a single set of definitions will suffice for all types of locomotion and organisms, or even for one particular case, but a critical terminology should not oversimplify fundamental biological differences which might exist and, indeed, it ought to help identify them. However, the participants of this Workshop have also come closer to recognizing the considerable extent of the difficulties inherent both in measuring and classifying directed behaviour. For example, the ability to formulate three parallel definitions of taxis (above) - each retaining at least some elements of validity - might indicate that a general fallacy of abstraction may be lurking within schemes to produce an authoritative terminology. Yet, perhaps the chief weakness in all the efforts to do so may be that the scientific focus has a tendency to become reversed, because the problematization and identification of patterns of behaviour become mediated through the scheme of characterization, whereas the intention is (or should be) to derive them from behaviour itself.

2. Biological Time

Some of the discrepancies between cell behaviour and the way it is portrayed may be accounted for by the feedback and temporal natures of reactions in living systems. The present conception of taxis and kinesis, and for that matter most other biological processes, is based on the premise of dose-response reactions; e.g. the number of attractant molecules bound to

cell surface receptors is thought to determine both the kinetic
response in an isotropic field and taxis in a gradient field
(Zigmond 1978). However, one of the most salient and remarkable
features of organisms is that they are more effectively considered as essentially non-linear systems. Their molecular responses to signals and the behavioural repertoire are governed by
feed-back networks in receptor physiology involving, e.g.,
signal-receptor desensitization and the adaptation of biochemical and behavioural responses, which have not been adequately
considered in the present concepts of directed locomotion. For
example, organisms may become desensitized or adapted to constant dose signals, even in so-called "chemotactic gradients".
Most significantly, stimuli capable of driving reactive networks beyond equilibrium may be expected to induce
qualitatively different responses from those involved in linear
reactions.

Organisms may also demonstrate oscillatory properties
intrinsic to their directional behaviour, which is inexplicable
on the basis of linear reactions. For example, the flagellar
motor of the "phototactic" *Halobacterium* expresses a switching
rhythm with a period of 20 s (Schimz and Hildebrand 1985). A
limit cycle has been used to model actomyosin dynamics in
eukaryotes (Alt 1984; Pohl, section I/1), which may be a key
determinant of directed-stimulus reading. The existence of
adaptive and feedback reactions and oscillatory behaviour
justifies conclusions that temporal signals are the most effective stimuli of particular systems, such as the cell-surface
reception of hormone signals. Their sensitivity to directional
signals may depend on the eigen-frequencies of their reception
and response systems (for details see Vicker *et al.* 1984,
Vicker and Rensing 1987). The consequences of biological time
evidently places limits on the sorts of signals that organisms
are capable of perceiving and indicates an hitherto unsuspected
dimension in the classification of directional signals and
behaviour.

3. Chemotaxis and Kinesis in Cell Populations

Tranquillo and Lauffenburger are confronted with an
extraordinarily complex and dynamic system in which the attrac-

tive signal is changing continually during the entire phase of the motility assay. At some positions the rate of development of the signal field is faster than that of the cell's locomotory response, and at others the inverse relationship holds. Their analysis is focussed on taking these changes of the signal into account, and it may offer a major benefit to researchers as well as clinicians, since it allows them the opportunity to master the uncertainties surrounding stimulus form, a topic usually ignored in most treatments of the subject.

Nevertheless, it is the dynamic aspect and identity of developing tactic stimuli which forms the main element of difference between the contributions by **Tranquillo and Lauffen-burger** and **Vicker**. The former base their analysis upon the assumption that the tactic signal is a spatial concentration gradient of a taxin, and they apply their analysis under the conditions that "the receptor-mediated motility and gradient sensing mechanisms are linearly dependent on the CA gradient". The latter author has attempted to identify the exclusive significance of the temporal stimulus (i.e., the attractant impulse) as the stimulus for directed turning. The two views are, indeed, controversial.

Compared to taxis, the topic of cell orthokinesis has been seldom addressed. Here, Tranquillo and Lauffenburger develop a useful mathematical tool for its analysis and both they and Vicker demonstrate its existence during the accumulative behaviour of chemotactic cells. However, the role of klinokinesis has only been briefly approached by both authors: in the mathematical treatment by Tranquillo and Lauffenburger in this section and by Vicker *et al.* (1984) as to its effect on spreading and evenly distributed cell populations. Yet a detailed analysis is still lacking. It is to be hoped that there will eventually be more opportunity to extend the theoretical and experimental investigations on these matters in the near future.

4. Perspectives

The work included in this section presents a brief glimpse of the direction of future research in the directed locomotion

of cells. New concepts of the dynamics of directional signals,
signal molecule receptors and the cytoskeleton (Pohl, Section
I/1) are being integrated with those of cell and population
behaviour. The result is an increasingly complex and differentiated picture of directed behaviour compared to that available
only a few years ago. These principles may contribute to the
understanding of the role of the directed behaviour of cells in
pathogenesis, e.g. inflammation and the diseases attributable
to phagocyte malfunctions, and in critical morphogenetic processes especially during the cell movements shaping the early
embryo.

REFERENCES

Alt, W. (1984) Contraction and oscillations in a simple model
   for cell plasma motion. In: Temporal Order. (eds. L. Rensing
   and N. Jaeger), pp. 163-174. Springer-Verlag, Berlin, Heidelberg, New York.
Diehn, B., Feinleib, M., Haupt, W., Hildebrand, E., Lenci, F.
   and Nultsch, W. (1977) Terminology of behavioral responses of
   motile organisms. Photochem. photobiol. 26, 559-560.
Gerisch, G., Hülser, D., Malchow, D. and Wick, U. (1975) Cell
   communication by periodic cyclic-AMP pulses. Phil. Trans. R.
   Soc. Lond. B. 272, 181-192.
Fisher, P.R., Merkl, R. and Gerisch, G. (1989) Quantitative
   analysis of cell motility and chemotaxis in *Dictyostelium
   discoideum* by using an image processing system and a novel
   chemotaxis chamber providing stationary chemical gradients.
   J. Cell Biol. 108, 973-984.
Fraenkel, G.S. and Gunn, D.L. (1961) The Orientation of
   Animals: Kinesis, Taxis and Compass Reactions. Dover, New
   York.
Harris, H. (1954) Role of chemotaxis in inflammation. Physiol.
   Rev. 34, 529-562.
Keller, H.U., Wilkinson, P.C., Abercrombie, M., Becker E.L.,
   Hirsch, J.G., Miller, M.E., Ramsey, W.S., and Zigmond, S.H.
   (1977) A proposal for the definition of terms related to
   locomotion of leukocytes and other cells. Clin. exp. Immunol.
   27, 377-380.
Koshland, D.E.Jr. (1977) A response regulator model in a simple
   sensory system. Science 196, 1055-1063.
Kühn, A. (1919) Die Orientierung der Tiere im Raum. Fischer,
   Jena.
Lauffenburger, D.A. (1982) Influence of external gradient con-
   centration fluctuations on leukocyte chemotactic orientation.
   Cell Biophys. 4, 177-209.

Schimz, A. and Hildebrand, E. (1985) Response regulation and sensory control in *Halobacterium halobium* based on an oscillator. Nature (Lond.) 317, 641-643.

Vicker, M.G., Schill, W. and Drescher, K. (1984) Chemoattraction and chemotaxis in *Dictyostelium discoideum*: myxamoeba cannot read spatial gradients of cyclic adenosine monophosphate. J. Cell Biol. 98, 2204-2214.

Vicker, M.G. and Rensing, L. (1987) Oscillations and the regulation of spatial order in developing systems. In: Temporal Disorder in Human Oscillatory Systems. (eds. L. Rensing, U. an der Heiden and M.C. Mackey), pp. 24-29. Springer-Verlag, Berlin, Heidelberg, New York.

Wilkinson, P.C. (1982) Chemotaxis and Inflammation. Churchill Livingstone, Edinburgh.

Zigmond, S.H. (1978) Chemotaxis by polymorphonuclear leukocytes. J. Cell Biol. 77, 269-287.

PART III

COLLECTIVE MOTION

Part III

COLLECTIVE MOTION

(Coordinator: Leah Edelstein-Keshet)

Introduction

In this chapter we present several examples of collective motion and illustrate theoretical approaches and techniques that can be applied to understanding aspects of the phenomena. The selection of contributions that make up this section illustrate how biological systems at extremes of morphological diversity fall into a common phenomenological category. We encounter here examples drawn from insect societies, from groups of cells, and from schools of fish. In each of these papers, the question addressed is how patterns or structures arise spontaneously in a setting of collective motion. How social behavior is coordinated, and what biological factors influence the type of structure or collective motion are also topics of discussion in these papers.

The fact that collective motion and pattern formation are ubiquitous hint that there may be common underlying principles in disparate examples. The individual mechanisms clearly differ from one case to another, but some aspects of the phenomena have analogs in other settings and even in other size scales. A possible role for theoretical work is to uncover such principles and, where feasible, to place disparate observations into a common framework.

A challenge in making a connection between the level of an individual and that of the population stems from the fact that population behavior depends on many factors, including properties of the individuals, interactions of the individuals with each other, with the environment, with chemical substances, or with other kinds of organisms. Experience with classical physics or chemistry teaches that even when the rules governing the behavior of individual particles are well known, an understanding of the collective properties of systems of particles may be quite difficult or intractable. The same observation can also be drawn from computer experiments with cellular automata (a famous example being Conway's "life game") which demonstrate that a collection of interacting units, even very simple ones, can exhibit complex behavior. This conclusion applies to many systems since, in general, interactions are nonlinerar (i.e., not simply cumulative). When the system is compounded by the complexity of biological organisms, the problem of collective behavior becomes even more challenging. The remarkable photographs reproduced in this introduction illustrate that we can talk about aggregate properties or collective motion in macroscopic as well as in microscopic settings.

Figure 1: Several hundred thousand wildbeest in a moving front, grazing on the Serengeti. From A.R.E. Sinclair (1977), reproduced by permission of the author and publisher.

In the photograph of wildbeest grazing on the Serengeti Plains of Africa, hundreds of thousands of animals form a moving front, a sharp transition in the density of the population which resembles protruding fingers, see Fig.1. A similary dense flock of flamingos on an African lake demonstrates cohesiveness, a well defined margin, as well as internal structure such as directionality, see Fig.2. The roughly circular wave of density emanating from a gap in the side of the flock may suggest some disturbance that initiated an outwards movement of escaping birds. A final example, that of branching patterns in the foraging columns of army ants illustrates that self-organization, and formation and preservation of pattern occur in the insect societies, see Fig.3. Other examples of collective motion and associated phenomena on the cellular level are given in the pages of this chapter.

The chapter begins with a discussion of the organization of traffic patterns in the ant, *Iridomyrmex humilis* (**Aron, Deneubourg, Goss and Pasteels**). The authors show experimentally that traffic between artificial nests has the tendency to select certain configurations. A mechanism underlying self-organization in the ant is that of trail following. Individuals secrete a chemical pheromone which marks their route, but which slowly fades away. Collectively, there is a tendency to select routes which are most heavily marked i.e.

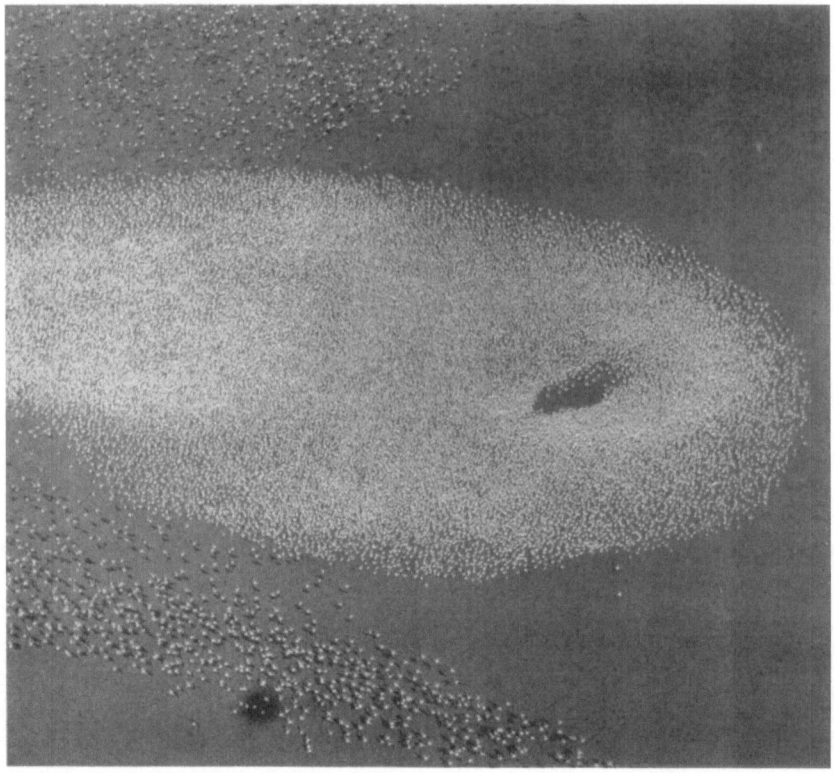

Figure 2: A flock of flamingos on an African lake. Note variations in density and in the orientations of the birds. A fairly distinct front is maintained.

which have carried the most traffic. The authors demonstrate that a model for pheromone and ant traffics accounts for experimentally observed traffic patterns.

Several contributions in this chapter further develop the theme of trail following at the microscopic scale. Two sections, one by Stevens and one by Pfistner present the swarming behavior of myxobacteria, microorganisms which alternate between free-living and multicellular stages. In the cellular form, each cell secretes slime trails on which gliding motion takes place. **Stevens** presents a computer simulation of the motion and trail following in a group of cells moving on a plane. She shows how changes in the assumptions about individual behavior modify the collective patterns and aggregations that are produced. **Pfistner**, on the other hand, approaches the problem by formulating a mathematical model in which individual cells are replaced by a cell density distribution, described by differential equations. She displays how parameters such as the perception of density and of density gradients can influence the existence of a stationary swarm, and finally, displays the density distribution in the theoretical swarms by numerical analysis of the continuum models.

Figure 3: The development of a raiding swarm of army ants (*Eciton burchelli*) showing three stages, from left to right: (spreading at dawn from their bivouac, and expansion in one sector at two later stages). After T.C. Schneirla (1971) Army Ants, a Study in Social Organization, W.H. Freeman.

The parallels between the above three papers illustrate that functional similarities may exist in cases that do not, at first, seem biologically related. As a curious sideline, as shown in Fig.3, trail-following ants are known to exhibit foraging patterns resembling branching networks (see also Deneubourg and Goss, 1989). The analogy between pheromone trails as branching tracks and ants as the particles creating and following such tracks recalls other branching networks in which crisscrossing of branches results in merging of separate tracks (see Edelstein-Keshet and Ermentrout, 1989, and references cited there for other examples).

Further discussion of cellular behavior is given in a paper by **Edelstein-Keshet and Ermentrout**, this time of mammalian cells in tissue culture. The authors show that certain aspects of self-organization can be produced by direct inbuilt cell responses, without other mediating factors. The ability of these cells to form a tissue with directionality, i.e. to select a single axis of orientation, is predicted by a model based on the contact response of these cells. (Cells turn and align upon contact with a neighbor.) The modelling consists both of continuum approximations by differential equations (as in **Pfistner**), and discrete simulation of "cells" (as in **Stevens**).

In another paper which connects inbuilt responses of organisms to each other and the properties of their collective motion, **Huth and Wissel** discuss the case of fish schools. Here, simulations for the motion of individual fish are used to explore a variety of assumptions and reveal which lead to the formation of cohesive schools. Although the mechanisms of responses of fish to one another differ from those of cells (visual cues and muscle control replacing the cell-membrane chemistry and subcellular forces of the previous example), the functional similarity is clear.

FUNCTIONAL SELF-ORGANISATION ILLUSTRATED BY INTER-NEST TRAFFIC IN ANTS : THE CASE OF THE ARGENTINE ANT

S. Aron, J.L. Deneubourg, S. Goss, J.M. Pasteels

Unit of Behavioural Ecology, C.P. 231,

Université Libre de Bruxelles, B-1O5O Bruxelles

Abstract

In many ant species the colony is not a single structure but rather a number of decentralised nests linked together by a network of trails. This network can be extended to include trails that form between the nests and long-lasting food sources or rich foraging areas. The formation of inter-nest networks was studied with laboratory colonies of the Argentine ant *Iridomyrmex humilis*. Bridges were placed to link isolated nests, with branches of equal length arranged in a triangle and a square linking three and four nests respectively, and two branches of different length linking two nests. The resulting ant traffic connected all the nests, neglecting redundant bridges, forming a minimum spanning tree. Cutting a frequented bridge caused the traffic to divert to a neglected bridge. Visual cues appear not to be essential as similar networks were generated both in light and darkness. Rotating the bridges caused the traffic to change correspondingly indicating a primary role of chemical cues with respect to memory or visual cues. Where the bridges between two nests were of unequal length, the ants neglected the longer one.

A simple model of trail laying and trail following generated networks similar to the experimental ones, when saturation of the workers' pheromone perception was taken into consideration. The agreement was not, however, perfect, especially in the square where the model tended to generate two pairs of nests well connected within a pair but totally separated between pairs. This suggests that a supplementary mechanism may intervene in the ants' behaviour to avoid the isolation of pairs of nests. These results are discussed with reference to the classical travelling salesman or shortest network problem, and the general principles of organisation in social insects.

1. Introduction

A caricature of modern science is that there is a unique optimal solution to each problem, and that this solution is the most desireable. To achieve optimality, one central-unit has access to all the data necessary, and the algorithms it uses to treat this data are necessarily complex and therefore highly specific. As such they can tolerate neither internal errors, inexact or incomplete information, nor changes. The consequences of this approach are that each solution must be constantly monitored and overhauled to cope with unforseen events, leading to a spiral of mutually increasing complexity and fragility. This is of course an excessively pessimistic view of a type of solution that does function and will continue to be the most appropriate for the majority of actual problems. There is however room for an alternative and complementary approach, both for many existing classes of problem and for a number of problems for which one cannot at present envisage a solution.

What we propose is in many respects diametrically opposed to the above caricature. Rather than one solitary central control unit that is specific, complex and omniscient, and which has a solution programmed into it, we propose a team of simple, random and identical units that need only be locally informed, without being hierarchically organised. Distributing the team within the environment of the problem to be solved and introducing positive feedback interactions between the units allows the amplification of localised information found by one or a few of the units. Thus coordinated, the team's collective reaction to these local signals is the solution to the problem. While no one individual is aware of all the alternatives possible, and no one individual contains an explicitly programmed solution, together they reach an "unconscious" decision. We term this process functional self-organisation.

The solutions these teams generate are necessarily less efficient than optimally tailored ones. However this short term sacrifice can be outweighed by other long term benefits, the first being reliability. The more decentralised and less omniscient the unit, the simpler it can be. The simpler the units, the easier it is to program them and the less likely they are to "break down". Working as a team of generalists implies that even if many of the units fail, the rest continue. The second benefit is flexibilty. As no solution is explicitly imposed, the decentralised units can react to environmental heterogeneities, either temporal or spatial. Indeed as the units are intimately mixed with the problem and its environment, these heterogeneities contribute to the solution, allowing the team to cope with unforseen and unforseeable circumstances. The third benefit is fault tolerance. Individual simplicity implies a high degree of randomness and error. The positive feed back interactions allow this and easily coordinate random individuals into an efficient team. Far from being undesirable, individual randomness offers an escape route to individuals caught in a maze, and can help the team to reach collective solutions that would otherwise be overlooked.

The trade off between these somewhat overlapping and interwoven benefits and efficiency is especially revealing in situations where positive feed-back can block a team in a sub-optimal solution. To escape from the sub-optimal solution to a better one it can be necessary to increase the individual complexity, at the risk of reducing the benefits of a collective organisation. We are therefore lead to examine a completely new question, namely where should a group place its intelligence and complexity? To what extent should it be within each unit and to what extent in the interactions between them?

Rather than inventing such systems from scratch, we have chosen to start by examining existing ones, the best possible material being provided by group-living animals, and in particular the 20,000 social insect species. The self-organisation exhibited by social insects is functional and not abstract, and is used to solve real problems in a physical and biological environment whose principal characteristic is its heterogeneity. We shall use them to illustrate these ideas in the following context.

In most social insects, the basic ecological unit is the colony. Workers forage around the nest and bring food back to share with their nest-mates. A colony is in competition with other colonies of the same or different species, and workers from a "foreign" colony are generally not well received. In a number of ant species, however, the notion of a colony is not so clear. Groups of workers, larvae and sexuals may leave one nest to form a new nest closer to important food-sources for example, while retaining a strong connection with the parent nest. Food, workers and larvae are regularly exchanged, and if you mark workers of one nest over a number of days they can turn up in neighbouring nests even over considerable distances. In this way, the ensemble of nests or sub-colonies, sometimes called a super-colony, can spread out over a very large territory, and, by forming and abandoning nests, distribute its work-force in response to environmental cues (e.g. Hölldobler and Wilson, 1977; Hölldobler and Lumsden, 1980; Rosengren, 1985). *Formica lugubris*, for example, can form super-colonies of 20,000,000 individuals, covering several thousand square metres (Cherix, 1987).

Whereas Calenbuhr and Deneubourg (section II/2) discuss how one ant follows a trail, in this article we shall concentrate on how the trails between different nests are formed under laboratory conditions in the Argentine ant, *Iridomyrmex humilis*. In a series of previous articles, we have shown how this species forms collective exploratory trails (Aron et al., 1989; Deneubourg et al., 1990), selects the shortest path between nest and food (Goss et al., 1989), and efficiently organises its foraging area (Aron and Pasteels, 1989; Aron et al., in press). Central to these articles is the idea that much of the individual forager's or explorer's behaviour can be reduced to marking the ground with pheromone as they move, their direction of movement being influenced by the marks left by preceding workers. The interactions between a number of workers, each obeying this simple behavioural rule, are capable of organising their collective movements into complex spatial and spatio-temporal patterns (see

Deneubourg and Goss, 1989, and Goss et al., 1990, for a general discussion; Deneubourg et al., 1989 for army ant raid patterns; Goss and Deneubourg, 1989, for rotating trail pattern in a harvester ant). We shall compare the predictions of this model with a series of experimental results, and relate them to the classical "travelling salesman" problem, or "how to form the shortest network" (review in Bern and Graham, 1989).

2. Methods

Nests containing approximately 5-8,000 *I. humilis* workers were each placed in separate 20 x 30 x 4cm plastic containers, together with an abundant food and water supply. These containers were then connected by 2cm wide carboard bridges, in three different configurations (figs. 1-3). An ant leaving a nest was in each case faced with a choice between two branches leading to the adjacent nests. The angle between the two branches at the choice point was 60 degrees, with no spatial bias to one or the other branch. This acute angle discourages ants arriving at the end of one branch from turning 300 degrees and continuing on the other rather than entering the nest. Each nest being well supplied, there was no *a priori* reason for the ants of one nest to move to another nest. Three series of experiments were carried out.

1. Triangular network (fig. 1): Three nests were placed at the vertices of an equilateral triangle (4 expts.). After a period of more or less two weeks during which the colonies were left to settle down and establish the links between each other, the traffic in each direction on each of the branches was counted three times, at two day intervals, for 10 min. (Note that the qualitative structures observed in all three bridge configurations are very stable, appearing after only a few hours, and so more extensive measurement was not necessary). The bridge structure as a whole was then rotated by 120 degrees, the nests remaining where they were, so that each pair of nests became connected by a different branch than before (3 expts.). After 4 days, the traffic was counted once for 10 minutes. The branch with the heaviest traffic was then cut (3 expts.), and the traffic counted for 10 minutes after 4 days.

2. Square network (fig. 2): Four nests were placed at the vertices of a square (3 expts.). After two weeks, the traffic in each direction on each of the branches was counted three times, at two day intervals, for 10 minutes. The same experiment was made in complete darkness (2 expts.). After two weeks, the traffic in each direction on each of the branches was counted three times at two day intervals, for 10 min. The base of the U-shaped traffic system that formed (see results) was then cut, and the traffic counted for 10 minutes after 4 days. Finally,

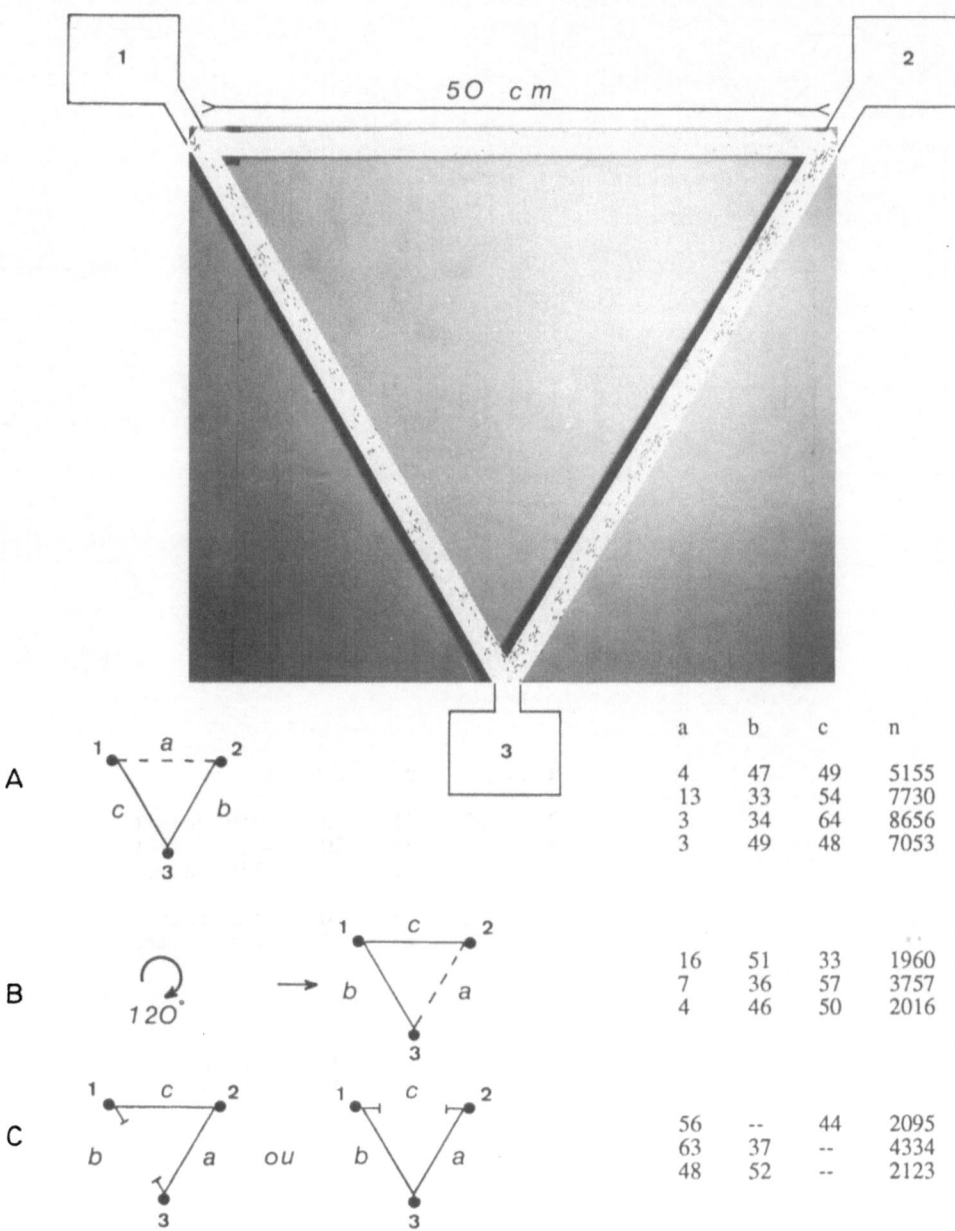

	a	b	c	n
A	4	47	49	5155
	13	33	54	7730
	3	34	64	8656
	3	49	48	7053
B	16	51	33	1960
	7	36	57	3757
	4	46	50	2016
C	56	--	44	2095
	63	37	--	4334
	48	52	--	2123

Figure 1. The experimental design (photo) and results for the triangular network. A: the triangular network was left for two weeks before counting the traffic. B: the entire bridge system was then rotated 120 degrees, the nests not being moved. C: the most frequented branch was then cut. The drawings indicate the qualitative solutions adopted. A solid line indicates heavy traffic, a dashed line very little traffic, and an interrupted line a cut branch. The numbers indicate the quantitative results for each experiment, with the % traffic on each branch and the total traffic (n) on the bridge.

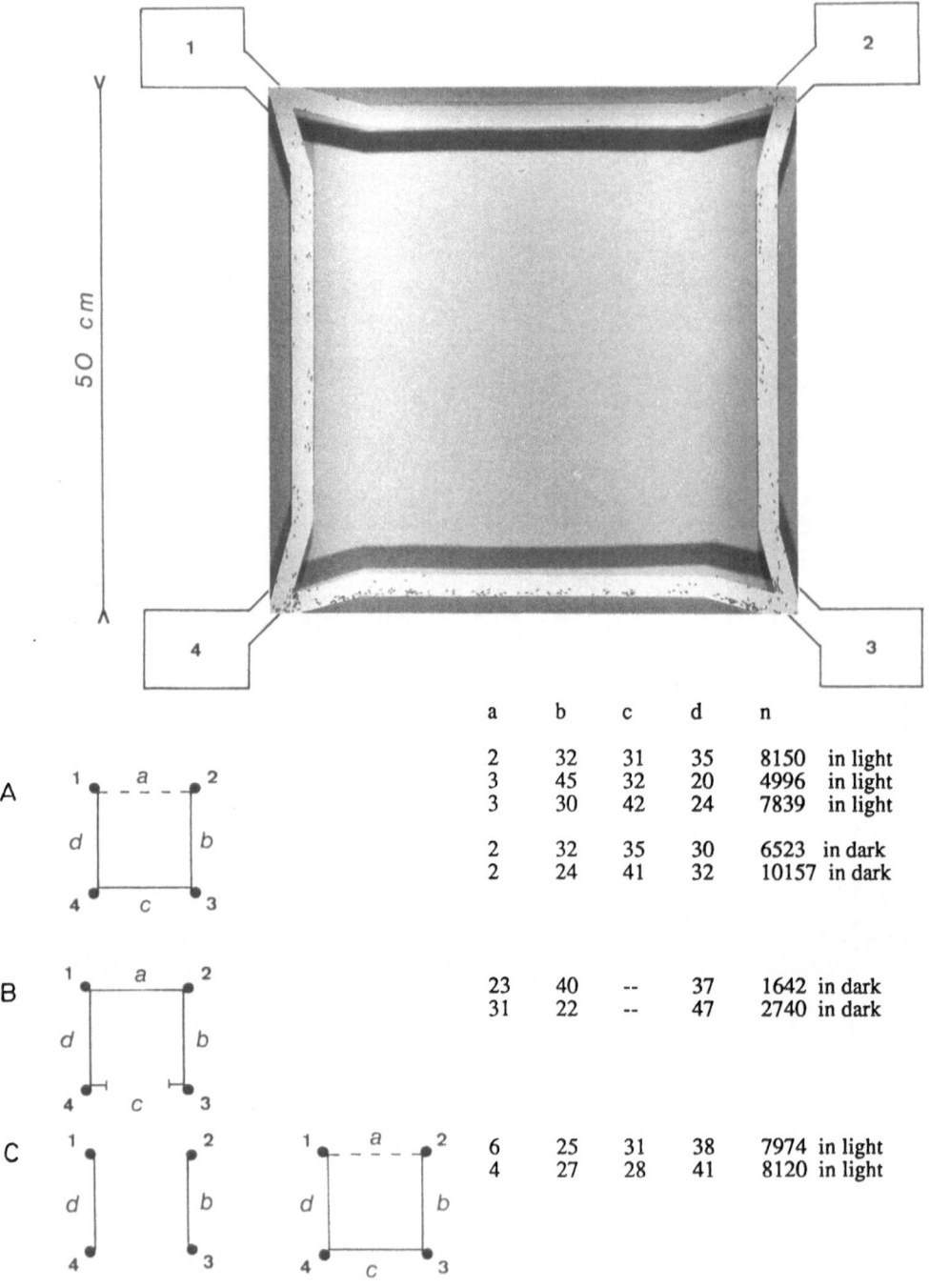

	a	b	c	d	n	
A	2	32	31	35	8150	in light
	3	45	32	20	4996	in light
	3	30	42	24	7839	in light
	2	32	35	30	6523	in dark
	2	24	41	32	10157	in dark
B	23	40	--	37	1642	in dark
	31	22	--	47	2740	in dark
C	6	25	31	38	7974	in light
	4	27	28	41	8120	in light

Figure 2. The experimental design (photo) and results for the square network. A: the square network was left for two weeks before counting the traffic. B: branch c (the base of the U-shaped solution in A) was then cut. C: branches b and d were placed for two weeks, branches a and c were then added. See legend of Figure 1.

four nests were again placed at the vertices of a square, but instead of four branches connecting them, only two opposite branches were placed (3 expts.). After two weeks, during which the traffic built up on these two branches, the missing two branches were added, and after one week the traffic was counted three times, at two day intervals, for 10 minutes.

3. Long / short network (fig. 3): Two colonies were placed at opposite ends of a bridge consisting of two alternative branches, 50cm and 70cm long (3 expts.). After two weeks, the traffic was counted three times, at two day intervals, for 10 minutes.

In each series, the orientation of the experimental set-up between replications was varied so as to avoid any systematic environmental biais. Similarily, the illumination was made as uniform as possible, and we avoided using the same nests in the same position in each series.

3. Results

The results of the experiments were as follows. In the case of the triangular network (fig. 1), after a brief period of more or less uniform traffic on each branch, two of the branches were used intensively and the third hardly at all. This formed a V-shaped traffic system, in which an ant can pass, for example, directly from nest 1 to nest 3, and from nest 2 to nest 3, whereas in order to pass from nest 1 to nest 2 it must transit via nest 3. All three nests are connected, either directly or indirectly, with the minimum of connections, one side of the triangle being in this sense redundant. (Note that all results have been re-ordered *a posteriori* to show the least used branch as branch a, whereas from experiment to experiment which branch was least used was random).

As we strongly suspected that chemical trails were important in this structure, we rotated the bridge system by 120 degrees while leaving the nests where they were. If the ants were following chemical cues, then the traffic would remain important on branches b and c and very low on branch a, even though these branches now connected different nests from before, i.e. nests 3-1, 1-2 and 2-3 respectively. If on the other hand the ants were following visual cues or, via some spatial memory, were specifically connecting nests 2-3 and nests 3-1 and neglecting to link nests 1-2, then the traffic would link the same nests as before although via different branches. In fact, the former hypothesis was confirmed , i.e. the "V" became a "<", chemical cues thus appearing to be predominant.

To see how the network reacted to a perturbation, we cut the branch carrying the most traffic. In each case the traffic on the hitherto neglected branch increased to more or less the

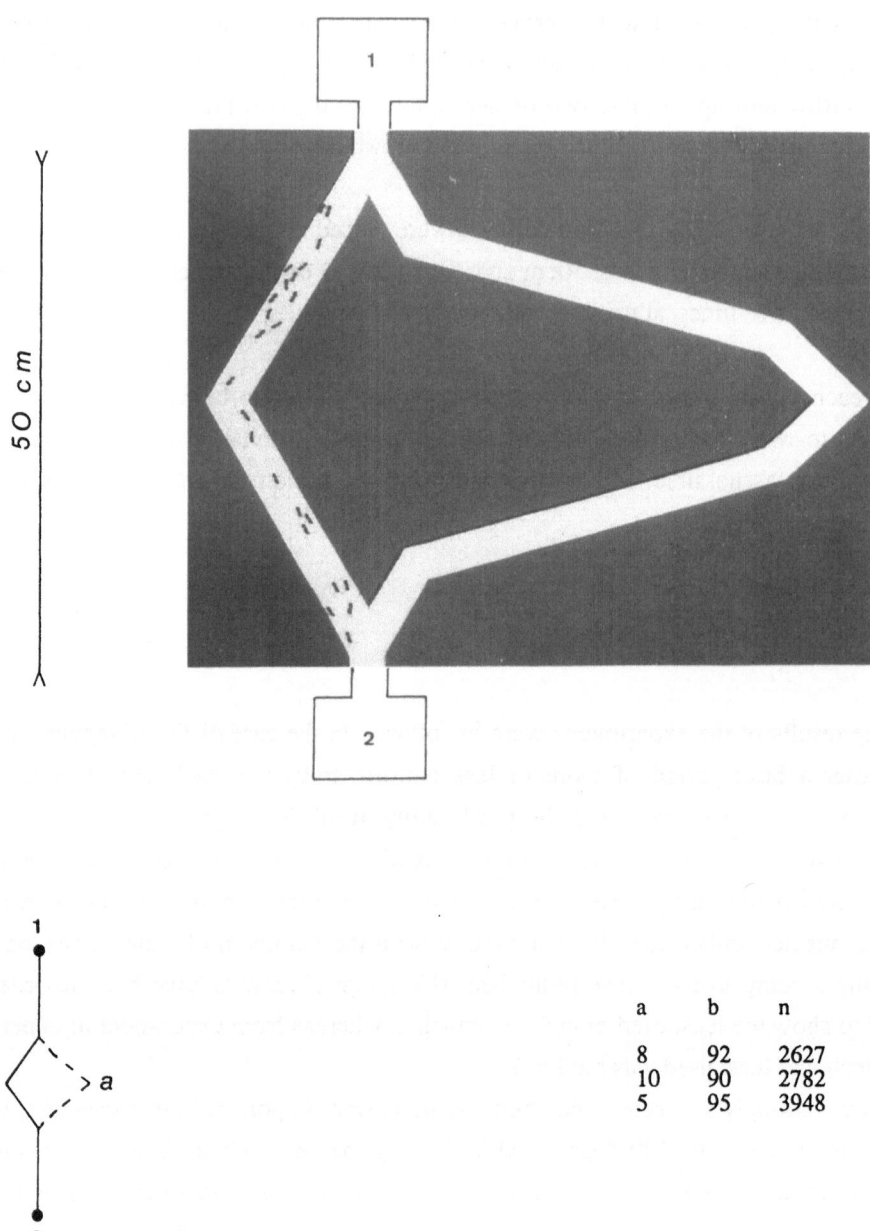

a	b	n
8	92	2627
10	90	2782
5	95	3948

Figure 3. The experimental design (photo) and results for the long / short network. See legend of Figure 1. (Note that the photo is taken out of context, in that unlike the experiments described, the bridge in the photo is 15cm long, the long branch is twice as long as the short, and it joins a nest to a food source).

same level as the remaining well-frequented branch. The network thus seems to respond so as to maintain the links between the nests.

The experiments on the square network gave similar results (fig. 2). Three sides were used intensively and one hardly at all, forming a U-shaped network. Again all the nests are interconected with a minimum of connections, the fourth branch being redundant. The experiments performed in darkness gave the same result, providing further proof that visual cues are not important in the formation of the traffic systems. When the base of these U-shaped sytems was cut, the ants form heavy traffic on the previously neglected fourth branch, thus reforming the U-system (or rather an inverted U-system).

This capacity to maintain and re-establish the links between all the nests is underlined by the last experiments of this series. By placing only two opposite sides of the square initially we tried to "force" the network into two independent sub-systems of two nests each. The idea was that the strong trails built up on these two parallel branches b and d would cause the ants to ignore the branches a and c when these were added after two weeks. In fact, after exploration on both of the newly added branches, one became heavily used and the other was neglected, recreating the U-system given when all four branches were present at the beginning.

Finally, in the long / short network (fig. 3), the ants' traffic was systematically concentrated on the shorter of the two branches connecting the two nests. This result corresponds to previous results, where an *I. humilis* colony was given access to a food source via a similar bridge and selected the shorter branch (Goss et al., 1989).

4. Model and Monte Carlo Simulations

We have previously performed a number of analyses showing that *I. humilis*' collective spatial organisation can result from a very simple individual trail laying and trail following behaviour (see introduction). Assuming, as indicated by the experiments in which the branches were rotated, that chemical cues are the principal ordering factor of these networks, their formation may be modelled as follows.

We will consider two similar sorts of networks. The circular networks consist of n nests linked together in a chain by n branches, such that each nest has two neighbours. Nest i is linked by branch i-1 to nest i-1 and by branch i to nest i+1. Nest n is linked by branch n-1 to nest n-1 and by branch n to nest 1, closing the circle. With n=3 and n=4, this corresponds to the triangular and square network studied experimentally.

The linear networks also consist of n nests but with only n-1 branches, and are as above except that there is no branch linking nests 1 and n, who therefore only have one neighbour. With n=3 or n=4, this corresponds to the triangular and square networks studied experimentally with one branch cut.

At any moment, nest i contains X_i ants. Each ant has a fixed probability, q, of leaving its nest at each time step that we consider here to be independent of the trail concentrations. At the fork just outside the nest, it chooses between the two branches (i, i-1), leading to its two neighbours (i+1, i-1), as a function of the quantity of pheromone (C_i, C_{i-1}) on them. Using the probability function evaluated experimentally for *I. humilis* (Deneubourg et al., 1990) and used in previous models (see introduction), an ant leaving nest i will have the following probability of choosing the branch i leading to nest i+1:

(1) $\text{prob}_{i,i+1} = (20 + C_i)^2 / ((20 + C_i)^2 + (20 + C_{i-1})^2)$

The complementary probability, $\text{prob}_{i,i-1}$, i.e. the probability that an ant leaving nest i moves to nest i-1 (via branch i-1), is given by $1-\text{prob}_{i,i+1}$. (The extreme nests of a linear network only having one neighbour, this choice does not apply).

Unlike most other ant species, Argentine ant foragers/explorers lay pheromone not just when returning to the nest with prey but more or less continuously as they move. Therefore, each ant that leaves nest i then crosses the branch it chose, adding one unit of pheromone to that branch, and enters the nest at the other end. In marking, each ant that crosses a branch increases the probability that the next ants leaving the nests at both ends of that branch will choose that branch. This positive feed-back is the key to the way the network becomes structured.

As the ants are moving from nest to nest, the number of ants in each nest, X_i, is a variable. Also, as the time scale is relatively long, a proportion, e, of the pheromone evaporates per step. The average equations are thus:

(2) $dX_i/dt = q(X_{i-1}\text{prob}_{i-1,i} + X_{i+1}\text{prob}_{i+1,i} - X_i)$ i=1,...,n

(3) $dC_i/dt = q(X_i\text{prob}_{i,i+1} + X_{i+1}\text{prob}_{i+1,i}) - eC_i$ i=1,...,n-1 or n

linear: $X_0=X_{n+1}=0$, $\text{prob}_{n+1,n}=\text{prob}_{0,1}=0$; circular: $X_0=X_n$, $X_{n+1}=X_1$, $C_0=C_n$, $C_{n+1}=C_1$

Analysis of the stationary state solutions of both linear and circular networks gives us the following rule:

(4) $C_i = C_{i+1}$ or $C_i = 400/C_{i+1}$

A pair of neighbouring branches can be either equal or inverse, with respect to the pheromone on them, but any other branch must therefore be equal to one or the other branch of that pair. In other words, either all the branches are equal, or each branch can take one of two values.

This generates a number of possibilities that depends on how many branches the network contains. If there are two branches, only two situations are possible, either the branches are equal or inverse. If there are three branches, either all the branches are equal, or two branches are equal and one is the inverse. If there are four branches, either all are equal, or three are equal and one is the inverse, or two branches have one value and the other two are also equal but have the inverse quantity of pheromone.

In general, the actual values of C_i at the stationary state depend on how many, let us say j, of the n branches equal one value, and how many, n-j, equal the other. Note that if n is an even number, j can vary from n/2 to n, whereas if n is odd j can vary from (n+1)/2 to n. C_i are the positive real solutions to the following quadratic equation:

$$(5) \qquad jC_i^2 - CC_i + 400(n-j) = 0 \qquad\qquad j=n/2 \text{ or } (n+1)/2,...,n$$

where C is the total amount of pheromone present at equilibrium, and is equal to qN/e, where N is the total number of ants. In the experimental conditions, the approximate values of the parameters are: N=6500n, where n is the number of nests, q=0.0002 per sec, and e=0.0005 per sec (estimated indirectly from Van Vorhis Key and Baker, 1982). C is thus of the order of 8000 for the triangular network, 10000 for the square network and 5000 for the long / short network.

The greater the value of C, the more solutions become possible, in the order j=n-1, j=n-2, etc. However, we are only interested in the solutions that are both stable and can be spontaneously reached from the initial condition $C_i=0$, $X_i=N/n$ (where n here is the number of nests).

Without going into all the details, the results of the model, confirmed by Monte Carlo simulations, give the following configurations. In all networks, the symmetrical solution, $C_i=C/n$, is only stable for low values of C, becoming unstable for experimental values of C.

For the triangular network with high values of C, corresponding to the experimental conditions, nearly all the traffic concentrates on one branch linking two large nests, with the third nest having been spontaneously "abandoned", in the passive sense that more ants have left it than have returned to it. This raises the interesting point that the network (as described by this model) is capable of reducing the number of nests and branches it contains, as happens during the period before winter. This occurs as a by-product of the ants individual behaviour, rather than being an "intentional" reorganisation. If you cut one of the branches of the triangle, the stable solution with the experimental value of C is such that nearly all the traffic is concentrated on one branch.

In the square network with mid-range values of C, nearly all the traffic is concentrated on one branch linking two large nests, the other two nests being more or less abandoned ($j=n-1$). With the higher experimental values of C, a second solution appears, where the traffic is divided equally between two opposite branches, and the number of ants in each nest is equal ($j=n-2$). This II-shaped solution dominates the previous one, in the sense that it is more systematically reached from the initial condition. With one branch cut, the solutions are as with the triangular network.

Finally, the model does not explicitly consider the time taken to cross the branches between the nests and is therefore inappropriate to analyse the long / short network (see Goss et al., 1989, for detailed analysis of this situation). With equal branches the model gives all the traffic concentrated on one branch.

These theoretical results evidently do not correspond to the experimental observations for the triangular and square networks. There is, however, one factor that could reasonably be introduced that modify the model's results, namely the idea that the ants' antennae become progressively saturated at high concentrations of pheromone (see also Calenbuhr and Deneubourg, section II/2). In other words, they are more sensitive to a given difference between two weakly marked branches, than to the same difference between two strongly marked branches. In mathematical terms, the only change in the model is that equation (1) becomes:

(6) $\mathrm{prob}_{i,i+1} = (20 + f(C_i))^2 / ((20 + f(C_i))^2 + (20 + f(C_{i-1}))^2)$

where $f(C_i) = gC_i/(g+C_i)$, g being a constant such that when $C_i=g$, the ants only "perceive" $C_i/2$. With a high value of g, the saturation effect is very small and equation (6) is approximately as equation (1). With low values of g, the saturation is important, and has the effect that, when the branches are well marked, the ants choose the lesser marked branch much more often than they would without saturation. This of course favours more dispersion of the ant traffic. The rule in equation (4) becomes, with $g>20$:

(7) $C_i = C_{i+1}$ or $g^2(1-C_iC_{i+1}/400+2(C_i+C_{i+1})/g) = -(C_i+C_{i+1}+C_iC_{i+1}/20)^2$

This generates some extra possibilities that were not present before, as well as modifying the values for those solutions that were. As the complete picture is rather complicated, we will only discuss the solutions obtained with the experimental values of C as estimated above.

The network with two branches now has two stable solutions, i.e. the previous solution, with all the traffic concentrated on one branch, and the symmetrical solution which has become stable.

The triangular network now has three stable solutions. These are the symmetrical solution which has become stable, the solution decribed above with the traffic concentrated on one branch, and a solution that corresponds to the experimentally observed V-shaped result, i.e. two branches equally well-used and the third neglected, with the nest in the middle becoming roughly twice the size of the the other nests. This solution existed without saturation, but was unstable. Monte Carlo simulations show that with low values of g, as one would expect the increased dispersion favours the symmetrical solution, which is the one most frequently reached from the initial condition. With medium values of g the experimental solution is the most frequently reached, and as one would expect with high values of g the model gives the concentrated solution that characterised the model without saturation.

Similarly, with experimental values of C, the square network now has three stable solutions, the symmetrical solution which has become stable, the II-solution described previously in which two opposite branches are equally used, and a solution that corresponds to the experimental U-shaped result, i.e. three equally well-used branches with the fourth neglected and the nests in the middle becoming roughly twice the size of the the other nests. This solution also existed without saturation, but was unstable. Again, with low values of g the symmetrical solution is the one most frequently reached from the initial condition, with medium values of g the experimental solution dominates, and with high values of g the concentrated solution with two separated pairs of branches dominates.

There is unfortunately no objective basis on which to estimate a value for g, although one could tentatively suggest a possible range of 500-5000. With a value of 2500, Monte Carlo simulations of the triangular network gave the experimentally observed V-shaped solution nine times out of ten, and the symmetrical solution once out of ten. When one of the well-used branches of the experimental V-solution was cut, the simulations recreated the experimentally observed reaction ten times out of ten, with the previously unused branch becoming as well-used as the other. For the square network the experimentally observed U-shaped solution was selected six times out of ten and the concentrated solution (two parallel branches heavily used) was selected four times. However, unlike what happened experimentally, when the base of the U-shaped solution was cut, the network reacted to form the concentrated II-solution ten times out of ten. Finally, smaller or larger values of g gave smaller proportions of the corresponding experimental solutions.

To summarise this theoretical section, let us say that the model now generates the experimentally observed traffic configurations. However these are not the only solutions to appear spontaneously in the Monte Carlo simulations. It is therefore necessary to enlarge the experimental base, in spite of the time required for each experiment, to see if these other so-lutions can appear experimentally, and with what frequency.

There is a second point of disagreement between the experiments and the model with saturation, namely that in the square network, when you start from two opposite well-marked branches, the model always stays on that II-configuration (for the parameter values investi-gated above), whereas the experimental network always reformed the U-shaped configura-tion.

This suggests there might be some additional factor in the ants' behaviour that pre-vents pairs of nests from becoming isolated, and which maintains a direct or indirect link between all the nests. While we have no clue at present as to its nature, it is certainly not ex-clusive of their trail-laying and trail-following behaviour, whose primary importance both in this and other contexts is now quite well established. Rather it must be something that has a corrective effect on the underlying structure generated by trail pheromone.

5. Discussion

How to link a number of points in 2-D space is a classical problem of operational analysis, often referred to as the travelling salesman or shortest network problem. In the lim-ited number of experimental configurations tested here, *I. humilis* appears to adopt a mini-mum spanning tree, that is to say links all nests together while neglecting superfluous bridges. Note that a minimum spanning-tree necessarily minimises total route length at the cost of an increased travel time. For example in the triangular configuration, the final network is two sides long but the ants can no longer pass directly from nest 1 to nest 2. However, when given the choice of two routes linking two points, the shortest route is selected.

While relatively modest algorithms are capable of achieving this, there is a very im-portant difference in the way a computer-aided analyst reaches his solution and the way the ants reach theirs. The analyst knows beforehand the number and position of each point, whereas each individual ant can be considered to be more or less unaware of this data, at least initially, and especially in the dark. How then do they achieve such a respectable solution?

At least part of the answer lies in their capacity for collective problem-solving or self-organisation, as illustrated by the above model. Solutions (in this example the traffic-configu-rations and the reorganisation of the nest sizes), are not coded explicitly into the individual ants, but rather emerge spontaneously from the interactions both between them and with the physical environment in which they are placed. This last point is important as it allows them to adopt in most situations an efficient trail system covering their particularly large territories, and to react rapidly and in generally appropriately to any modification in the environment. As nothing is specifically forseen, nothing can be unforseen.

Acknowledgements

This work is supported in part by La Fondation Schlumberger pour l'Education et la Recherche, the Ernst and Victor Hasselblad Foundation, the Belgian program on interuniversity attraction poles, Les Instituts Internationaux de Physique et de Chimie, IRSIA grant #86004 and the Belgian Fonds National de Recherche Scientifique.

REFERENCES

Aron S. and Pasteels J.M. (1989) Spatial organisation in the Argentine ant *Iridomyrmex humilis* (Mayr). Actes Coll. Insectes Soc. 5: 189-197

Aron S., Pasteels J.M. and Deneubourg J.L. (1989) Trail laying behaviour during exploratory recruitment in the Argentine ant *Iridomyrmex humilis* (Mayr). Biol. Behav. 14: 207-217

Aron S., Pasteels J.M. and Deneubourg J.L. (in press) Self-organising spatial patterns in the Argentine ant *Iridomyrmex humilis* (Mayr). In: K. Jaffe, A. Cedena and R.K. Vander Meer (eds.) Applied Myrmecology: A World Perspective. Westview Press, Boulder, Colorado

Bern M.W. and Graham R.L. (1989) The shortest-network problem. Scientific American January: 66-71

Cherix D. (1987) Relation between diet and polyethism in *Formica* colonies. In: J.M. Pasteels and J.L. Deneubourg (eds.) From Individual to Collective Behaviour in Social Insects: 93-116. Exp. Supp. 54, Birkhäuser Verlag, Basel

Deneubourg J.L. and Goss S. (1989) Collective patterns and decision making. Ecol. Ethol. Evol. 1: 295-311

Deneubourg J.L., Goss S., Franks N. and Pasteels J.M. (1989) The blind leading the blind: modelling chemically mediated army ant raid patterns. J. Ins. Behav. 2: 719-725

Deneubourg J.L., Aron S., Goss S. and Pasteels J.M. (1990) The self-organizing exploratory pattern of the Argentine ant. J. Ins. Behav. 3: 159-168

Goss S. and Deneubourg J.L. (1989) The self-organising clock-pattern of *Messor pergandei* (Formicidae, Myrmicinae). Insectes Soc. 36: 339-376

Goss S., Aron S., Deneubourg J.L. and Pasteels J.M. (1989) Self organised short cuts in the Argentine ant. Naturwissenschaften 76: 579-581

Goss S., Deneubourg J.L., Aron S., Beckers R. and Pasteels J.M. (1990) How trail laying and trail following can solve foraging problems for ant colonies. In : R.N. Hughes (ed.) Behavioural mechanisms of food selection: 661-678. NATO ASI Series, Vol G 20, Springer Verlag, Heidelberg

Hölldobler B. and Wilson E.O. (1977) The number of queens: an important trait in ant evolution. Naturwissenschaften 68: 8-15.

Hölldobler B. and Lumsden C.J. (1980) Territorial strategies in ants. Science 210: 732-739.

Rosengren R. (1985) Inter-nest relations in polydomous *Formica* colonies (Hym. Formicidae). Ges. Alg. Angew. Ent. 4: 288-291.

Van Vorhis Key S.E. ad Baker T.C. (1982) Trail following responses of the Argentine ant, *Iridomyrmex humilis* (Mayr), to a synthetic trail pheromone component and analogs. J. Chem. Ecol. 8: 3-14.

SIMULATIONS OF THE
GLIDING BEHAVIOR AND AGGREGATION
OF MYXOBACTERIA

Angela Stevens

Institut für Angewandte Mathematik, SFB 123

Im Neuenheimer Feld 294, D-6900 Heidelberg

Abstract

Up to now the social gliding and aggregation of myxobacteria has not been explained definitely. To get a better insight, several hypotheses were tested by Monte Carlo simulations, where the bacteria are represented by rod shaped figures moving in a square with periodic boundary conditions. Among these hypotheses are, for example, slime trail following and response to a diffusing chemo-attractant. The results of these simulations are the basis for a mathematical model.

1. Introduction

The gliding myxobacteria are ubiquitous soil bacteria which have the following life cycle (here the example of Myxococcus xanthus) :

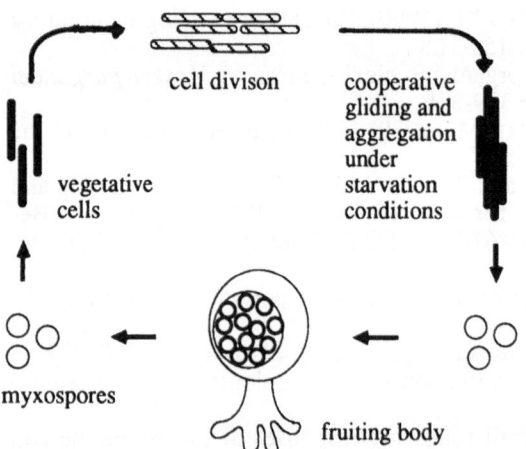

Figure 1. The rod shaped vegetative cells undergo cell division and are able to move on suitable surfaces. Under starvation conditions large population of cells cooperate and generate different patterns. Finally they aggregate to form fungi-like structures - so called fruiting bodies - whose stalks consist mainly of slimy material. Within these fruiting bodies the vegetative cells transform into metabolically dormant myxospores.

The myxobacteria are of special interest for biologists because their cooperative gliding, culminating in the formation of fruiting bodies, serves as a model for morphogenetic problems at the procaryotic level of organization (Reichenbach, 1986). Two metabolically different cell types are found among the myxobacteria: the Cystobacterinae and the Soranginae. In the following only the gliding of the Cystobacterinae is described because the Soranginae tend to sink in the agar and build ruts which restrict the freedom of movement.

2. Description of Movement

The phenomena of myxobacterial gliding are presented in great detail in the films by Reichenbach (Lehrfilme, IWF). The individual cells move on suitable surfaces parallel to their long axes, but they also stop and reverse direction in a irregular manner. They glide very slowly, about 5μm/min (Reichenbach, 1966). Nevertheless the gliding velocity is not constant. The myxobacteria produce extracellular slime, which they prefer to glide on. Once on a slime trail, they increase their gliding velocity. During cell division gliding motility stops.

The myxobacterial gliding behavior is mainly communal. The bacteria build tracks or swarms, which merge to form larger ones or split into smaller ones. They move in circles and spirals. Just before aggregation, patterns of rhythmic pulsating waves occur, where ripples in adjacent tracks move in opposite directions.

3. Hypotheses for myxobacterial gliding

Genes governing the gliding behavior of myxobacteria are the motility system A for adventurous single cell gliding and the motility system S for social gliding. Both systems are found in the wildtype. A^- mutants are only able to glide in groups and S^- mutants glide predominantely alone. It is interesting that rippling requires both gene systems (Hodgekin and Kaiser, 1979).

Dworkin et al. (1983) and Keller et al. (1983) suggested that the myxobacteria produce a chemical gradient which reduces the surface tension of the area they glide on. Their model predicts not only that two cells can pass each other but also that single cells can join a swarm. On the other hand their model does not explain the other movements nor does it account for the behavior of cells which continue to crawl on a surface even if immersed in water.

Myxobacteria feed on nutrients or on prey bacteria by response to chemical gradients. This and the fact that two neighboring swarms migrate towards each other with increasing velocity until they fuse, are a hint for the ability of the cells to sense and produce a diffusing chemo-attractant (Lauffenburger 1984 and Reichenbach 1965a). If this chemo-attractant exists the size of its diffusion coefficient determines the extent of accumulation. Kuner and Kaiser (1982) however found that fruiting body formation takes place even when the cells are immersed in water and Dworkin (1983) tested several substances for chemo-attraction without success.

Another idea is that the thread shaped cell projections - the pili - play an important role for communal gliding in the sense that the bacteria are attached to each other after they have collided.

Stanier (1942) found that the myxobacteria glide parallel to stress lines in the agar. So perhaps this behavior together with slime trail following is sufficient to account for aggregation (Reichenbach 1965b).

4. Simulations

4.1. Assumptions

The simulations are carried out under the following basic assumptions:

(a) The ratio of cell width to cell length is taken to be 1/8. This is a realistic average value.

(b) The region where the bacteria are gliding is a square with periodic boundary conditions. The density of bacteria is kept in a range between a low of 100 and a high of 6000 , see experiments of Shimkets and Dworkin (1981).

(c) Everytime a bacterium moves one cell width forward it produces slime over the whole cell length.

(d) If the conditions are suitable, a diffusing chemo-attractant is produced in the same way.

(e) Each cell decides where to move depending on four conditions in the following order of importance:

      (1) density of chemo-attractant,

      (2) density of slime,

      (3) preference for special direction,

      (4) escape probability.

A weighted average of the above four factors is used to compute the actual cell motion, see (Stevens 1990) for a detailed description.

(f) With increasing density of slime, the gliding velocity increases up to ten-fold.

4.2 Results of Simulations

In the figures the slime is marked by white dots. The colors for the myxo-bacteria change from light grey to black, depending on the number of bacteria gliding over each other. The chemo-attractant is chequered. Figure 2 shows the random walk of a single bacterium. The probability of walking left or right is 1/8 (nearly the ratio of cell width to cell length). This is the same in figure 3. No diffusing chemo-attractant is produced.

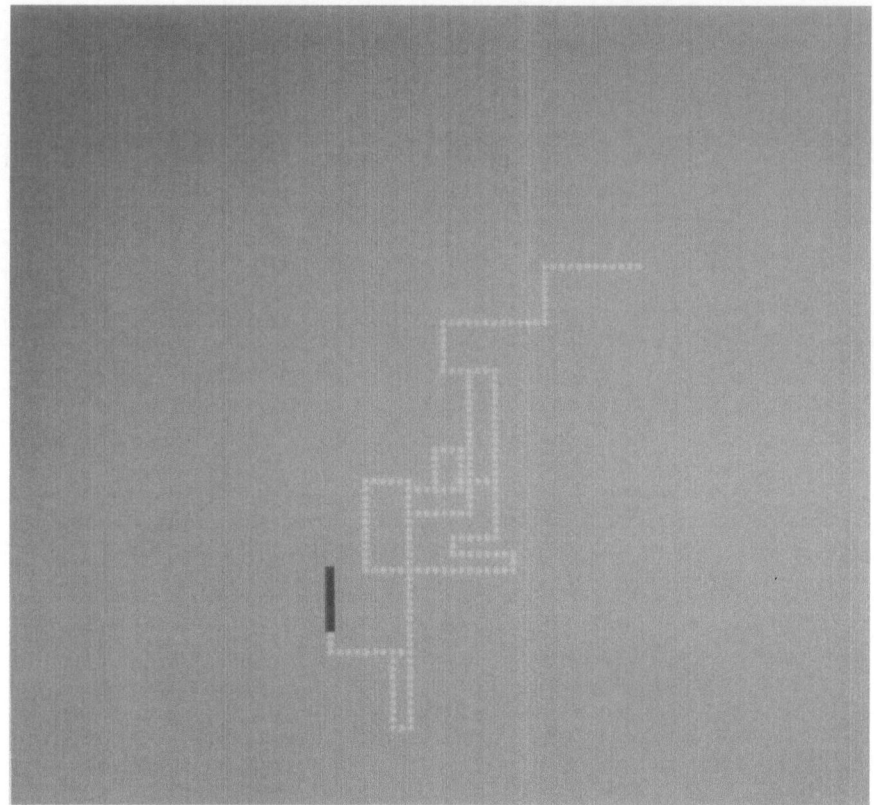

Figure 2. Typical random walk of a single bacterium (black).

The slime at position x at time t_{n+1}, $n \in \mathbf{N}$ is calculated by:

$$
(1) \quad
\begin{aligned}
slime(x, t_{n+1}) = (&slime(x, t_n) \cdot decomposition\ factor \\
&+ slime\ production(x, t_{n+1})) \cdot weight\ factor.
\end{aligned}
$$

It is important to choose the parameters such that memory is present, but not so strong that a single bacterium is caught on the slime trail forever.

Then 1000 bacteria are randomly distributed in the inspected area. After a while some closed loops can be seen where the bacteria concentrate. Some times later these loops disappear and are formed again at other places. These formations are comparable to temporary stable circles which occur in real myxobacterial cultures. If all bacteria are gliding with the same velocity, the loops do not

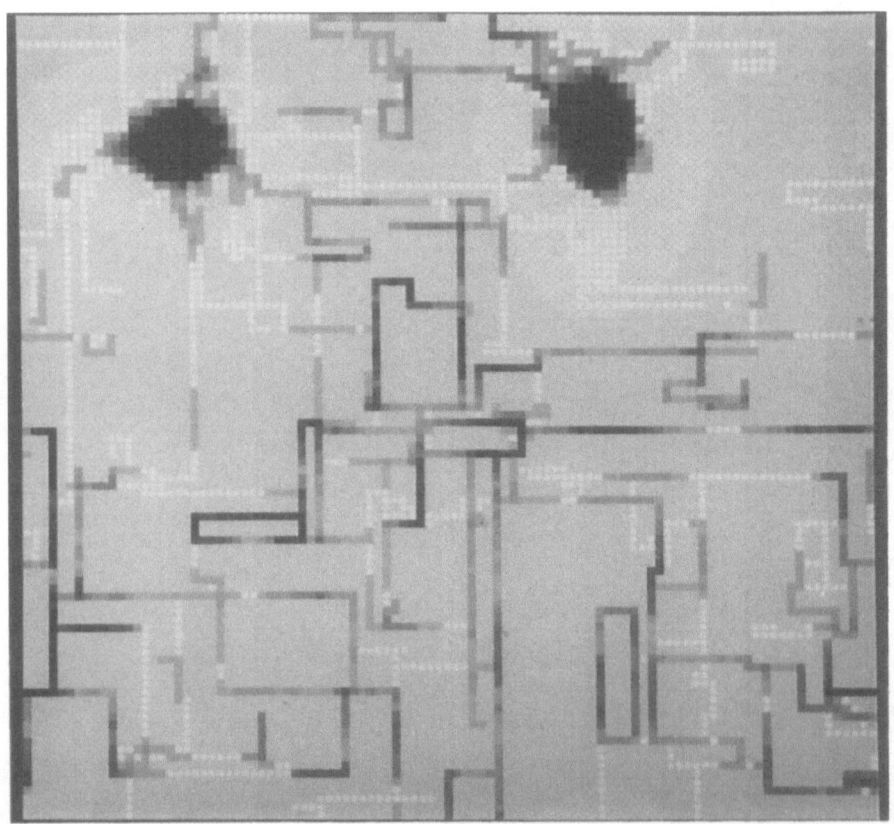

<u>Figure 3.</u> 1000 bacteria during the phase of aggregation (light grey for one
bacterium, black for more than five bacteria).

develop in this distinct way. So the different velocities seem to be important, but they are not sufficient to account for stable aggregation centres. Next, a threshhold for starting the production of the diffusing chemo-attractant is introduced. When the density of bacteria at some site exeeds a critical level, chemo-attractant is produced by one of the cells. Other bacteria sensing the attractant respond by producing more of this, but only when the ambient concentration is high enough. The result as seen in figure 3 is that the cells begin to aggregate. In areas where no aggregation takes place closed loops of bacteria and slime are formed.

Other biological ideas were tested:
When all bacteria produce the chemo-attractant simultaneously from the beginning, aggregation takes place, but too many fruiting bodies are formed and the patterns before aggregation do not appear. When the chemo-attractant is pulsed the following effect can be seen: the bigger the delay between two pulses of the attractant, the fewer fruiting bodies are found. The idea of slime trail following where the bacteria reduce the preference of moving straight after a while, see (White 1987), does not generate aggregation centres.

Some simulations were done for parallel gliding. This idea is realized by a coat-like adhesive structure around each bacterium (Stevens 1990).

5. Mathematical Model

In the silmulations the bacteria react to the density of slime and chemo-attractant at that points which are nearest neighbors to their positon. Keeping this in mind the following general system of partial differential equations can be derived at least heuristically from an interacting stochastic many particle system (compare Oelschläger 1987, 1989) to describe the myxobacterial gliding:

Define $\Omega \subset \mathbf{R}^2$ a suitable bounded domain with $x \in \Omega, t > 0$

$b(x, t)$: density of bacteria,
$s(x, t)$: density of slime,
$c(x, t)$: density of chemo-attractant at position x at time t,

$$
\begin{aligned}
\partial_t b &= \nabla(f(b, s)\nabla b - g(b, s)\nabla s - b\nabla c) \\
\partial_t s &= \beta_s(b, s) \\
\partial_t c &= \Delta c + \beta_c(b, c) \quad,
\end{aligned}
$$

(2)

$$b(x, 0) = b_0(x), s(x, 0) = s_0(x), c(x, 0) = c_0(x).$$

No flux conditions are fulfiled on the boundary. The celluar flux due to the gliding of myxobacteria on the slime trails is described by $-f(b, s)\nabla b + g(b, s)\nabla s$ which also contains the undirected random motion of the bacteria, $b\nabla c$ is the chemotactic flux, $\beta_s(b, s), \beta_c(b, c)$ describe production and degradation of slime and chemo-attractant. The introduction of a very small diffusion coefficient for the slime might also be plausible for the continuous model.

The connection between the mathematical model and the simulations will be discussed elsewhere. Compare Tranquillo and Lauffenburger, Sect II/3 for derivation and analysis of the chemotaxis equations.

6. Discussion

The Monte Carlo simulations suggest that following the slime trails with different velocities plays a role for pattern formation before the final aggregation has taken place but does not account for aggregation. Thus an additional attracting effect seems to be necessary which has to work in a small range of the future fruiting body. Different possibilities for this effect are described in paragraph 3. Here a diffusing chemo-attractant with cell density dependent production is introduced in order to generate aggregation centers after a while. Of course all theoretical suggestions have to be supported by biological experiments.

Acknowledgement

This work has been supported by the Deutsche Forschungsgemeinschaft as part of the research program of the Sonderforschungsbereich 123 (project C1) at the University of Heidelberg.

REFERENCES

Dworkin M. and Eide D. (1983) Myxococcus xanthus does not respond chemotactically to moderate concentration gradients. J. Bacteriol. 154: 437-442.

Dworkin M. and Keller K.H. and Weisberg D. (1983) Experimental observations consistent with surface tension model of gliding motility of Myxococcus xanthus. J. Bacteriol. 155, No 3: 1367-1371.

Hodgekin J. and Kaiser D. (1979) Genetics of gliding motility in Myxococcus xanthus (Myxobacterales): two gene systems control movement, Mol. Gen. Genet. 171: 177-191.

Keller K.H. and Grady M. and Dworkin M. (1983) Surface tension gradients: Feasible models for gliding motility in Myxococcus xanthus. J. Bacteriol. 155, No 3: 1358-1366.

Kuner J.M. and Kaiser D. (1982) Fruiting body morphogenesis in submerged cultures of Myxococcus xanthus. J. Bacteriol. 151: 458-461.

Lauffenburger D. (1984) An hypothesis for approaching swarms of Myxobacteria. J. theor. Biol. 110: 257-274.

Oelschläger K. (1987) A fluctuation theorem for moderately interacting diffusion processes. Probab. Th. Rel. Fields 74: 591 - 616.

Oelschläger K. (1989) On the derivation of reaction-diffusion equations as limit dynamics of systems of moderately interacting stochastic processes. Probab. Th. Rel. Fields 82, No. 4: 565 - 586.

Reichenbach H. (1965a) Rhythmische Vorgänge bei der Schwarmentfaltung von Myxobakterien. Berichte der deutschen Botanischen Gesellschaft 78: 102-105.

Reichenbach H. (1965b) Untersuchungen an Archangium violaceum. Ein Beitrag zur Kenntnis der Myxobakterien, Berichte der deutschen Botanischen Gesellschaft, Arch. Microbiol. 52: 376-403.

Reichenbach H.: Lehrfilme, Myxobakterien. Institut für den wissenschaftlichen Film, Göttingen: C893, E777, E779, E1582, E1583, E2421.

Reichenbach H. (1966) Myxococcus spp. (Myxobacterales): Schwarmentwicklung und Bildung von Protocysten. Publikationen zu wissenschaftlichen Filmen, Göttingen 1A: 557-578.

Reichenbach H. (1986) The myxobacteria: common organisms with uncommon behaviour. Microbiol. Sc. 3, No 9: 268-274.

Shimkets L.J. and Dworkin M. (1981) Execreted adenosine is a cell density signal for the initiation of fruiting body formation in Myxococcus xanthus. Dev. Biol. 84: 51.

Stanier R.Y. (1942) A note on elasticotaxis in myxobacteria, J. Bacteriol. 44: 405-412.

Stevens A. (1990) A model for gliding and aggregation of myxobacteria. Nonlinear wave processes in excitable media. Eds. A. Holden, M. Markus, H.G. Othmer. Manchester Univ. Press. To appear.

White D. (1987) Cell interactions and the control of development in myxobacteria populations. Intl. Rev. Cytol. 71: 203-227.

A ONE DIMENSIONAL MODEL FOR THE SWARMING BEHAVIOR OF MYXOBACTERIA

Beate Pfistner

Abteilung Theoretische Biologie, Universität Bonn, SFB 256

Kirschallee 1, D-5300 Bonn 1

Abstract

The gliding myxobacteria show complex phenomena of swarm building and aggregation to fruiting bodies. This article presents a one dimensional model, relevant for the behavior in 'streets' of swarms. The stationary state is analytically discussed and simulated. It results in a criterion for the facility of swarm building. A sketch of a two dimensional formulation is presented in Box 15 (Pfistner/Alt).

1. Introduction

The film clips (Reichenbach and Kühlwein, 1964-68), underlying the model, were taken from an *Archangium violaceum* swarm. This species belongs to the suborder Cysto-bacterineae. Their swarm edges are typically fringed and flame like, see Figure 1a. The whole swarm surface usually shows radial veins of different density (Reichenbach and Dworkin, 1981; Keller, 1983).

In low density regions of swarms, such as the streets in Figure 1b, the moving behavior of single cells can be observed. They tend to join other cells. Either they glide past each other or they move parallel in the same direction ('social gliding', see also Stevens, this part). Cells at the swarm edge rarely leave the swarm totally (Kaiser, 1979). If they venture from the swarm edge they stop after a short distance, reverse their direction and enter the swarm again (White, 1981).

In addition to the A- and S-gene systems responsible for adventurous and social glid-ing (for details see Stevens, this part) Blackhart and Zusman (1985) found the frz-genes ('frizzy' genes) to be part of a control mechanism of direction reversal. *Myxococcus xanthus* cells reverse their direction about every 6.8 min. Net movement occurs since the interval between reversals can vary widely. Perhaps a decrease in motility rate in the forward direction triggers a reversal in direction. Blackhart and Zusman proposed chemoresponse to macromolecular signals or cell-bound signals to trigger reversal. Nevertheless the rea-son for the decrease is unknown (White, 1981).

a) 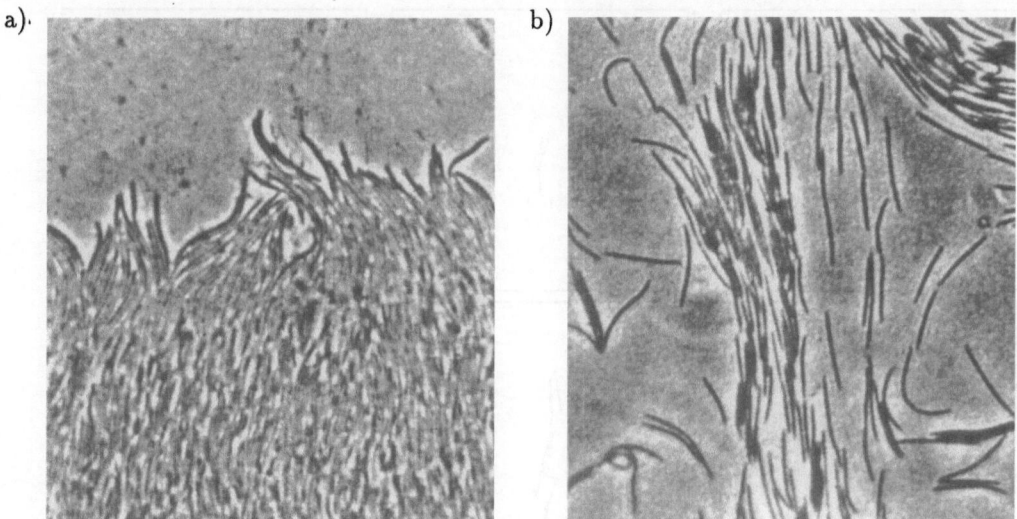 b)

Figure 1: Parts of an Archangium violaceum swarm: a) swarm edge, b) 'streets' in the swarm. (Reichenbach and Kühlwein, 1964-68)

2. Analysis of Myxobacterial paths

The paths of cells at the outest swarm edge were tracked (Figure 1a). For that purpose the film clip of an *Archangium violaceum* swarm edge (Reichenbach and Kühlwein, 1964-68) was projected onto a Summagraphics digitizing tablet MM1201, which was connected with an IBM AT personal computer, programmed with Sigma Scan for recording points with a stylus. The tablet had a resolution of 40 lines per mm. Cartesian coordinates were introduced in order to measure cell positions. The film was projected at a magnification producing a mean cell length of 14.4 mm. In the starting scene the top of a cell was defined as the end pointing to the moving direction of the swarm front and was tracked as long as the cell was clearly visible. Sometimes cells disappeared into the agar and could not be observed any longer. Nevertheless it was possible to track some cells over 350 pictures, of which every 8th was recorded.

Each picture of Figure 2 shows the path of the top of a cell. These 6 typical examples of cells, lying next to each other, show, that for most of the cells, a change of moving direction at the swarm edge is a reversal of direction. Only some cells leave the initial direction of their longitudinal axis. Therefore the gliding behavior at the swarm edge is similar to the behavior in the streets of the swarm (Figure 1b). There the cells are gliding almost in one dimension. So in this case change of direction is always a reversal. The difference between the interior and the edge behavior is the number of reversals per time, called <u>turning frequency</u>.

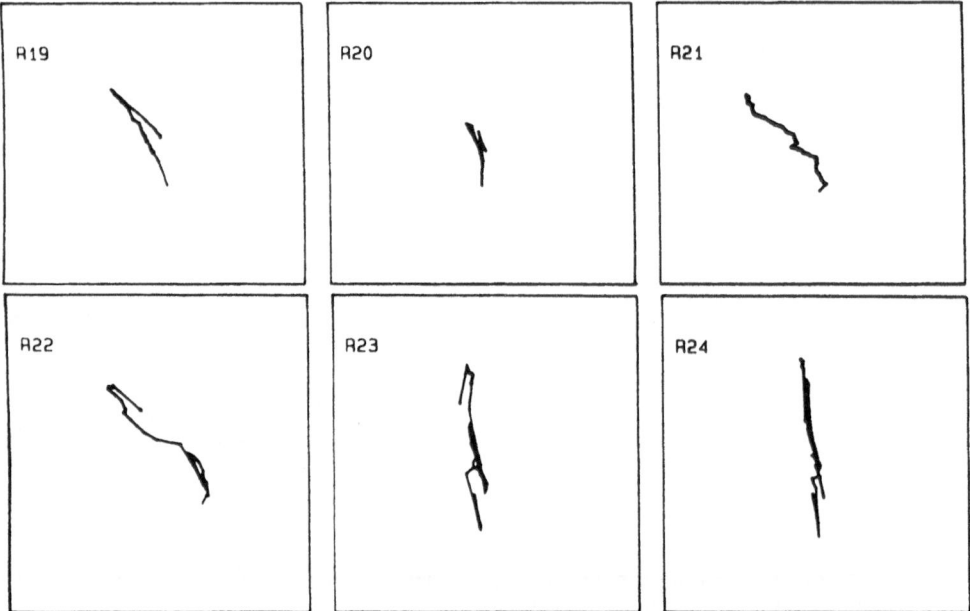

Figure 2: Paths of 6 neighbouring cells from the outest swarm edge.

3. Assumptions for the model

The differences in the turning frequencies within the streets and at the swarm edge must depend on signals the cells perceive from surrounding cells. So they depend on the density of surrounding cells. For the following model of cell interactions in a swarm not temporal but spatial sensing of cell densities is assumed. This means that each cell perceives the density distribution of surrounding cells up to a maximal radius R, but has no memory of past density distributions.

Cell density change is caused by turning frequencies of the individual cells (Othmer, Dunbar and Alt, 1988). The probability for a change of direction of one cell depends only on the present density distribution. The direction changes are assumed to be Poisson-distributed with mean waiting time T, which becomes constant $T = \tau$ for single cells without interaction. Also the average gliding speed s is assumed to be constant.

4. The one dimensional model

Within the streets (Figure 1b) cells are gliding almost in one dimension. So in this case change of direction is always a reversal and it is possible to introduce a positive $(+)$ and a negative $(-)$ moving direction.

Define the density $p^+(x,t)$ of cells moving in positive direction at point x and time t and the density $p^-(x,t)$ for the negative moving direction. The density change for one direction per time step is calculated as the difference between cells turning out of and into this direction. The rate of cells turning out of or into a direction is given by the product of their density, $p^+(x,t)$ or $p^-(x,t)$, and the corresponding turning frequency λ_+ or λ_- respectively. From that it follows a PDE system for the one dimensional model (Greenberg and Alt, 1987):

$$
\begin{aligned}
(1) \qquad \partial_t p^+(x,t) + s \cdot \partial_x p^+(x,t) &= -\lambda_+ \cdot p^+(x,t) + \lambda_- \cdot p^-(x,t) \\
\partial_t p^-(x,t) - s \cdot \partial_x p^-(x,t) &= \lambda_+ \cdot p^+(x,t) - \lambda_- \cdot p^-(x,t) \quad ,
\end{aligned}
$$

where s denotes the constant average moving speed. The total density at point x and time t is given by $p(x,t) = p^+(x,t) + p^-(x,t)$.

The function Λ defining the turning frequencies λ_+, λ_- depends on the density profile in the perception interval $[x - R, x + R]$ of cells at point x. The surrounding densities are weighed with $\alpha(r)$ or $\beta(r)$ as weight functions for the same or the opposite direction. So λ_+ and λ_- can be defined as follows:

$$
(2) \qquad \lambda_+ \; := \; \Lambda \left(\int_{-R}^{+R} \alpha(r) \cdot p^+(x+r,t) + \beta(r) \cdot p^-(x+r,t)\, dr \right)
$$

$$
\lambda_- \; := \; \Lambda \left(\int_{-R}^{+R} \alpha(r) \cdot p^-(x-r,t) + \beta(r) \cdot p^+(x-r,t)\, dr \right) \quad .
$$

The turning frequency function Λ has to be monotone growing and positive. From a biological view it is meaningful that the surrounding density of cells moving in the same direction has a negative influence on the turning frequency, i.e. $\alpha(r) < 0, \forall\, r \in \mathbb{R}$. Correspondingly $\beta(r)$, the weight for the opposite direction, has a positive influence on Λ, i.e. $\beta(r) > 0, \forall\, r \in \mathbb{R}$. Further, it is observed that cells prefer not to change their direction if in the neighborhood both directions of motion are equally distributed. Therefore $|\alpha(r)| > \beta(r), \forall\, r \in \mathbb{R}$.

By approximating the integral terms in (2) through local Taylor expansion for small R the turning frequencies λ_+, λ_- are found to be

$$
\begin{aligned}
(3) \qquad \lambda_+ &= \Lambda \left((\alpha_0 p^+ + \alpha_1 \partial_x p^+) + (\beta_0 p^- + \beta_1 \partial_x p^-) \right) \\
\lambda_- &= \Lambda \left((\alpha_0 p^- - \alpha_1 \partial_x p^-) + (\beta_0 p^+ - \beta_1 \partial_x p^+) \right) \quad ,
\end{aligned}
$$

where the weights α_i, β_i are:

$$
(4) \qquad
\begin{aligned}
\alpha_0 &:= \int_{-R}^{+R} \alpha(r)\, dr & \qquad \alpha_1 &:= \int_{-R}^{+R} r\alpha(r)\, dr \\[2mm]
\beta_0 &:= \int_{-R}^{+R} \beta(r)\, dr & \qquad \beta_1 &:= \int_{-R}^{+R} r\beta(r)\, dr \quad .
\end{aligned}
$$

5. The stationary state

The stationary state for system (1) occurs if $\partial_t p(x,t) = \partial_t p^+(x,t) + \partial_t p^-(x,t) \equiv 0$ and the flux $p^+(x,t) - p^-(x,t) \equiv 0$. It follows that for the stationary solution: $p^+(x,t) = p^-(x,t) = \frac{1}{2}p(x,t)$. With that, system (1) reduces to the following implicite ordinary differential equation:

$$(5) \qquad\qquad s \cdot \partial_x p(x,t) = (\lambda_- - \lambda_+) \cdot p(x,t) \quad,$$

The turning frequencies have the following form:

$$(6) \qquad\qquad \begin{aligned} \lambda_+ &= \Lambda(-Ap - B\partial_x p) \\ \lambda_- &= \Lambda(-Ap + B\partial_x p) \quad. \end{aligned}$$

Setting

$$(7) \qquad\qquad \begin{aligned} A &:= -\frac{1}{2}(\alpha_0 + \beta_0) \quad > 0 \\ B &:= -\frac{1}{2}(\alpha_1 + \beta_1) \quad > 0 \end{aligned}$$

$A > 0$ in accordance with the choice of the weight functions $\alpha(r), \beta(r)$, while $B > 0$ is demanded.

The observed swarms show clear, sharp edges. This can be modeled by demanding that $\lambda_\pm \to +\infty$ for $p \to 0$ and $\partial_x p \to \mp\infty$. A simple turning frequency function Λ, which meets the above conditions is

$$(8) \qquad\qquad \Lambda(y) = \frac{1}{\tau}e^y,$$

where τ is the constant average time period between two turns of a cell. Using equation (5) leads to:

$$(9) \qquad\qquad \frac{\sinh(B \cdot \partial_x p)}{\partial_x p} = \frac{s\tau}{2p} \cdot e^{Ap} \quad.$$

By rescaling with $\bar p := Ap, \bar x := 2x/s\tau$ and then dropping the bars this can be reduced to

$$(10) \qquad\qquad \frac{\sinh(k \cdot \partial_x p)}{\partial_x p} = \frac{1}{p} \cdot e^p \quad,$$

with the remaining parameter

$$(11) \qquad\qquad k := \frac{2B}{s\tau A} \quad.$$

A and B are weighting coefficients for the perceived density and the perceived density gradients, respectively; $s\tau$ is the average distance covered between two direction changes and k gives a measurement for perception of density gradients relative to the perceived density through half of the mean distance $s\tau$.

It is found that a swarm extremum exists only for $k > e$. For this case, depending on the value of the extremum, both a minimum (for $p_{ex} > 1$) or a maximum (for $p_{ex} < 1$) are possible (Figure 3).

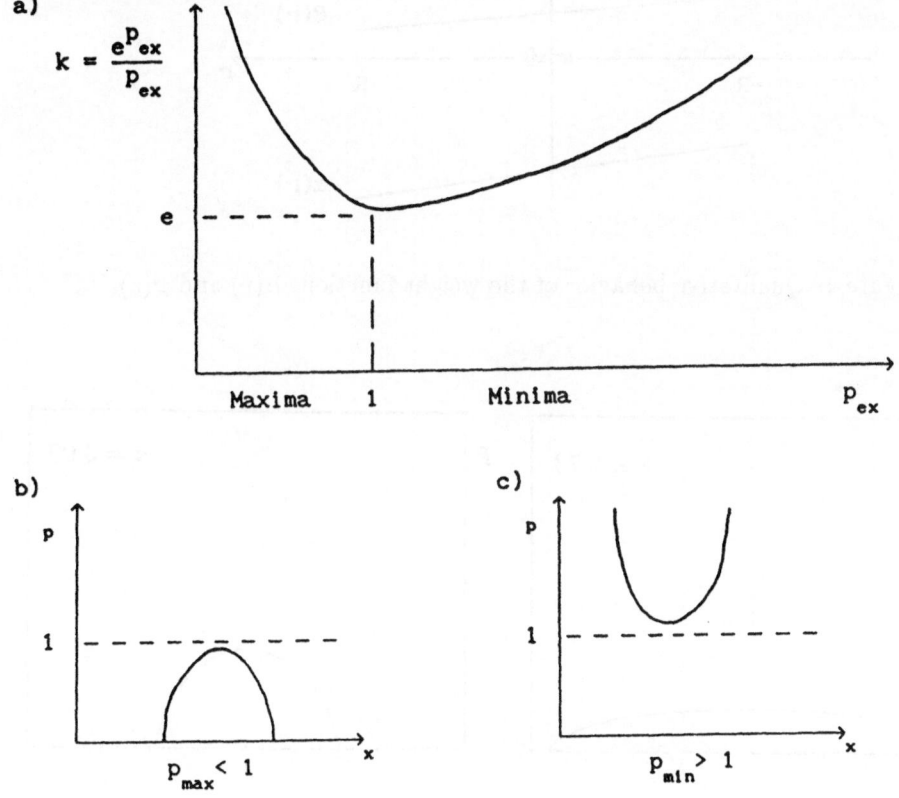

Figure 3: a) Qualitative course of k depending on the extremum p_{ex}, derived from function (9). b) Qualitative form of the swarm for $p_{ex} < 1$, i.e. existence of a maximum. c) Qualitative form of the swarm for $p_{ex} > 1$, i.e. existence of a minimum.

$k > e$ only can be achieved for $B > 0$, because $A > 0$. So for example if no density gradients are measured and therefore $B = 0$, no swarm forming is possible. $B > 0$ and $k > e$ leads to a sufficient condition for the existence of a stationary swarm with finite swarm mass:

(12)
$$\alpha(r) < \alpha(-r) \quad \text{and} \quad \beta(r) < \beta(-r) \qquad 0 \leq r \leq R$$
$$\text{and} \quad R < \frac{s\tau e}{2} \ .$$

The easiest way to fulfil condition (12) is to choose linear functions for $\alpha(r)$ and $\beta(r)$. Their qualitative shape is shown in Figure 4. So the density of cells moving in the same direction are more important in front of the cell than behind it. But the density of cells in the opposite direction are more heavily weighed behind the cell than in front of it.

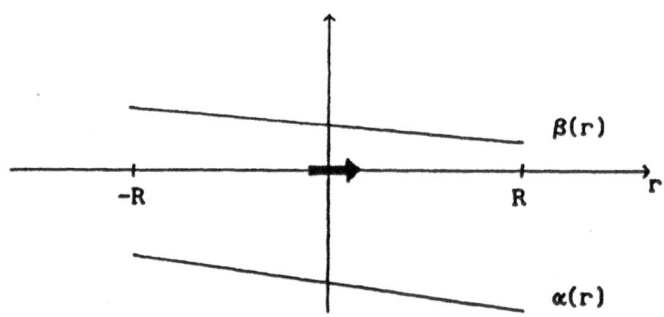

Figure 4: Qualitative behavior of the weight functions $\alpha(r)$ and $\beta(r)$.

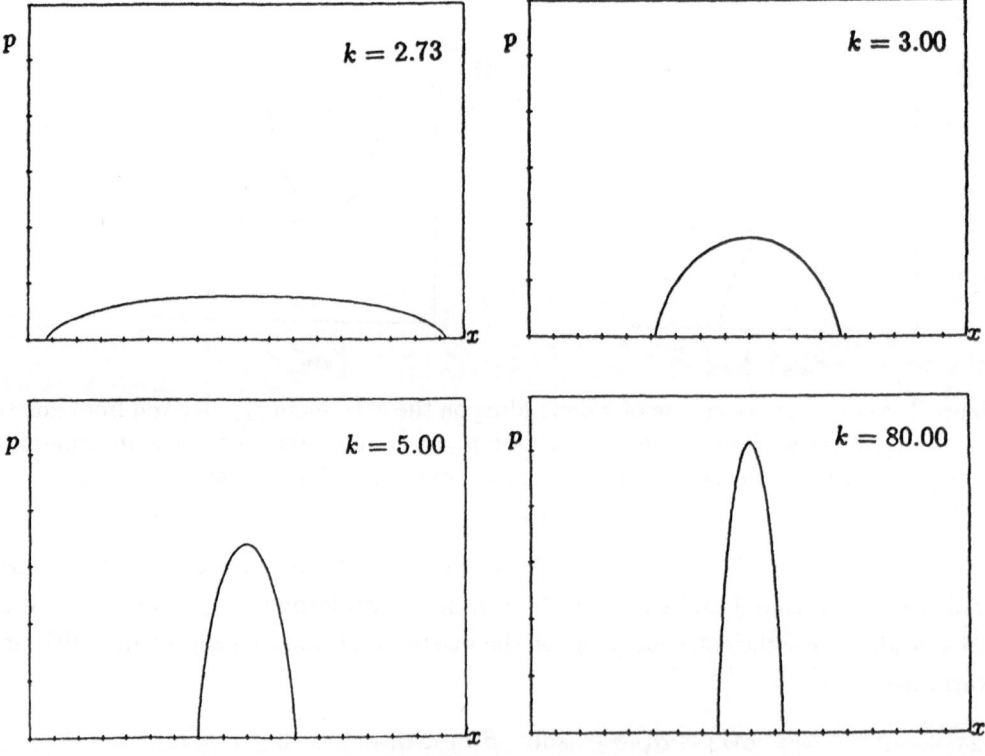

Figure 5: Simulations of stationary swarms for various k and step length $h = 0.01$.

6. Simulations of stationary swarms

The stationary swarm was simulated for several values of $k > e$. The maximum was calculated with a Newton method and then used as the starting value for the upwind method to simulate the density of the swarm. This way only one half of the symmetrical swarm had to be calculated.

The results are shown in Figure 5 where the total density of the swarm is normalized to 1 for better comparison of swarm shapes.

If k is increased, i.e. cells perceive density gradients better, we find that the swarm form concentrates. For $k \to e$ the swarm cohesion is weak, while for $k \to \infty$ the swarm theoretically contracts to one point. But then also $p_{max} \to 1$ (compare Figure 3a), i.e. no swarm develops.

REFERENCES

Blackhart B.D. and Zusman D.R. (1985) 'Frizzy' genes of *Myxococcus xanthus* are involved in control of frequency of reversal of gliding motility. Proc. Natl. Acad. Sci. USA Vol. 82: 8767-8770.

Greenberg J.M. and Alt W. (1987) Stability results for a diffusion equation with functional drift approximating a chemotaxis model. Trans AMS 300: 235-258.

Kaiser D. (1979) Social gliding is correlated with the presence of pili in *Myxococcus xanthus*. Proc. Natl. Sci. USA Vol.76, 11: 5952-5956.

Keller K.H., Grady M. and Dworkin M. (1983) Surface Tension Gradients: Feasible Model for Gliding Motility of Myxococcus xanthus. J. Bacteriology 1358.

Othmer H.G., Dunbar S.R. and Alt W. (1988) Models of Dispersal in Biological Systems. J. Math. Biology 26: 263.

Reichenbach H. and Kühlwein H. (1964-1968) Myxobakterien. Lehrfilm-Archivmaterial. Produced by the Institut für den wissenschaftlichen Film - Göttingen. Botanisches Institut Karlsruhe.

Reichenbach H. and Dworkin M. (1981) The Order Myxobacteriales. In: Prokaryotes, Starr M.P., Stolp H. et al. (Ed.). Vol. 1, Cap. 20. Springer Heidelberg, New York.

White D. (1987) Cell Interactions and the Control of Development in Myxobacteria Populations. Intl. Rev. Cytol. 71: 203-227.

BOX 15

A TWO DIMENSIONAL RANDOM WALK MODEL
FOR SWARMING BEHAVIOR

Beate Pfistner, Wolfgang Alt

This general model is based on a piecewise straight random walk with constant speed s of organisms, which interact only with other members of the swarm.

Model derivations

The position of an organism is characterized by the point of its tip $x = (x_1, x_2)$ and by the angle of its orientation $0 \leq \phi \leq 2\pi$. So $p(x, \phi, t)$ denotes the density of swarm organisms at state (x, ϕ) and time t.

The turning frequency $\lambda(x, \phi, t)$ gives the probability of directional change at state (x, ϕ) and time t and depends on the weighted densities at the surrounding states $(\tilde{x}, \tilde{\phi})$, at time t, up to a maximal perception radius R. Each perceived density is weighed relative to the state (x, ϕ). Such a weight function is given by $g_l(D_\phi(\tilde{x} - x), \tilde{\phi} - \phi)$, where D_ϕ denotes the turn matrix

$$(1) \qquad\qquad D_\phi = \begin{pmatrix} \cos\phi & \sin\phi \\ -\sin\phi & \cos\phi \end{pmatrix} .$$

This leads to the following general functional for the turning frequency

$$(2) \quad \lambda(x, \phi, t) = \Lambda \left(\int\limits_0^{2\pi} \int\limits_{\|\tilde{x}-x\|\leq R} g_l(D_\phi(\tilde{x} - x), \tilde{\phi} - \phi) \cdot p(\tilde{x}, \tilde{\phi}, t)\, d\tilde{x}\, d\tilde{\phi} \right) ,$$

g_l suitable chosen and Λ e.g. similar to equation (6) in (Pfistner, this part).

Under the condition, that at state (x, ϕ) the organism turns, $h(x, \phi, \phi_1)$ represents the probability of choosing ϕ_1 as a new direction. It also depends on the weighted densities at surrounding states $(\tilde{x}, \tilde{\phi})$, at time t, up to the maximal radius R, but now the weight function $g_h(D_\phi(\tilde{x} - x), \tilde{\phi} - \phi, \phi_1 - \phi)$ has to depend on the chosen angle, ϕ_1, too. If there is no organism in the perception area $g_h^o(\phi_1 - \phi)$ gives the probability of choosing ϕ_1 as a new direction. Thus the following direction choice function, or turn angle distribution, is proposed

BOX 15

(3) $\quad h(x,\phi,\phi_1) \;=\; \int\limits_{0}^{2\pi} \int\limits_{\|\tilde{x}-x\|\le R} g_h(D_\phi(\tilde{x}-x),\tilde{\phi}-\phi,\phi_1-\phi)\cdot p(\tilde{x},\tilde{\phi},t)\,d\tilde{x}\,d\tilde{\phi}$

$$+ g_h^\circ(\phi_1-\phi)\;,$$

$$\text{with}\qquad \int\limits_{0}^{2\pi} g_h(.\,,.\,,\varphi)\,d\varphi = 0 \quad,\qquad \int\limits_{0}^{2\pi} g_h^\circ(\varphi)\,d\varphi = 1 \quad.$$

Therefore the density change at state (x,ϕ) and time t, given by the difference of cells turning out of and into state (x,ϕ), satisfies the following (hyperbolic) differential integral equation, a transport equation of Boltzman-type, which has similarities to equation (1b) in (Edelstein-Keshet, Ermentrout, this part):

(4) $\quad \partial_t p(x,,\phi,t) + s e^{i\phi}\nabla_x p(x,\phi,t) \;=\; -\lambda(x,\phi,t)p(x,\phi,t)$

$$+ \int h(x,\phi_1,\phi)\lambda(x,\phi_1,t)p(x,\phi_1,t)\,d\phi_1\,,$$

with $se^{i\phi}$ as the velocity vector.

One dimensional case

The reduction of this general two dimensional model to one dimension directly leads e.g. to the model for the swarming behavior of myxobacteria presented by Pfistner (this part).

In this case the moving directions are set either $\phi = +\frac{\pi}{2}$ or $\phi = -\frac{\pi}{2}$. Therefore $p(x,\pm\frac{\pi}{2},t)$ in the notation above corresponds to $p^\pm(x,t)$ in the one dimensional model. Analogously $\lambda(x,\pm\frac{\pi}{2},t) = \lambda_\pm$, where $g_l(r,0) = \alpha(r)$ and $g_l(r,\pi) = \beta(r)$. Setting $g_h^\circ(\pm\pi) = 1$ and $g_h \equiv 0$ reduces equation (4) to the one dimensional system equation (1) in (Pfistner, this part).

Applications

Several applications of this general swarming model might be envisaged in this Part III, e.g. to fish schools (Huth/Wissel), myxobacteria swarming (Pfistner) or epithelial cell movement (Edelstein-Keshet/Ermentrout).

FROM CELL TO TISSUE;

CONTACT MEDIATED ORIENTATION SELECTION IN A POPULATION

Leah Edelstein—Keshet
Dept. of Mathematics
University of British Columbia
Vancouver, B.C.
Canada V6T 1Y4

G. Bard Ermentrout
Dept. of Mathematics
University of Pittsburgh
Pittsburgh, PA 15260
USA

Abstract

We model the motion and contact-response of cells such as fibroblasts using integro-partial differential equations, with orientation angle, θ, as the main independent variable. The model describes free cells that reorient randomly or when they come into contact and bind with other cells. The model predicts that a population initially randomly distributed over all possible orientations will eventually exhibit a single axis of orientation, i.e. that a single parallel array of cells will form.

1. Introduction.

This paper describes a mechanism for pattern formation stemming from responses of individuals to contact with one another. It is motivated by a phenomenon observed in cultures of fibroblasts (Elsdale 1972, 1973; Elsdale and Wasoff 1976) which also occurs in other cellular systems such as insect epidermal cells (Nübler-Jung, 1987) and myxobacteria (Shapiro, 1988). In these systems, cells tend to aggregate and form parallel arrays––clusters in which cells have a common axis of orientation. These structures develop from arbitrary initial conditions and are reestablished if perturbed. (See figures 1 and 2 below).

To model these systems we abstracted key features––random cell motion, contact-mediated responses, binding and unbinding of cells in cohesive groups––while omitting details of the underlying chemical mechanisms and detailed treatment of the motion of individual cells. We present a brief summary of this approach below. Other details appear in Edelstein-Keshet and Ermentrout (1989).

2. Definitions and notation.

We used the following quantities in the equations of our model:

θ = angle of orientation with respect to fixed axis, x = spatial position, t = time,

$P(\theta,x,t)$ = density of bound (aggregated) cells at x,t whose orientation is θ,

$C(\theta,x,t)$ = density of free cells at x,t whose direction of motion is θ,

$v = (s\cos\theta, s\sin\theta)$ = velocity vector of cells moving in direction θ,

$K(\varphi)$ = probability that contact between two cells at relative angle φ leads to
     realignment and binding of one cell,

The reader should refer to Box 16 for a detailed description of contact response kernels $K(\varphi)$ and their role in changes in the population of cells at a given angle.

3. Assumptions and equations of the model.

We make the following assumptions: (1) Bound cells do not move, reorient or interact directly with other bound cells. There is a fixed rate of unbinding, γ. (2) Free cells can move and reorient randomly or through contact with another cell. In contacts between free cells it is equally likely that either cell determines the net orientation of the pair. In contacts between a free and a bound cell, the orientation is always determined by the bound cell. (3) For random reorientation, the turning angle distribution $\alpha(\varphi)$ is sharply peaked and symmetric about $\theta = 0$ and waiting times are equally distributed. This approximately leads to a diffusion-like random reorientation operator [see Edelstein-Keshet and Ermentrout, 1989]. (4) Cell division is a slow process on the timescale of these interactions. Then a set of equations describing aggregates is:

(1a)
$$\frac{\partial P}{\partial t} = \beta_1 CK * C + \beta_2 PK * C - \gamma P,$$

(1b)
$$\frac{\partial C}{\partial t} = -\nabla_x \cdot Cv + \mu \partial^2 C / \partial\theta^2 - \beta_1 CK*C - \beta_2 CK*P + \gamma P,$$

where β_i's are constants that we describe below and K is the reorientation probability for free and bound cells. The convolution terms (*'s) depict contact that leads to reorientation and $\mu C_{\theta\theta}$ represents random reorientation of free cells. (See Box 16.)

We found data for the angle dependence of the contact kernels, K in a paper by Erickson (1978). Briefly, K is positive for contact angles smaller than some critical value a. [See Box 16 for several representative examples of functions depicting K.] Two cases

BOX 16

CONTACT RESPONSES, KERNELS, AND THEIR FOURIER TRANSFORMS

L. Edelstein–Keshet

We define below the probabilities that contact between two elements leads to alignment and binding. (Note the dependence on the contact response kernel, K).

ELEMENTS INTERACTING		RESULTS
A	B	probability (A aligns with B)
cell at angle θ	cell at angle θ'	$K(\theta - \theta')$
"	population of cells $C(\theta')$ at angle θ'	$K(\theta - \theta')C(\theta')$
"	population of cells at all angles $-\pi < \theta' < \pi$	$\int_{-\pi}^{\pi} K(\theta-\theta')C(\theta',t)d\theta' = K*C$
population of all cells at angle θ	"	$C(\theta)\int_{-\pi}^{\pi} K(\theta-\theta')C(\theta',t)d\theta'$ $= C\ K*C$

In analyzing the stability of the steady state (4a,b) to small perturbations of the form (5) we encounter a set of linearized equations whose coefficients (entries in the Jacobian J) depend on the Fourier transform of K, i.e. on

$$\hat{K}(k) = \int_{-\pi}^{\pi} e^{ik\varphi}K(\varphi)d\varphi.$$

Below is experimental data about the function $K(\varphi)$, redrawn from Erickson (1978).

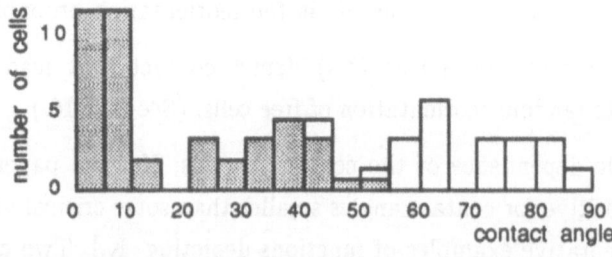

Figure. Cells meeting at angles smaller than 55 deg. aligned (dotted boxes) but at larger angles they did not (white boxes). From Erickson (1978). This result is used in our assumptions about the kernel $K(\theta)$.

BOX 16

Erickson's data were approximated by several convenient functional representations, (1) Assuming K is symetric about 0 (2) assuming K is symetric about both 0 and $\pi/2$. The functional form assumed on the basic interval $(0,a)$ is $f(\theta) = (\pi/4a) \cos(\pi\theta/2a)$ where a is the critical angle. A graph of $K(\theta)$ for each type, 1s and 1d is shown on the left as a function of θ and the fourier transform of K, $\hat{K}(k)$ is given. On the right, we show the graph of the function $\hat{K}(k)(1-\hat{K}(k))$ on which the stability of the steady state (9) depends. The parabola Ak^2 is also shown on the same graph for comparison. The configuration shown is consistent with instability at either k=1 or k=2.

Shape of Kernel $K(\theta)$ (graph)
and its Fourier transform $\hat{K}(k)$

Graph of
$\hat{K}(1-\hat{K})$

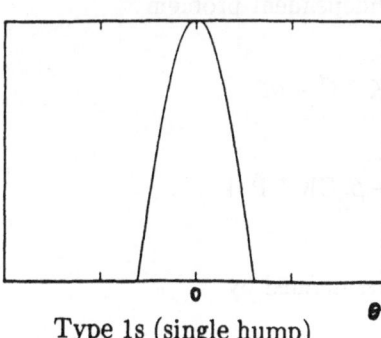

Type 1s (single hump)

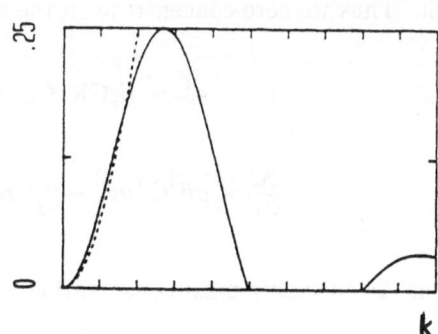

$$\hat{K}(k) = -(\frac{\pi}{2a})^2 \frac{\cos ka}{k^2 - (\pi/2a)^2}$$

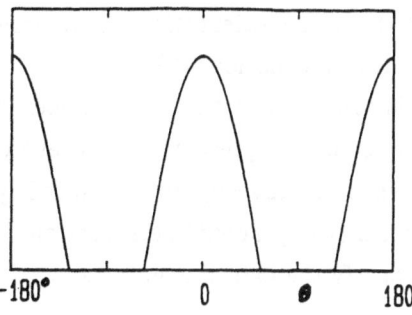

Type 1d (double humped)

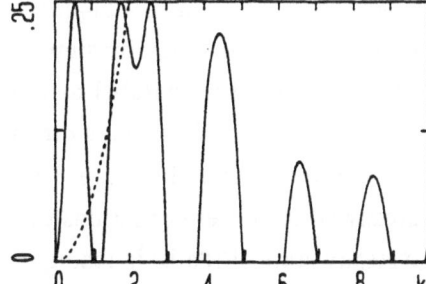

$$\hat{K}_{1d}(k) = \frac{1}{2} \hat{K}_{1s}(1 + \cos k\pi)$$

with different outcomes arise depending on whether (a) Cells adhere to each other only if facing the same way, (b) Cells adhere whether they face in opposite directions or in the same direction. In case (a) $K = K_s$ has a single hump on $[-\pi,\pi]$ whereas in case (b) $K = K_d$ has two humps on this interval.

We do not analyze these equations directly, but rather make a number of simplifications that lead to analytically tractable submodels.

4. A space–independent model.

To focus on the phenomenon of orientation selection we assume at first that densities are spatially homogeneous so that $C = C(\theta,t)$, $P = P(\theta,t)$, $\nabla_x \cdot Cv = 0$. The complete space-angle-time model is treated separately using a cellular automaton approach. Thus we here concentrate on the space-independent problem

(2a)
$$\frac{\partial P}{\partial t} = \beta_1 CK * C + \beta_2 PK * C - \gamma P,$$

(2b)
$$\frac{\partial C}{\partial t} = \mu \partial^2 C / \partial \theta^2 - \beta_1 CK * C - \beta_2 CK * P + \gamma P.$$

Equations (2a,b) satisfy mass conservation with mass defined by

(3)
$$M(t) = \frac{1}{2\pi} \int_{-\pi}^{\pi} [P(\theta,t) + C(\theta,t)]d\theta .$$

Later, we allow M to be a slowly increasing parameter to model cell division. A dividing cell does not necessarily have daughters with the same orientation so it would be inaccurate to model cell division by simple birth terms in equations (2).

To determine whether parallel arrays of cells can occur in the above model, we ask whether a state in which all angular directions are equally represented could evolve into one in which some orientations are favored. Thus we investigate the stability of a uniform steady state of these equations to nonuniform (θ-dependent) perturbations.

The steady state of (2a,b) is readily seen to satisfy the pair of equations

(4a,b)
$$\bar{P} = \frac{\beta_1 \ \bar{C}^2}{\gamma - \beta_2 \bar{C}} , \qquad M = \bar{P} + \bar{C} .$$

By graphical analysis in the PC plane we find that (4a,b) describes a single steady state $(\bar{P}, \bar{C})$ in the first quadrant. Examining small angle-dependent perturbations of this steady state, i.e.

(5)
$$\begin{pmatrix} P \\ C \end{pmatrix} = \begin{pmatrix} \bar{P} \\ \bar{C} \end{pmatrix} + \begin{pmatrix} P_0 \\ C_0 \end{pmatrix} e^{ik\theta + \lambda t},$$

for P_0, C_0 small, k an integer wavenumber, λ the growth rate, leads to the Jacobian

(6)
$$J = \begin{bmatrix} -\epsilon & \xi \\ \eta & -\delta \end{bmatrix},$$

whose individual elements are specified in Table I. Here it suffices to note that ϵ and δ are positive, so that $\mathrm{Tr}(J) = -(\epsilon + \delta)$ is always negative. The uniform steady state would thus be stable whenever $\mathrm{Det}(J) > 0$ and unstable when $\mathrm{Det}(J) < 0$. Because the expression for $\mathrm{Det}(J)$ is exceedingly cumbersome, we focus on three relevant subcases which are analytically simpler.

5. Behavior under limiting cases.

Case 1. *Free cells form bound pairs;* $(\beta_1 = \beta,\ \beta_2 = 0)$.

If only cell-cell interactions occur, i.e. cells bind in pairs, then $\beta_2 = 0$ in equations (2a,b). This can be viewed as an initial step in the development of clusters from a population of free cells. With this simplification we find a unique positive θ-independent steady state satisfying

(7)
$$\bar{P} = \beta \bar{C}^2 / \gamma, \qquad M = \bar{P} + \bar{C},$$

whose stability depends on the Jacobian (6) with $\beta_2 = 0$, [see Table I]. We find that $\mathrm{Det}(J) = \gamma \mu k^2$ which is nonnegative for all k. (The fact that $\mathrm{Det}(J) = 0$ for $k = 0$ is a consequence of mass conservation.) This precludes the possibility of instability to any wavelength k. Thus the uniform steady state, which consists of a combination of free and paired cells, is always stable, meaning that paired and free cells coexist at all possible orientations. Cell pairing cannot lead to parallel arrays in the population.

Case 2. *Free cells interact only with multicellular clumps;* $(\beta_1 = 0, \beta_2 = \beta)$

If free cells bind to multicellular clusters but do not interact with other free cells then $\beta_1 = 0$. Setting $\beta_1 = 0$ in equations (2) leads to the uniform steady state

$$(8) \qquad\qquad \bar{C} = \gamma/\beta \qquad \bar{P} = M - \gamma/\beta .$$

Its stability now hinges on the Jacobian (6) with $\beta_1 = 0$, [see Table I] so that $\text{Det}(J) = -\beta\gamma \hat{K}(1-\hat{K})$. This quantity is negative whenever $0 < \hat{K} < 1$, an inequality that is satisfied simultaneously at many wavelengths. Thus instability will occur but linear theory does not have the power to predict which wavelength will predominate, since many values of k satisfy the above condition.

Case 3. *Free and bound cells interact equally;* $(\beta_1 = \beta_2 = \beta > 0)$;

If cell-cell collisions have the same outcome whether one cell is bound or both are free then both β_1 and β_2 are nonzero. We assume that they are equal. Then a single positive uniform steady state exists, satisfying

$$(9) \qquad\qquad \frac{\bar{P}}{\bar{C}} = \frac{\beta M}{\gamma} ,$$

Stability hinges on the Jacobian (6) with $\beta_1 = \beta_2$, [see table I]. After extensive algebraic

Table I: Elements in the Jacobian (6)

	Case 0 $\beta_1 \neq \beta_2 > 0$	Case 1 $\beta_1 = \beta, \beta_2 = 0$	Case 2 $\beta_1 = 0, \beta_2 = \beta$	Case 3 $\beta_1 = \beta_2 = \beta$
ϵ	$\gamma - \beta_2 \bar{C}$	γ	0	$\gamma - \beta\bar{C}$
η	$\gamma - \beta_2 \bar{C}\hat{K}$	γ	$\gamma(1-\hat{K})$	$\gamma - \beta\bar{C}\hat{K}$
ξ	$\beta_1(1+\hat{K}) + \beta_2\bar{P}\hat{K}$	$\beta\bar{C}(1+\hat{K})$	$\beta\bar{P}\hat{K}$	$\beta\bar{C}(1+\hat{K}) + \beta\bar{P}\hat{K}$
δ	$\mu k^2 + \beta_1\bar{C}(1+\hat{K}) + \beta_2\bar{P}$	$\mu k^2 + \beta\bar{C}(1+\hat{K})$	$\mu k^2 + \beta\bar{P}$	$\mu k^2 + \beta\bar{C}(1+\hat{K}) + \beta\bar{P}$
Inst-abil.		never	too many wavenumbers	when $Ak^2 < \hat{K}(1-\hat{K})$ (see text)

manipulation we find that $\text{Det}(J) = \frac{\beta\gamma\bar{C}^2}{\bar{P}}[\frac{\mu k^2}{\gamma} + (\hat{K} - 1)\hat{K}(\frac{\beta M}{\gamma})^2]$. Thus, the uniform

steady state can be destabilized by any wavelength satisfying

$$(10) \qquad\qquad\qquad\qquad Ak^2 < \hat{K}(1 - \hat{K}) \,,$$

where $A = \frac{\mu}{\gamma}(\frac{\gamma}{\beta M})^2$. We examine this instability condition by comparing the parabola y
$= Ak^2$ and the graph of $y = \hat{K}(1 - \hat{K})$. To do so, first observe that $\hat{K}(0) = 1$ by
normalization. Box 16 gives several examples of K which qualitatively approximate the
data of Erickson (1978) reasonably well. We note again the distinction between single
humped and double humped kernels previously discussed, and find that the latter all
contain the multiple $\{1 + \cos k\pi\}$. Thus for these kernels, the quantity $\hat{K}(1 - \hat{K})$ is zero
at $k = 0$, at $k = 1,3,5...$ and all odd integer wave numbers.

Now consider the bifurcation phenomenon that would occur if the quantity A,
initially large, gradually decreased: for example this could happen if the rate of shedding
of cells, γ decreased, the stickiness of cells for one another increased causing β to be
greater, or the mass of the population, M, increased slowly by cell division. As the
parabola $y = Ak^2$, initially steep, becomes shallower, the possibility arises that for some
integer k, the inequality (10) is satisfied. From the graphs of $\hat{K}$ $(1 - \hat{K})$ in Box 16 we see
that the first such integer at which this can occur is $k = 2$ for double-humped kernels
(provided that $a < \pi/2$ so that $\hat{K}(2) > 0$). Thus, as the threshold is crossed, the
uniform steady state loses stability to any perturbation of wavelength $k = 2$. Since this
is the first wavelength to trigger instability , the new patterned state will be characterized
by $k = 2$. This means that two orientations, $180°$ apart will be reinforced in the
population – –i.e. a parallel array of intercalated cells will be formed. The case of single
humped kernels is similar, but $k = 1$ is the first destabilizing wavelength. Figure 1
provides an example of numerical solutions of equations 2a,b in which the kernel is K_{1s},
but with the total mass large enough that $k = 2$ eventually overtakes other modes.

6. Simulations with cellular automata.

In a second approach to the problem, we simulated the behavior of the cells directly,
without the approximating continuum representation. Taking a discrete number of angles
(in one case, 16 angles at multiples of 22.5 degrees) and a grid with a discrete set of
spatial positions, we represented cells by moving line segments, and implemented rules for
the way that motion, reorientation, sticking, and shedding of cells occurs. Free cells move

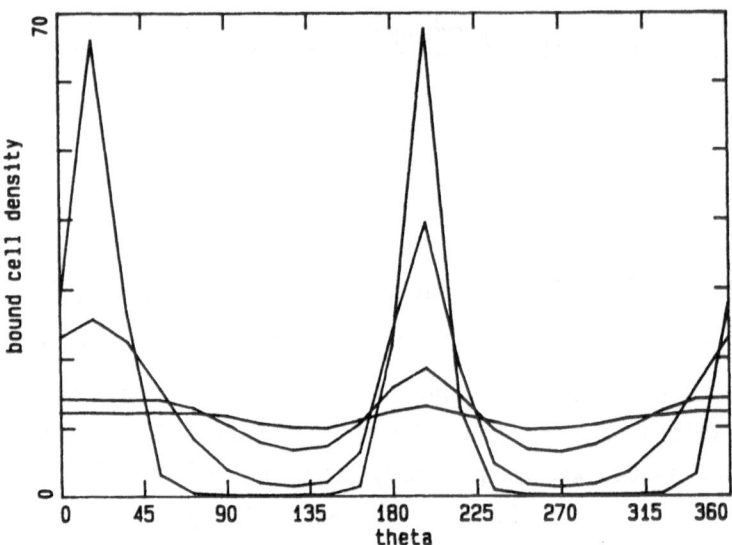

Figure 1. Typical development of solutions to equations 2(a,b) in case 3 with $\beta = 0.3$, $\gamma = 0.3$, $\mu = 0.3$, a = 55 degrees and kernel of type 1s. Numerical iterations with $\Delta\theta = 0.34$ radians, $\Delta t = 0.01$, are shown starting from 8,000 iterations in increments of 4000. Initial conditions were 10% random noise superimposed on $P(\theta, 0) = C(\theta, 0) = 6.0$. The total mass $M \approx 12$ is large enough that the mode with k = 2 eventually wins. Two peaks, 180 degrees apart are produced, indicating that most cells are aligned in parallel, and facing opposite directions.

in the direction of their axis, undergo random reorientation (called by a random number generator), and upon encounter with another cell at a small enough angle, align and bind. The bound cells are assumed non motile, but are shed randomly. The probabilities governing sticking, reorientation, and shedding can be changed, as can the size of the domain and the total density of the cells. We explored several shapes of the domain, and several types of boundary conditions. In Figure 2 we illustrate a typical development of stages in the formation of parallel arrays. This development occurs only for densities above the critical threshold. Otherwise, for low densities, no stable clusters form.

7. Discussion.

Our analysis reveals that contact-mediated cell reorientation does lead to spontaneous formation of structure in the population; i.e. to cells lying in parallel groups. The model further leads to several conclusions: (A) Cohesiveness is an essential feature of

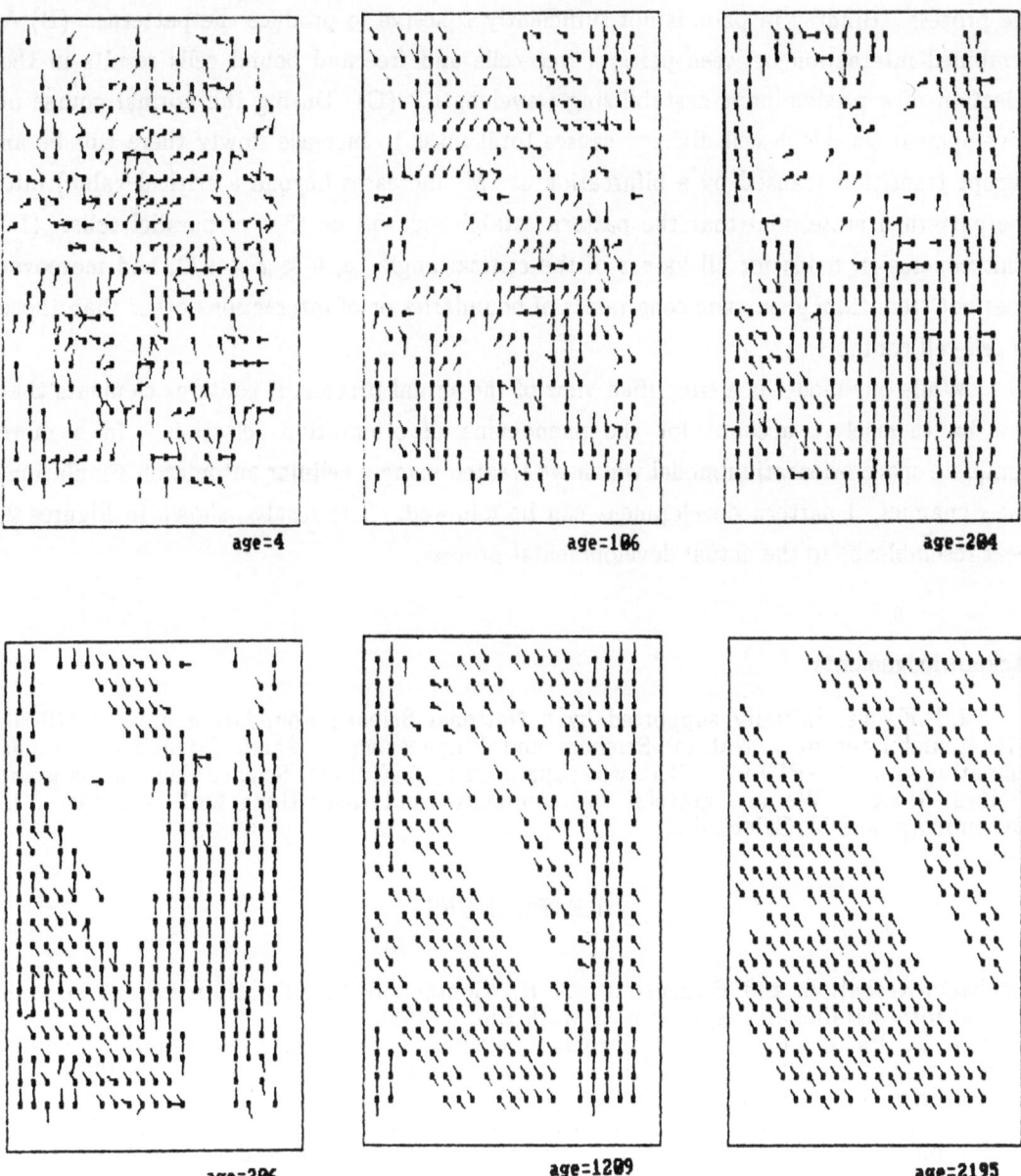

Figure 2. A sequence of stages in a computer simulation of cells, represented by line segments. The cells are initially oriented at random. They move at constant speed, reorienting randomly and due to contact with other cells. (Parameter values were $\mu=0.1$, $\beta=0.2$, $\gamma=0.1$.) Boundaries are periodic in both directions. The cells assemble into groups of parallel arrays. After some time only two to three directions are left, and these clusters compete with each other for recruiting free cells. After a long struggle, one direction wins.

the process. Binding in pairs is not sufficiently cohesive to produce the patterns. (B) A combined interaction between pairs of free cells and free and bound cells results in the selection of a predominant destabilizing wavelength. (C) During the normal course of development, in which cell division causes total mass to increase slowly there will be an abrupt transition (caused by a bifurcation as M increases beyond a critical value) into the patterned state, and that the pattern established will be that of parallel cells. (D) This conclusion holds for all values of the critical angle a, $0 < a < \pi/2$, and moreover does not stem from geometric constraints of boundaries or of interactions other than those of cell-cell contacts.

While our model is a simplified view of the actual process, it contains elements that can by themselves account for the phenomena of orientation selection. In a more complete space-orientation model which we treated using a cellular automaton simulation, the dynamics of pattern development can be followed. The results, shown in Figures 2, bear resemblence to the actual developmental process.

Acknowledgements

L.E.K. was initially supported by a National Science Foundation grant DMS-86-01644; and later by a Natural Sciences and Engineering Research Council of Canada operating grant OGPIN 021. B.E. was supported by a National Science Foundation grant DMS-87-01405. We are grateful for comments and discussions with W. Alt, J.L. Deneubourg, and V. Calenbuhr.

REFERENCES

Edelstein-Keshet, L. and Ermentrout, G. B., Contact mediated pattern formation; cells that form parallel arrays, in press J.Math Biol, (1990).

Elsdale, T., Pattern formation in fibroblast cultures, an inherently precise morphogenetic process, in Towards a Theoretical Biology, Vol. 4 (C.H. Waddington, Ed.), Edinburgh University Press, Edinburgh, (1972).

Elsdale, T., The generation and maintenance of parallel arrays in cultures of diploid fibroblasts, in Biology of Fibroblast, (E. Kulonen and J. Pikkarainen, eds.), Academic Press, New York, (1973).

Elsdale, T. and Wasoff, F., Fibroblast cultures and Dermatoglyphics: the topology of two planar patterns, Wilhelm Roux's Archives 180, 121–147 (1976).

Erickson, C. A., Analysis of the formation of parallel arrays by BHK cells in vitro, Experimental Cell Res. 115, 303–315 (1978).

Nübler-Jung, K., Tissue polarity in an insect segment: denticle patterns resemble spontaneously forming fibroblast patterns, Development 100, 171–177 (1987).

Shapiro, J. A., Bacteria as multicellular organisms, Sci. Amer., June 1988, 82–89.

THE MOVEMENT OF FISH SCHOOLS: A SIMULATION MODEL

Andreas Huth, Christian Wissel
Fachbereiche Physik und Biologie, Universität Marburg
Renthof 6, D-3550 Marburg

Abstract

Fish schools represent a biological form of self organization (synergetics): Fish don't need a leader or external stimuli for their school organization.

With the help of computer simulations we investigate the individual behaviour patterns which give rise to the self-organized movement of fish schools. Following the biology we modelled several basic behaviour patterns for the single fish in the school: attraction, repulsion, parallel orientation.

Connecting these patterns with an averaging concept for the mixing of the influences of fish's neighbours, we find simulated fish schools with the typical characteristics of real fish schools: a high degree of parallel orientation and a strong cohesion. Moreovers the nearest neighbour distance distribution of our model fish school is found to agree with that of real schools.

1. Introduction

One of the most important capabilities of fish schools is to form schools with a high parallel orientation of its members, known as polarized schools (Fig.1) (Breder 1951, 1959; Shaw 1970; Partridge et al. 1980a; Aoki et al. 1986). This formation is important for allowing effective group motion. Today, we know that fish do not need a leader to form

polarized schools. The fish in front of the school change
continually (Breder 1951; Radakov 1971). Nevertheless, the
possibility that the leader is not positioned in front of the
school should be considered. Aoki (1980) and Partridge et al.
(1980a) proved by measuring cross correlation functions that
many fish schools have no leaders.

We must distinguish between two sorts of different locomotions for fish schools in order to understand this phenomenon.
First there are locomotions which are influenced by external
stimuli. For example some fish species migrate seasonally to a
certain warmer region for spawning (Balchen 1975). In this case
it is easy to explain the organization of the school. Very
often it is not clear which stimuli are reasonable in the
special case (Balchen 1976).

On the other hand fish schools also show a high degree of
polarization in the absence of external stimuli. That is the
case for example when a food searching school swims through a
region without markings. Here the school behaviour represents a
biological form of self organization (synergetics).

It is unknown how fish manage this self organization. The
aim of our project is to investigate the synergetic behaviour
of fish schools. We want to understand with the help of
experimental data and computer simulations which behavioural
patterns of the individual fish produce polarized schools.

2. Which are the methods we use?

On the basis of the biological facts we designed various
models for the individual fish in a moving school. With the
help of computer simulations we "test" these models. If the
model fish group shows the characteristics typical of real fish
schools - a strong cohesion and high polarization - over many
time steps we can conclude that the modelled behaviour patterns
could possibly explain schooling of real fish. In this way we
hope we can contribute to revealing the secrets of a school
organization without a leader and get a better understanding of

the structure and organization of fish schools.

3. Basic behaviour patterns of swarming fish

With the help of several experimental observations we can
formulate three basic behaviour patterns for the individual
fish in a polarized school:

a) Repulsion

Normally the fish in a school do not collide. The probability distribution of the distance to the nearest neighbour
shows that the neighbours of fish never come closer than a
certain distance (fig.2, Partridge et al. 1980a). Thus we
assume that: If a neighbour fish is too close, the fish
tries to avoid a collision.

b) Attraction

If a fish is too far away from the school, it swims
towards the school (Keenleyside 1955). Moreover two fish of
the same species try to aggregate (Pitcher 1979, Aoki
1984). Shaw (1970) called this behaviour "biosocial mutual
attraction". We can formulate for our model: If the neighbour is too far away, the fish swims towards its neighbour.

c) Parallel Orientation

If a fish in a moving school has the preferred distance from
its neighbours (fig.2), it normally shows a high degree of
parallel orientation to its neighbours as well (Hunter
1966, Aoki et al. 1986, Breder 1959). This means for our
model: If the neighbour fish is situated in a certain
preferred range (fig.2), the fish tries to swim in the same
direction as its neighbour.

4. The model

In the following we present our 2-dimensional model for the
motion of a fish in a polarized school.

The velocity of the i-th fish $\vec{v}_i$ depends on the positions

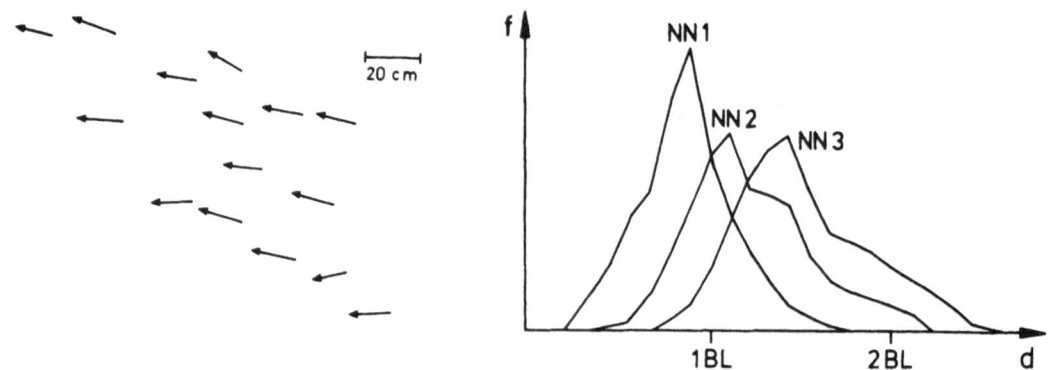

<u>Figure 1.</u> (left side) Positions and orientations of the indivi-
dual fish in a school of mackerels. After a photograph by Aoki
et al. (1986) taken at a depth of 75m at night.

<u>Figure 2.</u> (right side) Frequency f of the nearest neighbour
(NN1), the second nearest neighbour (NN2) and the third nearest
neighbour (NN3) at various distances d from the reference
fish. Data are for a saithe school. BL = body length. After
Partridge et al. (1980a).

and orientations of its neighbours in a complex way. Okubo
(1980) and Inagaki (1976) used the physical force concept for
modelling fish schools. We prefer the biological model concept
of Aoki (1982). Here the velocity changing is devided in two
parts: a change of the direction (described by α_i) and a change
of the amount of the velocity (described by g_i)

(1) $\vec{v}_i(t+\Delta t) = g_i \cdot D(\alpha_i) \cdot \vec{v}_i{}^o(t)$

(2) $\vec{x}_i(t+\Delta t) = \vec{x}_i(t) + \Delta t \cdot \vec{v}_i(t+\Delta t)$

           with $\alpha_i = \alpha(\vec{x}_i - \vec{x}_j, \vec{v}_i, \vec{v}_j, \xi)$ and $g_i = g_i(\xi)$

($\vec{x}_i(t), \vec{v}_i(t)$ position and velocity of the i-th fish at the
time t; $\vec{x}_j, \vec{v}_j$ position and velocity of its neighbour; $g_i =$
$|\vec{v}_i(t+\Delta t)|$ the new velocity of the i-th fish; $D(\alpha_i)$ is a
matrix rotating the vector of the swimming direction of the
fish $\vec{v}_i{}^o(t)$ ($=\vec{v}_i(t)/|\vec{v}_i(t)|$) through an angle α_i; ξ sto-
chastic variable).
Strictly speaking equation (1) discribes the motion of a fish
as a velocity jump process at regular time steps (Δt) with

uncorrelated speeds (g_i) and one step correlated directions $(D(\alpha_i) \cdot \vec{v}_i(t))$. In the following we will explain the construction of α_i and g_i.

a) The turning angle α_i

$\alpha_i(t)$ the turning angle of the i-th fish at the time t is randomly chosen from a normal distribution with standard deviation s and mean $m(\{\beta_{ij}\})$:

$$\alpha_i = \text{chance } (p(\alpha_i))$$

$$(3) \qquad p(\alpha_i) = \frac{1}{\sqrt{2\pi} \, s} \quad \exp(\, (\alpha_i - m(\{\beta_{ij}\}))^2 \, / \, 2 \, s^2 \,)$$

Here β_{ij}, the influence angle, is the mean turning angle of fish i under the influence of fish j alone, and m() is a mixing function (explanation follows later). β depends on $(\vec{x}_i - \vec{x}_j, \vec{v}_i, \vec{v}_j)$ in the following fashion:

$$(4) \qquad \beta_{ij} = \begin{cases} \measuredangle \, (\vec{v}_i, \vec{x}_j - \vec{x}_i) & \text{attraction} \\ \min \{ \measuredangle \, (\vec{v}_i, \vec{v}_{j\perp}) \} & \text{repulsion (with } (\vec{v}_j \cdot \vec{v}_{j\perp}) = 0) \\ \measuredangle \, (\vec{v}_i, \vec{v}_j) & \text{parallel orientation} \\ \text{randomly chosen} & \text{searching} \\ \text{between } -180° \\ \text{and} \qquad +180° \end{cases}$$

($\measuredangle$ $(\vec{a}, \vec{b})$ angle between two vectors; $\vec{v}_\perp$ vector perpendicular to $\vec{v}$, ° = DEG)

if $(r = |\vec{x}_j - \vec{x}_i|, \, \vartheta = \measuredangle \, (\vec{v}_i, \vec{x}_j - \vec{x}_i)) \, \epsilon$

$\{(r, \vartheta) | r < r_1$ and $|\vartheta| < 180° - \omega\}$ attraction area

$\{(r, \vartheta) | r_1 < r < r_2$ and $|\vartheta| < 180° - \omega\}$ repulsion area

$\{(r, \vartheta) | r_2 \leq r < r_3$ and $|\vartheta| < 180° - \omega\}$ parallel area

$\{(r, \vartheta) | r \geq r_3$ and $|\vartheta| \geq 180° - \omega\}$ searching area

The last part we have to explain in (3) is the mixing function m( ). It indicates how a fish mixes the influences (influence angles) of its different neighbours. As we have no knowledge about this mixing we have to make reasonable assumptions. We tried different concepts: a decision and averaging concept with different priorities. Here we mainly

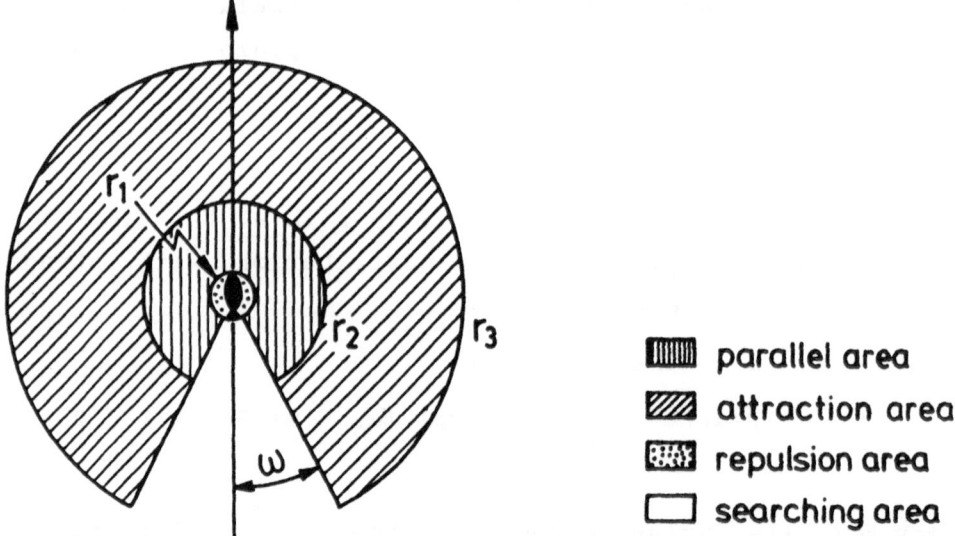

Figure 3. Ranges of the basic behaviour patterns. The black fish reacts with different basic behaviour pattern according to the range in which a neighbour is positioned: repulsion; parallel orientation, attraction and searching (see text for full explanation). The ranges are determined by the radius r_1, r_2, r_3 and the dead angle (typical values are $r_1 = 0.5$ BL, $r_2 = 2$ BL, $r_3 = 5$ BL and $\omega = 15$ DEG, BL=body length of a fish)

represent the averaging concept. It produces the most realistic results in our simulations.

$$(5) \qquad m(\{\beta_{ij}\}) = (1/n_{max}) \sum_{j=0}^{n_{max}} \beta_{ij} \qquad (n_{max} \le 4)$$

In the maximum case a fish averages over the influence angles of four neighbours. If a fish perceives more than four neighbours it selects four neighbours using a certain priority. For example a fish prefers the neighbours which are more in front of it ("front priority"), as Olst and Hunter (1970) Tembruck (1983), Bone and Marshall (1985) suspected on the basis of experimental observations. We include this in our model by renumbering the neighbour fish in the order of the relative angle γ under which a fish "sees" its neighbours.

$$(6) \qquad \gamma_1 < \gamma_2 < \gamma_3 < \gamma_4 \ldots \text{ with } \gamma_j = \measuredangle(\vec{v}_i, \vec{x}_j - \vec{x}_i).$$

We also investigate other priorities: "side priority" and
"distance priority". Using side priority we renumber the
neighbours with respect to their relative "side position"
$\sphericalangle(\vec{v}_i , \vec{x}_j - \vec{x}_i)$.

For the distance priority the nearest neighbours are preferred. In contrast to Aoki (1982) we attach importance also to
the fact that the senses of a fish have only a finite
reception range (Bone 1985) (The most important senses for
schooling are the eyes and the lateral lines (Shaw 1970,
Partridge et al. 1980b, Partridge 1981, Pitcher et al.
1976)). For our model this means that only neighbours
which swim in either the parallel, the attraction or the
repulsion areas of a fish are taken into account by equation
(5). If all the neighbours of a fish are positioned in its
dark area, the fish cannot perceive them and reacts by a
searching behaviour, modeled by a random turn.

In the decision-model (Aoki 1982) we calculate another
probability distribution for the turning angle α_i :

$$(7) \qquad p(\alpha_i) = \sum_{j=1}^{nmax} b_j \; \frac{1}{\sqrt{2\pi}\, s} \; \exp(-(\alpha_i - \beta_{ij})^2 / 2s^2)$$

with $b_{j+1} = 0.5 \cdot b_j$ and $\sum_{j=1}^{nmax} b_j = 1$ (j and nmax see (6)).

The consequence is that the fish selects for one neighbour
to which it reacts. Fig.4 shows an example which demonstrates the differences between the two model concepts.

b) The new velocity $g_i (t)$

To simplify our models the magnitude of the new velocity
g_i of the fish is chosen independtly of the other fish. The
new velocity $g_i(t)$ of every fish is calculated by chance
from an experimentally observed gamma distribution which is
found in experiments (fig.5, Aoki 1980, 1982)

$$g_i = chance \; (p(v)),$$

$$(8) \qquad p(v) = e^{-Av} v^{k-1} \cdot A^k / \Gamma(K)$$

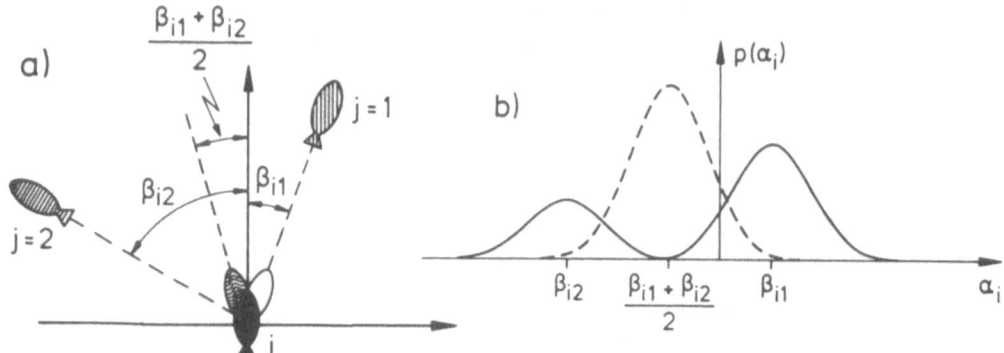

Figure 4. Example for the decision and averaging model.
a) The black fish has two neighbours (hatched) in its parallel orientation area. The white fish illustrates a new swimming direction which the black fish probably chooses after the next time step if the decision model (D) is used. The fish•with the undulated marking shows the probable new swimming direction for the averaging model (A).
b) Probability distribution $p(\alpha_i)$ for the turning angle α_i of the black fish (i). For the A-model the probability distribution (dashed line) consists only of one normal distribution around the mean angle $(\beta_{i1}+\beta_{i2})/2$. The black fish averages the influences of its neighbours. It swims roughly in a direction located between its two neighbours. For the D-model the distribution (full line) consists of two differently weighted normal distributions with standard deviations $s=15$ DEG. The weight of the normal distribution of the right neighbour, which is positioned more in front of the black fish is double the weight of the other distribution (front priority, right neighbour $j=1$, left neighbour $j=2$, black fish = i-th fish). In 2 of 3 cases the black fish swims roughly in direction of its right neighbour.

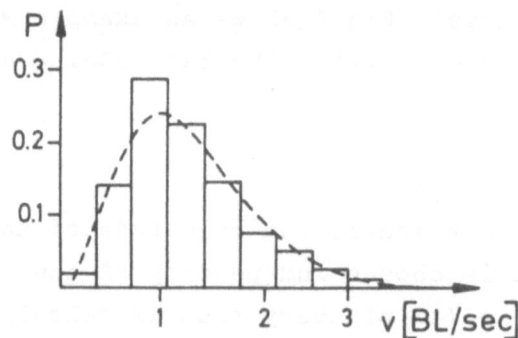

Figure 5. Typical velocity frequency distribution in a fish school determined by experiments and fitted by a Gamma-distribution (histo-gram datas from experiments, parameters of the Gamma distribution $A=4$, $K=33$, units in body length (BL/sec). After AOKI 1980.

(Γ(K) Gamma function, v velocity in BL/sec); K, A
parameters (K=4, A=3,3); the distance is expressed in body
lengths (BL); the average velocity of a fish is 1.2
BL/sec.).

5. Results

Usually we simulated 100 time steps with n=8 fish. At the
beginning the fish were set at random in a quadratic start area
(4.5 BL · 4.5 BL). The simulation parameters had the following
values: Δt=0.5 sec, s=15° (see also fig.3 and fig. 5).

Fig.6 shows a typical sequence of 6 successive time steps in
a simulation with the averaging model (A-model) and decision
model (D-model). The arrows represent the fish. The A-model
fish school is highly polarized (like real fish schools)
(fig.2) and also shows a strong cohesion. The D-model fish
school does not show polarization. We say the model school is
in confusion.

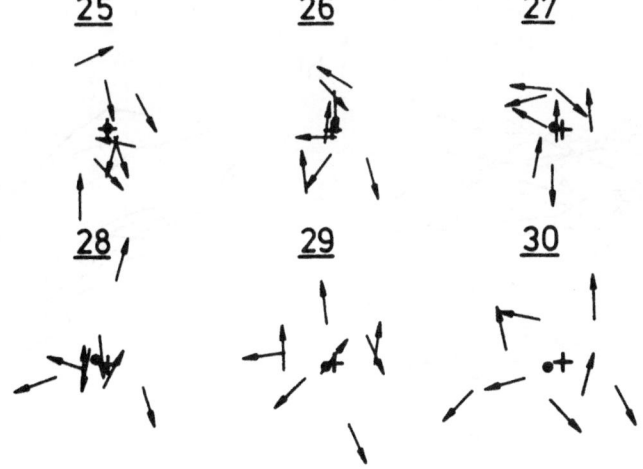

Figure 6a. Sequence from a typical simulation run with the
decision model (D). The arrows represent the fish. Time steps
25-30 are shown. The black dot marks the current centre of mass
of the school. The cross marks the position of the centre of
mass at the first illustrated time step. The polarization p of
the school is 57°, 74°, 70°, 80°, 80°, 77° (see text for
explanation, °=DEG).

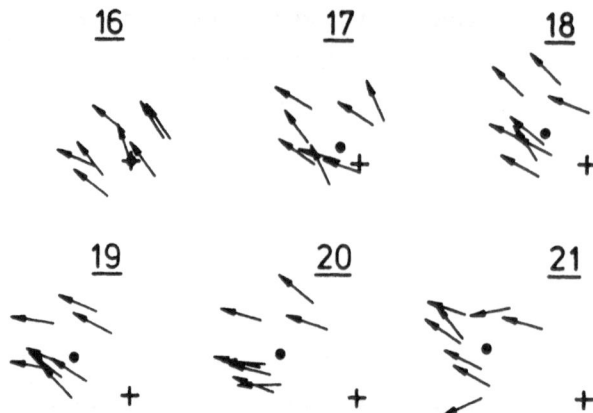

<u>Figure 6b.</u> Sequence from a typical simulation run with the averaging model (time steps 15-21), same presentation as in fig.6a. The polarization p of the school is 12°, 16°, 10°, 9°, 8°, 17° respectively.

   Fig.7b shows the same sequence of the A-model school in other representation. Fig.7a shows the motion of a real fish school for comparison. The similarity is clear.

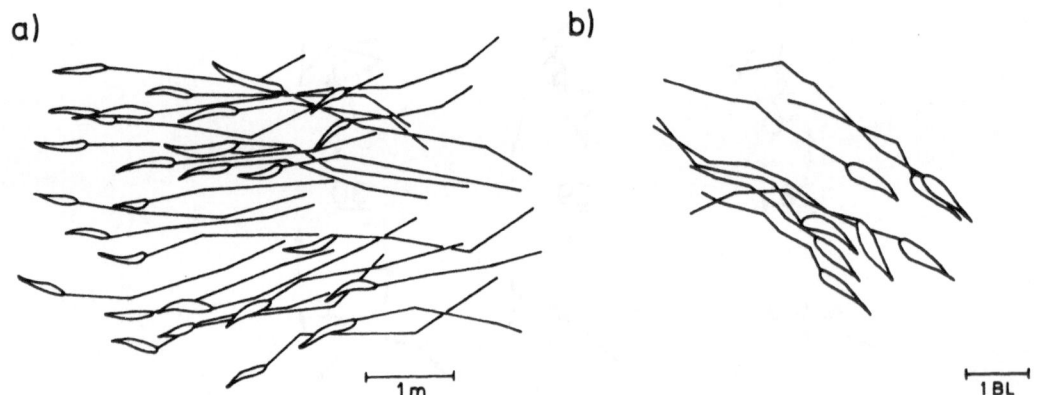

<u>Figure 7.</u> Paths of the members of a fish school
a) After experiments by Partridge (1981). The fish were traced by videotape and the positions of the fish are plotted every 1.4 sec.
b) Fish school of our simulations (A model). The positions of the fish are plotted every 0.5 sec. (BL body length)

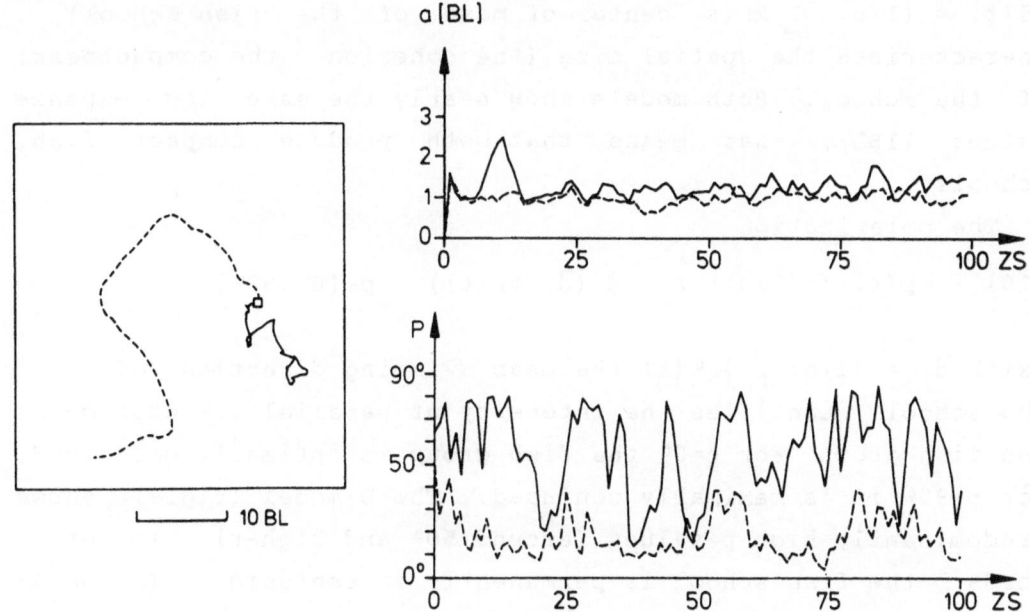

Figure 8. (left side) Path of the centre of mass for a typical simulation run with the averaging (dotted line) and decision model (full line). See text for the values of the model parameters. The initial point is marked by the little box. The duration of the simulation is 100 time steps.

Figure 9. (right side) Expanse a (in BL body length) and polarization p versus time steps (ZS = 0.5 sec) for the same simulations represented in fig.8 (A-model dotted line, D-model full line).

Fig.8 shows the paths of the center of mass of the above simulations. The path of the D-model (full line) is very twisted. The fish group changes its direction very often. On the other hand, the fish group of the A-model (dashed line) keeps a fixed direction for many time steps. Thus the fish group covers a good distance. The bends in the path arise from short periods of confusion.

Fig.9 gives an overview over the same simulations with the help of two variables which quantify the typical characteristics of real fish schools.

The <u>expanse</u>

$$(9) \qquad a(t) = \sqrt{\frac{1}{n} \sum_{i=1}^{n} (\vec{x}_i(t) - \vec{s}(t))^2}$$

($\vec{s}(t) = (1/n) \sum_{i=1}^{n} \vec{x}_i (t)$ center of mass of the fish school) characterises the spatial size (the cohesion, the compactness) of the school. Both models show nearly the same low expanse values (1BL), that means that both produce compact fish schools.

The <u>polarization</u>

(10) $p(t) = (1/n) \sum_{i=1}^{n} \not{\chi} (\vec{d}, \vec{v}_i (t))$ $p\epsilon [0^0, 90^0]$

(with $\vec{d} = (1/n) \sum_{i=1}^{n} \vec{v}_i{}^0 (t)$ the mean swimming direction of the school) quantifies the intensity of parallel orientation of the fish group. For $p=0^0$ the fish group is optimally polarized, for $p=90^0$ it is maximally confused. The D-model (fig.9) shows predominantly high p values (around 50^0 and higher). In other words, the fish school is permanently in confusion. In the A-model the p values are low (about 20^0). That means the fish group remains highly polarized nearly all the time.

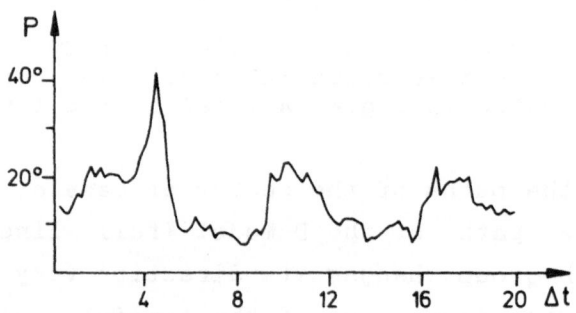

<u>Figure 10.</u> Polarization p versus time t ($\Delta t = 0.5$ sec) for a school of six jack mackerels. After Hunter (1966).

Fig.10 shows an experimental result by Hunter (1966): The polarization time sequence for a real fish school in a water tank (Hunter called p "mean angular deviation"). Comparing them with our model results we see that our model school shows the same low polarizations as real fish schools. The p values over 20^0 (fig.9, fig.10) arise from bends in the school path.

Next we investigated whether our model also reproduces the internal structure of fish schools in the form of the nearest

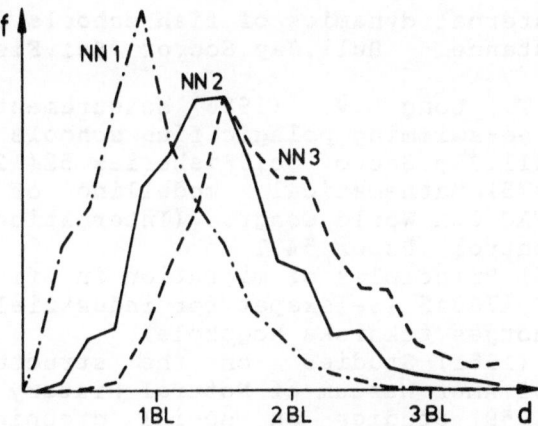

Figure 11. Frequency f of the neighbours at various distances d from the reference fish. Compare with fig.2. Data are from simulations with the A-model. Discussion see text.

neighbour distance (NN) distributions (fig.2). Fig.11 shows the NN distributions for our model school (with special parameters for the "behaviour areas"). They show the same typical charac-teristics of the NN distributions as real fish schools: for each neighbour a distinct maximum in the respective distribu-tion. The fact that the NN1 distribution does not decrease to zero at a finite distance (compare fig.2) demonstrates that the repulsion behaviour in our model is not sufficiently effective. This is surely a product of our model simplification that the velocity of each fish is chosen independently from its neigh-bours. We hope to get still a better conformity with the experimental NN distributions by dropping this simplification and designing a new repulsion behaviour.

REFERENCES

Aoki I.(1980) An analysis of the schooling behaviour of fish: Internal organization and communication process. Bull. Ocean. Res. Inst. No.12, Tokyo

Aoki I.(1982) A simulation study on the schooling mechanism in fish. Bull.Jap.Soc.Sci.Fish. 48: 1081-1088

Aoki I. (1984) Internal dynamics of fish schools in relation to inter-fish distance. Bull.Jap.Soc.of Sci.Fisheries 50(5): 751-758

Aoki I., Inagaki T., Long L.V. (1986) Measurements of the 3-D-structure of free-swimming pelagic fish schools in a natural environment. Bull.Jap.Soc.of Sci.Fisheries 52(12): 2069-2078

Balchen J.G. (1975) Mathematical modelling of fish behaviour. Proc.IFAC 6th World Congr. (International Federation of automatic control) Paper 54.1

Balchen J.G.(1976) Principles of migration in fishes. SINTEF Rapports STF 48 A76045 (Selskapet for industriell og tekmisk forskming ved norges tekniska hogskole)

Breder C.M.Jr. (1951) Studies on the structure of fish schools. Bull.of Amer.Museum of Natural History 98: 3-27

Breder C.M.Jr. (1959) Studies on social groupings in fishes. Bull.of Amer.Museum of Natural History 117: 395-481

Bone Q., Marshall N.B.(1985) Sinnesorgane der Fische. (insbes: 9.3 Sehen und Leuchten, 9.2 Elektrorezeptoren und elektr. Organe) in "Biologie der Fische" Stuttgart

Hunter J.R. (1966) Procedure for analysis of schooling behaviour. Journal Fish.Res.Bd.Canada 23(4): 547-562

Keenleyside M.H.A. (1955) Some aspects of the schooling behaviour. Behaviour (Leiden) 8: 183-247

Inagaki T.,Sakamoto W., Kuroki T. (1976) Studies on the schooling behaviour on fish II: Mathematical modeling of schooling form depending on the intensity of mutual force between individuals. Bull.Jap.Soc.of Sci.fisheries 42(3): 256-270

Okubo A. (1980) 'The dynamics of animal grouping' in: Diffusion and ecological problems: Mathematical models. Springer p.110-131

Olst J.C., Hunter J.R. (1970) Some aspects of the organization of fish schools. J.Fish.Res.Board Canad. 27: 1225-1238

Partridge B.L. (1981) Internal dynamics and the interrelations of fish schools. J.Comp.Physiol. 144: 313-325

Partridge B.L., Pitcher T.J., Cullen J.M., Wilson J. (1980a) The 3-D structure of fish schools. Beh.Ecol.Sociobiol. 6: 277-288

Partridge B.L., Pitcher T.J. (1980b) The Sensory Basis of Fish Schools: Relative Role of Lateral Line and Vision. 1980, J.Comp.Physiol. 135: 315-325

Pitcher T.J., Partridge B.L., Wardle C.S. (1976) A blind fish can school. Science 194: 963-965

Pitcher T.J. (1979) Sensory information and the organization of a behaviour in a schooling cyprint fish. Anim.Behav.27: p.126-149

Radakov D.V. (1971) Schooling in the ecology of fish. Halsted Press, New York

Shaw E. (1970) 'Schooling in fishes: critique and review' in: Aronson L.(ed) "Development and evolution of behaviour", Freeman, San Francisco

Tembrock G. (1983) 'Fische: Biosozialverhalten' in: Spezielle Verhaltensökologie. Bd.II p.584-594, Stuttgart

Conclusions

A comparison of the methods used in this chapter reveals a dichotomy between computer simulations of individuals and equations describing densities or other aspects of the population. The former approach can have the advantage of direct visual results. (For example, in simulations of fish, we obtain the actual position and relative orientations of the animals. In the case of cells, the simulations provide a record similar to a "time-lapse" movie of the aggregation and clustering of cells.) A disadvantage of simulations is the necessity for intensive parameter space exploration. This is especially forbidding when the system is governed by numerous parameters, each with a different effect.

In the latter approach, the details of individual position or orientation are averaged and replaced by spatially varying densities. This continuum approach has the advantages of being amenable to analysis (in some cases) by a toolkit of applied mathematical techniques. Even when only partial analysis is possible, the role of parameters and assumptions in the resulting predicted behavior is spelled out more clearly here than by computer simulations alone. The disadvantage is that assumptions made in formulating the equations and simplification required for analytic tractability may restrict the validity of the results.

Numerical analysis of the approximating continuum equations is often necessary. Ideally, direct simulations combined with analysis of a continuum approximation lead to complementary approaches which prove most effective when used together on the same problem.

Regardless of the particular biological situation modelled, and of the exact mathematical framework used, a key question is whether one can distinguish between a variety of interaction mechanisms (at the level of the motile organisms) based only on differences in the structure or dynamical properties of the swarms they produce. A subsidiary question is whether similarities and differences in the functional properties of disparate interactions show up in the mathematical models. One expects parallels in the formalisms used to model "direct" interactions (e.g. antennae contact of ants, lamellipodia contact or sliding cohesion of myxobacteria or fibroblasts, visual or acustic communication between fish). Similary, "indirect" interactions mediated by intermediates should also fall into related modelling categories. (Examples include deposited slime or pheromone trails, prey or nutrients, predation pressure, chemical signals and gradients, and others). Clearly, however, most models include components specific to the system being investigated, in which temporal and/or spatial kinetics of intermediates are related to the signal perception and motor response (for example, *kinesis* and *taxis*, see section II/3) and in which the abundance of intermediates in turn influences the motility of aggregative drift of the population (Tranquillo and Lauffenburger, Section II/3, Keller and Segel 1970, Alt and Lauffenburger 1987, Othmer and Monk 1988).

At the end of this chapter, a brief bibliography highlights a selection of literature on motion and collective behavior (both classic and recent). Among these are papers noted by diffusion approximations for random motion of organisms (Fisher 1937, Skellam

1951, Gurtin and MacCamy 1977, Kolmogorov et al. 1937, Okubo 1980, 1986, Kareiva 1983, Levin 1986, and Tranquillo and Lauffenburger in Section II/3), leading to parabolic equations. Hyperbolic transport equations appear in the contribution by Pfistner, and e.g. in (Greenberg and Alt 1987). Chemotaxis is treated in (Keller and Segel 1970, Alt and Lauffenburger 1987), and selections in (Jaeger and Murray 1984).

Aside from the implication of direct responses and responses to chemical or biological intermediates, in some systems an important role is played by physical or mechanical forces that guide motion and shape the population behavior. Recently, mechanochemical interactions between cells and other structural components of the environment have been studied in a variety of contexts by Oster, Murray and Odell, see (Oster et al. 1983, Murray 1989) for a review. Such mechanochemical theories have particular relevance to cellular and multicellular phenomena, although they may exist at some level in animal populations when close contact to individuals occur. Examples include "pushing" in crowds, close herding as a defense against predation and mating behavior in some African ungulates where males physically "round up" a harem of females, leading to local aggregation centers (A.R.E. Sinclair, personal communication).

As pointed out by Focardi and Magagnoli (preprint, personal communication and workshop discussion) in the ecological settings, two predominant factors influencing movement and social behavior are the pressures of predation and the distribution of resources. Hamilton (1971) indicated that herd geometry (namely ratio of herd circumference to the occupied area) can increase chances of survival in the face of predation. Availability of food, on the other hand, may limit herd size, and thus these two factors have competing effects.

According to Focardi and Magagnoli, in ecology one must carefully distinguish between cases in which interindividual communication alone suffices to explain collective behaviour versus cases in which such externally imposed factors are the major determinants. Further work in this direction is outlined in (Clark and Mangel 1985, Focardi and Toso 1987, Hoffmann et al. 1981, Sinclair 1977, and Thompson et al. 1974).

Other phenomena in which structure and self-organization arises are well known. A theory which spans several size scales is that of Turing for diffusive instability, originally applied to the problem of chemical prepattern formation in morphogenesis. The idea that a pair of diffusing, mutually interacting "activator-inhibitor" chemicals can spontaneously form patterns (Meinhardt 1982, Gierer and Meinhardt 1972) has also been applied to a population-interaction setting to explain the phenomenon of plankton patchiness by Segel and Jackson (1972), and Mimura and Murray (1978). Other topics of current interest appear in selected contributions in (Jaeger and Murray 1984, Teramoto and Yamaguti 1987) and (Goldbeter 1989). For predominantly mathematical treatments of the problems and questions of aggregating populations see (Alt 1985 b, Britton 1986, Shigesada 1980, Teramoto and Seno 1988).

Most of the approaches outlined above have led to continuum (partial differential) equations, which have yielded to linear stability analysis and computer simulations. In some cases, large systems of detailed equations have been studied. However, it is often worthwhile and adequate to reduce complexity and treat simpler approximating models. In particular, this is the case if characteristic *length* or *time scales* for signal transmission are relatively small compared to those for the locomotion process. For example, in the case of instantaneous transmission one obtains equations (e.g. degenerate diffusion equations), whose drift motion is described by a spatial integral term and which still reproduce the phenomena of aggregation and population waves (Nagai and Mimura 1983, Alt 1985 a, Mimura and Takigawa 1987). If, in addition, the radius of mutual interaction between individuals is negligably small, then local Taylor expansion condenses these functional integral terms to pointwise functions of the population density and its derivatives. While there are cases where Taylor series expansion up to degree 3 leading to fourth order parabolic equations have been used (Murray 1989), in other cases (e.g. Pfistner) the property of aggregation is reproduced by a Taylor localization of only degree one leading to a hyperbolic system of first order.

The concluding discussion during the workshop revealed that these mathematically interesting nonlinear differential equations or stochastic processes are of use for biologists only insofar as they characterize the quantitative effects on collective motion (e.g. *swarm size* or *shape, speed of advance* of fronts, *degree of alignment* within a swarm) which are induced by changing parameters or functional properties of the assumed interaction mechanisms. Applied mathematicians involved in analysis of collective motion will increasingly be asked to develop models, analyses, and numerical simulations hand in hand with quantitative criteria in order to estimate parameters, discriminate between competing theories or assumptions, and allow for feedback from empirical observations.

In particular, effective image processing and detailed correlation analysis of collectively moving cells or animals, for example cell motility imaging (Thurstin et al. 1988) or software for analysis of fish schools (Spieser 1989), could then help to detect and understand the detailed rules governing their mutual communication and the resulting adjustment of their motile activity.

REFERENCES

Alt W. (1985 a) Degenerate diffusion equations with drift functionals modelling aggregation. Nonlinear Analysis, Theor. Math. Appl. 9: 811-836

Alt W. (1985 b) Models for mutual attraction of motile individuals. Lecture Notes in Biomathematics Vol 57: 33-38

Alt W. and Lauffenburger D.A. (1987) Transient behavior of a chemotaxis system modelling certain types of tissue inflammation. J. Math. Biol. 24: 691-722

Britton N.F. (1986) Reaction diffusion equations and their applications to biology. Academic Press NY.

Clark C.W. and Mangel M. (1985) The evolutionary advantages of group foraging. Theor. Pop. Biol. 30: 45-75

Deneubourg J.L. and Goss S. (1989) Collective patterns and decision making. Ethology, Ecology, and Evolution 1: 295-311

Edelstein-Keshet L. and Ermentrout B. (1989) Models for branching networks in two dimensions. SIAM J. Appl. Math. 49 Nr.4: 1136-1157

Fisher R.A. (1937) The wave of advance of advantageous genes. Ann. Eugen. (London) 7: 355-369

Focardi S. and Toso S. (1987) Foraging and social behavior of ungulates: proposals for a mathematical model. pp. 295-304. In: (P.Ellen and C.Thinus-Bland eds.) Cognitive processes and spatial orientation in animal and man. Vol 1. Martinus Nijhoff Publishers, Boston

Gierer A. and Meinhardt H. (1972) A theory of biological pattern formation. Kybernetik, 12: 30-39

Goldbeter A. (1989) Cell to Cell Signalling: From Experiments to Theoretical Models. Academic Press, NY.

Greenberg J.M. and Alt W. (1987) Stability results for a diffusion equation with functional drift approximating a chemotaxis model. Trans. AMS 300: 235-258

Gurtin M.E. and MacCamy R.C. (1977) On the diffusion of biological populations. Math. Biosci. 33: 35-49

Hamilton W.D. (1971) Geometry of the selfish herd. J. Theor. Biol. 31: 295- 311

Hoffman W., Heinemann D. and Wiens J. (1981) The ecology of seabirds feeding flocks in Alaska. Auk. 98: 437-456

Jaeger W. and Murray J.D., eds. (1984) Modelling Patterns in Space and Time. Proceedings of a Workshop (Heidelberg 1983), Springer-Verlag, Lect. Notes in Biomath. Vol.55 Springer, NY.

Kareiva P.M. (1983) Local movement in herbivorous insects: applying a passive diffusion model to mark-recapture field experiments. Oecologia (Berlin) 57: 322-327

Keller E.F. and Segel L.A. (1970) Initiation of slime mold aggregation viewed as an instability. J. Theor. Biol. 26: 399-415

Kolmogorov A., Petrovsky I. and Piscunov (1937) Etude de l'equation de la diffusion avec croissance de la quantite a de la matiere et son application a un probleme biologique. Bull. Univ. Moscou. Ser. Intern. Sec A1, 6: 1-25

Lauffenburger D.A., Grady M. and Keller K.H. (1984) A hypothesis for approaching swarms of myxobacteria. J. Theor. Biol. 110: 257-274

Levin S.A. (1986 a) Ecological and evolutionary aspects of dispersal. In: Teramoto and Yamaguti, eds. (see below) pp. 80-87

Levin S.A. (1986 b) Random walk models of movement and their implications. In: (T.Hallam and S.A.Levin eds.) Mathematical Ecology: An Introduction. Springer, NY.

Meinhardt H. (1982) Models of Biological Pattern Formation. Acad.Press, London

Mimura M. and Murray J.D. (1978) On the diffusive predator-prey model which exhibits patchiness. J. Theor. Biol. 75: 249-262

Mimura M. and Takigawa S. (1987) A spatially aggregating population model involving site-distributed dynamics. In: Teramoto and Yamaguti, eds. (see below)

Nagai T. and Mimura M. (1983) Some degenerate diffusion equations related to population dynamics. J. Math. Soc. Japan 35: 539-562

Okubo A. (1980) Diffusion and Ecological Problems. Mathematical Models. Springer, NY.

Okubo A. (1986) Dynamical aspects of animal grouping: swarms, schools, flocks, and herds. Adv. in Biophys. 22: 1-94

Oster G.F., Murray J.D. and Harris A.K. (1983) Mechanical aspects of mesenchymal morphogenesis. J. Embryol. Exp. Morphol. 78: 83-125

Ricciardi L.M., ed. (1988) Biomathematics and Related Computational Problems. Kluwer Acad. Publ., Dordrecht

Segel L.A. and Jackson J.L. (1972) Dissipative structure. An explanation and an ecological example. J. Theor. Biol. 37: 545-559

Shigesada N. (1980) Spatial distribution of dispersing animals. J.Math.Biol. 9: 85-96

Sinclair A.R.E. (1977) The African buffalo. Univ. of Chicago Press

Skellam J.G. (1951) Random dispersal in theoretical populations. Biometrika 38: 196-218

Spieser D. (1989) Behavioquant. Software package for movement analysis. Ges. Strahlen- und Umweltforschung, München

Thompson W.A., Vertinsky I. and Krebs J.R. (1974) The survival value of flocking in birds: a simulation model. J. Animal Ecol. 43: 785-820

Thurston G., Jaggi B. and Palcic B. (1988) Measurement of cell motility and morphology with an automated microscope system. Cytometry 9: 411-417

Teramoto E. and Seno H. (1988) Modelling of biological aggregation patterns. In: Ricciardi, ed. (see above) pp. 409-419

Teramoto E. and Yamaguti M., eds. (1986) Mathematical Topics in Population Biology, Morphogenesis, and Neurosciences. Lect. Notes in Biomath. Vol. 71. Springer, NY.

LIST OF PARTICIPANTS

Workshop on

MODELING, ANALYSIS AND SIMULATION OF BIOLOGICAL MOTION

Königswinter, March 16-19,1989

Alt, Wolfgang, Abt. Theoretische Biologie, Botan. Inst. der Univ. Bonn, Kirschallee 1, D-5300 Bonn 1, Tel: *(49-)228-735577 E-mail: UNB11B@DBNRHRZ1

Baba, Shoji, Dept. of Biology, Ochanomizu University, 2-1-1 Otsuka, J-Tokyo 112, Japan

Behrend, Konstantin, Zoologisches Institut, Universität Mainz, Saarstr. 21, D-6500 Mainz, Tel: *(49-)6131-393379

Benhamou, Simon, C.N.R.S.-L.N.F. 2, 32 chemin J. Aiguier, F-13009 Marseille, Frankreich

Bereiter-Hahn, Jürgen, Kinem. Zellforschung, FB Biol. J.-W.-Goethe-Univ., Senckenberganlage 27, D-6000 Frankfurt am Main, Tel: *(49-)611-7982335

Bovet, Pierre, Inst. Neurophys.et Psychophys., CNRS-LNF 2, 32 chemin J. Aiguier, F-13009 Marseille, Frankreich, E-mail: BOVET@FRMOP11

Briegleb, Wolfgang, DLR, Postfach 906058, D-5000 Köln 90, Tel: *(49-)2203-6013088

Brokaw, Charles J., Division of Biology 156-29, California Inst. of Technology, Pasadena, Ca 91125, USA, Tel: *(1-)818- 356-4927

Brosteanu, Oana, Abt. Theoretische Biologie, Botan. Inst. der Univ. Bonn, Kirschallee 1, D-5300 Bonn 1, Tel: *(49-)228-737409

Calenbuhr, V., Service de Chimie Physique II, C.Postal n°231, C.Plaine U.L.B., Boulevard du Triomphe, B-1050 Bruxelles, Belgien, Tel: *(32-)2-6400015

Crenshaw, Hugh, Dept. of Zoology, Duke University, Durham, North Carolina 27706, USA, Tel: *(1-)919-684-3791

Cruse, Holk, Fakultät für Biologie, Postfach 8640, D-4800 Bielefeld 1

Deneubourg, Jean-Louis, Service de Chimie Physique II, C.Postal n°231, C.Plaine U.L.B., Boulevard du Triomphe, B-1050 Bruxelles, Belgien, Tel: *(32-)2-6400015 E-mail: ULBG028@BEARN

Doucet, Paul G., Biologisch Laboratorium, De Boele Laan 1087, NL-1007 MC Amsterdam, Niederlandc, Tel: *(31-)20-5842901

Dunn, Graham A., M.R.C., Cell Biophysics Unit, 26-29 Drury Lane, GB-London WC2B5RL, England, Tel: *(2-)1-8368851

Edelstein-Keshet, Leah, Dept. Mathematics, University of British Columbia, 121-1984 Math. Road, Vancouver, BC, Canada VGT 144, E-mail: USERKESH@UBCMTSG

Engelmann, U., Dermatolog. Klinik u. Poliklinik der Uni. München, Frauenlobstr. 9-11, D-8000 München 2

Focardi, Stefano, Istituto Nazionale di Biologia della Selvaggina, Via Ca'Fornacetta 9, I-40064 Ozzana Emilia, Italien, Tel: *(39-)51-798746

Gruler, Hans, Abteilung für Biophysik, Univ. Ulm, Oberer Eselsberg, Postfach 4066, D-7900 Ulm, Tel: *(49-)731-176-2426

Häder, Donat-P., Institut für Botanik und, Pharmazeutische Biologie, Staudtstr. 5, D-8520 Erlangen, Tel: *(49-)91331-85-8216 E-mail: HAEDER@DERRZE0

Haustein, Werner, MPI für Verhaltensphysiologie, D-8130 Seewiesen, Tel: *(49-)8157-29-353

Hemmersbach-Krause, Ruth, DLR, Postfach 906058, D-5000 Köln 90, Tel: *(49-)2203-6013094

Hoffmann, Gerhard, Zoolog. Institut II der Universität Würzburg, Röntgenring 10, D-8700 Würzburg, Tel: *(49-)931-31639

Huth, Andreas, FB Physik, Universität Marburg, Renthof 6, D-3550 Marburg, Tel: *(49-)6421-284233

Jackson, G. A., Inst. of Marine Resources, Univ. of California, San Diego, La Jolla, USA-92093 California, USA

Keller, Hans-Uli, Pathologisches Institut, Universität Bern, CH-3070 Bern, Schweiz, Tel: *(41-)31-643208

Kils, U., Institut für Meereskunde Kiel, Düsternbrocker Weg 20, D-2300 Kiel 1,

Levandowsky, Mike, Haskins Laboratories, Pace University, 41 Park Row, New York, NY 10038, USA, Tel: *(1-)212-488-1246

Machemer, Hans, Fakultät für Biologie, Ruhr-Univ.Bochum, Geb. ND6/28, Postfach 102148, D-4630 Bochum 1, Tel: *(49-)234-7004350

Marchetti, Federico, Univ.di Genova, Dipartimento di Matematica, v. L. B. Alberti 4, I-16132 Genova, Italien, Tel: *(39-)10-353-8717 E-mail: MARKETTI@IGECUNIV

Melkonian, Michael, Botanisches Institut, Gyrhofstr. 15, D-5000 Köln 41, Tel: *(49-)221-4702475

Mendes France, M., Laboratoire ass. CNRS n° 226, U.E.R. Math. Inform., Univ. de Bordeaux I, F-33405 Talence Cedex, Frankreich

Mogami, Yoshihoro, Fakultät für Biologie, Ruhr-Univ.Bochum, Geb. ND6/28, Postfach 102148, D-4630 Bochum 1

Müller, Uwe, Fakultät für Biologie, Postfach 8640, D-4800 Bielefeld 1

Noble, Peter B., Dept.Oral Biol.,Fac.of Dentist, McGill University, 3640 University Street, CDN-Montreal, Quebec, H3A 2B2, Kanada

Petermann, Harald, Inst. Allg. Angew. Geophysik, Universität München, Theresienstr. 41, D-8000 München 2, Tel: *(49-)89-23944226

Pfistner, Beate, Abt. Theoretische Biologie, Botan. Inst. der Univ. Bonn, Kirschallee 1, D-5300 Bonn 1, Tel: *(49-)228-737409 E-mail: UNB10D@DBNRHRZ1

Pohl, Thomas, Abt. Theoretische Biologie, Botan. Inst. der Univ. Bonn, Kirschallee 1, D-5300 Bonn 1, Tel: *(49-)228-735541

Reichard, Klaus, Inst. für Math.,Naturwiss. Fak, Universität Witten-Herdecke, Stockumer Str. 10, D-4630 Witten, Tel: *(49-)2302-669-182

Reize, I., Botanisches Institut, Gyrhofstr. 15, D-5000 Köln 41, Tel: *(49-)221-4702463

Scharstein, Hans, Zoolog. Institut, Lehrstuhl(II) Tierphysiologie, Weyertal 119, D-5000 Köln 44, Tel: *(49-)221-4703103

Schatz, Albrecht, DLR, Postfach 906058, D-5000 Köln 90, Tel: *(49-)2203-6013090

Schuber, Marianne, DFVLR, Postfach 906058, D-5000 Köln 90, Tel: *(49-)2203-6013270

Stevens, Angela, Inst. für Angewandte Math., Universität Heidelberg, Im Neuenheimer Feld 294, D-6900 Heidelberg 1, Tel: *(49-)6221-562980 E-mail: L39@DHDURZ1

Stokes, Cynthia, NIH Bldg. 31, Dept. Health + Human Sciences, Bethesda, Maryland 20892, USA, Tel: *(1-)301-496-4325 E-mail: STOKES@FCRFV1.NCIFCRF.GOV

Sugino, Kazuyuki, Dept. of Biol. Sciences, University of Tsukuba, Tennodai 1-1, J-Tsukuba 305, Japan

Tranquillo, Robert T., Dept. Chemical Eng., 1551 Amundson Hall, 421 Washington Ave S.E., Minneapolis, MN 55455, USA, E-mail: FQX6452@UMNACVX

Vicker, Michael, FB Biologie, Universität Bremen, Postfach 330440, D-2800 Bremen, Tel: *(49-)421-2183105

Wawrowsky, Kolja, Zoolog. Institut II der Universität Würzburg, Röntgenring 10, D-8700 Würzburg

Weber, Thomas, DLR, Bereich FF-ME-MUSC, Linderhöhe, D-5000 Köln 90

Weiss, Dieter G., Institut für Zoologie, Technische Univ. München, Lichtenbergstr. 4, D-8046 Garching, Tel: *(49-)89-3209-3669 E-mail: T321102@DM0LRZ01

Wissel, Ch., FB Physik, Philipps-Universität Marburg, Renthof 5, D-3550 Marburg

Wortmann, Michael, I. Zoologisches Institut, Universität Göttingen, Berliner Str. 28, D-3400 Göttingen, Tel: *(49-)551-395430

Zarnack, Wolfram, I. Zoologisches Institut, Universität Göttingen, Berliner Str. 28, D-3400 Göttingen, Tel: *(49-)551-395430

Zimmermann, Arthur, Pathologisches Institut, Universität Bern, CH-3070 Bern, Schweiz, Tel: *(41-)31-643208

Index

Journal of
Mathematical Biology

From a recent issue:

F. Cervantes-Pérez, M. A. Arbib:
Stability and parameter dependency analysis of a facilitation tectal column (FTC) model

L. Edelstein-Keshet, G. B. Ermentrout:
Models for contact-mediated pattern formation: cells that form parallel arrays

M. Notohara:
The coalescent and the genealogical process in geographically structured population

L. F. Murphy, S. J. Smith:
Optimal harvesting of an age-structured population

R. Sridhara:
Inference on system parameters in a model for interacting species

Covered by
Current Contents and
Zentralblatt für Mathematik

Springer-Verlag
Berlin
Heidelberg
New York
London
Paris
Tokyo
Hong Kong
Barcelona

*Subscription information
and other details are available
from the publisher at one
of the given addresses.*

☐ Heidelberger Platz 3, W-1000 Berlin 33 / F.R. Germany ☐ 175 Fifth Ave., New York, NY 10010, USA
☐ 8 Alexandra Rd., London SW19 7JZ, England ☐ 26, rue des Carmes, F-75005 Paris
☐ 37-3, Hongo 3-chome, Bunkyo-ku, Tokyo 113, Japan ☐ Citicorp Centre, Room 1603, 18 Whitfield Road,
Causeway Bay, Hong Kong ☐ Avinguda Diagonal, 468-4°C, E-08006 Barcelona

Biomathematics

Managing Editor: S. A. Levin

Editorial Board: C. DeLisi, M. Feldman, J. Keller, R. M. May, J. D. Murray, A. Perelson, L. A. Segel

Volume 18

S. A. Levin, Cornell University, Ithaca, NY;
T. G. Hallam, L. J. Gross, University of Tennessee, Knoxville, TN (Eds.)

Applied Mathematical Ecology

1989. XIV, 491 pp. 114 figs. Hardcover DM 98,–
ISBN 3-540-19465-7

This book builds on the basic framework developed in the earlier volume – "Mathematical Ecology", edited by T. G. Hallam and S. A. Levin, Springer 1986, which lays out the essentials of the subject. In the present book, the applications of mathematical ecology in ecotoxicology, in resource management, and epidemiology are illustrated in detail. The most important features are the case studies, and the interrelatedness of theory and application. There is no comparable text in the literature so far. The reader of the two-volume set will gain an appreciation of the broad scope of mathematical ecology.

Volume 19

J. D. Murray, Oxford University

Mathematical Biology

1989. XIV, 767 pp. 292 figs. Hardcover DM 98,–
ISBN 3-540-19460-6

This textbook gives an in-depth account of the practical use of mathematical modelling in several important and diverse areas in the biomedical sciences. The emphasis is on what is required to solve the real biological problem. The subject matter is drawn, for example, from population biology, reaction kinetics, biological oscillators and switches, Belousov-Zhabotinskii reaction, neural models, spread of epidemics.
The aim of the book is to provide a thorough training in practical mathematical biology and to show how exciting and novel mathematical challenges arise from a genuine interdisciplinary involvement with the biosciences. It also aims to show how mathematics can contribute to biology and how physical scientists must get involved.

Volume 20

J. E. Cohen, Rockefeller University, New York, NY;
F. Briand, Gland; **C. M. Newman,** University of Arizona, Tucson, AZ

Community Food Webs

Data and Theory

1990. XII, 308 pp. 46 figs. Hardcover DM 148,–
ISBN 3-540-51129-6

Contents: I. General Introduction: Food Webs and Community Structure. – **II. Empirical Regularities:** Untangling an Entangled Bank. – **A. General Regularities:** Ratio of Prey to Predators in Community Food Webs. – Community Food Webs have Scale-Invariant Structure. – Trophic Links of Community Food Webs. – Food Webs and the Dimensionality of Trophic Niche Space. – **B. Differential Regularities:** Environmental Control of Food Web Structure. – Environmental Correlates of Food Chain Length. – **III. A Stochastic Theory of Community Food Webs:** Theory: Circles of Complexity, Spherical Horses. – Models and Aggregated Data. – Individual Webs. – Predicted and Observed Lengths of Food Chains. – Theory of Food Chain Lengths in Large Webs. – Intervality and Triangulation in the Trophic Niche Overlap Graph. – **IV. Data on 113 Community Food Webs.** – Bibliography. – Subject Index. – Acknowledgements.

Springer-Verlag
Berlin
Heidelberg
New York
London
Paris
Tokyo
Hong Kong
Barcelona

Lecture Notes in Biomathematics